# Biofuels and Biorefineries

Volume 7

## Aims and Scope of the Series

The Biofuels and Biorefineries Series aims at being a comprehensive and integrated reference for biomass, bioenergy, biofuels, and bioproducts. The series provides leading global research advances and critical evaluations of methods for converting biomass to biofuels and chemicals. Scientific and engineering challenges in biomass production and conversion are covered that show technological advances and approaches for creating new bio-economies in a format that is suitable for both industrialists and environmental policy decision-makers.

The Biofuels and Biorefineries Series provides readers with clear and concisely written chapters that are peer-reviewed on significant topics in biomass production, biofuels, bioproducts, chemicals, catalysts, energy policy, economics, and processing technologies. The text covers major fields of plant science, green chemistry, economics and economy, biotechnology, microbiology, chemical engineering, mechanical engineering, and energy.

**Series Description**

Annual global biomass production is about 220 billion dry tons or 4,500 EJ, equivalent to 8.3 times the world's energy consumption in 2014 (543 EJ). On the other hand, world-proven oil reserves at the end of 2011 reached 1652.6 billion barrels, which can only meet just over 50 years of global production. Therefore, alternative resources are needed to both supplement and replace fossil oils as the raw material for transportation fuels, chemicals, and materials in petroleum-based industries. Renewable biomass is a likely candidate, because it is prevalent over the Earth and is readily converted to other products. Compared with coal, some of the advantages of biomass are (i) its carbon-neutral and sustainable nature when properly managed, (ii) its reactivity in biological conversion processes, (iii) its potential to produce bio-oil (ca. yields of 75%) by fast pyrolysis because of its high oxygen content, (iv) its low sulfur and lack of undesirable contaminants (e.g., metals, nitrogen content), (v) its wide geographical distribution, and (vi) its potential for creating jobs and industries in energy crop productions and conversion plants. Many researchers, governments, research institutions, and industries are developing projects for converting biomass including forest woody and herbaceous biomass into chemicals, biofuels, and materials, and the race is on for creating new "biorefinery" processes needed for future economies. The development of biorefineries will create remarkable opportunities for the forestry sector and biotechnology, materials, and chemical processing industry and stimulate advances in agriculture. It will help to create a sustainable society and industries that use renewable and carbon-neutral resources.

More information about this series at http://www.springer.com/series/11687

Zhen Fang • Richard L. Smith, Jr. • Xinhua Qi
Editors

# Production of Platform Chemicals from Sustainable Resources

Springer

*Editors*
Zhen Fang
Nanjing Agricultural University
Nanjing, China

Richard L. Smith, Jr.
Tohoku University
Sendai, Japan

Xinhua Qi
Agro-Environmental Protection Institute
(AEPI), Ministry of Agriculture
Tianjin, China

ISSN 2214-1537 ISSN 2214-1545 (electronic)
Biofuels and Biorefineries
ISBN 978-981-10-4171-6 ISBN 978-981-10-4172-3 (eBook)
DOI 10.1007/978-981-10-4172-3

Library of Congress Control Number: 2017938599



Printed on acid-free paper

This Springer imprint is published by Springer Nature
The registered company is Springer Nature Singapore Pte Ltd.
The registered company address is: 152 Beach Road, #21-01/04 Gateway East, Singapore 189721, Singapore

# Preface

With increasing concerns on environmental pollution and global warming that resulted from traditional fossil resource applications, much progress has been made in the past few years in developing catalytic reaction systems and chemistries for the conversion of various biomass resources into platform chemicals. This text provides state-of-the-art reviews, current research, prospects, and challenges of production of platform chemicals such as C6 sugars, 5-hydroxymethylfurfural, furfural, $\gamma$-valerolactone, xylitol, 2,5-furandicarboxylic acid, levulinic acid, ethanol, and others from sustainable biomass resources with processes that include heterogeneous catalysis, ionic liquid, hydrothermal/solvothermal, electrochemical, and fermentation methods. Reaction mechanism, methods for product separation and purification, and process integration are introduced. The application of these chemicals and their derivatives for synthesizing commodity chemicals via various routes is also covered.

This book is the seventh book of the series entitled "Biofuels and Biorefineries," and it contains 14 chapters contributed by leading experts in the field. The text is arranged into five key areas:

**Part I:** Production of Sugars (Chap. 1)
**Part II:** Production of Aldehydes (Chaps. 2, 3, and 4)
**Part III:** Production of Acids (Chaps. 5, 6, 7, and 8)
**Part IV:** Production of Alcohols (Chaps. 9, 10, 11 and 12)
**Part V:** Production of Lactones and Amino Acids (Chaps. 13 and 14)

**Chapter 1** presents a brief introduction into the characterization of lignocellulosic biomass and outlines some developing and promising pretreatment and hydrolysis methods for lignocellulosic biomass. **Chapter 2** provides state-of-the-art developments in the field of catalytic synthesis of furfural from C5 sugars and hemicellulose biomass, taking into consideration green chemistry principles, and gives critical analyses and perspectives of the development of sustainable furfural production processes. **Chapter 3** summarizes the catalytic production of 5-hydroxymethylfurfural from biomass-derived sugars and lignocelluloses and mainly focuses on the characteristics and superiority of different catalysts on the catalytic transformation of various feedstocks. **Chapter 4** provides an overview of the historical role of 5-(halomethyl)furfurals in the chemical investigation of carbohydrates and describes multiple approaches to their preparation. Commercial markets that can be unlocked by synthetic manipulation of 5-(chloromethyl)furfural and its immediate derivatives are highlighted. **Chapter 5** offers an overview on process technology studies of kinetic models and the status of large-scale production of levulinic acid from biomass. Levulinic acid derivatives and their application are also presented. **Chapter 6** gives a concise overview of up-to-date methods for the synthesis of 2,5-furandicarboxylic acid from 5-hydroxymethylfurfural or directly from carbohydrates, with special attention to catalytic systems, mechanistic insight, reaction pathway, and catalyst stability. **Chapter 7** introduces the chemical processes for the production of gluconic acid and glucaric acid from monosaccharides and polysaccharides with a focus on heterogeneous catalysts. **Chapter 8** reviews microorganism producers, cultivation, separation technologies, alternative substrates of lignocellulosic biomass, and integration strategies to provide analysis of the strategies and economics of 1,4-diacid commercial-scale production. **Chapter 9** analyzes sorbitol's current market and its potential as a platform chemical and describes sorbitol production methods by chemical, electrochemical, and biotechnological routes. Some prospects about the direction of future research for overcoming current bottlenecks for further development are discussed. **Chapter 10** describes biotechnological xylitol production from the selection and preparation of the raw material to fermentative process conditions, downstream strategies, and future perspectives. **Chapter 11** introduces the development of heterogeneous catalysts for the production of C2 to C6 diols by removal of OH groups, hydrogenation of COOH to $CH_2OH$, and/or ring-opening C-O hydrogenolysis. **Chapter 12** compiles recent advances in lignocellulosic ethanol production processes, from novel raw materials or fermenting microorganisms to new processing technologies and their present commercialization. **Chapter 13** surveys the methodology and recent advances in the production of γ-valerolactone (GVL) from different renewable biomass-derived sources, from the pioneering studies to the present state of the art. **Chapter 14** provides an overview of the microbial production of L-glutamic acid and L-lysine and their applications as building block chemicals from biomass for synthesizing commodity chemicals and non-protein amino acids.

The text should be of interest to students, researchers, academicians, and industrialists who are working in the areas of renewable energy, environmental and chemical sciences, engineering, resource development, biomass processing, sustainability, materials, biofuels, and chemical industries.

Nanjing, Jiangsu, China — Zhen Fang
Sendai, Japan — Richard L. Smith Jr.
Tianjin, China — Xinhua Qi

# Acknowledgments

First and foremost, we would like to cordially thank all the contributing authors for their great efforts in writing and revising the chapters and ensuring the reliability of the information given in their chapters. Their contributions have really made this project realizable.

Apart from the efforts of authors, we would also like to acknowledge the individuals listed below for carefully reading the book chapters and giving constructive comments that significantly improved the quality of many aspects of the chapters:

Dr. Pieter Bruijnincx, Utrecht University, Netherlands
Dr. Raghunath V. Chaudhari, University of Kansas, USA
Dr. KeKe Cheng, Tsinghua University, China
Dr. Attilio Converti, University of Genoa, Italy
Prof. Avelino Corma, Universitat Politècnica de València, Spain
Dr. Thomas J. Farmer, University of York, UK
Dr. Susana Ferreira de Gouveia, University of Vigo, Spain
Dr. Masayuki Inui, Research Institute of Innovative Technology for the Earth (RITE), Japan
Prof. Ning Li, Dalian Institute of Chemical Physics, Chinese Academy of Sciences, China
Prof. Min Jiang, Nanjing Technology University, China
Dr. Wei Jin, Institute of Process Engineering, Chinese Academy of Science, China
Prof. Michael R. Ladisch, Purdue University, USA
Dr. Angelos Lappas, Chemical Process Engineering Research Institute (CPERI)/ Center for Research and Technology Hellas (CERTH), Greece
Dr. J. J. Leahy, University of Limerick, Ireland
Prof. Zhimin Li, East China University of Science and Technology, China
Prof. Li Liu, Dalian University of Technology, China
Dr. Qian Liu, West Virginia University, USA
Dr. Mahdi Malmali, University of Minnesota, USA
Dr. Carlos Martín, Umeå University, Sweden
Dr. Tomoo Mizugaki, Osaka University, Japan

Prof. Dmitry Yu. Murzin, Åbo Akademi University, Finland
Dr. Yoshinao Nakagawa, Tohoku University, Japan
Dr. Manuel Fernando R. Pereira, University of Porto, Portugal
Dr. Marcus Rose, RWTH Aachen University, Germany
Dr. Héctor A. Ruiz, Autonomous University of Coahuila, Mexico
Dr. Mazloom Shah, Xishuangbanna Tropical Botanical Garden, Chinese Academy of Sciences, China
Dr. Krishna Vankudoth, Indian Institute of Chemical Technology (CSIR-IICT), India
Prof. Johannes Gerardus de Vries, Leibniz Institute for Catalysis, Germany
Dr. Hongliang Wang, Washington State University, USA
Prof. Yanqin Wang, East China University of Science and Technology, China
Prof. Dr. Volker F. Wendisch, Bielefeld University, Germany
Prof. Xianmei Xie, Taiyuan University of Technology, China
Prof. Youzhu Yuan, Xiamen University, China
Dr. Jianan Zhang, Tsinghua University, China
Prof. Junhua Zhang, Northwest Agriculture and Forestry University, China
Dr. Yuying Zhao, Taiyuan University of Science and Technology, China

We are also grateful to Ms. Becky Zhao (senior editor) and Ms. Abbey Huang (editorial assistant) for their encouragement, assistance, and guidance during the preparation of the book.

Finally, we would like to express our deepest gratitude toward our families for their love, understanding, and encouragement, which help us in the completion of this project.

Zhen Fang, January 24, 2017, in Nanjing (Zhen Fang)

Richard L. Smith, Jr., January 24, 2017, in Sendai (Richard L. Smith, Jr.)

Xinhua Qi, January 24, 2017, in Tianjin (Xinhua Qi)

# Contents

# Contributors

**Mário Antônio Alves da Cunha** Department of Chemistry, Federal University of Technology of Paraná, Curitiba, Brazil

**Pablo Alvira** LISBP, Université de Toulouse, CNRS, INRA, INSA, Toulouse, France

**Felipe Antonio Fernandes Antunes** Department of Biotechnology, Engineering School of Lorena, University of São Paulo, São Paulo, Brazil

**Larissa Pereira Brumano** Department of Biotechnology, Engineering School of Lorena, University of São Paulo, São Paulo, Brazil

**Andre M. da Costa Lopes** Laboratório Nacional de Energia e Geologia, I.P., Unidade de Bioenergia, Lisbon, Portugal

LAQV-REQUIMTE, Departamento de Química, Faculdade de Ciências e Tecnologia, Universidade NOVA de Lisboa, Caparica, Portugal

**Sai Swaroop Dalli** Lakehead University, Thunder Bay, ON, Canada

**Débora Danielle Virginio da Silva** Department of Biotechnology, Engineering School of Lorena, University of São Paulo, São Paulo, Brazil

**Silvio Silvério da Silva** Department of Biotechnology, Engineering School of Lorena, University of São Paulo, São Paulo, Brazil

**Júlio César dos Santos** Department of Biotechnology, Engineering School of Lorena, University of São Paulo, São Paulo, Brazil

**Thais Suzane dos Santos Milessi** Department of Biotechnology, Engineering School of Lorena, University of São Paulo, São Paulo, Brazil

**Swapnil Gaikwad** Department of Biotechnology, Engineering School of Lorena, University of São Paulo, São Paulo, Brazil

**Buana Girisuta** Institute of Chemical and Engineering Sciences, Jurong Island, Singapore

**Hero Jan Heeres** Department of Chemical Engineering, University of Groningen, Groningen, AG, The Netherlands

**Jianli Hu** Department of Chemical and Biomedical Engineering, West Virginia University, Morgantown, WV, USA

**David Ibarra** Forestry Products Department, INIA-CIFOR, Madrid, Spain

**Akihiko Kondo** Graduate School of Science, Technology and Innovation, Kobe University, Kobe, Hyogo, Japan

Department of Chemical Science and Engineering, Graduate School of Engineering, Kobe University, Kobe, Japan

RIKEN Center for Sustainable Resource Science, Yokohama, Kanagawa, Japan

**Bing-Zhi Li** Key Laboratory of Systems Bioengineering (Ministry of Education), School of Chemical Engineering and Technology, SynBio Research Platform, Collaborative Innovation Center of Chemical Science and Engineering (Tianjin), Tianjin University, Tianjin, People's Republic of China

**Qiang Li** National Key Laboratory of Biochemical Engineering, Institute of Process Engineering, Chinese Academy of Sciences, Beijing, China

University of Chinese Academy of Sciences, Beijing, China

**Wen-Chao Li** Key Laboratory of Systems Bioengineering (Ministry of Education), School of Chemical Engineering and Technology, SynBio Research Platform, Collaborative Innovation Center of Chemical Science and Engineering (Tianjin), Tianjin University, Tianjin, People's Republic of China

**Rafał M. Łukasik** Laboratório Nacional de Energia e Geologia, I.P., Unidade de Bioenergia, Lisbon, Portugal

**Huixia Luo** Department of Chemistry, Princeton University, Princeton, NJ, USA

**Mark Mascal** Department of Chemistry, University of California Davis, Davis, CA, USA

**Kelly Johana Dussán Medina** Department of Biochemistry and Technological Chemistry, Institute of Chemistry, State University of São Paulo – UNESP, São Paulo, Brazil

**Ana Rita C. Morais** Laboratório Nacional de Energia e Geologia, I.P., Unidade de Bioenergia, Lisbon, Portugal

LAQV-REQUIMTE, Departamento de Química, Faculdade de Ciências e Tecnologia, Universidade NOVA de Lisboa, Caparica, Portugal

**Antonio D. Moreno** Department of Energy, CIEMAT, Madrid, Spain

**Yoshinao Nakagawa** Department of Applied Chemistry, School of Engineering, Tohoku University, Sendai, Japan

**José R. Ochoa-Gómez** TECNALIA, Division of Energy and Environment, Biorefinery Department, Parque Tecnológico de Álava, Miñano, Spain

**Ayumu Onda** Research Laboratory of Hydrothermal Chemistry, Faculty of Science, Kochi University, Kochi, Japan

**Guilherme Fernando Dias Peres** Department of Biotechnology, Engineering School of Lorena, University of São Paulo, São Paulo, Brazil

**Lei Qin** Key Laboratory of Systems Bioengineering (Ministry of Education), School of Chemical Engineering and Technology, SynBio Research Platform, Collaborative Innovation Center of Chemical Science and Engineering (Tianjin), Tianjin University, Tianjin, People's Republic of China

**Tomás Roncal** TECNALIA, Division of Energy and Environment, Biorefinery Department, Parque Tecnológico de Álava, Miñano, Spain

**Masazumi Tamura** Department of Applied Chemistry, School of Engineering, Tohoku University, Sendai, Japan

**Ruly Terán-Hilares** Department of Biotechnology, Engineering School of Lorena, University of São Paulo, São Paulo, Brazil

**Elia Tomás-Pejó** Biotechnological Processes for Energy Production Unit, IMDEA Energía, Móstoles, Spain

**Keiichi Tomishige** Department of Applied Chemistry, School of Engineering, Tohoku University, Sendai, Japan

**Xinli Tong** Tianjin Key Laboratory of Organic Solar Cells and Photochemical Conversion, School of Chemistry and Chemical Engineering, Tianjin University of Technology, Tianjin, People's Republic of China

**Yota Tsuge** Institute for Frontier Science Initiative, Kanazawa University, Kanazawa, Ishikawa, Japan

**Jianmin Xing** National Key Laboratory of Biochemical Engineering, Institute of Process Engineering, Chinese Academy of Sciences, Beijing, China

University of Chinese Academy of Sciences, Beijing, China

**Song Xue** Tianjin Key Laboratory of Organic Solar Cells and Photochemical Conversion, School of Chemistry and Chemical Engineering, Tianjin University of Technology, Tianjin, People's Republic of China

**Kai Yan** School of Engineering, Brown University, Providence, RI, USA

**Ying-Jin Yuan** Key Laboratory of Systems Bioengineering (Ministry of Education), School of Chemical Engineering and Technology, SynBio Research Platform, Collaborative Innovation Center of Chemical Science and Engineering (Tianjin), Tianjin University, Tianjin, People's Republic of China

**Zehui Zhang** Key Laboratory of Catalysis and Materials Sciences of the State Ethnic Affairs Commission & Ministry of Education, College of Chemistry and Material Science, South-Central University for Nationalities, Wuhan, China

**Peng Zhou** Key Laboratory of Catalysis and Materials Sciences of the State Ethnic Affairs Commission & Ministry of Education, College of Chemistry and Material Science, South-Central University for Nationalities, Wuhan, China

**Jia-Qing Zhu** Key Laboratory of Systems Bioengineering (Ministry of Education), School of Chemical Engineering and Technology, SynBio Research Platform, Collaborative Innovation Center of Chemical Science and Engineering (Tianjin), Tianjin University, Tianjin, People's Republic of China

# About the Editors

**Zhen Fang** is professor and leader of the biomass group in Nanjing Agricultural University. He is the inventor of the "fast hydrolysis" process. He is listed in the "Most Cited Chinese Researchers" in energy for 2014, 2015 and 2016 (Elsevier-Scopus). Professor Fang specializes in thermal/biochemical conversion of biomass, nanocatalyst synthesis and its applications, and pretreatment of biomass for biorefineries. He obtained his PhDs from China Agricultural University (biological and agricultural engineering, Beijing) and McGill University (materials engineering, Montreal). Professor Fang is associate editor of *Biotechnology for Biofuels* and is serving on editorial boards of major international journals in energy.

**Richard L. Smith, Jr.** is professor of chemical engineering at the Graduate School of Environmental Studies, Research Center of Supercritical Fluid Technology, Tohoku University, Japan. Professor Smith has a strong background in physical properties and separations and obtained his PhD in chemical engineering from the Georgia Institute of Technology (USA). His research focuses on developing green chemical processes especially those that use water and carbon dioxide as solvents in their supercritical state. He has expertise in physical property measurements and in separation techniques with ionic liquids and has published more than 200 scientific papers, patents, and reports in the field of chemical engineering. Professor Smith is the Asia regional editor for the *Journal of Supercritical Fluids* and has served on editorial boards of major international journals associated with properties and energy.

**Xinhua Qi** is professor of environmental science at the Agro-Environmental Protection Institute (AEPI), Ministry of Agriculture, China. Professor Qi obtained his PhD from the Department of Environmental Science, Nankai University, and has worked there until 2015. Professor Qi has a strong background in environmental treatment techniques in water and in chemical transformations of biomass. His research focuses on the synthesis of functional carbonaceous materials from biomass and their applications in the catalytic conversion of biomass resources into chemicals and biofuels. Professor Qi has written/edited five books and three book chapters and has published more than 70 peer-reviewed scientific papers with a number of papers being in top-ranked international journals. These papers have been cited over 1700 times in ISI Web of Science.

# Part I
# Production of Sugars

# Chapter 1
# Hydrolysis of Lignocellulosic Biomass to Sugars

**Lei Qin, Wen-Chao Li, Jia-Qing Zhu, Bing-Zhi Li, and Ying-Jin Yuan**

**Abstract** Lignocellulosic biomass is a widely available resource that can be used to produce renewable chemicals. Conversion of lignocellulosic biomass into sugars is one of the major challenges in producing biofuels and chemicals, because inherent biomass recalcitrance hinders the efficient conversion. The most available method in industry is to combine thermochemical pretreatment with enzymatic hydrolysis treatment. In this chapter, a brief introduction into the characterization of lignocellulosic biomass is presented. Some developing and promising pretreatment methods are introduced. Process description, reaction mechanisms and developments in each pretreatment method are reviewed. Enzyme systems for lignocellulose hydrolysis and the influencing factors to enzymatic hydrolysis are presented and promising hydrolysis strategies are outlined.

**Keywords** Lignocellulose • Biomass • Cellulose • Hemicellulose • Lignin • Pretreatment • Enzymatic hydrolysis • Cellulase • Inhibition

## 1.1 Introduction

Lignocellulosic biomass is regarded as a potential resource to produce mixed sugars, of which more than half of the amount of material consists of carbohydrates. The most commonly used method for producing sugars from lignocellulosic biomass is enzymatic hydrolysis. Several related technologies have appeared in the past years so that this conversion process can be carried out. Now, the objective is to make it more cost-competitive in today's markets. Plants evolve the recalcitrance to resist microbial and enzymatic deconstruction. Several factors contributing to the recalcitrance have been well-studied in past years and these findings are instructive for enhancing discussion on enzymatic digestibility of lignocellulosic biomass.

L. Qin • W.-C. Li • J.-Q. Zhu • B.-Z. Li (✉) • Y.-J. Yuan
Key Laboratory of Systems Bioengineering (Ministry of Education), School of Chemical Engineering and Technology, SynBio Research Platform, Collaborative Innovation Center of Chemical Science and Engineering (Tianjin), Tianjin University,
Weijin Road 92, Nankai District, Tianjin 300072, People's Republic of China
e-mail: bzli@tju.edu.cn



Z. Fang et al. (eds.), *Production of Platform Chemicals from Sustainable Resources*, Biofuels and Biorefineries 7, DOI 10.1007/978-981-10-4172-3_1

## 1.1.1 *Lignocellulosic Biomass*

### 1.1.1.1 Biomass Cell Wall Structure

Plant cell walls are the primary materials of lignocellulosic biomass. Lignocellulosic materials include herbaceous plant (*e.g.* corn stover and switchgrass), softwood (*e.g.* spruce), hardwood (*e.g.* aspen), and other waste materials (*e.g.* distillers' grains and waste paper). The composition and structure of plant cell wall determine the design of downstream processes in a biorefinery. Plants consist of an orderly arrangement of cells with walls composed of varying amounts of a mixture of cellulose (~40%), hemicellulose (~15–35%), lignin (~15–30%) and other components (*e.g.* pectin, proteins, ash, lipids, wax, and other extractives) [1]. The plant body is like a building made of "osmotic bricks", which means that pressure fastens the cells by creating tension in the cell walls [2].

Plant cell wall structures in multi-scales are displayed in Fig. 1.1. There is a lumen in the supporting cell wall after the protoplast is dead, which has the function of transporting water and nutrients. The outer layer of the cell wall is primary cell wall which is a heterogeneous mixture of cellulose, hemicellulose and pectin. The secondary wall is inside of the primary cell wall and can be divided into three sublayers. The secondary wall accounts for the majority of cell wall mass, which mainly consists of cellulose, hemicellulose and lignin. Cellulose microfibrils are covered and cross-linked by amorphous materials (mainly hemicellulose and lignin) [3, 4]. Lignin also cross-links with hemicellulose and forms lignin-carbohydrate complexes (LCC).

### 1.1.1.2 Cellulose

Cellulose (*i.e.* glucan) is unbranched homopolymers of $\beta$-(1,4)-linked D-glucose. Native cellulose molecules in higher plants are in fibril and crystal form, which is named as cellulose I. The visually structures of cellulose are cellulose microfibrils with the diameter of 2~10 nm, which are cross-linked by other components such as glucoronoarabinoxylan and xyloglucans [2]. Microfibrils, synthesized at the plasma membrane, are unbranched fibrils and composed of 30~36 glucan chains aggregated

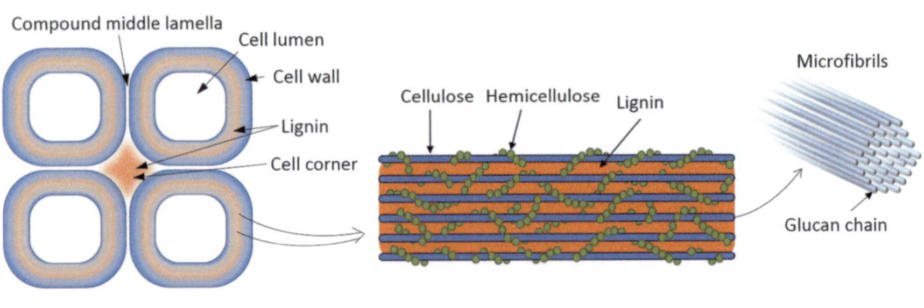


**Fig. 1.1** Visualization of plant cell wall structures from micrometer scale to nano scale

through intra- and inter-hydrogen bonding and van der Waals forces to form the crystalline structures [4]. Microcrystalline cellulose is divided into two different crystal phases: $I_\alpha$ and $I_\beta$. Cellulose $I_\alpha$ and $I_\beta$ are the primary allomorphic forms of cellulose in microorganisms and higher plants, respectively. Besides cellulose I, other allomorphs can be produced through various thermochemical treatments. Native cellulose can transform to cellulose II irreversibly by caustic mercerization or regeneration (such as ionic liquids) and to cellulose III by liquid ammonia or amine treatment. The lattice spacing distances of cellulose III and cellulose II are longer than that of cellulose I due to the alteration of the networks of intra- and inter-hydrogen bindings [5, 6].

#### 1.1.1.3 Hemicellulose

Hemicellulose is a class of heterogeneous polysaccharides, generally accounting for 15~35% of plant biomass. Hemicellulose may contain pentoses (*β*-D-xylose, *α*-L-arabinose), hexoses (*β*-D-mannose, *β*-D-glucose and *α*-D-galactose) and/or uronic acids (*α*-D-glucuronic, *α*-D-4-*O*-methylgalacturonic and *α*-D-galacturonic acids). Other sugars (*α*-L-rhamnose and *α*-L-fucose) may also exist in small amounts. Acetyl groups may partially substitute the hydroxyl groups of sugars. Xylan, which is an ideal polymer of xylose, is the most abundant relevant hemicellulose. Xylan is the main hemicellulose component in the secondary cell walls, accounting for about 20~30% in hardwoods and herbaceous plants. Mannan-type hemicelluloses (for example, glucomannan and galactoglucomannan) are the major types of hemicellulose in the secondary cell wall of softwoods. Different hemicellulose structures can be detected depending on the different biological origins [7].

#### 1.1.1.4 Lignin

Lignin is abundant in the middle lamella (20~40%) and the interlamination of secondary cell walls (60~80%). Lignin is a complex macromolecular polymer synthesized mainly from three *p*-hydroxycinnamyl alcohols: *p*-coumaryl, coniferyl and sinapyl alcohols. These monolignols produce the *p*-hydroxyphenyl (H), guaiacyl (G), and syringyl (S) subunits, respectively. Lignin subunits are cross-linked by a series of bonds (such as *β*-O-aryl ether and *β*-*β*' linkages). Lignin may cross-link to polysaccharides via coupling of feruloylated xylans [8, 9].

### *1.1.2 Conversion Pathways from Lignocellulose Biomass to Sugars*

There are three major hydrolysis processes used to produce sugars from lignocellulose: dilute-acid hydrolysis, concentrated-acid hydrolysis, and enzymatic hydrolysis.

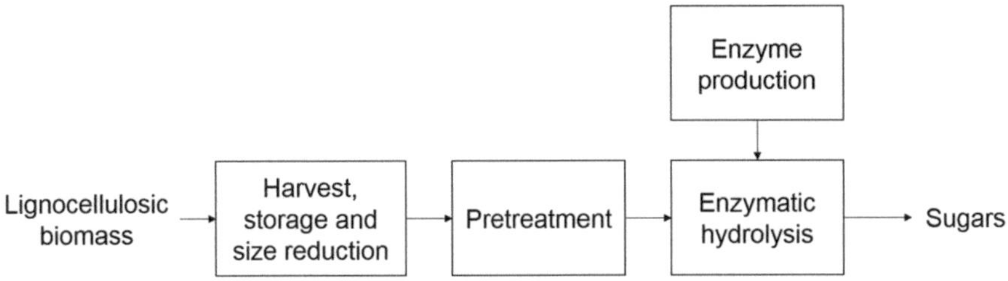


**Fig. 1.2** Bioconversion process from lignocellulosic biomass to sugars

In the dilute-acid process, the concentration of sulfuric acid can be over a wide range of values (0.4~20%). Hemicellulose can be easily hydrolyzed by dilute acids under moderate conditions, but much more severe conditions are needed for cellulose hydrolysis, namely, the reaction needs to be carried out at high temperatures (up to 190 °C) and pressures [10].

The concentrated-acid hydrolysis process can produce high glucose yields from lignocellulose. Sulfuric acid concentration used is in the range of 10~30%. Lower operating temperatures and pressures are adequate during the concentrated-acid hydrolysis process.

However, acid hydrolysis processes have several disadvantages limiting the application to industry, for example, acid hydrolysis may give a surprisingly large ratio of sulfuric acid to cellulose on dry weight basis (varying from 0.03 to 5), the degradation of sugars to by-products is hard to control, the acid is difficult to be separated and recovered from the sugar products, large amounts of acid may contaminate the environment, and dilute acid is corrosive to equipment although corrosion is less of an issue at very high acid concentrations.

Enzymatic hydrolysis, which is a bioconversion process, is the most common method of producing sugars from lignocellulosic biomass due to its moderate operating conditions, high specificity and affinity, and non-pollution to environment. Enzymes produced by various categories of microorganisms are enable to break down lignocellulose into sugars. This chapter focuses mainly on the enzymatic hydrolysis to produce sugars from lignocellulose. Prior to the enzymatic hydrolysis of lignocellulose, several operation units are required, including feedstock harvest and storage, enzyme production, and pretreatment (Fig. 1.2).

### *1.1.3 Biomass Recalcitrance*

The enzymatic digestibility of lignocellulosic biomass is hindered by many factors, including structural and compositional factors. The natural resistance of plant cell walls to enzymatic deconstruction is collectively known as "biomass recalcitrance". The biomass recalcitrance is mainly attributed to: (1) the mechanical defenses from the epidermal tissue (particularly the cuticle and epicuticular waxes), the sclerenchymatous tissue, and the compact cell wall connection; (2) the protection of lignin; (3) the high crystallinity and degree of polymerization of cellulose [4, 11, 12]. Biomass needs pretreatment so that the cellulose in the plant fibers can be exposed to enzymes. Increasing accessibility of cellulose to enzymes is the fundamental factor to break down the biomass recalcitrance and thus enhances enzymatic digestibility. The specific factors contributing to biomass recalcitrance studied in details during past years are stated as follows:

*Lignin, Hemicellulose and Pectin Content* In general, cellulose-rich substrates with more highly accessible cellulose require low enzyme loading to obtain efficient hydrolysis. The removal of hemicellulose and lignin relieves the physical barrier and thus increases the accessibility of the enzymes to the cellulose chains [13, 14]. Degradation of either hemicellulose or lignin can increase the cellulose conversion [15].

*Cellulose Crystallinity* Reduction of crystallinity remarkably increases the hydrolysis rate and reduces the enzyme dosage required to achieve high cellulose digestibility [16, 17]. Moreover, it is believed cellulose allomorph has a strong influence on cellulose digestibility, which decreases in the following order: amorphous cellulose > cellulose III > cellulose II > cellulose I [18, 19]. However, the variation of the degree of polymerization of cellulose has shown to be insensitive to enzymatic digestibility [20].

*Porosity* The increasing biomass fiber porosities has a significantly positive influence on the efficiency of enzymatic hydrolysis [20].

## 1.2 Pretreatment

Pretreatment is an important process for biological conversion of cellulosic biomass to sugars and for fermentation to biofuels and other chemicals. Pretreatment is aimed to break down the biomass recalcitrance, make cellulose more accessible to the enzymes, and improve enzymatic digestibility. The role of pretreatment in practice is generally the disruption of the naturally resistant lignin-carbohydrate shield and/or the crystalline structure of cellulose. Pretreatment is regarded as one of the most expensive processing steps in the bioconversion of lignocellulosic biomass. The average cost of pretreatment has been estimated to be about 20 dollar/ton biomass in 2005 [21]. Even so, pretreatment has great potential to improve enzymatic efficiency and lower enzyme dosage and cost. Pretreatment also has the pervasive impact on processing of all other operations, such as the handling of the liquid streams generated and solids from pretreatment, waste treatment, and the potential production of co-products [22]. Therefore, the choice of pretreatment technology should take into account not only the enzymatic digestibility of cellulose, but also the operation pattern and the compatibility with the overall process [22, 23]. There are several key aspects to be taken into consideration for low-cost and advanced pretreatment process (Fig. 1.3):

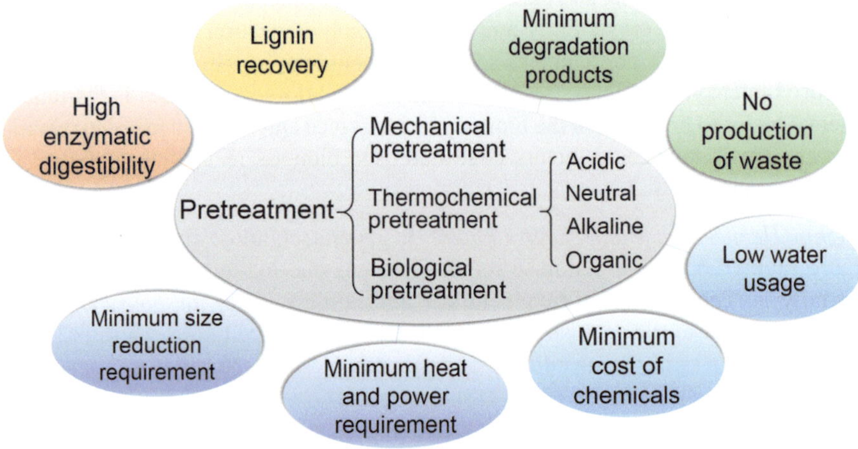


**Fig. 1.3** Classification and requirements of lignocellulose pretreatment

1. High enzymatic digestibility of pretreated solid. Generally, cellulose conversion of pretreated biomass needs to be as high as possible. In addition, relative low enzyme loading is desired due to the high cost of enzymes [24].
2. Lignin recovery. Lignin and the derived materials should be recovered for their conversions to valuable co-products in subsequent processes.
3. Minimum degradation products. Degradation products from pretreatment of lignocellulose materials can be divided into the following classes: carboxylic acids (*e.g.* acetic, formic and levulinic acid), furan derivatives (*e.g.* furfural and 5-HMF), and phenolic compounds, all of which have inhibition effects to enzymatic hydrolysis to some extent.
4. Minimum heat and power requirements. The heat and power for pretreatment should be low or compatible for thermal integration with other processes.
5. Low water usage. The use of raw biomass at high dry matter content can reduce energy consumption during pretreatment.
6. Minimum cost of chemicals. The need for chemicals in pretreatment and subsequent neutralization should be minimal and inexpensive. Otherwise, the chemicals should be recoverable and the recovery process should be inexpensive.
7. Minimum size reduction requirement. Milling of biomass to small particles is energy intensive and costly. Pretreatment technologies appropriate for large particle sizes are desirable.
8. No production of waste. The chemicals or materials produced should not present processing or disposal challenges.

Over the years, many technologies have been developed for realizing low cost and high sugar yields from both cellulose and hemicellulose, which can be categorized as mechanical, thermochemical and biological processes. In many cases, the combinations of these approaches are necessary to obtain high enzymatic digestion.

In this section, various pretreatment methods as well as the operation and mechanism are reviewed. Although some pretreatment methods have been applied in industrial processes, none are absolutely predominant among the methods. More understanding of the effect of pretreatment on structural and molecular level and more novel pretreatment technologies still need to be developed. Besides pretreatment methods, some preceding processes, such as harvest, storage, size reduction and fractionation that have considerable impact on pretreatment efficiency are discussed.

## *1.2.1 Thermochemical Pretreatment*

Thermochemical pretreatment employing high temperature with or without chemical catalysts to disrupt the biomass recalcitrance occupies the most important place in pretreatment processes, since most lignocellulosic feedstocks require high intensity to achieve satisfactory enzymatic digestibility.

Thermochemical pretreatment can be categorized according to catalyst property as acidic, neutral, alkaline and organic catalyzed pretreatment. Acidic pretreatment generally results in the hydrolysis of hemicellulose to monomeric sugars and other by-products. Neutral pretreatment hydrolyses hemicellulose through autohydrolysis. Alkaline pretreatment makes lignin and partial hemicellulose soluble [25]. A general function of acidic and alkaline pretreatment on lignocellulose is demonstrated in Fig. 1.4. Organic pretreatment selectively dissolves carbohydrates and/or lignin. Pretreatment can also be categorized as chemical and physicochemical pretreatment methods. Physicochemical pretreatment combines heat and rapid pressure release, such as steam explosion and ammonia fiber expansion. Various pretreatment methods listed in Table 1.1 are introduced.

### 1.2.1.1 Dilute Acid (DA) Pretreatment

Dilute acid (DA) pretreatment appears as a candidate method for industrial applications and for treating a wide range of lignocellulosic biomass. Dilute acid pretreatment has historically been conducted at temperatures of 120~180 °C, acid concentrations of 0.2~2% (w/w) and reaction times within 60 min. The most widely used approach is based on sulfuric acid. Nitric acid, hydrochloric acid, phosphoric acid and organic acids have also been shown to be effective.

Combined severity (CS), a function of pretreatment time, temperature and pH, is usually applied to evaluate the pretreatment severity. The combined severity is defined as [26, 27]:

$$CS = \log R_0 - pH \tag{1.1}$$

$$R_0 = t \times \exp[(T_H - 100)/\omega] \tag{1.2}$$

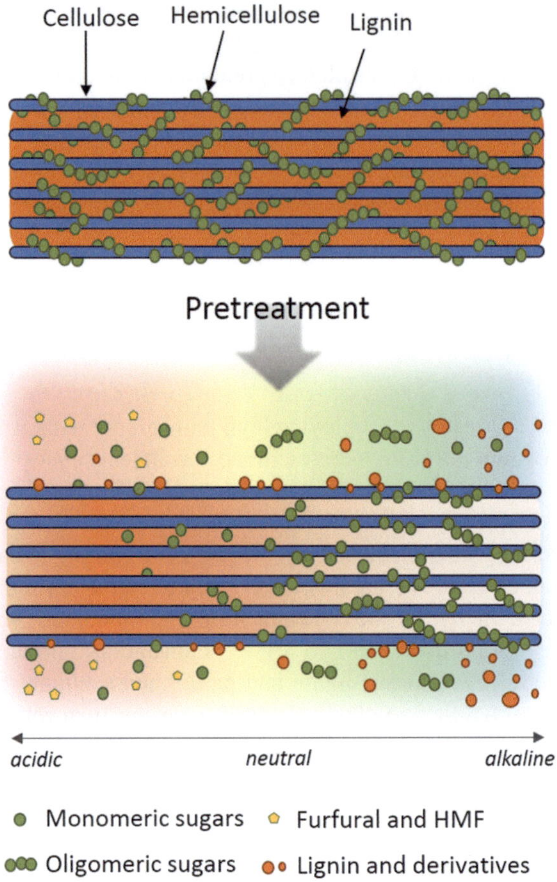


**Fig. 1.4** General effects of acidic and alkaline pretreatment on solubilization of hemicellulose and lignin in lignocellulose

in which $R_0$ is a severity factor; $t$ is reaction time in minutes; $T$ is the hydrolysis temperature in °C; $\omega$ is an empirical parameter, usually assigned to the value of 14.75, which equates to a reaction that doubles in reaction rate for every 10 °C increase in temperature [28].

The remarkable improvement of enzymatic digestion of dilute acid pretreated biomass is correlated to the removal of hemicellulose. Under acidic conditions, hemicellulose is more easily degraded to monomeric sugars and by-products than cellulose due to the amorphous structures. The majority of lignin and cellulose components are retained in the acid-insoluble residues. Soluble sugars hydrolyzed by acid during pretreatment can be further degraded to furfural (which is produced from pentoses) and 5-hydroxymethylfurfural (5-HMF, which is produced from hexoses). The degradation routine of lignocellulosic biomass during acidic pretreatment is elucidated in Fig. 1.5. Acid hydrolysis is also an important method to

**Table 1.1** Summary of thermochemical pretreatment methods for lignocellulosic biomass

| Pretreatment method | Typical pretreatment condition | Advantages | Disadvantages |
|---|---|---|---|
| DA[a] | 160 °C, 0.5% sulfuric acid, 15 min, solid to liquid ratio of 1:10 [38] | Low cost; reduces hemicellulose content | Equipment erosion; chemical recovery; formation of toxic components |
| SE[b] | 200 °C, 10 min [49] | Cost-effective; solubilizes hemicellulose; no chemical used | Equipment cost; formation of toxic components |
| $SO_2$ explosion | 190 °C, 6% $SO_2$, 5 min [43] | Cost-effective; solubilizes hemicellulose | Equipment cost; equipment erosion; formation of toxic components |
| LHW[c] | 190 °C, 15 min, solid to liquid ratio of 1:6 [50] | Solubilizes hemicellulose; few degradation products | Equipment cost |
| Flowthrough | 170 °C, 25 mL/min, 35 min [54] | Solubilizes hemicellulose and removes lignin | Equipment cost |
| AFEX[d] | 135 °C, water to solid ratio of 1:1; ammonia to solid ratio of 1:1, 45 min [60] | Low degradation of carbohydrates; low formation of toxic components; low water usage | Pressure requirement; cost of ammonia; chemical recovery |
| EDA[e] | 120 °C, EDA to solid ratio of 1:1, 60 min [19] | Low water usage; low equipment cost; low pressure | Cost of EDA; chemical recovery |
| SAA[f] | 75 °C, 15% aqueous ammonia, 48 h, solid to liquid ratio of 1:12 [68] | Reduce lignin and hemicellulose content; low pressure; low formation of toxic components | Long pretreatment period; chemical recovery |
| ARP[g] | 170 °C, 15% aqueous ammonia, 5 mL/min, 10 min [69] | Reduce lignin and hemicellulose content; low formation of toxic components | Equipment cost; chemical recovery |
| Lime | 55 °C, 0.5 g $Ca(OH)_2$/g solid, 4 weeks [70] | Reduce lignin content; low cost of chemicals | Long pretreatment period; chemical recovery |
| ILs[h] | 160 °C, solid to [C2mim][Ac] ratio of 1:33 [74] | Extracts cellulose and reduce crystallinity | Cost of ILs; chemical recovery |
| CELF[i] | 150 °C, 50% (v/v) THF and 0.5% sulfuric acid, solid to liquid ratio of 1:20, 25 min [81] | Reduces lignin and hemicellulose content; separation and recovery of lignin | Cost of organic solvents; chemical recovery |

(continued)

**Table 1.1** (continued)

| Pretreatment method | Typical pretreatment condition | Advantages | Disadvantages |
|---|---|---|---|
| COSLIF[j] | 50 °C, 85% $H_3PO_4$, solid to liquid ratio of 1:8, 60 min, washing by ethanol [82] | Low temperature and pressure; effective for tough feedstocks | Cost of acid and recovery |
| SPORL[k] | 180 °C, 9% sodium bisulfite and 1.8% sulfuric acid, solid to liquid ratio of 1:20, 30 min [83] | Reduce lignin and hemicellulose content; low formation of toxic components | Cost of chemicals; chemical recovery; Equipment erosion |
| Wet oxidation | 185 °C, 0.6 MPa $O_2$, 10 min, solid to liquid ratio of 1:6 [85] | Reduce lignin content; low formation of toxic components | Equipment cost |

[a]Dilute acid. [b]Steam explosion. [c]Liquid hot water. [d]Ammonia fiber expansion. [e]Ethylenediamine. [f]Soaking in aqueous ammonia. [g]Ammonia recycled percolation. [h]Ionic liquids. [i]Co-solvent-enhanced lignocellulosic fractionation. [j]Cellulose and organic-solvent-based lignocellulose fractionation. [k]Sulphite pretreatment to overcome recalcitrance of lignocellulosics

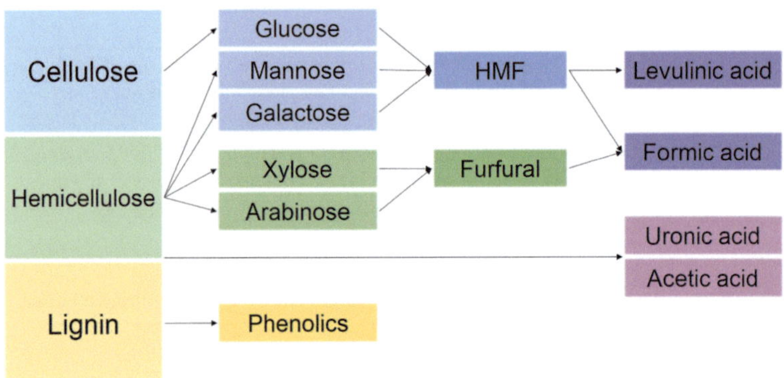


**Fig. 1.5** Degradation of lignocellulosic biomass in acidic pretreatment

produce xylose and/or furfural in the industry. The removal of hemicellulose increases when the combined severity increases. Xylose yield increases to a maximum between a combined severity of 2~3 [29–31]. The yields of sugar degradation compounds (furfural and 5-HMF) and lignin degradation compounds increase with increasing combined severity.

With increasing combined severity (>2), a progressive increase in the Klason lignin content of the acid pretreated material is observed, which is attributed to the acid catalyzed dehydration polymerization of carbohydrates, termed as "pseudo-lignin" [32]. The droplets of pseudo-lignin cover a range of sizes from 0.3 to 8.0 μm and significantly retard cellulose hydrolysis [33]. The addition of DMSO can effectively suppress pseudo-lignin formation, since DMSO preferentially solvates and stabilizes furans which is the key intermediate to form pseudo-lignin [34].

Besides removal of hemicellulose, dilute acid pretreatment results in the cellulose fibril swelling, reduction in the molecular weight of cellulose, and the increase in the crystallinity index of pretreated biomass [35–37].

During pretreatment, redistribution of lignin location and degradation of cross-linking between aromatic subunit also occurs, resulting in the loss in matrix between neighboring cell walls and increasing porosity [38].

Dry dilute acid (DDA) pretreatment process has promising features compared with traditional dilute acid pretreatment and employs high solid loading (up to 70%) with a helically agitated impeller. The dry dilute acid pretreatment can achieve high hydrolysis digestion with several advantages, such as low steam consumption, no pre-hydrolysate solution generation, less sugar loss and less by-product generation [39, 40].

Organic acids, such as maleic acid and oxalic acid, have some advantages compared with sulfuric acid. Organic acids result in lower concentrations of degradation products and higher concentrations of oligomers due to their higher dissociation constants ($pK_a$) compared with mineral acids [31, 41, 42].

#### 1.2.1.2 Steam Explosion (SE) Pretreatment

Steam explosion (SE) pretreatment is one of the most widely applied technologies for pretreating lignocellulosic biomass. In this physicochemical process, biomass is subjected to pressurized steam (160~220 °C) for a short period of time and then suddenly depressurized by the explosive decompression.

During pretreatment, biomass is heated up by the condensation of steam leading to the microporous structure being filled with liquid hot water. Water acts as a weak acid, which lowers the pH to 3~4 and initiates the depolymerization of hemicellulose. The hydrolysis of acetyl groups in hemicellulose results in further acidic environment leading to cleavage of more glycosidic bonds. This process is termed as "autohydrolysis".

Impregnating biomass in sulfuric acid or sulfur dioxide before pretreatment, known as acid explosion and $SO_2$ explosion respectively, decreases pretreatment time and temperature and improves enzymatic digestion [43].

After explosive decompression, biomass particle size, fiber length and cellulose crystallinity are reduced, while surface area, porosity and cellulose accessibility are increased [44–46], resulting in much higher enzymatic digestibility compared with that without the explosion process [47].

After pretreatment, hemicellulose can degrade to xylooligomer, xylose, or furfural. Lignin is redistributed and minorly degraded under acidic conditions to some extent. A pretreatment severity factor between 3~4 is optimal for maximum sugar yields. Similar to dilute acid pretreatment, higher temperatures result in the increased removal of hemicellulose and enhanced cellulose enzymatic digestibility, but also promote for generation of furfural and 5-HMF [48, 49].

#### 1.2.1.3 Liquid Hot Water (LHW) Pretreatment

Liquid hot water (LHW) pretreatment is a hydrothermal process which does not employ any catalyst or chemicals. Pressure is applied to maintain water in the liquid state at elevated temperatures (160~240 °C).

The main effect of liquid hot water is solubilization and degradation of hemicellulose, making the cellulose more accessible. Similar to steam explosion, autohydrolysis occurs with hemicellulose in liquid hot water. The pH can be controlled between 4 and 7 during the pretreatment. At this pH range, hemicellulosic sugars are retained in oligomeric form and the formation of monomeric sugars is minimized. Hence, the formation of by-products is also very low [50]. LHW pretreated biomass presents higher crystallinity and thermal stability than untreated biomass due to the removal of the non-crystalline materials [51]. Lignin depolymerization facilitates lignin extraction and the extractability of syringyl groups is higher than guaiacyl groups [52]. The condensation reactions of lignin molecules also occurs during pretreatment and leads to the formation of spherical lignin droplets (diameters are 1~10 μm) on the cell wall surface [53]. The lignin migration increases the plant cell wall porosity [54].

Researchers found that the lower $\omega$ value corresponding to the increased relative contribution of temperature to severity factor (Eq. 1.2) is better to predict the pretreatment responses (*e.g.* hemicellulose solubilization and enzymatic digestibility of cellulose) [55].

Flow-through LHW pretreatment systems can remove more hemicellulose and lignin than batch mode systems. Increasing temperature during flow-through pretreatment enhances lignin removal to almost 100%. Dilute acid can also be applied in flow-through systems [56].

Furthermore, a multi-stage LHW pretreatment has been explored [55]. The first low-severity pretreatment is to partially degrade hemicellulose and minimize the formation of furans. The subsequent pretreatment is performed at over 200 °C to increase cellulose accessibility to enzymes.

#### 1.2.1.4 Ammonia Fiber Expansion (AFEX) Pretreatment

Ammonia fiber expansion (AFEX) process treats lignocellulose (mainly herbaceous plants) with high-pressure liquid ammonia at temperatures between 60 and 160 °C and then releases the pressure, resulting in rapid expansion of ammonia gas that causes swelling and physical disruption of biomass fibers. Other parameters affecting enzymatic digestibility of pretreated biomass include ammonia loading, biomass moisture and reaction time [57]. The combined chemical and physical effect increases lignocellulose susceptibility to enzymatic attack. AFEX is a dry-to-dry process (pretreatments can be efficient for a wide range of moisture content, typically 0~100%), which does not produce liquid fraction after pretreatment. Ammonia recovery and recycle is feasible and thus this diminishes total cost.

AFEX pretreatment completely preserves carbohydrates and no furans are generated. Only a small amount of hemicellulose and lignin becomes solubilized. Hemicellulose is mainly converted to xylooligomers and other oligomeric sugars. Cellulose crystallinity after pretreatment is decreased [16].

Ammonia breaks down the lignin-carbohydrate complex bonds and ester linkages in lignin, resulting in the formation of acetamide and various phenolic amides [58, 59]. The formation of amides provides nutrients in the subsequent fermentation [60, 61]. As the ammonia evaporates, degradation products (rich in lignin, hemicellulose and cell wall decomposition products) redeposit on outer cell wall surfaces. As a result, nanoporous tunnel-like networks are formed within the cell walls, which greatly enhances cellulose accessibility [62].

Pre-soaking biomass in hydrogen peroxide before AFEX pretreatment can enhance delignification and enzymatic digestibility of both cellulose and hemicellulose [63].

A modified liquid ammonia pretreatment named as extractive ammonia (EA) pretreatment has been developed, which employs much higher ammonia loading (2~3 g ammonia/g biomass). This method converts native crystalline cellulose $I_\beta$ into a more highly digestible allomorph, cellulose III, and simultaneously extracts up to nearly half the amount of lignin from lignocellulosic biomass while preserving all polysaccharides. EA pretreated corn stover requires low enzyme dosage and give high sugar yields compared with traditional AFEX pretreatment [64].

#### 1.2.1.5 Ethylenediamine (EDA) Pretreatment

Ethylenediamine (EDA) pretreatment is a novel pretreatment method that can be operated at ambient pressure and temperatures of 40~180 °C due to the particular physical and chemical properties of ethylenediamine. It is also a dry-to-dry process without the requirement of water addition. Cellulose enzymatic digestibility of pretreated corn stover can reach 95% with the optimal condition (150 °C and 0.8 mL EDA/g solid).

Similar to AFEX pretreatment, EDA can also break down and ammonolyze ester bonds of *p*-coumarate and ferulate units in lignin. Lignin is relocated to the cell wall surface of biomass resulting in cell wall delamination. Crystal cellulose I is transformed into cellulose III or amorphous cellulose according to pretreatment conditions. After pretreatment, about 20% of hemicellulose and 40% of lignin are solubilized [19, 65].

Cadoxen, which is a solvent made by dissolving 5% cadmium oxide in 28% aqueous EDA, can also treat various lignocellulosic materials under ambient conditions. This method exhibits higher efficiency compared with the individual use of aqueous EDA. Cadoxen can be recovered and reused [66].

#### 1.2.1.6 Aqueous Ammonia (AA) Pretreatment

Aqueous ammonia (AA) pretreatment is a suitable method for many herbaceous materials. Aqueous ammonia (usually 5~20%, w/w) is known to react primarily with lignin and cause depolymerization of lignin and cleavage of lignin-carbohydrate linkages. After AA pretreatment, dozens of phenolics derived from lignin (such as vanillin and syringaldehyde) are solubilized in aqueous ammonia, which may have an inhibitory effect on enzymatic hydrolysis. Acetyl groups from hemicellulose are also removed, and partial hemicellulose is solubilized to oligomers in the liquid fraction, while cellulose remains intact in a solid fraction [67]. Biomass crystallinity increases after pretreatment due to the removal of the amorphous portion of biomass. The enzymatic digestibility is related with the removal of lignin and hemicellulose.

AA pretreatment can be undertaken at either low temperature/long pretreatment times (60~100 °C for several days, known as soaking in aqueous ammonia, SAA) [68] or high temperature/short pretreatment times (120~180 °C within 1 hour) [67].

Pretreatment in a flow-through system involves putting ammonia solution through a column reactor packed with biomass at elevated temperatures (160~180 °C). This method is known as ammonia recycled percolation (ARP) process since ammonia is separated and recycled [69]. The flow-through system removes more lignin and hemicellulose content than batch mode.

#### 1.2.1.7 Lime Pretreatment

Lime (calcium hydroxide) pretreatment removes amorphous materials such as lignin and hemicellulose and increases the crystallinity index, which conforms to the mechanism of alkaline pretreatment (such as aqueous ammonia and sodium hydroxide) [70]. Lime has been proven to be effective at temperatures from 50~150 °C. Lime pretreatment is much less expensive and has fewer safety requirements compared with sodium hydroxide or potassium hydroxide pretreatment.

Non-oxidative conditions during pretreatment are effective for the biomass with low lignin content (below approximately 18% lignin), whereas oxidative conditions are required for higher lignin content materials [70].

#### 1.2.1.8 Ionic Liquid (IL) Pretreatment

Ionic liquids (ILs) are salts, typically composed of large organic cations (*e.g.* 1-ethyl-3-methylimidazolium [Emim] and 1-n-butyl-3-methylimidazolium [$C_4$min]) and small inorganic anions (*e.g.* chloride and acetate), which exist as liquids at relatively low temperatures (<100 °C). Pretreatment is conducted at 60~160 °C for several hours with relatively low solids loading (usually 5%). Carbohydrates and lignin can be simultaneously dissolved in ILs. The solvent properties, dissolving capacity and pretreatment efficacy can be varied by adjusting the anion and the alkyl constituents of the cation [71]. There are two alternative operating modes: (1) dissolving biomass in IL solutions and precipitating cellulose by adding anti-solvent (*e.g.* water); (2) treating biomass with IL-water mixture directly [72, 73]. Both approaches can result in the similar outcomes. After the pretreatment, the ionic liquid liquor contains the majority of lignin and hemicellulose.

The pretreated cellulosic biomass exhibits extremely high enzymatic digestibility due to significant reduction of cellulose crystallinity and lignin content, and the increase of surface area [74]. IL anion can act with the inter- and intra-molecular hydrogen bonds of cellulose and cellulose I is transformed to cellulose II and amorphous cellulose with the removal of IL anion [75]. The formation of cellulose II greatly depends on the biomass type and pretreatment severity [76]. The depolymerization degree of hemicellulose and lignin in the ILs depends on the nature of the anion and cation.

Novel types of ILs that are easily prepared and elicit high cellulose conversion have been investigated, such as DBU-MEA-$SO_2$ and [Ch][AA] (cholinium amino acids) [77, 78].

#### 1.2.1.9 Organosolv Pretreatment

Organosolv pretreatment can be carried out in a large number of organic solvent systems with or without additional catalyst in a temperature range of 100~250 °C, while peracid pretreatment can be conducted under milder conditions, even at room temperature. The organic solvents that have frequently been used are divided into three classes: alcohols with low boiling points (*e.g.* methanol and ethanol), alcohols with high boiling points (*e.g.* ethylene glycol and glycerol), and other organic compounds (*e.g.* DMSO). The use of acid catalysts (*e.g.* hydrochloride acid or sulfuric acid) in organic solvents can reduce the temperature required and increase the removal of lignin and hemicellulose.

After organosolv pretreatment, cellulose remains in the solid fraction. Organic solvents can be recovered by distillation and recycled for pretreatment. The recovery of organic chemical after pretreatment can isolate lignin as a solid material [79]. The isolated lignin demonstrates the decreased molecular weight and increased polydispersity due to the cleavage of $\beta$-O-4 linkage [80].

A novel organosolv pretreatment called co-solvent-enhanced lignocellulosic fractionation (CELF) employing tetrahydrofuran miscible with aqueous dilute acid can obtain up to 95 % theoretical yield of fermentable sugars from corn stover and extremely reduce the enzyme dosage (2 mg enzyme/g glucan) [81].

#### 1.2.1.10 COSLIF Pretreatment

The cellulose and organic-solvent-based lignocellulose fractionation (COSLIF) process uses concentrated phosphoric acid to dissolve biomass at low temperatures (~50 °C) and then adds ethanol to precipitate the cellulose. It generates highly reactive amorphous cellulose, high substrate accessibility and high enzymatic cellulose digestibility. COSLIF is considered as a feedstock-independent pretreatment that is suitable for handling diverse kinds of biomass by only adjusting pretreatment reaction time [82].

#### 1.2.1.11 Sulfite Pretreatment

Sulfite pretreatment to overcome recalcitrance of lignocellulose (SPORL) is a method derived from the sulfite pulping process in the pulp and paper industry, which consists of sulfite treatment of wood chips with sodium bisulfite and sulfuric acid at 180 °C for 30 min followed by mechanical size reduction using disk refining. Size reduction after pretreatment significantly reduces electric energy consumption compared with pre-milling [83].

Enzymatic digestibility of sulfite pretreated biomass increases that is attributed to the lignin removal and lignin sulfonation. Hemicellulose is partially removed. The liquid fraction from pretreatment mainly contains lignosulfonate and degraded hemicellulose.

#### 1.2.1.12 Wet Oxidation Pretreatment

Wet oxidation is an oxidative pretreatment method that employs oxygen or air as catalyst. The oxidation is performed for 10~15 min at temperatures from 150 to 200 °C. The pretreatment process is exothermic when temperature is above 170 °C and oxygen is added, which reduces the total energy demand. It is an efficient method for solubilization of hemicelluloses and lignin to increase digestibility of cellulose. The main reaction during wet oxidation is the formation of acids from hydrolytic processes and oxidative reactions. Phenolic compounds derived from lignin are not end-products during wet oxidation because they are further degraded to carboxylic acids. Sodium carbonate addition can decrease the formation of furfural and 5-HMF by maintaining pH in the neutral to alkaline range [84, 85].

### *1.2.2 Biological Pretreatment*

Lignocellulosic materials are recycled and decomposed by microorganisms under ambient conditions in nature. The naturally occurring biological process can be adopted to address the major barriers in lignocellulose conversion. Biological pretreatments have mostly depended on the capacity of producing enzymes that can degrade lignin in fungi. White rot fungi have been considered as the most effective species for biological pretreatment, which are able to degrade all components of plant cell walls, especially lignin. Degradation of lignin enables white rot fungi to gain access to holocellulose, which is their real carbon and energy resource. Some white rot fungal species selectively degrade lignin or hemicellulose without decomposing cellulose. Various enzymes for lignin biodegradation are secreted by white rot fungi, such as laccases, lignin peroxidases (LiPs), manganese peroxidases (MnPs), manganese-independent peroxidases (MiPs) and versatile peroxidases (VPs). Laccases, LiPs and MnPs oxidize phenolic compounds to create phenoxy radicals. These radicals subsequently function as lignin-degrading peroxidase substrates. LiPs act in the oxidation of non-phenolic phenylpropanoid units, which results in cleavage of C-C and C-O linkages and degradation of lignin polymer [86]. White rot fungi possessing high lignolytic activities include *Phanerochaete chrysosporium*, *Phlebia radiata*, *Dichmitus squalens*, *Rigidosporus lignosus* and *Jungua separabilima* [87].

The rate of biological pretreatment is much lower than that of most thermochemical pretreatment methods. However, biological pretreatment could be exploited as a step in combination with another pretreatment method. For example, laccase treatment performed on steam exploded wheat straw reduces the toxic effect of phenolic compounds in enzymatic hydrolysis by laccase polymerization [88].

### 1.2.3 Biomass Harvest and Storage

Plant cell wall structures and compositions are considered to be changed throughout the growing seasons and growing areas. The differences in harvest date and region of growth can impact optimal pretreatment conditions, enzymatic digestibility and overall achievable sugar yields [89, 90]. Therefore, the best harvest time and scenario should be adopted.

The storage method of harvested biomass is also regarded as an influencing factor on pretreatment and enzymatic hydrolysis efficiency [91]. Storage of biomass for a long period changes structural and compositional properties compared with fresh biomass, which may have positive or negative effects on biomass digestion. Dry storage results in the shrinkage of fibrils, collapse of cell wall pores and increased crystallinity due to the loss of water. Wet storage (ensilage) can be regarded as a biological pretreatment process, during which the micro-structure maintains porosity and permeability. Microorganisms naturally grow on biomass and thus spoil the structures, part of cell wall structure is likely to be extensively loosened. Further, the carbohydrates may be metabolized by microorganisms and their content may be reduced. Size reduction prior to storage also has influence on biomass properties and pretreatment efficiencies.

### 1.2.4 Mechanical Comminution

Different types of physical comminution processes such as chipping, grinding, and milling (*e.g.* ball, two-roll, hammer, colloid and vibro energy milling) can be used to improve the enzymatic digestibility of lignocellulosic biomass [23]. Mechanical comminution itself is insufficient to attain economically feasible biomass conversion, which is always prior to thermochemical pretreatment. Moreover, the energy consumption of mechanical processes is relatively high and depends on the final particle size required and the initial biomass characteristics.

Mechanical comminution influences the efficiency of thermochemical pretreatment and enzymatic hydrolysis by altering the specific surface area, pore size and even the cellulose crystallinity of biomass. Virtually, all of the thermochemical pretreatment processes require significant size reduction of the feedstock to fine particles to achieve satisfactory cellulose enzymatic conversion. The advisable particle size or maximal size, defined as the particle size below which no increase in pretreatment effectiveness, measured in terms of the enzymatic conversion resulting from the pretreatment, needed for thermochemical pretreatment exhibits a wide range from micrometer to centimeter level [92]. The maximal size is dependent on the feedstock species and pretreatment method. Herbaceous biomass exhibits lower optimal particle size ranges than woody biomass. Steam explosion and liquid hot water pretreatments show the higher maximal size range than dilute acid and alkaline pretreatments. For example, higher enzymatic digestibility of steam exploded corn stover is achieved with larger particle size (1~2.5 cm) compared with smaller particle size (<1 cm) [93]. The larger particle size fractions (>0.85 mm) of ammonia fiber expansion pretreated corn stover are found to be more recalcitrant to enzymatic hydrolysis compared with the smaller sizes (<0.15 mm) [94].

### *1.2.5 Fractionation*

Plant biomass, especially C4 grasses (such as corn stover and sorghum), has different functional morphological fractions, which is termed as "heterogeneity". The structural properties of plant tissues vary in each morphological fraction, which affects the pretreatment and enzymatic efficiency, although the composition is very similar in different tissues. Considering the heterogeneity, each morphological fraction has its optimum pretreatment method and conditions to obtain a satisfactory use of the entire mass [95, 96]. For example, branch and boll shell of cotton stalks require lower pretreatment severity than stems to achieve similar enzymatic digestibility [97]. The lower hardness fractions in corn stover (corncob, stem pith and leaf) result in higher enzymatic digestibility compared with higher hardness fractions (stem node and rind) pretreated at identical conditions [98].

## 1.3 Enzymes for Lignocellulose Hydrolysis

Although extensive research has been made to improving enzyme production and enzyme performance, highly efficient and cost-effective enzymatic saccharification of lignocellulosic biomass still remains a challenge. Hydrolysis of lignocellulose by cellulases is considered as a complex, heterogenous multi-enzyme process.

The kinetics are ambiguous owing to the complexity and interference factors, such as the changing substrate properties and inhibitory mechanisms. This section introduces the basic concepts and categories of enzymes for lignocellulose hydrolysis and the interaction between enzymes and pure cellulose. The adsorption profile, rate-limiting factors, and synergistic effect play important roles in enzymatic hydrolysis of crystalline cellulose.

## *1.3.1 Classification of Enzymes*

There are three general categories of enzymes to hydrolyze biomass cell wall: cellulases, hemicellulases, and accessory enzymes such as hemicellulose debranching, phenolic acid esterase, lignin degrading and modifying enzymes. To date, commercial cellulase and hemicellulase preparations are mainly supplied by Novozyme (*e.g.* Cellic CTec/HTec and Celluclast) and Genencor (*e.g.* Spezyme and Accelarase) [99]. Most of the cellulases and hemicellulases exert maximum activities at about 50 °C and pH 4.5~5.5.

Cellulases are cellulolytic enzymes that can be categorized as exoglucanases (also named cellobiohydrolases, CBHs), endoglucanases (EGs) and *β*-glucosidases (BGs, also named cellobiase) (Table 1.2). The narrow definition of cellulase refers to the assembly of CBH and EG, excluding BG. The well-characterized cellulolytic system, produced by industrially important filamentous fungus *Trichoderma reesei* (a clonal derivative of *Hypocrea jecorina*), consists of at least two CBHs, TrCel7A (formerly CBH I) and TrCel6A (CBH II); five EGs, TrCel7B (EG I), TrCel5A (EG II), TrCel12A (EG III), TrCel61A (EG IV), and TrCel45A (EG V); and two BGs [100].

CBHs are exo-active processive enzymes which hydrolyze cellulose chain from the reducing and non-reducing end (by CBH I and CBH II, respectively) to produce cellobiose and other glucooligomers. CBH has a modular structure consisting of a catalytic domain (CD) connected by a linker peptide to a carbohydrate binding module (CBM). The catalytic amino acids of the CD are located in a relatively long tunnel formed by surface loops extending from the central fold. The three-dimensional structure of Cel7A reveals the 50 Å-long tunnel shaped active site of

**Table 1.2** Cellulases

| Enzyme activities | Abbreviations | Substrates | Proteins |
|---|---|---|---|
| Cellobiohydrolase I | CBH I | Reducing end of cellulose | Cel7A |
| Cellobiohydrolase II | CBH II | Non-reducing of cellulose | Cel6A |
| Endoglucanase | EG | Amorphous cellulose | Cel7B, Cel5A, Cel5B, Cel12A, Cel61A, Cel45A |
| *β*-Glucosidase | BG | Cello-oligomers | BGI, BGII |

**Table 1.3** Hemicellulases

| Enzyme activities | Abbreviations | Substrates |
|---|---|---|
| Endoxylanase | XYN (or EX) | $\beta$-1,4-Xylan |
| $\beta$-Xylosidase | $\beta$-XYL (or BX) | $\beta$-1,4-Xylooligomers |
| $\alpha$-D-Glucuronidase | $\alpha$-GLU | 4-*O*-Methyl-$\alpha$-glucuronic acid (1→2) xylooligomers |
| $\alpha$-L-Arabinofuranosidase | $\alpha$-AF | $\alpha$-Arabinofuranosyl (1→2) or (1→3) xylooligomers |
| $\alpha$-D-Galactosidase | $\alpha$-GAL | a-Galactopyranose (1→6) mannooligomers |
| Acetylxylan esterase | AXE | 2- or 3-*O*-Acetyl xylan |
| Feruloyl esterase | FE | Ester bond between arabinose substitutions and ferulic acid |

TrCel7A resides in CD and accommodates 10 consecutive glucose subsites (at least seven substrate-binding subsites and two product subsites) along cellulose chain [101]. It has been shown that separated CBM and CD of TrCel7A can both bind to cellulose. The proposed mechanism of CBM includes: (1) adsorption of the CBM at cellulose micro-cracks; (2) penetration of CBM into the interfibrillar space and insertion of mechanical pressure on pore walls; (3) penetration of water molecules into the interfibrillar space, cleavage of hydrogen bonds and formation of free chain ends [102].

Compared with CBH, the active site of EG is more open and groove-shaped. CBHs are more efficient in the degradation of crystalline cellulose. Whereas, EGs preferentially target amorphous cellulose regions, producing oligosaccharides and a little glucose. EG reacts rapidly on the amorphous cellulose and reduces its degree of polymerization to 30~60, after which this reaction with amorphous cellulose ceases [103]. Prior to hydrolysis, TrCel7B can swell the bundles of microfibrils and thus loosen surface fibrils and expose microfibril ends [104].

BGs react and hydrolyze cellobiose and soluble oligosaccharides to glucose, which thereby relieves the product inhibition to CBHs.

Hemicellulases are a diverse group of enzymes that hydrolyze hemicellulose, including $\alpha$-D-glucuronidase, endoxylanase, $\alpha$-L-arabinofuranosidase, $\beta$-xylosidase, feruloyl esterase, acetylxylan esterase, $\alpha$-D-galactosidase, and other activities (Table 1.3) [100, 105]. Hemicellulose degradation involves the synergistic action of these diverse enzyme activities.

Large enzymatic assemblies called cellulosomes, which recruit multiple catalytic activities to protein scaffolds has been found in some bacteria (*e.g. Caldicellulosiruptor bescii*). One of these, CelA, consisting of a glycoside hydrolase family 9, a family 48 CD and three type CBMs, outperforms mixtures of commercially relevant CBH and EG [106].

### 1.3.2 Enzyme-Cellulose Interaction

The mechanism of cellulases action on cellulose can be summarized into several steps, although they are not fully elucidated yet [107]. First, cellulases bind to cellulose and possibly disrupt its local crystalline structure through the CBM or CD. Cellulases primarily contact with the hydrophobic cellulose surface [1]. Then, it is hypothesized that the individual cellulose chains are decrystallized from the surface by the bound cellulases and discretely slide into the catalytic site tunnel of cellulases, eventually leading to the formation of the enzyme-substrate complexes. Finally, the hydrolysis of the complex glucan chains proceed to produce either cellobiose or other glucooligomers.

Enzymatic hydrolysis of cellulosic materials takes place at the solid-liquid interface. Therefore, a pre-requisite for catalysis is the adsorption of enzymes onto the solid surface. Enzymes (CBHs and EGs) adsorb quickly at the early stage of hydrolysis, and start to release as the degree of hydrolysis increases or remains bonded throughout the hydrolysis depending on the substrate characteristics [108]. For example, alteration of the crystalline structure from native cellulose I to cellulose III exerts 40~50% a lower binding coefficient, but it enhances hydrolytic efficiency on the latter allomorph [107]. At low concentrations of free cellulase, the binding is exclusively active-site mediated and conforms to Langmuir's one binding site model. With the increased concentrations, the isotherm gradually deviates from the Langmuir's one binding site model. Furthermore, the isotherm depends on the cellulose concentration: more efficient binding is observed at low cellulose concentrations [102].

The precipitous slowing down in the enzymatic hydrolysis rate of cellulose is one of the major limitations to the commercialization of lignocellulose. It is known that the hydrolysis rate declines drastically during the early stage and then declines slowly and steadily throughout the rest of hydrolysis time. The reasons for the reduced reactivity for pure crystalline cellulosic substrate is probably attributed to: (1) permanent denaturation of enzyme activity in the reaction, (2) reversible non-productive binding or product inhibition by sugars, and (3) a decrease in the hydrolyzability of the substrate [14, 109]. Besides these hypotheses, "jamming effect" of adjacent cellulases can explain the dramatically reduced rate of hydrolysis at high degrees of conversion. In previous studies, CBH molecules have been observed by atomic force microscopy to slide undirectionally along the surface of crystalline cellulose. The cellulose fibrils are in parallel with a characteristic lateral distance to each other which is on the same order as the diameters of Cel7A and Cel6A molecules. At high enzyme concentrations, several binding enzyme molecules adjoin to one another and the CBH molecules crowd onto the small surface area of cellulose and restrain the moving speed. This effect is termed as "jamming" and results in the reduction of overall hydrolysis rate [110, 111]. It has been further proven by a "restart" experiment [112]. Changing the crystalline polymorphic form of cellulose by means of an ammonia treatment can increase the apparent number of accessible lanes on the crystalline cellulose surface and consequently the number of moving cellulase molecules [113].

Several kinetic models have been developed to represent the actions of cellulases in enzymatic hydrolysis, such as intrinsic kinetic model [110], functionally based model [114], cellular automata model [115], and rate-constraining model [14].

The complementary action of cellulolytic enzymes has been observed and is ascribed to synergistic effects, as the combined enzymatic activity of the mixture of two or more enzymes being much higher than the sum of their individual activities [116]. The jamming effect can be remarkably eliminated by simultaneously employing Cel7A and Cel6A due to their synergistic effect. EGs prefer to hydrolyze obstacle-like amorphous cellulose, therefore assisting the processive movement of adsorbed CBH and accelerating the release of trapped CBH [117]. The synergistic effect of CBHs and EGs constitutes the major pattern in the degradation of cellulose. An appropriate and balanced combination of these cellulase activities determines the efficiency of cellulose hydrolysis.

## 1.4 Factors Affecting Enzymatic Hydrolysis of Lignocellulose

Compared with the enzymatic hydrolysis of pure cellulose substrate, enzymatic hydrolysis of thermochemical pretreated lignocellulosic biomass is affected by many more factors. Besides overcoming the nature biomass recalcitrance, diminishing other inhibitory factors from pretreatment processes exerts significant improvement in enzymatic hydrolysis of lignocellulose. The inhibitory factors from pretreated biomass consist of inhibition from lignin non-productive binding, lignin derived phenolics, oligomeric sugars, end-products, and high solid loading.

Although enzymatic hydrolysis for high solids loading that produces high sugar concentrations is needed and has many advantages, it brings about reduced enzymatic efficiency. Reducing enzyme loading and ensuring the efficiency of enzymatic saccharification are still challenges for commercialization of the lignocellulose bioconversion processes.

### *1.4.1 Inhibitors to Enzymatic Hydrolysis*

#### 1.4.1.1 Lignin Non-productive Adsorption

Besides the steric hindrance of lignin, non-productive enzyme adsorption onto lignin inhibits enzymatic hydrolysis of lignocellulosic biomass. For many pretreatment methods, such as hydrothermal and acidic pretreatments, lignin is marginally modified and remains in the pretreated solid residues. Increasing pretreatment severity enhances accessibility of the enzymes not only to cellulose but also to lignin. The lignin can reversibly or irreversibly adsorb cellulase proteins resulting in the loss of cellulase activity. More enzymes are needed to compensate for this loss in activity but this increases enzyme loading [118, 119].

Increasing temperature results in increased protein adsorption capacity onto lignin [120, 121]. Cellulase adsorption onto lignin at 50 °C is slower than adsorption on cellulose and takes over 12 h to reach equilibrium, but the maximum adsorption capacity of lignin is comparable to cellulose and is ten times higher than the adsorption capacity of lignin at 4 °C [122].

The nature of lignin obtained after pretreatment significantly influences enzymatic hydrolysis. Guaiacyl-rich lignin has higher adsorption capacities on enzymes than syringyl-rich lignin. Lignin with low S/G ratio and highly uniform fragment size contributes the high adsorption capacity [123]. Decreasing the carboxylic acid groups or increasing phenolic hydroxyl groups in lignin can also increase the adsorption capacity [120, 123, 124]. Molecular dynamics simulation of a model containing cellulose, lignin and TrCel7A has found that hydrophobic interactions between lignin and the enzyme play important roles [125]. The exposed hydrophobic clusters on lignin surface is demonstrated to preferentially adsorb to proteins [126]. Adding hydrophilic or negatively charged residues to Cel7A CBM and linker by site-directed mutagenesis generates a mutant with 2.5-fold less lignin affinity and fully retaining cellulose affinity [127]. An elevated pH (5.5~6.0) can significantly increase the negative charge on the lignin surface, making lignin more hydrophilic and consequently reducing its coordination affinity to cellulase [128, 129].

Various cellulases differ by up to 3.5-fold in their inhibition degree by lignin, while xylanases and BG show less affected by lignin [130]. EGs have higher isolated lignin affinity than CBHs. Both CD and CBM in CBHs and EGs have markedly adsorption capacities [131]. CBM can additionally bind to the specific residues of lignin [125].

To minimize non-productive lignin binding and improve enzyme specific activity, developing pretreatment technologies for reducing or modifying lignin content is one of the most important ways. Exploring low-lignin binding enzymes by protein engineering can possibly eliminate non-productive adsorption [119].

#### 1.4.1.2 Lignin Derived Phenolics

Unfortunately, both macromolecular lignin and soluble lignin derivatives (phenolics) can restrain enzyme hydrolysis and reduce sugar yields. After most pretreatment processes, lignin generally degrades into more than forty kinds of phenolics and carboxylic acids [132]. The molecular structure and concentration of phenolics depend on the biomass species and pretreatment methods. Phenolics are regarded as the most inhibitory components to enzymes among all the potential soluble materials (*e.g.* soluble sugars, furan derivatives and organic acids) in enzymatic hydrolysate [133].

The inhibition mechanism of phenolic compounds to enzymes is different from that of the insoluble lignin. The inhibition of phenolics is unable to be entirely eliminated by increasing enzyme dosage or by adding blocking proteins due to the dispersity of phenolics compared with insoluble lignin [134]. Phenolics, such as vanillin, can deactivate enzymes by both reversible and irreversible manners. Phenolics can cause protein denaturation [133, 135]. Some researchers suggest the reversible inhibition exists and is competitive [136], while some researchers have demonstrated that the inhibition type is non-competitive [137]. CBHs and EGs are found more susceptible to be inhibited than BGs [138]. The chemical groups of phenolics, such as hydroxyl, carbonyl and methoxy groups affect the inhibition degree [134, 137].

The inhibition degree of phenolics, defined as the ratio of cellulose conversion without inhibitor ($A_0$) to that with inhibitor ($A$) has a linear relationship with inhibitor concentration according to the following expression [139]:

$$A_0 / A = 1 + \beta I \tag{1.3}$$

where $I$ is the concentration of inhibitor (g/L) and $\beta$ is the inhibitor-binding constant (L/g). The higher $\beta$ value means a stronger combining capacity of inhibitor to the enzyme. The inhibitor-binding constant, which is exponentially related to the cellulose and cellulase concentrations, can be lowered by increasing cellulose or cellulase concentration [134]. Thus, by this means, the inhibition can be retarded, but it cannot be eliminated. Washing of pretreated biomass prior to enzymatic hydrolysis is effective for lowering inhibition caused by soluble phenolics.

#### 1.4.1.3 Oligo-saccharides

Xylose, xylan and xylooligomers can dramatically decrease enzymatic digestion of cellulose. Xylooligomers are more inhibitory than xylose and xylan under the same equivalent concentration and cause a decrease in initial hydrolysis rate and in final glucose yield [140]. Furthermore, the addition of monomeric sugars accompanied with oligosaccharides increases the inhibitory degree of oligosaccharides on cellulases [141]. Commercial cellulase preparations contain activities enabling them to hydrolyze non-cellulosic polysaccharides (primarily hemicellulose and pectin). However, xylanase and $\beta$-xylosidase activities detected in many commercial enzyme complexes have been shown to be insufficient to hydrolyze xylan completely, leading to high xylooligomer concentration in the hydrolysate of xylan-rich pretreated biomass. It has been reported that about 10~30% of xylooligomers with high degrees of polymerization are difficult to be degraded by cellulase or xylanase and about 5% of xylooligomers with low degrees of polymerization (mainly xylobiose) are resistant to hydrolysis by cellulase or $\beta$-glucosidase [142].

It has been found that the xylooligomer competitively impedes CBH activity, but it does not significantly affect EG [143]. Xylooligomer structures, such as xylotriose, xylotetraose and xylopentaose, are able to bind at the entrance of the substrate-binding tunnel of Cel7A, in which xylose residues are shifted ~2.4 Å towards the catalytic center compared with the binding of glucan or glucooligomers. The occupancy of two consecutive xylose residues at subsites -2 and -1 shows the different binding mode in the vicinity of the catalytic center, preventing the hydrolysis of cellulose. Moreover, the -1 xylosyl unit may exhibit the open aldehyde conformation or the ring-closed pyranoside conformation, which interferes with cellulase activity [144].

Other hemicellulose-derived oligosaccharides, such manno-oligomers, can inhibit CBH, but not BG. The lower degrees of polymerization of manno-oligomers (*e.g.* mannobiose) have higher inhibitory effects than higher degrees of polymerization [145].

#### 1.4.1.4 Products Inhibition

Cellobiose is the dominant product of CBHs and is highly inhibitory, while EGs are less affected by cellobiose. Cellobiose is found to bind almost exclusively to the +1 and +2 subsites of the catalytic tunnel of CBH, which competitively inhibits the hydrolysis of cellulose [146]. BG activity is important to release the inhibition of cellobiose to cellulases. Cel7A mutants display improved tolerances without the loss of enzyme activity [147].

Glucose inhibits *T. reesei* cellulase with a non-competitive mode [139].

### *1.4.2 Additives to Improve Enzymatic Hydrolysis*

Table 1.4 summarizes the effective additives for improving enzymatic digestion of lignocellulose.

#### 1.4.2.1 Non-hydrolytic Proteins

Several kinds of proteins, such as expansins, expansin-like proteins and swollenin, can disrupt or loosen the inaccessible or crystalline regions of cellulose microfibrils in a process known as amorphogenesis), thereby increasing the cellulose surface area [148]. Swollenin is naturally produced by *Trichoderma reesei* in low yields.

**Table 1.4** Effective additives to improve enzymatic hydrolysis

| Additive types | Specific types | Functions |
|---|---|---|
| Proteins | Expansin, swollenin, BSA | Swelling cellulose or blocking lignin |
| Surfactants | Tween, PEG, sterol, sodium lignosulfonate | Reducing lignin adsorption |
| Metal ions | Calcium, magnesium | Reducing lignin adsorption |

The recombinant swollenin causes deagglomeration and dipersion of cellulose microfibrils, and decreases the particle size and crystallinity, resulting in an increased maximum cellulase adsorption and increased hydrolysis rate [149].

Another kind of non-productive proteins is lignin blocking proteins, such as bovine serum albumin (BSA) and soybean protein. BSA and soybean protein non-specific competitive and irreversible adsorbs on lignin, and thus reduces adsorption of cellulase and particularly EGs and BGs on lignin [150, 151].

#### 1.4.2.2 Surfactants

Many kinds of surfactants have been examined for their ability to improve enzymatic hydrolysis. Non-ionic surfactants, such as Tween, Triton and polyethylene glycol (PEG) are found to be the most effective. Non-ionic surfactants can both increase cellulase temperature stability and reduce the non-productive enzyme adsorption to lignin by changing the hydrophobic interaction [152]. One study shows that PEG enhances the hydrolysis efficiency of CBH, without influencing EG or BG, although thermostability of all enzymes is increased and suggests that PEG increasing enzyme activity is related to the increased water availability [153].

Sterol in wood extractives is found to enhance enzymatic hydrolysis by affecting the adsorption of enzymes onto the crystalline cellulose surface [154]. Sodium lignosulfonate with high molecular weight has strong blocking effects on non-productive cellulase adsorption onto lignin [155].

#### 1.4.2.3 Metal Ions

Enzymatic hydrolysis is promoted by some metal ions at low concentrations (10 mM), such as $Ca^{2+}$ and $Mn^{2+}$, in which these ions are shown to increase the adsorption tightness and affinity of enzymes to cellulose [156]. $Mg^{2+}$ was shown to improve enzymatic hydrolysis of acid pretreated biomass, through reducing the non-productive adsorption of cellulase to lignin [157]. However, some metal ions, such as $Fe^{3+}$ and $Cu^{2+}$, have inhibitory effects on cellulase [158].

### *1.4.3 Synergistic Effect*

The synergistic effect on cellulose between CBHs and EGs has been introduced in the above section, which is regarded as intramolecular synergism. Hemicellulase and accessory enzymes (*e.g.* pectinase) also exhibit synergistic effects with cellulases, which is an important issue for the hemicellulose-rich biomass [159, 160]. This effect is defined as intermolecular synergism [161, 162]. The addition of glucuronidase, $\beta$-xylosidase, $\alpha$-L-arabinofuranosidase, acetyl esterase and pectinase are essential in achieving complete degradation of heteroxylans and high cellulose digestion. For example, xylanases are found to efficiently hydrolyze AFEX pre-treated biomass in an intermolecular synergistic manner. The optimal mass ratio of xylanases to cellulases is 1:3. The $\beta$-xylosidase is crucial to obtaining monomeric xylose, while $\alpha$-arabinofuranosidase and $\alpha$-glucuronidase can both further increase xylose yield. Cellulases supplemented with accessory hemicellulases not only increase both glucose and xylose yields but also decrease the total enzyme loading needed for equivalent yields [163].

### *1.4.4 High Solids Loading*

To achieve high sugar concentrations and reduce the cost of downstream purification, high solids loading (commonly ≥15% solids, w/w) in the enzymatic hydrolysis process is required [164]. Enzymatic hydrolysis at high solids loading is identified as a bottleneck affecting overall sugar yield, since it results in substantial reduction in the conversion of polysaccharides to monomeric sugars. Although the exact cause of the solid effect is not clear, there are several hypotheses that have been suggested, including: the lack of available water to promote mass transfer, increased substrate viscosity, increased concentration of inhibitors, and increased product inhibition.

The concentration of inhibitors (*e.g.* phenolics and xylooligomers) increases proportionately with the cellulose and enzyme concentration [67]. However, the impact of increased inhibitor concentration is more dramatic than the increased enzyme concentration [134]. Therefore, pretreated biomass containing a minimum amount of inhibitors is required for enzymatic hydrolysis with high solids loading.

Reactors capable of handling high-solids loadings are being developed for research purposes and used on the bench- and pilot-scales to study mass and heat transfer [165]. Previous works have shown that helical impeller and plate-and-frame impeller can result in more consistent mixing and higher substrate conversions than other impellers, indicating the free-fall mixing is effective [166, 167]. Periodic peristalsis reactors can improve mixing efficiency by releasing constrained water [168].

## 1.5 Hydrolysis Strategy

To meet industrial production requirements, process design and optimization of enzymatic hydrolysis are necessary. Low enzyme dosage and high solids loading in hydrolysis can be fulfilled by some strategies, such as enzyme recycling, pelletization, and consolidation with downstream process (*e.g.* fermentation). These promising strategies are able to achieve high sugar yields and concentrations, as introduced in this section.

### 1.5.1 Enzyme Recycling

The high cost of cellulolytic enzyme is one of the main challenges to industrial application of the lignocellulosic conversion process. A potential solution to alleviate this problem is the recycling of enzymes. There are three strategies to recycle enzymes:

1. Recovering enzymes in the supernatant by ultrafiltration. Through this way, recycled enzymes remain active for three rounds of recycle [169]. This method can retain most $\beta$-glucosidase due to its low adsorption capacity [170].
2. Desorbing enzymes from solid residues and then recycling after being separated [171].
3. Recycling bound enzymes together with solid residues. The enzyme dosage can be decreased by about 30% compared with non-recycling hydrolysis using the biomass pretreated by AFEX. This method results in an increase in total solids concentrations, reaction volumes and lignin content. However, hydrolysis efficiency is only slightly influenced by the increased lignin concentration or solid effect throughout the cycles [172–174].

The choice of the recycling process depends on the enzyme distribution between solid residue and hydrolysate. In other words, the choice of the recycling process depends on the enzyme adsorption capacity on the solid residue. Lignin content of pretreated biomass, as well as temperature, pH and surfactant addition are the major factors governing enzyme desorption from the residual substrate [170, 171].

### 1.5.2 Pelletization

Pelletization of pretreated biomass prior to enzymatic hydrolysis has many advantages that benefits the logistic handling of biomass (*e.g.* storage and transportation) and facilitates enzymatic hydrolysis at high solids loadings. Pelletization is reported to increase the initial enzymatic hydrolysis rate and final sugar yields compared with unpelletized biomass. The lower water absorption and retention capacity of pelletized biomass allows the slurries to remain well mixed in enzymatic hydrolysis without the fed-batch operation [175]. The bulk density of the pelletized AFEX pretreated biomass can increase by 1.2~6 fold compared with unpelletized biomass [176]. Pretreated biomass is shown to be easier to compress compared with untreated biomass. High moisture content (~20%) can be compressed at relatively low pressure to produce highly compacted pellets [177].

### *1.5.3 Application of Bioconversion from Lignocellulose to Sugars—SSF Process and Fed-Batch for Bioethanol Production*

In general, sugars produced from lignocellulosic biomass are usually used for fermentation to produce biofuels or chemicals. In comparison to classical separate hydrolysis and fermentation (SHF), the simultaneously saccharification and fermentation (SSF) process integrates both the enzymatic hydrolysis of pretreated lignocellulosic material and the fermentation of released sugars into one step. SSF avoids end-product inhibition, reduces operating costs, decreases risk of contamination, and improves production efficiency. The combination of hydrolysis and fermentation decreases the number of vessels and operating units. The decrease in capital investment has been estimated to be up to 20% [178]. Furthermore, sugars do not need to be separated from the solid residue following a separate enzymatic hydrolysis step, thereby avoiding potential sugar loss.

Inevitably, there are also disadvantages of SSF compared with SHF process. The optimal conditions for enzymatic hydrolysis and fermentation are different. For example, temperature in SSF should be lower than the separate enzymatic hydrolysis, as the optimum temperature for enzymatic hydrolysis is around 50 °C while for fermentation is 30~37 °C [179].

For the production of lignocellulosic ethanol, both high ethanol yields and high concentrations are important targets, because the distillation cost significantly decreases for high ethanol concentrations. To increase the ethanol concentration, high solids loading is necessary. Effective strategies, such as fed-batch operation, can be employed [180, 181]. Through adding feedstocks in fed-batch mode, the concentrations of inhibitors can be kept lower, exerting less inhibition to enzymatic hydrolysis and fermentation. A suitable feed rate allows the continuous degradation of inhibitors by microorganism [182, 183].

## 1.6 Conclusions and Future Outlook

Lignocellulosic biomass is the most abundant renewable resource which contains plentiful amounts of polysaccharides, mainly cellulose and hemicellulose. The strategy of combining enzymatic hydrolysis with pretreatment effectively converts the polysaccharides in biomass to monomeric or oligomeric sugars. Pretreatment breaks down the biomass recalcitrance and increases the cellulose accessibility to enzymes. Thermochemical pretreatments, classified into acidic, neutral, alkaline and organic method, are considered as integral and influential to the subsequent operations, such as solid-liquid separation, detoxification, neutralization and enzymatic hydrolysis. Several thermochemical pretreatment technologies seem promising for industrial application, although all of them have virtues and faults. Mechanical and biological pretreatments also play roles in enhancing enzymatic digestibility of biomass. Enzymes for lignocellulose hydrolysis involves cellulases, hemicellulases and accessory enzymes, which react collaboratively on the lignocellulose. The complete degradation of pure cellulose is well understood and depends on three kinds of cellulase: cellobiohydrolase, endoglucanase and $\beta$-glucosidase, whereas, the enzymatic hydrolysis of lignocellulosic materials actually is confronted with several obstacles, such as inhibitors to enzyme activities, the lack of accessory enzyme activities in commercial enzyme preparations, and difficulties in increasing solid loading in the hydrolysate. Several materials, including lignin, phenolics and xylooligomers, are strong inhibitors to enzymes in reversible or irreversible manners in the hydrolysate of pretreated biomass. To prevent inhibition and improve enzymatic hydrolysis, some additives, such as non-productive proteins, non-ionic surfactants and some metal ions as well as the accessory enzymes have been demonstrated to be effective. Enzyme recycling is beneficial to reduce enzyme dosage, which has important implication for saving cost. Pelletization of pretreated biomass and integration enzymatic hydrolysis with subsequent fermentation process can facilitate the operation process and the high-solid enzymatic hydrolysis.

The sugar yield, concentration and productivity from lignocellulose enzymatic hydrolysis are still far below amylohydrolysis. On top of that, the refinery cost of lignocellulose is much higher than starchy feedstocks, which makes the process less competitive in business. Both new cognitions and novel technologies in pretreatments and enzymatic hydrolysis are required to improve the sugar concentration, yield and rate of the enzymatic hydrolysis. To reduce the production cost and improve the efficiency of bioconversion of lignocellulose, further research is needed on: (1) the development of new pretreatment technologies with strong merits; (2) genetic engineering of lignocellulosic biomass with reduced biomass recalcitrance; (3) new cognition about the relationship between enzymatic digestibility and biomass structure and composition; (4) new enzymes possessing higher activity, specificity and inhibitor-tolerance, developed by protein engineering; (5) the production of hemicellulases and accessory enzymes; (6) technologies and ways for removing the inhibition in enzymatic hydrolysis; (7) consolidated operation processes and reactor design for increasing solid loading and sugars concentration.

## References


1. Ding SY, Liu YS, Zeng YN, Himmel ME, Baker JO, Bayer EA. How does plant cell wall nanoscale architecture correlate with enzymatic digestibility? Science. 2012;338(6110):1055–60.
2. Somerville C, Bauer S, Brininstool G, Facette M, Hamann T, Milne J, Osborne E, Paredez A, Persson S, Raab T, Vorwerk S, Youngs H. Toward a systems approach to understanding plant cell walls. Science. 2004;306(5705):2206–11.
3. Brandt A, Gräsvik J, Hallett JP, Welton T. Deconstruction of lignocellulosic biomass with ionic liquids. Green Chem. 2013;15:550.
4. Himmel ME, Ding SY, Johnson DK, Adney WS, Nimlos MR, Brady JW, Foust TD. Biomass recalcitrance: engineering plants and enzymes for biofuels production. Science. 2007;315(5813):804–7.

5. Wada M, Chanzy H, Nishiyama Y, Langan P. Cellulose $III_I$ crystal structure and hydrogen bonding by synchrotron x-ray and neutron fiber diffraction. Macromolecules. 2004;37(23):8548–55.
6. Zugenmaier P. Conformation and packing of various crystalline cellulose fibers. Prog Polym Sci. 2001;26(9):1341–417.
7. Gírio FM, Fonseca C, Carvalheiro F, Duarte LC, Marques S, Bogel-Łukasik R. Hemicelluloses for fuel ethanol: A review. Bioresour Technol. 2010;101(13):4775–800.
8. Ralph J, Lundquist K, Brunow G, Lu F, Kim H, Schatz PF, Marita JM, Hatfield RD, Ralph SA, Christensen JH, Boerjan W. Lignins: natural polymers from oxidative coupling of 4-hydroxyphenylpropanoids. Phytochem Rev. 2004;3:29–60.
9. Ragauskas J, Beckham GT, Biddy MJ, Chandra R, Chen F, Davis MF, Davison BH, Dixon RA, Gilna P, Keller M, Langan P, Naskar AK, Saddler JN, Tschaplinski TJ, Tuskan GA, Wyman CE. Lignin valorization: improving lignin processing in the biorefinery. Science. 2014;344(6185):1246843.
10. Bienkowski PR, Ladisch MR, Narayan R, Tsao GT, Eckert R. Correlation of glucose (dextrose) degradation at 90 to 190 °C in 0.4 to 20% acid. Chem Eng Commun. 1987;51:179–92.
11. Zhao XB, Zhang LH, Liu DH. Biomass recalcitrance. Part I: The chemical compositions and physical structures affecting the enzymatic hydrolysis of lignocelluloses. Biofuels Bioprod Bioref. 2012;6:465–82.
12. Liu ZH, Qin L, Li BZ, Yuan YJ. Physical and chemical characterizations of corn stover from leading pretreatment methods and effects on enzymatic hydrolysis. ACS Sustain Chem Eng. 2014;3(1):140–6.
13. Arantes V, Saddler JN. Cellulose accessibility limits the effectiveness of minimum cellulase loading on the efficient hydrolysis of pretreated lignocellulosic substrates. Biotechnol Biofuels. 2011;4:3.
14. Pihlajaniemi V, Sipponen MH, Kallioinen A, Nyyssölä A, Laakso S. Rate-constraining changes in surface properties, porosity and hydrolysis kinetics of lignocellulose in the course of enzymatic saccharification. Biotechnol Biofuels. 2016;9:18.
15. Ohgren K, Bura R, Saddler J, Zacchi G. Effect of hemicellulose and lignin removal on enzymatic hydrolysis of steam pretreated corn stover. Bioresour Technol. 2007;98(13):2503–10.
16. Laureano-Perez L, Teymouri F, Alizadeh H, Dale BE. Understanding factors that limit enzymatic hydrolysis of biomass: characterization of pretreated corn stover. Appl Biochem Biotechnol. 2005;121-124:1081–99.
17. Zhu L, O'Dwyer JP, Chang VS, Granda CB, Holtzapple MT. Structural features affecting biomass enzymatic digestibility. Bioresour Technol. 2008;99(9):3817–28.
18. Cui T, Li J, Yan Z, Yu M, Li S. The correlation between the enzymatic saccharification and the multidimensional structure of cellulose changed by different pretreatments. Biotechnol Biofuels. 2014;7:134.
19. Qin L, Li WC, Zhu JQ, Liang JN, Li BZ, Yuan YJ. Ethylenediamine pretreatment changes cellulose allomorph and lignin structure of lignocellulose at ambient pressure. Biotechnol Biofuels. 2015;8:174.
20. Ju X, Grego C, Zhang X. Specific effects of fiber size and fiber swelling on biomass substrate surface area and enzymatic digestibility. Bioresour Technol. 2013;144:232–9.
21. Mosier N, Wyman C, Dale B, Elander R, Lee YY, Holtzapple M, Ladisch M. Features of promising technologies for pretreatment of lignocellulosic biomass. Bioresour Technol. 2005;96(6):673–86.
22. Yang B, Wyman CE. Pretreatment: the key to unlocking low-cost cellulosic ethanol. Biofuels Bioprod Bioref. 2008;2(2):26–40.
23. Alvira P, Tomás-Pejó E, Ballesteros M, Negro MJ. Pretreatment technologies for an efficient bioethanol production process based on enzymatic hydrolysis: a review. Bioresour Technol. 2010;101(13):4851–61.
24. Klein-Marcushamer D, Oleskowicz-Popeil P, Simmons BA, Blanch HW. The challenge of enzyme cost in the production of lignocellulosic biofuels. Biotechnol Bioeng. 2012;109(14):1083–7.

25. Garlock RJ, Balan V, Dale BE, Pallapolu VR, Lee YY, Kim Y, Mosier NS, Ladisch MR, Holtzapple MT, Falls M, Sierra-Ramirez R, Shi J, Ebrik MA, Redmond T, Yang B, Wyman CE, Donohoe BS, Vinzant TB, Elander RT, Hames B, Thomas S, Warner RE. Comparative material balances around pretreatment technologies for the conversion of switchgrass to soluble sugars. Bioresour Technol. 2011;102(24):11063–71.
26. Overend RP, Chornet E. Fractionation of lignocellulosics by steam-aqueous pretreatments. Phil Trans R Soc Lnd. 1987;A321:523–36.
27. Chum HL, Johnson DK, Black SK. Organosolv pretreatment for enzymatic hydrolysis of poplars. 2. Catalyst effects and the combined severity parameter. Ind Eng Chem Res. 1990;29:2.
28. Chen SF, Mowery RA, Chambliss CK, Walsum GP. Pseudo reaction kinetics of organic degradation products in dilute-acid-catalyzed corn stover pretreatment hydrolysates. Biotechnol Bioeng. 2007;98:1135–45.
29. Larsson S, Palmqvist E, Hahn-Hägerdal B, Tengborg C, Stenberg K, Zacchi G, Nilvebrant NO. The generation of fermentation inhibitors during dilute acid hydrolysis of softwood. Enzyme Microb Technol. 1999;24:151–9.
30. Jensen JR, Morinelly JE, Gossen KR, Brodeur-Campbell MJ, Shonnard DR. Effects of dilute acid pretreatment conditions on enzymatic hydrolysis monomer and oligomer sugar yields for aspen, balsam, and switchgrass. Bioresour Technol. 2010;101(7):2317–25.
31. Qin L, Liu ZH, Li BZ, Dale BE, Yuan YJ. Mass balance and transformation of corn stover by pretreatment with different dilute organic acids. Bioresour Technol. 2012;112:319–26.
32. Sannigrahi P, Kim DH, Jung S, Ragauskas A. Pseudo-lignin and pretreatment chemistry. Energy Environ Sci. 2011;4:1306.
33. Kumar R, Hu F, Sannigrahi P, Jung S, Ragauskas AJ, Wyman CE. Carbohydrate derived-pseudo-lignin can retard cellulose biological conversion. Biotechnol Bioeng. 2013;110(3):737–53.
34. Hu F, Ragauskas A. Suppression of pseudo-lignin formation under dilute acid pretreatment conditions. RSC Adv. 2014;4:4317.
35. Foston M, Ragauskas AJ. Changes in lignocellulosic supramolecular and ultrastructure during dilute acid pretreatment of Populus and switchgrass. Biomass Bioenerg. 2010;34(12):1885–95.
36. Pingali SV, Urban VS, Heller WT, McGaughey J, O'Neill H, Foston M, Myles DA, Ragauskas A, Evans BR. Breakdown of cell wall nanostructure in dilute acid pretreated biomass. Biomacromolecules. 2010;11(9):2329–35.
37. Kshirsagar SD, Waghmare PR, Loni PC, Patil SA, Govindwar SP. Dilute acid pretreatment of rice straw, structural characterization and optimization of enzymatic hydrolysis conditions by response surface methodology. Rsc Adv. 2015;5(58):46525–33.
38. Ji Z, Zhang X, Ling Z, Zhou X, Ramaswamy S, Xu F. Visualization of *Miscanthus x giganteus* cell wall deconstruction subjected to dilute acid pretreatment for enhanced enzymatic digestibility. Biotechnol Biofuels. 2015;8:103.
39. Zhang J, Wang X, Chu D, He Y, Bao J. Dry pretreatment of lignocellulose with extremely low steam and water usage for bioethanol production. Bioresour Technol. 2011;102(6):4480–8.
40. He Y, Zhang L, Zhang J, Bao J. Helically agitated mixing in dry dilute acid pretreatment enhances the bioconversion of corn stover into ethanol. Biotechnol Biofuels. 2014;7:1.
41. Mosier NS, Sarikaya A, Ladisch CM, Ladisch MR. Characterization of dicarboxylic acids for cellulose hydrolysis. Biotechnol Prog. 2001;17:474–80.
42. Kim Y, Kreke T, Ladisch MR. Reaction mechanisms and kinetics of xylo-oligosaccharide hydrolysis by dicarboxylic acids. AIChE J. 2013;59(1):188–99.
43. Bura R, Mansfield SD, Saddler JN, Bothast RJ. $SO_2$-catalyzed steam explosion of corn fiber for ethanol production. Appl Biochem Biotechnol. 2002;100(1-9):59–72.
44. Nitsos CK, Matis KA, Triantafyllidis KS. Optimization of hydrothermal pretreatment of lignocellulosic biomass in the bioethanol production process. ChemSusChem. 2013;6:110–22.

45. Kang Y, Bansal P, Realff MJ, Bommarius AS. $SO_2$-catalyzed steam explosion: the effects of different severity on digestibility, accessibility, and crystallinity of lignocellulosic biomass. Biotechnol Prog. 2013;29(4):909–16.
46. Diop CI, Lavoie JM, Huneault MA. Structural changes of *Salix miyabeana* cellulose fibres during dilute-acid steam explosion: impact of reaction temperature and retention time. Carbohydr Polym. 2015;119:8–17.
47. Pielhop T, Amgarten J, von Rohr PR, Studer MH. Steam explosion pretreatment of softwood: the effect of the explosive decompression on enzymatic digestibility. Biotechnol Biofuels. 2016;9:152.
48. Liu ZH, Chen HZ. Xylose production from corn stover biomass by steam explosion combined with enzymatic digestibility. Bioresour Technol. 2015;193:345–56.
49. Agudelo RA, García-Aparicio MP, Görgens JF. Steam explosion pretreatment of triticale (*x Triticosecale Wittmack*) straw for sugar production. N Biotechnol. 2016;33(1):153–63.
50. Mosier N, Hendrickson R, Ho N, Sedlak M, Ladisch MR. Optimization of pH controlled liquid hot water pretreatment of corn stover. Bioresour Technol. 2005;96:1986–93.
51. Michelin M, Teixeira JA. Liquid hot water pretreatment of multi feedstocks and enzymatic hydrolysis of solids obtained thereof. Bioresour Technol. 2016;216:862–9.
52. Trajano HL, Engle NL, Foston M, Ragauskas AJ, Tschaplinski TJ, Wyman CE. The fate of lignin during hydrothermal pretreatment. Biotechnol Biofuels. 2013;6:110.
53. Ko JK, Kim Y, Ximenes E, Ladisch MR. Effect of liquid hot water pretreatment severity on properties of hardwood lignin and enzymatic hydrolysis of cellulose. Biotechnol Bioeng. 2015;112(2):252–62.
54. Reddy P, Lekha P, Reynolds W, Kirsch C. Structural characterisation of pretreated solids from flow-through liquid hot water treatment of sugarcane bagasse in a fixed-bed reactor. Bioresour Technol. 2015;183:259–61.
55. Kim Y, Kreke T, Mosier NS, Ladisch MR. Severity factor coefficients for subcritical liquid hot water pretreatment of hardwood chips. Biotechnol Bioeng. 2014;111(2):254–63.
56. Zhang L, Yan L, Wang Z, Laskar DD, Swita MS, Cort JR, Yang B. Characterization of lignin derived from water-only and dilute acid flowthrough pretreatment of poplar wood at elevated temperatures. Biotechnol Biofuels. 2015;8:203.
57. Murnen HK, Balan V, Chundawat SPS, Bals B, da Costa SL, Dale BE. Optimization of Ammonia Fiber Expansion (AFEX) Pretreatment and Enzymatic Hydrolysis of *Miscanthus x giganteus* to Fermentable Sugars. Biotechnol Prog. 2007;23:846–50.
58. Krishnan C, da Costa SL, Jin M, Chang L, Dale BE, Balan V. Alkali-based AFEX pretreatment for the conversion of sugarcane bagasse and cane leaf residues to ethanol. Biotechnol Bioeng. 2010;107(3):441–50.
59. Chundawat SPS, Vismeh R, Sharma LN, Humpula JF, da Costa SL, Chambliss CK, Jones AD, Balan V, Dale BE. Multifaceted characterization of cell wall decomposition products formed during ammonia fiber expansion (AFEX) and dilute acid based pretreatments. Bioresour Technol. 2010;101(21):8429–38.
60. Lau MJ, Lau MW, Gunawan C, Dale BE. Ammonia Fiber Expansion (AFEX) Pretreatment, Enzymatic Hydrolysis, and Fermentation on Empty Palm Fruit Bunch Fiber (EPFBF) for Cellulosic Ethanol Production. Appl Biochem Biotechnol. 2010;162:1847–57.
61. Li BZ, Balan V, Yuan YJ, Dale BE. Process optimization to convert forage and sweet sorghum bagasse to ethanol based on ammonia fiber expansion (AFEX) pretreatment. Bioresour Technol. 2010;101(4):1285–92.
62. Chundawat SPS, Donohoe BS, da Costa SL, Elder T, Agarwal UP, Lu F, Ralph J, Himmel ME, Balan V, Dale BE. Multi-scale visualization and characterization of lignocellulosic plant cell wall deconstruction during thermochemical pretreatment. Energy Environ Sci. 2011;4:973–84.
63. Shao Q, Cheng C, Ong RG, Zhu L, Zhao C. Hydrogen peroxide presoaking of bamboo prior to AFEX pretreatment and impact on enzymatic conversion to fermentable sugars. Bioresour Technol. 2013;142:26–31.

64. da Costa SL, Jin MJ, Chundawat SPS, Bokade V, Tang X, Azarpira A, Lu F, Avci U, Humpula J, Uppugundla N, Gunawan C, Pattathil S, Cheh AM, Kothari N, Kumar R, Ralph J, Hahn MG, Wyman CE, Singh S, Simmons BA, Dale BE, Balan V. Next-generation ammonia pretreatment enhances cellulosic biofuel production. Energy Environ Sci. 2016;9(4):1215–23.
65. Qin L, Liu L, Li WC, Zhu JQ, Li BZ, Yuan YJ. Evaluation of soluble fraction and enzymatic residual fraction of dilute dry acid, ethylenediamine, and steam explosion pretreated corn stover on the enzymatic hydrolysis of cellulose. Bioresour Technol. 2016;209:172–9.
66. Ladisch MR, Ladisch CM, Tsao GT. Cellulose to sugars: new path gives quantitative yield. Science. 1978;201:743–5.
67. Qin L, Liu ZH, Jin M, Li BZ, Yuan YJ. High temperature aqueous ammonia pretreatment and post-washing enhance the high solids enzymatic hydrolysis of corn stover. Bioresour Technol. 2013;146:504–11.
68. Kim TH, Taylor F, Hicks KB. Bioethanol production from barley hull using SAA (soaking in aqueous ammonia) pretreatment. Bioresour Technol. 2008;99(99):5694–702.
69. Kim TH, Kim JS, Sunwoo C, Lee YY. Pretreatment of corn stover by aqueous ammonia. Bioresour Technol. 2003;90(1):39–47.
70. Kim S, Holtzapple MT. Lime pretreatment and enzymatic hydrolysis of corn stover. Bioresour Technol. 2006;96(18):1994–2006.
71. Zavrel M, Bross D, Funke M, Büchs J, Spiess AC. High-throughput screening for ionic liquids dissolving (ligno-)cellulose. Bioresour Technol. 2009;100(9):2580–7.
72. Brandt A, Ray MJ, To TQ, Leak DJ, Murphy RJ, Welton T. Ionic liquid pretreatment of lignocellulosic biomass with ionic liquid–water mixtures. Green Chem. 2011;13:2489.
73. Fu D, Mazza G. Aqueous ionic liquid pretreatment of straw. Bioresour Technol. 2011;102(13):7008–11.
74. Li C, Knierim B, Manisseri C, Arora R, Scheller HV, Auer M, Vogel KP, Simmons BA, Singh S. Comparison of dilute acid and ionic liquid pretreatment of switchgrass: Biomass recalcitrance, delignification and enzymatic saccharification. Bioresour Technol. 2010;101(13):4900–6.
75. Remsing RC, Swatloski RP, Rogers RD, Moyna G. Mechanism of cellulose dissolution in the ionic liquid 1-n-butyl-3-methylimidazolium chloride: A $^{13}C$ and $^{35/37}Cl$ NMR relaxation study on model systems. Chem Commun (Camb). 2006;12:1271–3.
76. Cheng G, Varanasi P, Li C, Liu H, Melnichenko YB, Simmons BA, Kent MS, Singh S. Transition of cellulose crystalline structure and surface morphology of biomass as a function of ionic liquid pretreatment and its relation to enzymatic hydrolysis. Biomacromolecules. 2011;12(4):933–41.
77. Hou XD, Smith TJ, Li N, Zong MH. Novel renewable ionic liquids as highly effective solvents for pretreatment of rice straw biomass by selective removal of lignin. Biotechnol Bioeng. 2012;109(10):2484–93.
78. Soudham VP, Raut DG, Anugwom I, Brandberg T, Larsson C, Mikkola JP. Coupled enzymatic hydrolysis and ethanol fermentation: ionic liquid pretreatment for enhanced yields. Biotechnol Biofuels. 2015;8:135.
79. Zhao X, Cheng K, Liu D. Organosolv pretreatment of lignocellulosic biomass for enzymatic hydrolysis. Appl Microb Biotechnol. 2009;82(5):815–27.
80. Hu G, Cateto C, Pu Y, Samuel R, Ragauskas AJ. Structural characterization of switchgrass lignin after ethanol organosolv pretreatment. Energy Fuels. 2011;26(1):740–5.
81. Nguyen TY, Cai CM, Kumar R, Wyman CE. Co-solvent pretreatment reduces costly enzyme requirements for high sugar and ethanol yields from lignocellulosic biomass. ChemSusChem. 2015;8(10):1716–25.
82. Sathitsuksanoh N, Zhu Z, Zhang YHP. Cellulose solvent- and organic solvent-based lignocellulose fractionation enabled efficient sugar release from a variety of lignocellulosic feedstocks. Bioresour Technol. 2012;117(10):228–33.
83. Zhu JY, Pan XJ, Wang GS, Gleisner R. Sulfite pretreatment (SPORL) for robust enzymatic saccharification of spruce and red pine. Bioresour Technol. 2009;100(8):2411–8.

84. Klinke HB, Ahring BK, Schmidt AS, Thomsen AB. Characterization of degradation products from alkaline wet oxidation of wheat straw. Bioresour Technol. 2002;82(82):15–26.
85. Biswas R, Uellendahl H, Ahring BK. Wet explosion pretreatment of sugarcane bagasse for enhanced enzymatic hydrolysis. Biomass Bioenerg. 2014;61(2):104–13.
86. Singh D, Chen SL. The white-rot fungus *Phanerochaete chrysosporium*: conditions for the production of lignin-degrading enzymes. Appl Microbiol Biotechnol. 2008;81:399–417.
87. Sun S, Sun S, Cao X, Sun R. The role of pretreatment in improving the enzymatic hydrolysis of lignocellulosic materials. Bioresour Technol. 2016;199:49–58.
88. Jurado M, Prieto A, Martínez-Alcalá A, Martínez AT, Martínez MJ. Laccase detoxification of steam-exploded wheat straw for second generation bioethanol. Bioresour Technol. 2009;100(24):6378–84.
89. Garlock RJ, Chundawat SP, Balan V, Dale BE. Optimizing harvest of corn stover fractions based on overall sugar yields following ammonia fiber expansion pretreatment and enzymatic hydrolysis. Biotechnol Biofuels. 2009;2:29.
90. Garlock RJ, Balan V, Dale BE. Optimization of AFEX™ pretreatment conditions and enzyme mixtures to maximize sugar release from upland and lowland switchgrass. Bioresour Technol. 2012;104:757–68.
91. Liu ZH, Qin L, Jin MJ, Pang F, Li BZ, Kang Y, Dale BE, Yuan YJ. Evaluation of storage methods for the conversion of corn stover biomass to sugars based on steam explosion pretreatment. Bioresour Technol. 2013;132(2):5–15.
92. Vidal Jr BC, Dien BS, Ting KC, Singh V. Influence of feedstock particle size on lignocellulose conversion--a review. Appl Biochem Biotechnol. 2011;164(8):1405–21.
93. Liu ZH, Qin L, Pang F, Jin MJ, Li BZ, Kang Y, Dale BE, Yuan YJ. Effects of biomass particle size on steam explosion pretreatment performance for improving the enzyme digestibility of corn stover. Ind Crops Prod. 2013;44(2):176–84.
94. Chundawat SPS, Balan V, Dale BE. Effect of particle size based separation of milled corn stover on AFEX pretreatment and enzymatic digestibility. Biotechnol Bioeng. 2007;96(2):219–31.
95. Zeng M, Ximenes E, Ladisch MR, Mosier NS, Vermerris W, Huang CP, Sherman DM. Tissue-specific biomass recalcitrance in corn stover pretreated with liquid hot-water: enzymatic hydrolysis (Part 1). Biotechnol Bioeng. 2011;109(2):390–7.
96. Zeng M, Ximenes E, Ladisch MR, Mosier NS, Vermerris W, Huang CP, Sherman DM. Tissue-specific biomass recalcitrance in corn stover pretreated with liquid hot-water: enzymatic hydrolysis (Part 2). Biotechnol Bioeng. 2011;109(2):398–404.
97. Jiang W, Chang S, Li H, Oleskowicz-Popiel P, Xu J. Liquid hot water pretreatment on different parts of cotton stalk to facilitate ethanol production. Bioresour Technol. 2015;176:175–80.
98. Liu ZH, Chen HZ. Mechanical property of different corn stover morphological fractions and its correlations with high solids enzymatic hydrolysis by periodic peristalsis. Bioresour Technol. 2016;214:292–302.
99. Jin M, da Costa SL, Schwartz C, He Y, Sarks C, Gunawan C, Balan V, Dale BE. Toward lower cost cellulosic biofuel production using ammonia based pretreatment technologies. Green Chem. 2015;18(4):957–66.
100. Berlin A. Microbiology. No barriers to cellulose breakdown. Science. 2013;342(6165):1454–6.
101. Divne C, Ståhlberg J, Teeri TT, Jones TA. High resolution crystal structures reveal how a cellulose chain is bound in the 50-Å-long tunnel of cellobiohydrolase I from *Trichoderma reesei*. J Mol Biol. 1998;275:309–25.
102. Jalak J, Väljamäe P. Multi-mode binding of Cellobiohydrolase Cel7A from *Trichoderma reesei* to cellulose. PLoS One. 2014;9(9):e108181.
103. Gupta R, Lee YY. Mechanism of cellulase reaction on pure cellulosic substrates. Biotechnol Bioeng. 2009;102(6):1570–81.

104. Wang J, Quirk A, Lipkowski J, Dutcher JR, Hill C, Mark A, Clarke AJ. Real-time observation of the swelling and hydrolysis of a single crystalline cellulose fiber catalyzed by cellulase 7B from *Trichoderma reesei*. Langmuir. 2012;28:9664–72.
105. Shallom D, Shoham Y. Microbial hemicellulases. Curr Opin Microbiol. 2003;6:219–28.
106. Brunecky R, Alahuhta M, Xu Q, Donohoe BS, Crowley MF, Kataeva IA, Yang SJ, Resch MG, Adams MW, Lunin VV, Himmel ME, Bomble YJ. Revealing nature's cellulase diversity: the digestion mechanism of *Caldicellulosiruptor bescii* CelA. Science. 2013;342(6165):1513–6.
107. Gao D, Chundawat SP, Sethi A, Balan V, Gnanakaran S, Dale BE. Increased enzyme binding to substrate is not necessary for more efficient cellulose hydrolysis. Proc Nat Acad Sci. 2013;110(27):10922–7.
108. Várnai A, Viikari L, Marjamaa K, Siika-aho M. Adsorption of monocomponent enzymes in enzyme mixture analyzed quantitatively during hydrolysis of lignocellulose substrates. Bioresour Technol. 2011;102(2):1220–7.
109. Bansal P, Vowell BJ, Hall M, Realff MJ, Lee JH, Bommarius AS. Elucidation of cellulose accessibility, hydrolysability and reactivity as the major limitations in the enzymatic hydrolysis of cellulose. Bioresour Technol. 2012;107:243–50.
110. Peri S, Karra S, Lee YY, Karim MN. Modeling intrinsic kinetics of enzymatic cellulose hydrolysis. Biotechnol Prog. 2007;23(3):626–37.
111. Bommarius AS, Katona A, Cheben SE, Patel AS, Ragauskas AJ, Knudson K, Pu Y. Cellulase kinetics as a function of cellulose pretreatment. Metab Eng. 2008;10(6):370–81.
112. Yang B, Willies DM, Wyman CE. Change in the enzymatic hydrolysis rate of avicel cellulose with conversion. Biotechnol Bioeng. 2006;94(6):1122–8.
113. Igarashi K, Uchihashi T, Koivula A, Wada M, Kimura S, Okamoto T, Penttilä M, Ando T, Samejima M. Traffic jams reduce hydrolytic efficiency of cellulase on cellulose surface. Science. 2011;333(6047):1279–82.
114. Zhang YHP, Lynd LR. A functionally based model for hydrolysis of cellulose by fungal cellulase. Biotechnol Bioeng. 2006;94:889–98.
115. Eibinger M, Zahel T, Ganner T, Plank H, Nidetzky B. Cellular automata modeling depicts degradation of cellulosic material by a cellulase system with single-molecule resolution. Biotechnol Biofuels. 2016;9:56.
116. Wang J, Quirk A, Lipkowski J, Dutcher JR, Clarke AJ. Direct *in situ* observation of synergism between cellulolytic enzymes during the biodegradation of crystalline cellulose fibers. Langmuir. 2013;29(48):14997–5005.
117. Jalak J, Kurasin M, Teugjas H, Valjamae P. Endo-exo synergism in cellulose hydrolysis revisited. J Biol Chem. 2012;287:28802–15.
118. Kim Y, Kreke T, Ko JK, Ladisch MR. Hydrolysis-determining substrate characteristics in liquid hot water pretreated hardwood. Biotechnol Bioeng. 2015;112(4):677–87.
119. Gao D, Haarmeyer C, Balan V, Whitehead TA, Dale BE, Chundawat SP. Lignin triggers irreversible cellulase loss during pretreated lignocellulosic biomass saccharification. Biotechnol Biofuels. 2014;7(1):175.
120. Pareek N, Gillgren T, Jönsson LJ. Adsorption of proteins involved in hydrolysis of lignocellulose on lignins and hemicelluloses. Bioresour Technol. 2013;148:70–7.
121. Rahikainen JL, Moilanen U, Nurmi-Rantala S, Lappas A, Koivula A, Viikari L, Kruus K. Effect of temperature on lignin-derived inhibition studied with three structurally different cellobiohydrolases. Bioresour Technol. 2013;146:118–25.
122. Zheng Y, Zhang S, Miao S, Su Z, Wang P. Temperature sensitivity of cellulase adsorption on lignin and its impact on enzymatic hydrolysis of lignocellulosic biomass. J Biotechnol. 2013;166(3):135–43.
123. Guo F, Shi W, Sun W, Li X, Wang F, Zhao J, Yu Y. Differences in the adsorption of enzymes onto lignins from diverse types of lignocellulosic biomass and the underlying mechanism. Biotechnol Biofuels. 2014;7:38.

124. Nakagame S, Chandra RP, Kadla JF, Saddler JN. Enhancing the enzymatic hydrolysis of lignocellulosic biomass by increasing the carboxylic acid content of the associated lignin. Biotechnol Bioeng. 2011;108(108):538–48.
125. Vermaas JV, Petridis L, Qi X, Schulz R, Lindner B, Smith JC. Mechanism of lignin inhibition of enzymatic biomass deconstruction. Biotechnol Biofuels. 2015;8:217.
126. Sammond DW, Yarbrough JM, Mansfield E, Bomble YJ, Hobdey SE, Decker SR, Taylor LE, Resch MG, Bozell JJ, Himmel ME, Vinzant TB, Crowley MF. Predicting enzyme adsorption to lignin films by calculating enzyme surface hydrophobicity. J Biol Chem. 2014;289(30):20960–9.
127. Strobel KL, Pfeiffer KA, Blanch HW, Clark DS. Engineering Cel7A carbohydrate binding module and linker for reduced lignin inhibition. Biotechnol Bioeng. 2016;113:1369–74.
128. Lou H, Zhu JY, Lan TQ, Lai H, Qiu X. pH-induced lignin surface modification to reduce nonspecific cellulase binding and enhance enzymatic saccharification of lignocelluloses. Chemsuschem. 2013;6(5):919–27.
129. Rahikainen JL, Evans JD, Mikander S, Kalliola A, Puranen T, Tamminen T, Marjamaa K, Kruus K. Cellulase-lignin interactions-the role of carbohydrate-binding module and pH in non-productive binding. Enzyme Microb Technol. 2013;53(5):315–21.
130. Berlin A, Balakshin M, Gilkes N, Kadla J, Maximenko V, Kubo S, Saddler J. Inhibition of cellulase, xylanase and beta-glucosidase activities by softwood lignin preparations. J Biotechnol. 2006;125(2):198–209.
131. Palonen H, Tjerneld F, Zacchi G, Tenkanen M. Adsorption of *Trichoderma reesei* CBH I and EG II and their catalytic domains on steam pretreated softwood and isolated lignin. J Biotechnol. 2004;107(1):65–72.
132. Du B, Sharma LN, Becker C, Chen SF, Mowery RA, van Walsum GP, Chambliss CK. Effect of varying feedstock-pretreatment chemistry combinations on the formation and accumulation of potentially inhibitory degradation products in biomass hydrolysates. Biotechnol Bioeng. 2010;107(3):430–40.
133. Kim Y, Ximenes E, Mosier NS, Ladisch MR. Soluble inhibitors/deactivators of cellulase enzymes from lignocellulosic biomass. Enzyme Microb Technol. 2011;48(4-5):408–15.
134. Qin L, Li WC, Liu L, Zhu JQ, Li X, Li BZ, Yuan YJ. Inhibition of lignin-derived phenolic compounds to cellulase. Biotechnol Biofuels. 2016;9:70.
135. Ximenes E, Kim Y, Mosier N, Dien B, Ladisch M. Deactivation of cellulases by phenols. Enzyme Microb Technol. 2011;48(1):54–60.
136. Jing X, Zhang X, Bao J. Inhibition performance of lignocellulose degradation products on industrial cellulase enzymes during cellulose hydrolysis. Appl Biochem Biotechnol. 2009;159(3):696–707.
137. Li Y, Qi B, Wan Y. Inhibitory effect of vanillin on cellulase activity in hydrolysis of cellulosic biomass. Bioresour Technol. 2014;167:324–30.
138. Ximenes E, Kim Y, Mosier N, Dien B, Ladisch M. Inhibition of cellulases by phenols. Enzyme Microb Technol. 2010;46:170–6.
139. Holtzapple M, Cognata M, Shu Y, Hendrickson C. Inhibition of *Trichoderma reesei* cellulase by sugars and solvents. Biotechnol Bioeng. 1990;36(3):275–87.
140. Qing Q, Yang B, Wyman CE. Xylooligomers are strong inhibitors of cellulose hydrolysis by enzymes. Bioresour Technol. 2010;101(24):9624–30.
141. Xue S, Uppugundla N, Bowman MJ, Cavalier D, Da Costa SL, Dale BE, Balan V. Sugar loss and enzyme inhibition due to oligosaccharide accumulation during high solids-loading enzymatic hydrolysis. Biotechnol Biofuels. 2015;8:195.
142. Qing Q, Wyman CE. Hydrolysis of different chain length xylooliogmers by cellulase and hemicellulase. Bioresour Technol. 2011;102(2):1359–66.
143. Zhang J, Viikari L. Xylo-oligosaccharides are competitive inhibitors of cellobiohydrolase I from *Thermoascus aurantiacus*. Bioresour Technol. 2012;117(4):286–91.

144. Momeni MH, Ubhayasekera W, Sandgren M, Ståhlberg J, Hansson H. Structural insights into the inhibition of cellobiohydrolase Cel7A by xylo-oligosaccharides. FEBS J. 2015;282(11):2167–77.
145. Xin D, Ge X, Sun Z, Viikari L, Zhang J. Competitive inhibition of cellobiohydrolase I by manno-oligosaccharides Enzyme Microb Technol. 2014;68:62–8.
146. Gruno M, Väljamäe P, Pettersson G, Johansson G. Inhibition of the *Trichoderma reesei* cellulases by cellobiose is strongly dependent on the nature of the substrate. Biotechnol Bioeng. 2004;86(5):503–11.
147. Atreya ME, Strobel KL, Clark DS. Alleviating product inhibition in cellulase enzyme Cel7A. Biotechnol Bioeng. 2016;113(2):330–8.
148. Arantes V, Saddler JN. Access to cellulose limits the efficiency of enzymatic hydrolysis: the role of amorphogenesis. Biotechnol Biofuels. 2010;3:4.
149. Jäger G, Girfoglio M, Dollo F, Rinaldi R, Bongard H, Commandeur U, Fischer R, Spiess AC, Büchs J. How recombinant swollenin from *Kluyveromyces lactis* affects cellulosic substrates and accelerates their hydrolysis. Biotechnol Biofuels. 2011;4:33.
150. Yang B, Wyman CE. BSA treatment to enhance enzymatic hydrolysis of cellulose in lignin containing substrates. Biotechnol Bioeng. 2006;94(4):611–7.
151. Florencio C, Badino AC, Farinas CS. Soybean protein as a cost-effective lignin-blocking additive for the saccharification of sugarcane bagasse. Bioresour Technol. 2016;221:172–80.
152. Eriksson T, Börjesson J, Tjerneld F. Mechanism of surfactant effect in enzymatic hydrolysis of lignocellulose. Enzyme Microb Technol. 2002;31:353–64.
153. Hsieh CW, Cannella D, Jørgensen H, Felby C, Thygesen LG. Cellobiohydrolase and endoglucanase respond differently to surfactants during the hydrolysis of cellulose. Biotechnol Biofuels. 2015;8:52.
154. Leskinen T, Salas C, Kelley SS, Argyropoulos DS. Wood extractives promote cellulase activity on cellulosic substrates. Biomacromolecules. 2015;16(10):3226–34.
155. Lou H, Wang M, Lai H, Lin X, Zhou M, Yang D, Qiu X. Reducing non-productive adsorption of cellulase and enhancing enzymatic hydrolysis of lignocelluloses by noncovalent modification of lignin with lignosulfonate. Bioresour Technol. 2013;146:478–84.
156. Kim DW, Jang YH, Kim CS, Lee NS. Effect of metal ions on the degradation and adsorption of two cellobiohydrolases on microcrystalline cellulose. Bull Korean Chem Soc. 2001;22(7):716–20.
157. Akimkulova A, Zhou Y, Zhao X, Liu D. Improving the enzymatic hydrolysis of dilute acid pretreated wheat straw by metal ion blocking of non-productive cellulase adsorption on lignin. Bioresour Technol. 2016;208:110–6.
158. Tejirian A, Xu F. Inhibition of cellulase-catalyzed lignocellulosic hydrolysis by iron and oxidative metal ions and complexes. Appl Environ Microbiol. 2010;76(23):7673–82.
159. Zhang J, Pakarinen A, Viikari L. Synergy between cellulases and pectinases in the hydrolysis of hemp. Bioresour Technol. 2013;129(2):302–7.
160. Hu J, Arantes V, Pribowo A, Saddler JN. Synergistic action of accessory enzymes enhances the hydrolytic potential of a "cellulase mixture" but is highly substrate specific. Biotechnol Biofuels. 2013;6:112.
161. Selig MJ, Knoshaug EP, Adney WS, Himmel ME, Decker SR. Synergistic enhancement of cellobiohydrolase performance on pretreated corn stover by addition of xylanase and esterase activities. Bioresour Technol. 2008;99(11):4997–5005.
162. Várnai A, Huikko L, Pere J, Siikaaho M, Viikari L. Synergistic action of xylanase and mannanase improves the total hydrolysis of softwood. Bioresour Technol. 2011;102(19):9096–104.
163. Gao D, Uppugundla N, Chundawat SP, Yu X, Hermanson S, Gowda K, Brumm P, Mead D, Balan V, Dale BE. Hemicellulases and auxiliary enzymes for improved conversion of lignocellulosic biomass to monosaccharides. Biotechnol Biofuels. 2011;4:5.

164. Modenbach AA, Nokes SE. The use of high-solids loadings in biomass pretreatment-a review. Biotechnol Bioeng. 2012;109(6):1430–42.
165. Koppram R, Nielsen F, Albers E, Lambert A, Wännström S, Welin L, Zacchi G, Olsson L. Simultaneous saccharification and co-fermentation for bioethanol production using corncobs at lab, PDU and demo scales. Biotechnol Biofuels. 2013;6:2.
166. Zhang J, Chu DQ, Huang J, Yu ZC, Dai GC, Bao J. Simultaneous saccharification and ethanol fermentation at high corn stover solids loading in a helical stirring bioreactor. Biotechnol Bioeng. 2010;105(4):718–28.
167. Wang W, Zhuang X, Yuan Z, Yu Q, Qi W, Wang Q, Tan X. High consistency enzymatic saccharification of sweet sorghum bagasse pretreated with liquid hot water. Bioresour Technol. 2012;108:252–7.
168. Liu ZH, Chen HZ. Periodic peristalsis releasing constrained water in high solids enzymatic hydrolysis of steam exploded corn stover. Bioresour Technol. 2016;205:142–52.
169. Lu Y, Yang B, Gregg D, Saddler JN, Mansfield SD. Cellulase adsorption and an evaluation of enzyme recycle during hydrolysis of steam-exploded softwood residues. Appl Biochem Biotechnol. 2002;98-100:641–54.
170. Qi B, Chen X, Su Y, Wan Y. Enzyme adsorption and recycling during hydrolysis of wheat straw lignocellulose. Bioresour Technol. 2011;102(3):2881–9.
171. Tu M, Zhang X, Paice M, MacFarlane P, Saddler JN. The potential of enzyme recycling during the hydrolysis of a mixed softwood feedstock. Bioresour Technol. 2009;100(24):6407–15.
172. Jin MJ, Gunawan C, Uppugundla N, Balan V, Dale BE. A novel integrated biological process for cellulosic ethanol production featuring high ethanol productivity, enzyme recycling and yeast cells reuse. Energy Environ Sci. 2012;5(5):7168–75.
173. Weiss N, Börjesson J, Pedersen LS, Meyer AS. Enzymatic lignocellulose hydrolysis: Improved cellulase productivity by insoluble solids recycling. Biotechnol Biofuels. 2013;6:5.
174. Visser EM, Leal TF, de Almeida MN, Guimarães VM. Increased enzymatic hydrolysis of sugarcane bagasse from enzyme recycling. Biotechnol Biofuels. 2015;8:5.
175. Bals BD, Christa G, Janette M, Farzaneh T, Dale BE. Enzymatic hydrolysis of pelletized AFEX™-treated corn stover at high solid loadings. Biotechnol Bioeng. 2014;111(2):264–71.
176. Karki B, Muthukumarappan K, Wang Y, Dale B, Balan V, Gibbons WR, Karunanithy C. Physical characteristics of AFEX-pretreated and densified switchgrass, prairie cord grass, and corn stover. Biomass Bioenerg. 2015;78:164–74.
177. Sundaram V, Muthukumarappan K, Kamireddy SR. Effect of ammonia fiber expansion (AFEX™) pretreatment on compression behavior of corn stover, prairie cord grass and switchgrass. Ind Crops Prod. 2015;74:45–54.
178. Wingren A, Galbe M, Zacchi G. Techno-economic evaluation of producing ethanol from softwood: Comparison of SSF and SHF and identification of bottlenecks. Biotechnol Prog. 2003;19:1109–17.
179. Park I, Kim I, Kang K, Sohn H, Rhee I, Jin I. Cellulose ethanol production from waste newsprint by simultaneous saccharification and fermentation using *Saccharomyces cerevisiae* KNU5377. Process Biochem. 2010;45:487–92.
180. Zhu JQ, Qin L, Li BZ, Yuan YJ. Simultaneous saccharification and cofermentation of aqueous ammonia pretreated corn stover with an engineered *Saccharomyces cerevisiae* SyBE005. Bioresour Technol. 2014;169:9–18.
181. Zhu JQ, Qin L, Li WC, Zhang J, Bao J, Li BZ, Yuan YJ. Simultaneous saccharification and co-fermentation of dry diluted acid pretreated corn stover at high dry matter loading: overcoming the inhibitors by non-tolerant yeast. Bioresour Technol. 2015;198:39–46.
182. Taherzadeh MJ, Niklasson C, Liden G. Conversion of dilute-acid hydrolyzates of spruce and birch to ethanol by fed-batch fermentation. Bioresour Technol. 1999;69:59–66.
183. Tengborg C, Galbe M, Zacchi G. Reduced inhibition of enzymatic hydrolysis of steam-pretreated softwood. Enzyme Microb Technol. 2001;28:835–44.

# Part II
# Production of Aldehydes

# Chapter 2
# Sustainable Catalytic Strategies for $C_5$-Sugars and Biomass Hemicellulose Conversion Towards Furfural Production

**Andre M. da Costa Lopes, Ana Rita C. Morais, and Rafał M. Łukasik**

**Abstract** Furfural has been identified as one of the most important biomass-based platform chemicals and has the potential to be used as a substitute of petrochemical-derived building blocks in the production of chemicals and advanced biofuels. Despite that the current industrial production technology of furfural is well established, it is characterised by moderate production yield and selectivity which reduces its competitiveness with crude oil-based alternatives. Furthermore, conventional furfural production requires high energy and generates acidic waste streams. Thus, research on more economic and environmentally benign furfural production strategies from hemicellulose biomass and pentose sugar feedstocks has become of worldwide interest in the scientific community and chemical industry. The present chapter aims to provide state-of-the-art developments in the field of catalytic synthesis of furfural from $C_5$-sugars and hemicellulose biomass, taking into the consideration of green chemistry principles. Among the many advances, the employment of homogeneous catalysts *i.e.* metal halides, ionic liquids and high-pressure $CO_2$ is presented. Application of heterogeneous catalysts is addressed briefly. The performance and efficiency of each catalytic approach in terms of catalyst reactivity, furfural yield and selectivity, as well as the sustainability of furfural production are analysed. Finally, critical outlook and perspectives of the development of sustainable furfural production processes are provided.

**Keywords** Furfural • Catalyst • Biorefinery • Pentoses • Biomass • Ionic liquids • Supercritical fluids • Metal halides

---

A.M. da Costa Lopes • A.R.C. Morais
Laboratório Nacional de Energia e Geologia, I.P., Unidade de Bioenergia,
Estrada do Paço do Lumiar 22, 1649-038 Lisbon, Portugal

LAQV-REQUIMTE, Departamento de Química, Faculdade de Ciências e Tecnologia,
Universidade NOVA de Lisboa, Caparica, Portugal

R.M. Łukasik (✉)
Laboratório Nacional de Energia e Geologia, I.P., Unidade de Bioenergia,
Estrada do Paço do Lumiar 22, 1649-038 Lisbon, Portugal
e-mail: rafal.lukasik@lneg.pt



Z. Fang et al. (eds.), *Production of Platform Chemicals from Sustainable Resources*, Biofuels and Biorefineries 7, DOI 10.1007/978-981-10-4172-3_2

## 2.1 Introduction

Furfural (furan-2-carbaldehyde) is a heterocyclic aldehyde discovered almost 2 centuries ago, in 1821, by Döbereiner who distilled bran with dilute sulphuric acid [1]. The empirical formula of furfural ($C_5H_4O_2$) was revealed in 1840 by Stenhouse, and in 1845, Fownes proposed the name furfurol that came from bran (furfur) and oil (ol). In 1845, it was discovered that furfural has aldehyde functions hence the name furfur-ol was changed to furfur-al.

Furfural is an oily product with a characteristic almond-like scent. The furfural molecule contains an aldehyde group linked to a furan ring. In the presence of oxidative agents (including air), the colourless furfural solution turns to yellow, then brown and finally can reach even a black colour. The reason for this is the formation of polymeric structures, due to the presence of conjugated double bonds, that can occur even at a very low concentrations ($10^{-5}$ M) [2]. Due to the present of the aldehyde group as well as the furan ring, furfural has the ability to form conjugated double bonds with other unsaturated molecules. Because of this, furfural is a very selective solvent and can be used for extraction of aromatic compounds from fuels or vegetable oils [2]. Besides, the wide natural presence of furfural in fruits, coffee, tea, brown bread [2] allows furfural can be used in agrochemistry with low risk for human and animal health. Despite that, the main use of furfural is its employment as a building block for different chemicals. According to the US Department of Energy, furfural and its sister compound, 5-hydroxymethylfurfural (5-HMF), are considered to be in the *Top 30 Building Blocks* derived from biomass [3]. Furfural is widely used directly as a solvent or converted into dozens of types of chemicals. Furfural alcohol is one of the most important furfural-based chemicals that are used in the production of polymers and plastics. Other important commodities that can be obtained from furfural are methylfuran and tetrahydrofuran which are widely used for the synthesis of polymers containing a furan ring. The potential conversion pathways of furfural to several hundred classes of chemicals are given in Fig. 2.1.

### 2.1.1 *Mechanistic Considerations of Furfural Formation*

As it was aforementioned, furfural was discovered by Döbereiner and was produced from biomass and more specifically from its hemicellulosic fraction. The hemicellulose, mainly pentosans, after the acid hydrolysis, produce saccharide in either oligomeric or monomeric forms [5, 6]. Under more severe reaction conditions, xylose undergoes further dehydration and cyclisation to produce furfural with a theoretical yield of approximately 73 mol%. The kinetics of biomass hydrolysis in the presence of acid catalysts shows that the hydrolysis of hemicellulose to saccharides is faster than their further dehydration [2, 7, 8]. Therefore, the limiting factor for furfural production seems to be the dehydration reaction which involves a series of elementary steps. Nowadays, three different theories exist regarding the

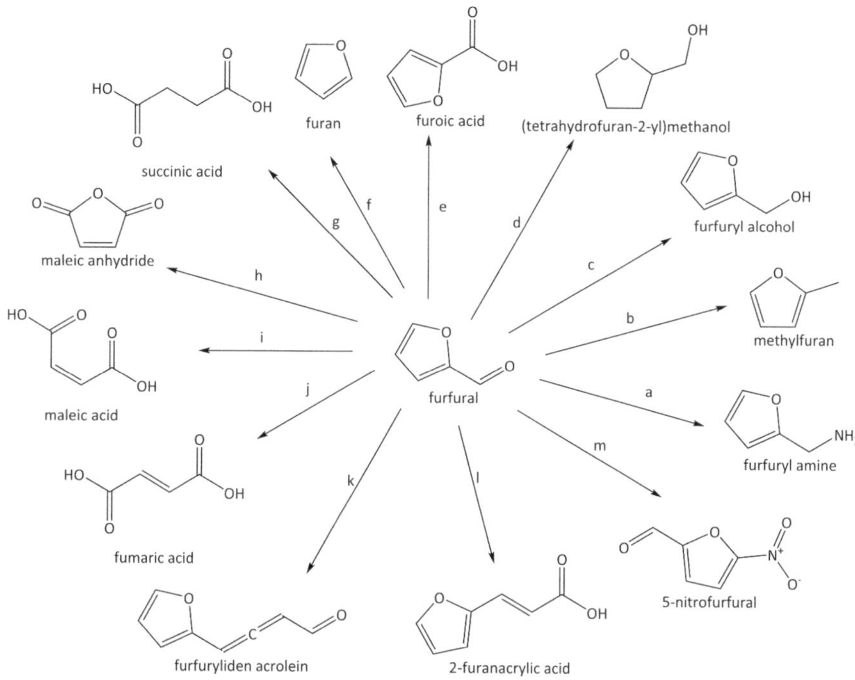


**Fig. 2.1** Furfural-based chemical product family tree. Examples of products obtained from furfural via (*a*) reductive amination, (*b*) catalytic hydrogenation, (*c*) reduction over Ni- and CuCrO-catalysts, (*d*) catalytic vapour phase hydrogenation, (*e*) Cannizarro reaction, (*f*) catalytic decarbonylation, (*g*) catalytic oxidation with $O_2$, (*h*) oxidation, (*i*) catalytic oxidation, (*j*) catalytic oxidation with NaClO3 in presence of $V_2O_5$ catalyst, (*k*) condensation with acetylaldehyde, (*l*) (a) + malonic acid in presence of pyridine (Perkin reaction), (*m*) nitration with $HNO_3$ in $(CH_3CO)_2O$, For full caption of figure please consult Ref. [4] (Adapted from Biorefineries – Industrial Processes and Products. Status Quo and Future Directions. Vol. 2 Edited by Birgit Kamm, Patrick R. Gruber, Michael Kamm Copyright © 2006 WILEY-VCH Verlag GmbH & Co. KGaA, Weinheim with permission)

mechanisms of furfural production from pentoses. The first possible mechanism for furfural formation from pentoses involves the reaction through a 1,2-enediol formation and subsequent triple dehydration [9, 10] as shown in Fig. 2.2.

Alternatively, the second possible mechanism goes through a 2,3-(α,β)-unsaturated aldehyde [11] according to reaction chains depicted in Fig. 2.3.

A third mechanism assumes the furfural formation from the pyranose form of pentoses and later due to acidic attack leads to a first dehydration and structural reformulation to form a furanose ring, followed by a subsequent dehydrations to furfural [8, 12, 13] according to the mechanism demonstrated in Fig. 2.4.

The literature data on the $^{14}C$ isotope tracking of dehydration of xylose [11] as well as other works show a significantly higher proportion of acyclic being formed in water than aldopentoses [14]. This confirms that the first possible mechanism (Fig. 2.2) is more likely than the other mechanisms (Figs. 2.3, and 2.4). Besides,

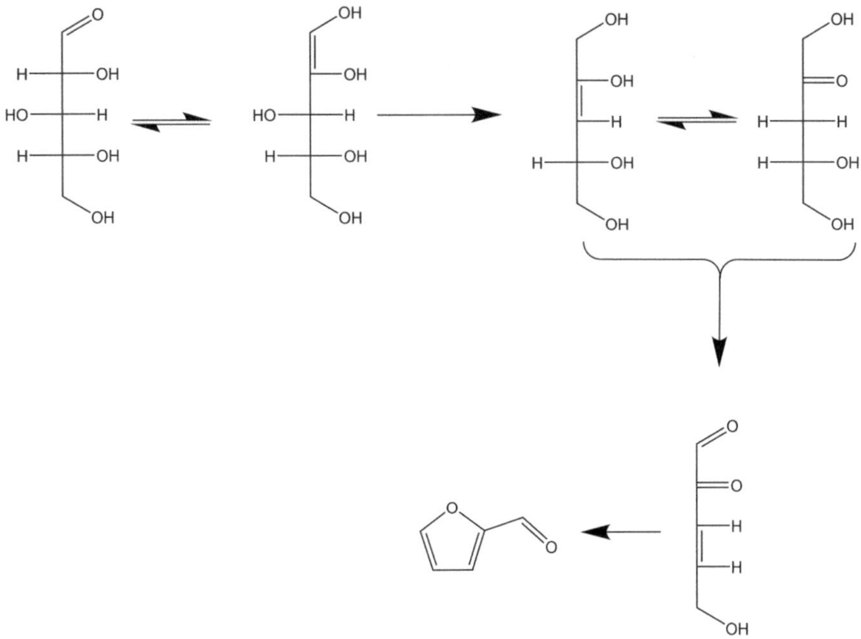


**Fig. 2.2** Possible mechanism (proposal 1) of pentoses dehydration to furfural though 1,2-enediol intermediate

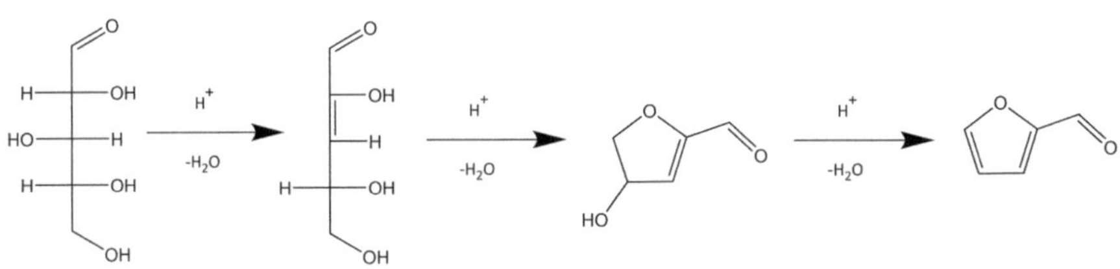


**Fig. 2.3** Possible mechanism (proposal 2) of xylose dehydration to furfural *via* β-elimination

taking into account that the dehydration of xylose to furfural is the limiting step, it can be considered that its formation should proceed through 1,2-enediol intermediate because enolisation reactions are generally reversible hence the 1,2-enediol formation from xylose can be considered to be rate limiting step in the furfural formation [10]. Despite these diverse scientific evidence published in literature [8–24], it is difficult to indicate one mechanism and rather all aforementioned mechanisms are possible and reaction conditions and catalytic/solvent system probably determine the importance of one mechanism over the other mechanisms.

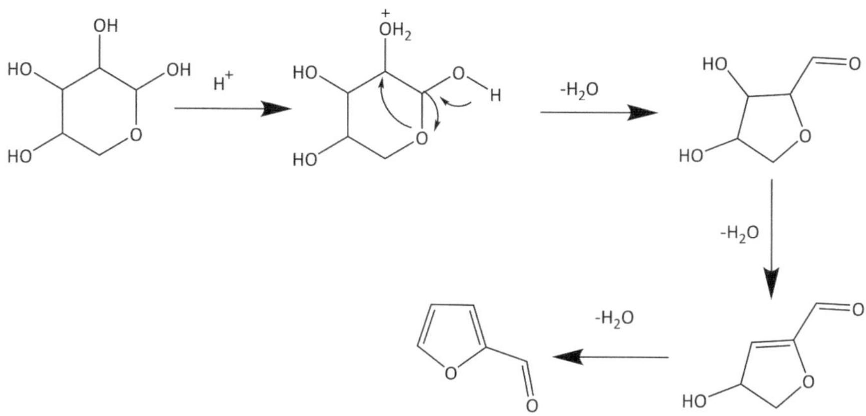


**Fig. 2.4** Possible mechanism (proposal 3) of furfural formation *via* cyclic intermediates

### 2.1.2 *Industrial Furfural Manufacturing and Their Recent Updates*

During World War I, the USA was driven to be self-efficient. Because of that, furfural became one of the most important pivotal chemicals in its aim to achieve this target. For this reason, between 1914 and 1918 the intense research on industrial valorisation of agricultural wastes took place. The National Bureau of Chemistry (US) studied corncob as a feedstock for furfural production. The huge Quaker Oats stockpiles of cereal wastes were considered as a potential source of furfural, and in 1921, the Quaker Oats company commenced production of furfural from oat hulls using "left over" reactors [1, 2]. The bath process of several tonnes per month involved the use of concentrated sulphuric acid.

Nowadays worldwide production of furfural is estimated at 300,000 tonnes/a [25] and China is the main furfural producer. Chinese producers contribute to half of the annual furfural global production capacity, however the installations are widespread and use inefficient (~50% of the theoretical yield) small-scale fixed bed reactors. On the contrary, South Africa and Dominican Republic producers use large-scale plants with higher efficiency than in China [26].

Nonetheless, to reach the significant levels of annual productions several industrial processes have been developed during last decades. One of the pioneers was BIOFINE process patented by Fitzpatrick in 1990 [27]. The BIOFINE process involves the formation of furfural as an intermediate product in levulinic acid production. The BIOFINE process uses two reactors in which biomass is processed with dilute sulphuric acid (1.5–3% depending on the alkalinity of the raw material). Then, the biomass is treated with steam (25 bar) at a temperature above 200 °C for a short time to promote cellobiose hydrolysis. The solution is fed into a second reactor in which the temperature is varied around 200 °C to have a reaction time of 20 min at 14 bar pressure. This allows levulinic acid to be formed, however any previously formed furfural can be recovered separately. The furfural yield obtained from this process is close to 70% of theoretical and as the rest is incorporated into char [28]. LeCalorie S.p.A. in Italy operated this process with 3 ktonnes/a for processing paper mill sludge and tobacco plant wastes [29].

Gravitis and Vedernikov from Latvian State University of Wood Chemistry developed an alternative process to enhance furfural production yield and minimise cellulose degradation [30]. Their process involves the use of combined aliquots of strong acid catalysts and salts to promote hydrolysis and dehydration reactions. The advantage of their process is the relatively high furfural yields (between 55 and 75% of the theoretical value), and a five-fold reduction in cellulose degradation compared with other processes. It is possible to achieve selective hydrolysis of pentosan-rich materials in two-steps.

In the CIMV (Compagnie Industrielle de la Matiere Vegetale) process, continuous fractionation of biomass with organic acid organosolv is used [31]. The main products of this process are whitened paper pulp, sulphur-free lignin (Biolignin™) and syrup containing xylose. Starting from 1 tonne of biomass, 270 kg of Biolignin™, 490 kg of cellulose pulp and 220 kg of sugar syrup with 48% of pentoses can be obtained. The pentoses can be converted to furfural. The actual processing capacity of the CIMV plant is based on 70 kg/h of straw, and the plant located in France started operation at the beginning of 2007 [32].

The Lignol Innovations Corporation developed a continuous biorefinery process employing the organosolv treatment to remove lignin, hemicellulose and extractives. The main product of this technology is cellulosic ethanol, while the liquor obtained from the organosolv pre-treatment is processed to produce furfural, xylose, acetic acid, lipophilic extractives and lignin. The concept of this process is based on the economic feasibility of ethanol production that can be achieved by the co-production of value-added products especially that Lignol biorefinery technology processes ∽100 tonnes/day of dry wood [33].

In 2000, Zeitsch patented the SupraYield process [2, 34]. The acid hydrolysis pre-treatment process allows simultaneous production of furfural and its immediate removal from the reacting, acid phase. The removal occurs by maintaining the reaction mixture to a continuously boiling state. In 2011, International Furan Technology started to explore sugarcane bagasse as a feedstock in the modified SupraYield process [35] in Central Romana factory in Dominican Republic, which is the largest in the world, and in Illovo Sugar, which is the second largest in the world [36].

Delft University developed MTC (Multi-Turbine-Column) process that is focused on the pre-treatment of biomass and the subsequent hydrolysis to produce a gaseous furfural stream and a solid residue. The furfural purification occurs using alkaline solutions and toluene as the organic phase. Among the claimed advantages of the process are high furfural yield, due to continuous removal of furfural from the reacting mixture, less by-product formation, and lower energy demands compared to batch processes [26].

Current industrial batch and continuous processes use mineral acids as catalysts, which leads to corrosion, issues with product and catalysts recuperation from reaction mixture, and risks to health and environment. In order to maximise the furfural production yield, diverse approaches have been proposed in the literature involving new types of catalysts in monophasic and biphasic reaction systems. Ionic liquids, high-pressure fluids, metal halides as well as heterogeneous catalysts are some of the strategies proposed to achieve higher furfural yield with lower negative environmental impact. All of these novel and greener concepts are discussed in the following sections.

## 2.2 Emerging Strategies of Furfural Production

### 2.2.1 *Homogeneous Catalysis*

Presently used methods for furfural production have technological or environmental issues. Alternatives to wasteful, energy intensive processes will require safe and environmentally friendly catalysts accordingly to green chemistry principles. In view of active research in alternative homogeneous catalysts, metal halides, supercritical $CO_2$, and ionic liquids (ILs) have gained increased attention in this field. The present contribution has been aimed to highlight alternative homogeneous catalysts that can significantly contribute to the development of advanced catalysts to be used in future furfural production processes.

#### 2.2.1.1 Metal Halides

Fundamentals and Mechanism

Metal halides are inexpensive catalysts and show great catalytic activity [37]. Metal halides can be categorised in four main groups accordingly to the periodic table: alkali metals, *e.g.* Li, Na and K; alkaline earth metals *e.g.* Mg and Ca; transition metals, *e.g.* Fe, Cr, Cu, Mn, Co, Zn and group IIIA metals. For the last group, phosphate, sulphate and nitrate anions were also tested as well as halides.

Numerous works have demonstrated efficient application of metal halides as catalyst [38, 39] for the conversion of lignocellulosic biomass to furfural, namely in aqueous systems, and often also coupled with other types of processes, such as microwave irradiation or ionic liquids [40]. Despite the great potential of metal halides in furfural production, there is still a lack of detailed information about their mechanism of hydrolysis and dehydration of pentoses into furfural. In general, the conversion of hemicelluloses into furfural with metal halides, especially metal chlorides, is related to their Lewis acid character, due to their ability to attract electron pairs. The addition of metal chlorides *e.g.* $AlCl_3$, $CuCl_2$, to water medium leads to its dissociation into complex ions. This drives to the formation of coordinate covalent bonds composed by several molecules of water [41]. In general, the nomenclature of metal ions ligand complexes is demonstrated as $[M(H_2O)_n]^{z+}$, where M is the metal ion, n is the solvation number (in the range of 4 to 9) and $z$ is the cation oxidation state [41, 42]. Water molecules bond to the central metal cation as monodentate ligands, *i.e.* only one donor atom is used to bond to the central metal cation, whereas the metal cation becomes polarised due to the water molecules. Then, addition of metal chlorides to water promotes hydrolysis reactions, which consequently, results in complex formation of cations. Near the metal cation, water molecules form a primary hydration sphere. Water molecules that bond to the primary hydration sphere also form a secondary hydration sphere. Bonding of hydronium to naturally more acidic hydrated cations results in relatively strong hydrogen bonds and forms a secondary hydration sphere [43]. Then, the metal cations that remain in the sphere of hydration promote the release of electrons from water molecules [41, 43]. Thus, the metal cations act as Lewis acids having a positive effect in the hydrolysis of glycosidic linkages of xylan polymer, whereas the coordinate water molecules from the hydrated cation act as nucleophiles to bond with xylose [41]. However, the application of high temperatures and pressures can destroy the hydration structure, whereby the metal cations remain in their elementary states [43, 44].

On other hand, other types of metal chlorides, *e.g.* $FeCl_3$, have a Brønsted acid character. For instance, it is well-known that aqueous solutions of $FeCl_3$ have Brønsted acid behaviour once the hydrolysis of $Fe^{3+}$ leads to the formation of different kinds of complexes as shown in Eqs. 2.1 and 2.2 [45, 46]:

$$\left[Fe\left(\mathrm{H_2O}\right)_6\right]^{3+} \leftrightarrow \mathrm{H}^+ + \left[Fe\left(OH\right)\left(\mathrm{H_2O}\right)_5\right]^{2+} \tag{2.1}$$

$$2\left[Fe\left(OH\right)\left(\mathrm{H_2O}\right)_5\right]^{2+} \leftrightarrow \left[\left(\mathrm{H_2O}\right)_5\mathrm{FeOFe}\left(\mathrm{H_2O}\right)_5\right]^{+4} + \mathrm{H_2O} \tag{2.2}$$

The pH value and the initial iron concentration play an important role in the equilibrium of these reactions [46]. The pH value of the solution drops dramatically when aqueous solution of $FeCl_3$ is boiled to form a dark precipitate [46] that consists mainly of iron oxides such as $\alpha$-$Fe_2O_3$ and $\beta$-FeOOH [47]. Thus, the concentration of $H^+$ in the solution increases according to the Eq. 2.3.:

$$3Fe^{3+} + 5\mathrm{H_2O} \leftrightarrow 9\mathrm{H}^+ + Fe_2\mathrm{O}_3 \downarrow + \mathrm{FeOOH} \downarrow \tag{2.3}$$

The presence of $FeCl_3$ in the aqueous solution acts as a source of HCl and iron oxide, decreasing the pH of the reactional medium [46]. Similar effect is foreseen from other trivalent metal chlorides [48]. Liu et al. reported that $FeCl_3$ in water had a greater catalytic effect on hemicellulose removal than strong acid solutions at a similar pH values [49]. Additionally, Mao et al. reported that $Fe^{3+}$ promoted seawater-based production of furfural through hydrolysis of biomass in acetic acid steam [50].

The role of halide salts in the dehydration of xylan type sugars to furfural has been investigated thoroughly [22, 51]. For instance, Marcotullio and de Jong determined the role of halide salts in assisting the enolisation of protonated acyclic xylose and consecutive dehydration steps (Fig. 2.5) [51].

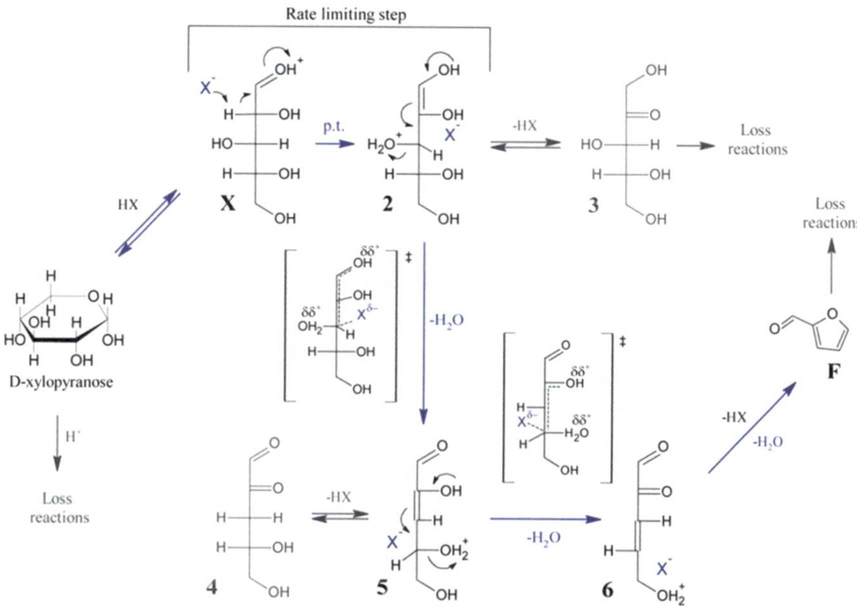


**Fig. 2.5** Role of halide salts in proposed mechanism of xylose conversion to furfural in acidic media by 1,2-enediol formation (2) and dehydration (5 and 6), according to ref. [51]. $X^-$ indicates aqueous halide (Reprinted from Carbohydrate Research, Vol 346, Marcotullio G, de Jong W, Furfural formation from D-xylose: the use of different halides in dilute aqueous acidic solutions allows for exceptionally high yields, Pages 1291–1293. Copyright © (2011), with permission from Elsevier)

According to the authors, $Cl^-$ had a greater effect in promoting enolisation reaction than $Br^-$ and $I^-$, possibly due to its nucleophilic character. Enslow and Bell found that the selectivity of xylose dehydration to furfural followed the same trend, however the conversion of xylose in aqueous solution increased in the following order: $Cl^- > Br^- > I^-$ [52].

### Interaction of Metal Halides with Water

The interaction of metal halides with a solvent has a great influence on activity as indicated by the solvent electron pair acceptor number. For example, water is characterised as having one of the highest acceptor numbers, *i.e.* 54.8 [42]. After the formation of a metal ion in aqueous solution, *i.e.* aqua ion, it is subject to hydrolysis producing hydroxide and oxide species, which in consequence, leads to a loss of its Lewis acid character. However, hard pre-transition metal ions or lanthanide ions and $Al^{3+}$ are highly resistant cations for preserving their Lewis acid character [42]. Furthermore, the acid strength of a metal cation in solution is highly depended on pKa. A lower pKa of a metal cation gives the ion a higher acidic character. Table 2.1 shows pKa values of some metal cations. As can be seen, $Fe^{3+}$, $Al^{3+}$ and $Cu^{2+}$ are the metal cations with stronger acid character than $Fe^{2+}$ and $Ca^{2+}$. Because of this, acidic

**Table 2.1** Selected metal cations pKa and hardness values

| Metal cation | pKa | Hardness value (eV) [54] |
|---|---|---|
| $Fe^{3+}$ | 2.46 | 13.1 |
| $Al^{3+}$ | 4.85 | 45.8 |
| $Cu^{2+}$ | 6.50 | 8.3 |
| $Fe^{2+}$ | 9.49 | 7.3 |
| $Ca^{2+}$ | 12.7 | 19.7 |



metal chlorides, such as $FeCl_3$, $AlCl_3$, $CrCl_2$, have been widely used due to their ability to catalyse the dehydration of $C_5$-sugars into furfural [17, 53].

Liu and Wyman reported that aqueous solutions containing 0.8 wt. % $CaCl_2$ (pH = 5.59) or 0.8 wt. % $MgCl_2$ (pH = 5.87) are acidic and that the pH value of 0.8 wt. % $FeCl_3$ solutions is particularly low (pH = 1.86) [55]. Cai et al. reported the influence of several metal halides, $FeCl_3$, $AlCl_3$, $CrCl_3$, $CuCl_2$, and $ZrOCl_2$ in THF-water system, on the conversion of different lignocellulosic biomasses, such as maple wood and corn stover, into furfural [56]. All pH values of metal solutions were determined in 1:1 THF-water system. The $FeCl_3$ solution (pH = 1.90) allowed the highest theoretical furfural yield of 95 mol% from maple wood at 170 °C for 60 min of reaction time and at 4:1 THF-water system. Its high Brønsted character permitted slowing down the xylose conversion rates and enhancing the furfural selectivity at longer reaction times. On the other hand, $CuCl_2$ (pH = 2.78) and $AlCl_3$ (pH = 2.88) had a moderate effect resulting in furfural yields in the range of 81–89 mol% and 58–76 mol%, respectively, probably due to their moderate Brønsted acidity. The effect of $ZrOCl_2$ (pH = 1.65) on maple wood was characterised by a low furfural yield (44 mol%) and a poor selectivity (37 mol%), despite its strong Brønsted acidity. The $CrCl_3$ solution (pH = 3.13) allowed high xylose conversion (100% in 5 min), however it had a negative impact on furfural yield as furfural losses quickly overtook the furfural production. The authors proposed that the strong Lewis acid character of $CrCl_3$ played an important role in furfural losses [56].

### Furfural from $C_5$-Sugars and Lignocellulosic Feedstocks

#### *Monophasic Aqueous and Non-aqueous Systems*

Converting xylan polymer to furfural is a critical step in the utilisation of lignocellulosic biomass, since β-D-xylose is the main constituent of hemicellulose. The ability to produce furfural with a high yield and selectivity implies a low production of humins as conversion of xylose to furfural is crucial to limit the production of side products and humins.

The kinetics of fundamental steps of main and loss reactions in the conversion mechanism of $C_5$-sugars play a crucial role in achieving a high yield of furfural. Liu and Wyman examined the influence of various metal halides, KCl, NaCl, $CaCl_2$, $MgCl_2$ and $FeCl_3$ on the degradation of xylose and xylotriose into furfural in aqueous solutions at 180 °C [55]. Adding 0.8 wt. % $FeCl_3$ to the aqueous solution resulted in the greatest increase of xylose and xylotriose degradation rate (6- and 49 fold, respectively) in comparison to those obtained in hot water. The improvement of xylose degradation rate was found to be in the following order: NaCl < KCl < $CaCl_2$ < $MgCl_2$ < $FeCl_3$. However, the formation of other degradation products and unidentified compounds increased in the presence of $FeCl_3$. Those authors found that unidentified products could be produced from furfural resinification and condensation reactions [55]. Yang et al. have investigated the conversion of commercial xylan-derived beech wood to furfural in aqueous reactional medium containing various types of metal chlorides, such as $CrCl_3$, $CrCl_2$, $AlCl_3$, $FeCl_3$, $SnCl_4$, $SnCl_2$, $GeCl_4$ and $InCl_3$, as catalyst [57]. They found that the catalytic activity of those metal chlorides was highly dependent on cation with $Sn^{4+}$ being the most effective. The effectiveness of $SnCl_4$ can be attributed to the hydrolysis of $Sn^{4+}$ in aqueous solution and the equilibrium with colloidal tin oxide and protons [58], as represented by Eq. 2.4, which shows that the pH value decreases and promotes hydrolysis of xylan into xylose.

$$Sn^{4+} + (\mathrm{x}+2)\mathrm{H_2O} \leftrightarrow Sn\mathrm{O_2} \cdot \mathrm{xH_2O} \downarrow + 4\mathrm{H^+} \tag{2.4}$$

The formed hydronium ion is mostly responsible for the hydrolysis of hemicellulose from the cell wall of lignocellulosic biomass promoting the release of xylose, whereas both $Sn^{4+}$ and $SnO_2 \bullet xH_2O$ act as Lewis acids promoting the xylose dehydration reactions, besides the acidic property of $Sn^{4+}$ [57]. Wang et al. found that mixtures of two metal chlorides, such as $SnCl_4$/LiCl, can improve by 8% the dehydration of xylose into furfural in comparison to $SnCl_4$ alone [59]. These data are in good agreement with that reported by Binder et al. who stated that $CrCl_2$ along with LiBr as additive, led to xylose to furfural yields up to 56 mol% at 100 °C for 4 h [17].

Some studies have focused on the use of non-aqueous systems, namely ionic liquids (ILs). The selective transformations of pentoses, native hemicellulose and biomass into furfural in reaction media composed of ILs as solvents (or co-solvents) and metal halides have been reported in many works [38, 40, 60, 61]. To reach this transformation, 1-butyl-3-metylimidazolium chloride ([bmim]Cl) has been chosen in the majority of those studies, since it has demonstrated to have a high ability to dissolve carbohydrates, as well as it can stabilise furfural at high temperatures [38]. The incorporation of metal halides ($CrCl_3$ and $AlCl_3$) in the IL medium was demonstrated to be more effective for converting xylose into furfural than $H_2SO_4$ [38, 60, 62]. For instance, Peleteiro et al. reached ~50 mol% of furfural yield from xylose using [bmim]Cl/$CrCl_3$ system at 120 °C for 30 min. Nevertheless, practically half of the initial xylose was converted into other products (soluble and insoluble)

[60]. Higher furfural yield (82.2 mol% from xylose) was obtained with [bmim]Cl/$AlCl_3$ system, although this enhancement was also promoted by microwave irradiation [38]. In fact, microwave irradiation improves the ability of metal halides in the presence of ILs to transform diverse feedstocks, such as pentoses, xylan and biomass into furfural [38, 40]. Just a few minutes or even seconds are needed to achieve high furfural yields from xylan. For instance, using 400 W irradiation power of a 63 mol% of xylan was converted into furfural in the presence of [bmim]Cl/$CrCl_3$ system within 2 min [40]. On the other hand, higher furfural yield (~85 mol%) was obtained with the microwave-assisted transformation of xylan in [bmim]Cl/$AlCl_3$ system in only 10 seconds at 170 °C [38]. Among several examined metal halides ($CrCl_3$, $AlCl_3$, $FeCl_3{\cdot}6H_2O$, $CuCl_2{\cdot}2H_2O$, LiCl and CuCl), $AlCl_3$ was demonstrated to be the most efficient catalyst in the production of furfural from xylan assisted by microwave irradiation. Furthermore, a correlation between the number of coordinated chlorides in the catalysts and the catalytic efficiency was observed. Considering the efficiency in the producing furfural, examined catalysts could be ordered as follows: $AlCl_3 > FeCl_3{\cdot}6H_2O > CrCl_3 > CuCl_2{\cdot}2H_2O > CuCl > LiCl$ [38]. The presence of chloride anions enhances the dehydration of xylose by acting as weak bases promoting the formation of the 1,2-enediol from the acyclic form of xylose. Subsequently, they accelerate the three dehydration steps and ring closure to form furfural [38, 51]. The cation $Al^{3+}$ is more efficient in furfural synthesis than $Cr^{3+}$ and $Fe^{3+}$ cations. However, when lignocellulosic biomass is used as substrate, the efficiency of the [bmim]Cl/metal halide reaction system drastically decreases [38, 40]. Corn stalk, rice straw and pine wood have been treated in [bmim]Cl/$CrCl_3$ system under 3 min of microwave irradiation (400 W) and produced furfural in 23–31 mol% yields [40]. Similar ranges of furfural yields (16–33 mol%) were obtained for corncob, grass and pine wood with the [bmim]Cl/$AlCl_3$ reaction system [38]. The main reasons for the very low yields obtained in comparison to those reactions with xylose and xylan as raw materials are dictated by the complex interactions between biomass components as well as by the presence of more complex parallel reactions between formed xylose, furfural and other highly reactive compounds composing biomass (*e.g.* phenolic compounds originated from lignin). A possible solution may lie in a pre-extraction of hemicellulose from biomass to improve the selectivity of catalysts. Peleteiro et al. performed a hydrothermal treatment of pine wood and hemicellulose in polymeric and oligomeric forms were obtained [61]. Nevertheless, the application of the [bmim]Cl/$CrCl_3$ reaction system to this substrate allowed to reach only 37.7 mol% of furfural yield [61]. Despite the relatively low furfural yields, developing processes that allow conversions at mild reaction conditions is highly desirable.

Notwithstanding the number of metal halides investigations performed with aqueous or non-aqueous systems, the direct comparison of their performance varies widely, due to conditions such as temperature, reaction time, catalyst loading and composition of reactional mixture. Table 2.2 depicts an overview of strategies used for conversion of $C_5$-sugars and hemicelluloses to furfural using metal halides as catalysts in both aqueous and non-aqueous systems.

**Table 2.2** Selected examples on metal halides strategies for furfural production in aqueous and non-aqueous solutions

| Starting material | Catalyst | Solvent(s) | Conditions | | FUR yield/ mol% | Ref. |
|---|---|---|---|---|---|---|
| | | | T/°C | t/min | | |
| **Pentose** | | | | | | |
| Xylose | $AlCl_3$ | Choline chloride-citric acid-$H_2O$[e] | 140 | 15 | 59 | [63] |
| Xylose | $CrCl_2$ | DMA[a] | 100 | 240 | 56 | [17] |
| Xylose | $CrCl_3$ | [bmim]Cl | 120 | 30 | 50 | [60] |
| Xylose | $AlCl_3$ | [bmim]Cl[b,c] | 160 | 1.5 | 82 | [38] |
| **Hemicellulose polysaccharides** | | | | | | |
| Xylan | $FeCl_3$ | GVL/water[d] | 170 | 35 | 69 | [64] |
| Xylan | $AlCl_3$ | Choline chloride-citric acid-$H_2O$ | 140 | 25 | 54 | [63] |
| Commercial beech xylan | $SnCl_4$ | Water | 150 | 120 | 49 | [57] |
| Xylan | $CrCl_3$ | [bmim]Cl[b] | 200 | 2 | 63 | [40] |
| Xylan | $AlCl_3$ | [bmim]Cl[b,c] | 170 | 1/6 | 85 | [38] |
| **Lignocellulosic biomass** | | | | | | |
| Corncob | $FeCl_3$ | GVL/water | 185 | 100 | 80 | [64] |
| Maple wood | $FeCl_3$ | THF/water | 170 | 80 | 95 | [56] |
| Corn stover | $FeCl_3$ | THF/water | 170 | 80 | 95 | [56] |
| Corncob | $AlCl_3$ | [bmim]Cl[b,c] | 160 | 3 | 19 | [38] |
| Pinewood | $AlCl_3$ | [bmim]Cl[b,c] | 160 | 3 | 34 | [38] |
| Corn stalk | $CrCl_3$ | [bmim]Cl[b] | 100 | 3 | 23 | [40] |
| Rice straw | $CrCl_3$ | [bmim]Cl[b] | 100 | 3 | 25 | [40] |
| Pine wood | $CrCl_3$ | [bmim]Cl[b] | 100 | 3 | 31 | [40] |

Ref.: References, FUR: furfural, GVL: γ-valerolactone, THF: tetrahydrofuran, [bmim]Cl: 1-butyl-3-methylimidazolium chloride, DMA: N,N-dimethylacetamide

[a]LiBr used as additive. [b]Under microwave irradiation. [c]10 μL of water was added to the reactional system. [d]10 wt. % water was added to the reaction system. [e]10 mg of water was added to the reaction mixture.

*Biphasic Systems*

Water is one of the most economical solvents for the dehydration reaction of pentoses to furfural. However, water speeds up some degradation reactions of furfural since it can also act as reactant [65]. In the search to improve furfural yields and selectivity, recent studies in aqueous/organic biphasic systems have demonstrated new directions. Although many organic solvents have low solubility in an aqueous phase, the limited furfural yield and selectivity can be overcome by addition of salts, such as NaCl. NaCl has a double positive influence on the production of furfural. Firstly, NaCl additive in water increases furfural selectivity and xylose to furfural rate (in acidic systems) [22], and secondly it acts as a phase splitting inducer of furfural in biphasic systems, due to an increase in the internal pressure caused by the action of the salt [39, 53]. Specifically, a biphasic system composed of an aqueous phase, constituted by $FeCl_3$ and NaCl (or seawater), and an organic phase constituted by a biomass-derived organic solvent *i.e.* 2-methyltetrahydrofuran (2-MTHF) has been acknowledged as a highly effective solvent system for furfural production [39]. This strategy led to a maximum xylose to furfural yield of 71 mol% for 20 wt. % of NaCl as additive at 140 °C for 4 h of reaction. Adding 2-MTHF as extraction solvent, 98 mol% of furfural could be extracted from the aqueous/reactional phase. Interestingly is that the direct use of seawater, which in this case acts as a source of various salts, resulting in an improvement in furfural production. However, the conversion of non-purified xylose effluents obtained from beech wood fractionation in water/2-MTHF system using $FeCl_3$ as catalyst resulted in a furfural yield of only 41 mol% [39].

Yang et al. reported on a case where $AlCl_3$ was used in large amount hence it was not considered as catalyst in the traditional sense in a biphasic water/tetrahydrofuran (THF) system at 140 °C for 45 min to form furfural from xylose and xylan [53]. Their approach gave under microwave irradiation, a xylose and xylan to furfural yields of 75% and 64%, respectively. $AlCl_3$ was also shown to be a good performer in the effective conversion of various types of lignocellulosic biomasses, *e.g.* corn stover, pinewood, switchgrass and poplar, resulting in different furfural yields depending on their recalcitrance. Biphasic systems have been used to make progress in improving the furfural yield and selectivity, however the conversion of lignocellulosic biomass into furfural using eco-friendly solvents and routes remains a big challenge that must still be addressed. Table 2.3 shows a wide-range of works reporting the use of metal halides to produce furfural from xylose and different biomass hemicelluloses in biphasic systems.

#### 2.2.1.2 Supercritical Fluids

A supercritical fluid is often referred to as a pure substance found at greater temperatures and pressures than the critical values ($T_c$ and $p_c$, respectively) as shown in Fig. 2.6. Under supercritical conditions, liquid-like densities are approached, allowing solvation of many compounds, whereas viscosity is comparable with that of ordinary gases, and diffusivity is two magnitudes higher than those of liquids [66]. At supercritical conditions, slight changes in temperature and pressure result in tremendous variations of density, without experiencing a phase change which plays an important role in the ability to dissolve other compounds, *i.e.* solvation power. Thus, the low surface tension coupled with high diffusivities of supercritical fluids allow the diffusion of supercritical fluids into the recalcitrant structure of biomass, while the liquid-like density of supercritical fluids permits the dissolution of components present in biomass. Among all of the chemicals that can be used as supercritical fluids, carbon dioxide and water are the fluids considered as potentially interesting regarding to the increasing concerns of green technologies for furfural production [23, 67].

**Table 2.3** Selected examples on metal halides for furfural production in biphasic systems

| Starting material | Catalyst | Solvent(s) | Conditions | | FUR yield/ mol% | Ref. |
|---|---|---|---|---|---|---|
| | | | T/°C | t/min | | |
| **Pentoses** | | | | | | |
| Xylose | $AlCl_3$[a] | THF/water[b,c] | 140 | 45 | 75 | [53] |
| Xylose | $SnCl_4$/LiCl | DMSO/water | 130 | 360 | 63 | [59] |
| Non-purified xylose from beech wood | $FeCl_3$ | 2-MTHF/water[c] | 140 | 240 | 68 | [39] |
| **Hemicellulose Polysaccharides** | | | | | | |
| Xylan | $AlCl_3$[a] | THF/water[b,c] | 140 | 60 | 64 | [53] |
| Commercial beech xylan[e] | $SnCl_4$ | 2-MTHF/water | 150 | 120 | 78 | [57] |
| **Lignocellulosic Biomass** | | | | | | |
| Corn stover | $AlCl_3$[a] | THF/water[b,c] | 160 | 60 | 55 | [53] |
| Pinewood | $AlCl_3$[a] | THF/water[b,c] | 160 | 60 | 38 | [53] |
| Switchgrass | $AlCl_3$[a] | THF/water[b,c] | 160 | 60 | 56 | [53] |
| Poplar | $AlCl_3$[a] | THF/water[b,c] | 160 | 60 | 64 | [53] |
| Corncob[d] | $SnCl_4$ | 2-MTHF/water | 150 | 120 | 69 | [57] |
| Bagasse[d] | $SnCl_4$ | 2-MTHF/water | 150 | 120 | 67 | [57] |
| Corncob[e] | $FeCl_3$ | Sea water/acetic acid | 190 | 30 | 73 | [50] |

Ref.: References, THF: tetrahydrofuran, FUR: furfural, 2-MTHF: 2-methyltetrahydrofuran, DMSO: dimethyl sulphoxide
[a]$AlCl_3$ was used in large amount hence it was not considered as catalyst in the classical meaning. [b]Under microwave irradiation. [c]NaCl used as additive. [d]Corncob and bagasse hemicelluloses were previously treated by oven drying. [e]Semi-bath reaction mode

### Supercritical Carbon Dioxide

Carbon dioxide is not only inexpensive and readily available, but also safe to handle, does not contaminate the product and leads to less production of wastes that require neutralisation [69, 70]. The energy demand to bring $CO_2$ to supercritical condition ($T_c$ = 31.0 °C and $p_c$ = 74 bar) is relatively low in comparison to other solvents, *e.g.* water ($Tc$ = 374.2 °C and $Pc$ = 221 bar). Consequently, it can "help" in the development of more energy-efficient furfural processes. Additionally, $CO_2$ has attracted increasing interest as a substitute of environmentally unattractive catalyst and, when being recovered and recycled, it does not contribute to increase of greenhouse gas (GHG) levels in the atmosphere. Recently, supercritical $CO_2$ has been used in valorisation of lignocellulosic biomass due to the enhancement of hemicelluloses hydrolysis, followed by the improvement of enzymatic digestibility of the processed biomass, over those obtained in conventional hydrothermal biomass processes [5, 6, 71, 72].

The extraction of hemicelluloses from biomass occurs generally in aqueous stream hence, the ideal process of furfural synthesis would have the reaction taking place in the aqueous phase along with catalyst. Besides the low solubility of $CO_2$ in

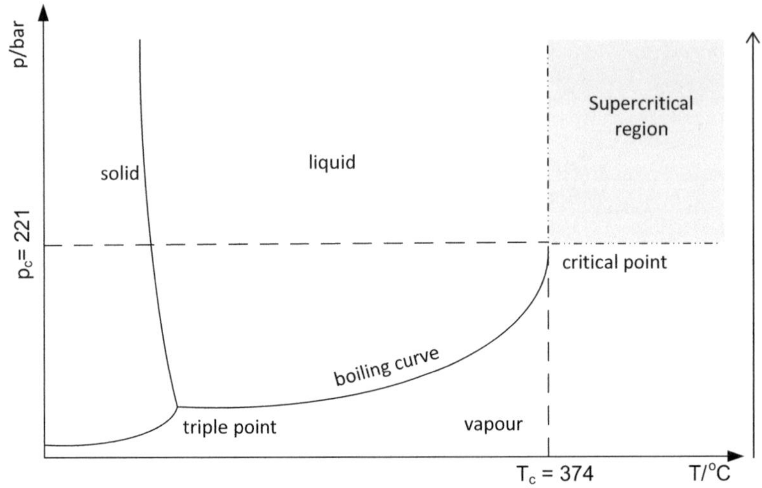


**Fig. 2.6** Phase diagram of water (Adapted from Bioresource Technology, Vol. 101, Girio FM, Fonseca C, Carvalheiro F, Duarte LC, Marques S, Bogel-Lukasik R, Hemicelluloses for fuel ethanol: A review. Pages 4775–4800, Copyright © (2010), with permission from Elsevier [68])

water, the dissolution of $CO_2$ in aqueous medium leads to the formation *in-situ* of carbonic acid according to Eqs. (2.5), (2.6), (2.7) and (2.8):

$$CO_2(g) \leftrightarrow CO_2(aq.) \tag{2.5}$$

$$CO_2 + H_2O \leftrightarrow H_2CO_3 \tag{2.6}$$

$$H_2CO_3 + H_2O \leftrightarrow H_3O^+ + HCO_3^- \tag{2.7}$$

$$HCO_3^- + H_2O \leftrightarrow 2H_3O^+ + CO_3^{2-} \tag{2.8}$$

The two-stage dissociation and liberation of hydronium ions promotes acid catalysed dehydration of hemicellulosic sugars into furans, in a similar way to mineral acids. The use of supercritical $CO_2$ as catalyst in aqueous medium has a great benefit since the *in-situ* formed carbonic acid does not constitute an environmental problem, as $CO_2$ can be easily removed during the depressurisation step, increasing the pH value of the solution [6]. The used $CO_2$ can also act as an extraction solvent, and due to the high solubility of furfural in supercritical $CO_2$, furfural can be easily extracted [73, 74].

*Furfural Formation from Pentoses and Biomass in Supercritical $CO_2$*

Gairola and Smirnova explored the conversion of L-arabinose, D-xylose and hemicellulose into furfural coupled with simultaneous extraction of furfural using supercritical $CO_2$ in six batch reactors [75]. Firstly, the kinetic studies of dehydration of

**Table 2.4** Supercritical $CO_2$ strategies for the conversion of xylose and lignocellulosic biomasses hydrolysates into furfural

| Starting material | Solvent(s) | Conditions T/°C | p/bar | t/min | FUR yield/ mol% | Ref. |
|---|---|---|---|---|---|---|
| Xylose | Water | 230 | 120[a] | 24 | 68 | [75] |
| Xylose | Water/THF | 180 | 50 | 60 | 69 | [23] |
| Xylose | Water/THF/MIBK | 180 | 50 | 60 | 57 | [67] |
| Wheat straw[b] | Water | 230 | 80[a] | 25–27 | 29 | [75] |
| Wheat straw[b] | Water/THF/MIBK | 180 | 50 | 60 | 43[c] | [67] |
| Brewery waste[b] | Water | 230 | 80[a] | 25–27 | 13[c] | [75] |

Ref.: References, FUR: furfural, THF: tetrahydrofuran, MIBK: methyl isobutyl ketone
[a]$CO_2$ flow rate of 3.6 g/min. [b]Lignocellulosic biomass hydrolysate. [c]With simultaneous supercritical $CO_2$ extraction

L-arabinose into furfural and the associated loss reactions in aqueous media, over a temperature range of 180–260 °C up to 60 min of reaction time at an initial $CO_2$ pressure of 30 bar with initial arabinose concentrations of 0.2, 1 and 5 wt. % were addressed. According to results obtained in those experiments, reaction temperatures greater than 220 °C were needed to achieve almost complete conversion of L-arabinose within 60 min of reaction time. The authors reported that for optimal modelling, the kinetic parameters for furfural degradation should be calculated separately from other kinetic parameters, because furfural degradation increases in more complex reaction mixtures. So, in an effort to determinate separately the kinetic parameters for furfural degradation, the declining furfural concentration in those assays, in which a total disappearance of pentoses was obtained within certain reaction time, was used. Additionally, the estimated Arrhenius parameters for arabinose were in good agreement with those reported by Jing and Xyuyang [76]. Gairola and Smirnova proposed that it is justified the solely use of xylose as the model compound for conversion of hemicellulose into furfural. Danon et al. stated that the above mentioned author's assumption seems to be "coarse", firstly because this assumption was not validated with experimental data, and secondly, the xylose degradation was around 30% faster than arabinose [48]. Besides the arabinose conversion experiments, Gairola and Smirnova also studied the optimal conditions for the formation of furfural from xylose, taking into account the temperature, pressure, reaction time, $CO_2$ flow rate and xylose concentration [75]. The highest overall xylose to furfural yield of 51 mol% was achieved at 230 °C, 80 bar for 24 min of reaction time. Increasing $CO_2$ pressure from 80 to 120 bar resulted in an improvement of furfural yield from 51 mol% to 68 mol%. Further increasing of pressure up to 160 bar did not result in an enhancement of furfural yield [75]. Additionally, those authors reported experimental data for the conversion of wheat straw and brewery waste hydrolysates using simultaneous supercritical $CO_2$ extraction of furfural, as shown in Table 2.4. Lower furfural yields and higher content of humins were obtained in comparison to pure xylose. These data were mainly explained by the presence of organic acids and phenolic compounds in the hydrolysates because furfural is well-known to promote the crosslinking of phenolic compounds and lignin under hydrothermal conditions [45, 77].

Morais and Bogel-Lukasik investigated the degradation of xylose in aqueous solutions using supercritical $CO_2$ and THF as solvent [23]. Over a temperature range of 160 and 180 °C, the conversion of xylose to furfural was studied in pure water (without $CO_2$ and THF), in water and $CO_2$ (without THF), in water with various amounts of THF (without $CO_2$) and in water/THF/$CO_2$ systems. They concluded that supercritical $CO_2$ plays an important role as catalyst in the improvement of furfural yield and selectivity. For instance, by using $CO_2$ as the catalyst in the system only with water, a 43 mol% yield of furfural was achieved, whereas in the absence of $CO_2$ the highest furfural yield was only 25 mol%, at 180 °C for 60 min. Besides the substantial improvement in furfural yield, it is still relatively low due to the occurrence of degradation reactions leading to the formation of high amounts of insoluble solid by-products. In an effort to prevent furfural degradation reactions in aqueous phase, it was proposed the use of THF aiming to "protect" furfural from undesired degradation reactions. In fact, the addition of THF to the system (composed of water and $CO_2$) improved both the yield and selectivity of furfural to 69 mol% and 83%, respectively. One possible explanation for the improvement of furfural yield in the presence of THF might be due to the observation reported by Pollet et al. who found that $CO_2$ acts as a phase splitting inducer that is able to separate THF from the aqueous phase [78]. Gaseous $CO_2$ has a very low solubility in water, while it has an infinite solubility in THF, and consequently the presence of $CO_2$ in water/THF mixtures promotes phase separation. The $CO_2$ promotes the transition of a substantial amounts of THF into a separate liquid phase ($CO_2$-expanded THF liquid phase), and a later increase of pressure leads to the formation of a gas or liquid phase (depending on the reaction geometrics) that is rich in $CO_2$ and THF (Fig. 2.7). Therefore, under the investigated operation conditions, $CO_2$ acts as an acid catalyst and contributes to phase spitting allowing the *in-situ* extraction of produced furfural from the aqueous phase, protecting it from degradation reactions. This approach benefits from the use of easily recovery of low-boiling THF by room temperature vacuum distillation and does not require additional steps in downstream processing of the salt-saturated aqueous phase [79]. Yet, Morais et al. investigated the conversion of xylose into furfural in a system composed of water, THF, MIBK and supercritical $CO_2$, where MIBK acted as a water immiscible extracting solvent [67]. The reactions at temperatures between 160–200 °C and for up to 120 min of reaction time under an initial $CO_2$ pressures of 50 bar with a xylose feed streams of 12.5 g/L in water were studied in a batch reactor. At temperatures below of 180 °C, the dehydration of xylose into furfural was shown to be very limited even for prolonged reaction times, resulting in a furfural yield of only 28 mol%. Interestingly, increasing temperature up to 180 °C resulted in a great increase in furfural yield up to 57% within 60 min of reaction time. Further increase of temperature promoted furfural degradation reactions reducing the amount of furfural recovered. The same authors also investigated the role of initial $CO_2$ pressure in the conversion of xylose into furfural. The addition of only 20 bar $CO_2$ increased the xylose conversion up to 85%, however similar effects were not observed for furfural yield as it remained

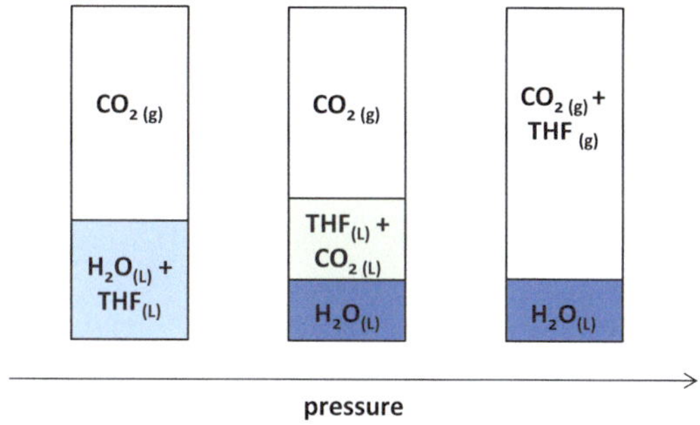


**Fig. 2.7** Phase separation of water/THF system induced by $CO_2$ [67] (Reproduced by permission of The Royal Society of Chemistry)

low, *i.e.* similar to the $CO_2$-free reaction (< 40 mol%). It was stated that the above-referred pressure is enough to promote the dehydration of xylose, but without occurrence of THF-water phase separation. Initial $CO_2$ pressures above 50 bar were enough to promote sufficient splitting phase, and thus prevented the further degradation of furfural from side reactions that resulted in an improvement in furfural yield up to 57 mol%. Also, Morais et al. studied wheat straw hydrolysates conversion. Similar to those results obtained by Gairola and Smirnova, the furfural yield (43 mol%) and selectivity (44 mol%) were significantly lower than when xylose was used as feedstock [67].

The analysis of the data presented in Table 2.4 permits the conclusions that the use of supercritical $CO_2$ as an alternative catalyst in the production of furfural is still in its development. However, the extraordinary properties of supercritical $CO_2$ either acting as a catalyst or as a co-solvent in mixtures makes it a possible solution to tackle the well-known challenges found in the industrial furfural production processes.

#### 2.2.1.3 Ionic Liquids

Ionic liquids are organic salts composed solely of an anion and a cation and by definition they have melting points below 100 °C. The wide range of existing anions and cations allows multiple possibilities in the synthesis of ILs to have different physicochemical properties for specific applications. One example of the potential of ILs as solvents or catalysts to assist or to mediate lignocellulosic biomass transformation into a wide range of products is given in a recent text by one of the authors [80]. The selectivity of ILs to form a desired product is a key advantage, surpassing conventional acid-catalysed reactions in aqueous medium for similar purposes. The production of furfural from pentoses, native hemicellulose or directly from biomass, assisted or mediated by ILs, is one of those successful examples. Nowadays, three different IL concepts for production of furfural have been studied and are characterised as: i) IL solvent (or co-solvent) with an external catalyst (*e.g.* metal halide as discussed above); ii) IL catalyst; or iii) IL as both solvent and

**Table 2.5** Selected examples of furfural (FUR) production with ionic liquids (furfural yield ≥50 mol%)

| Starting material | IL/co-solvent(s) | Conditions | | FUR/mol% | Ref. |
|---|---|---|---|---|---|
| | | T/°C | t/h | | |
| **Pentoses** | | | | | |
| Xylose | [bmim][$HSO_4$]/Dioxane | 140 | 6 | 82 | [24] |
| Xylose | [bmim][$HSO_4$]/MIBK | 140 | 6 | 80 | [24] |
| Xylose | [bmim][$HSO_4$]/Toluene | 140 | 4 | 74 | [24] |
| Xylose | [$bSO_3$Hmim][$HSO_4$]/ Water/MIBK | 150 | 25/60[a] | 92 | [81] |
| Xylose | [$bSO_3$Hmpyr][$BF_4$]/ Water/THF | 180 | 1[c] | 85 | [82] |
| Xylose | [emim][$HSO_4$]/ Toluene | 100 | 6 | 84 | [83] |
| Xylose | IL-$SiO_2$/DMSO[b] | 110 | 2 | 50 | [84] |
| **Hemicellulose polysaccharides** | | | | | |
| Extracted hemicellulose from eucalyptus wood | [bmim][$HSO_4$]/Dioxane | 160 | 4 | 59 | [85] |

[a]Under microwave irradiation. [b]IL- 3-sulphobutyl-1-(3-propyltriethoxysilane)-imidazolium hydrogensulphate

catalyst. Table 2.5 represents optimal conditions for furfural production for each research work using these IL concepts.

## Ionic Liquids Used as Acidic Catalysts

A strategy based on the use of ILs with their acidic properties is to use them to catalyse the production of furfural in aqueous medium [81, 82, 84]. In this case, selectivity is not provided by the high stability of furfural in ILs, such as [bmim]Cl, but by the catalytic activity of acidic sites present in the ILs. The main example is 1-(4-sulphonic acid)-butyl-3-methylimidazolium hydrogensulphate ([$bSO_3$Hmim][$HSO_4$]), which presents two Brønsted acid sites, sulphonic acid ($SO_3H$) attached to imidazolium cation and hydrogensulphate anion. The conversion of xylose mediated by this IL in aqueous medium resulted in approximately of 92 mol% furfural yield at 150 °C for 25 min [81]. Furthermore, the selectivity was as high as 96% at these conditions. An optimal dosage of 0.5 g IL was found for the conversion of 1.0 g of xylose, which is a considerable economic saving in comparison to the IL being used as solvent. Nevertheless, it is worth to mention that the application of immiscible methyl isobutyl ketone (MIBK) allowed continuous extraction of furfural from the aqueous phase during reaction, avoiding its consumption into undesired products. These results show the potential of [$bSO_3$Hmim][$HSO_4$]/water/MIBK system for the selectively converting pentoses into furfural with high yields [81].

In other work, the effect of other $SO_3$H-based ILs and other extracting organic solvents in the dehydration reaction of pentoses and oligomers to furfural was studied [82]. Among several organic solvents screened for furfural extraction, tetrahydrofuran (THF) was the most suitable solvent for this purpose. Ethyl acetate showed low stability, while toluene and MIBK produced high amount of black deposits at the same reaction conditions (microwave irradiation – 100 W). Afterwards, four different $SO_3H$-based ILs, namely 1-(4-sulphonylbutyl)pyridinium methanesulphonate ([b$SO_3$Hmpyr][Me$SO_3$]), 1-(4-sulphonylbutyl)triethylammonium methanesulphonate ([b$SO_3$HEt$_3$N][Me$SO_3$]), 1-(4-sulphonylbutyl)pyridinium tetrafluoroborate ([b$SO_3$Hmpyr][$BF_4$]) and 1-(4-sulphonylbutyl)triethylammonium tetrafluoroborate ([b$SO_3$HEt$_3$N][$BF_4$]) were examined. Among them, [b$SO_3$Hmpyr][$BF_4$] was the most efficient for converting 90 mol% of xylose into 75 mol% of furfural at 150 °C under 100 W microwave irradiation for 1 h. Increasing temperature to 180 °C, xylose conversion reached almost 100 mol% with a furfural yield of 85 mol%. The high performance of [b$SO_3$Hmpyr][$BF_4$] was associated with not only the acidic $SO_3H$ group but also with the *in-situ* formation of hydrofluoric acid (through $BF_4$ hydrolysis in the presence of water) that aided dehydration of xylose into furfural. In comparison to $BF_4$ anion, Me$SO_3$ has higher stability in the aqueous medium at the examined conditions. Regarding the screening of cations, those authors suggested that pyridinium enhances the acidity of $SO_3H$ group via stabilisation of the deprotonated sulphonic group, showing higher efficiency in the conversion than its counterpart triethylammonium. The [b$SO_3$Hmpyr][$BF_4$]/water/THF system was tested in the conversion of agricultural lignocellulosic waste derived syrup containing xylose and arabinose in mono and oligomeric forms, however, much lower furfural yields were obtained (30–45 mol%). This shows the same difficulties as those observed with IL/additional catalyst systems reviewed above [82].

The use of ILs as catalysts can be approached by immobilising acidic ILs in silica resins [84]. In this context, 3-sulphobutyl-1-(3-propyltriethoxysilane)-imidazolium hydrogensulphate was synthesised and applied to the conversion of xylose. The novelty lies in easier recovery of silica supported IL catalysts after the reaction. Nevertheless, only 44.5 mol% of furfural yield was obtained, albeit with a 94 mol% conversion of xylose. This moderate furfural yield results from a process that used dimethylsulphoxide (DMSO) as solvent without a second extraction solvent phase [84]. As mentioned before, organic solvents, such as THF and MIBK, allow recovery of furfural from the reaction system (in this case: IL and DMSO) avoiding unwanted side reactions.

#### Ionic Liquids Used as Both Solvents and Acidic Catalysts

The ability of acidic ILs to have a role as both solvents and catalysts without need of additional external catalysts has been addressed in the literature [24, 83, 85–87]. In the previous section, ILs containing acidic $SO_3H$ groups were used in catalytic amounts (due to their high acidity) to convert xylose into furfural in aqueous medium. However, less acidic [$HSO_4$]-based ILs, such as 1-ethyl-3-methylimidazolium hydrogensulphate ([emim][$HSO_4$]) and 1-butyl-3-methylimidazolium hydrogensulphate ([bmim][$HSO_4$]), can be used as solvents for biomass feedstocks and simultaneously as selective catalysts for the hemicellulose fraction to produce pentoses and/or furfural [83, 87, 88]. In this context, Lima et al. demonstrated high conversion of xylose into furfural using [emim][$HSO_4$] as both solvent and catalyst. After 6 h at only 100 °C, 84 mol% of xylose was converted to furfural that was assisted with toluene as extraction solvent. Without toluene, only 33 mol% of furfural yield was obtained [83]. Similar yields were obtained by Peleteiro et al. when using [bmim][$HSO_4$] for the reaction system [24, 86]. These results clearly show the crucial role of organic solvents in extraction of furfural from the reaction system to avoid unwanted side reactions that involve furfural. Therefore, screening of organic solvents, such as MIBK and dioxane, was performed in the conversion of xylose with [bmim][$HSO_4$] [24]. The results showed that high furfural yields were obtained with MIBK and this was ascribed to its higher ability of MIBK to extract/solubilise furfural than that of toluene. Although similar yields were obtained with MIBK and dioxane, faster production of furfural was observed when the latter was applied [24].

Nevertheless, applying [emim][$HSO_4$]/toluene biphasic system for more complex substrates, such as xylan, was shown to give approximately 29 mol% of furfural yield at 100 °C [83]. In this case, higher temperatures would increase the production of furfural, although a previous hydrolysis step is needed to release xylose from xylan. In other work, temperatures between 120 and 160 °C were examined for [bmim][$HSO_4$]/dioxane biphasic system in the conversion of hemicellulose fraction from eucalyptus wood (hot compressed water processing) [85]. The extracted hemicellulose containing xylose and arabinose in monomeric, oligomeric and polymeric forms was converted with a maximum of 59.1 mol% of furfural at 160 °C in 4 h reaction time. For this case, between 88.6 and 89.9% of produced furfural was found in the organic phase, showing the high ability of dioxane as an extracting solvent [85].

The application of lignocellulosic biomass, such as Miscanthus and wheat straw, as feedstock for furfural production using solely [bmim][$HSO_4$] has been studied, although only moderate furfural yields (up to 33 mol%) have been obtained [87, 88]. The lack of efficiency in producing furfural from raw biomass in comparison to pentose or hemicellulose could be associated to the high complexity of the lignocellulosic matrix as well as to the reactivity of some biomass components that possess strong chelating groups, which could cause leaching or poisoning of the catalysts [89].

### *2.2.2 Heterogeneous Catalysis*

Heterogeneous catalysis of carbohydrates for the selective production of furfural has been widely explored in the last few years. The main advantages of using heterogeneous catalysts over homogeneous catalysts are that they are less hazardous and are easier to recover, hence they can be reused. The most used and studied solid acid catalysts that have been applied to the production of furfural are zeolites, mesoporous materials and metal oxides, which show potential to be used in industrial processes and to substitute current toxic and corrosive mineral acids. The most promising studies approaching the application of this type of catalysis towards furfural production will be briefly reviewed. Table 2.6 presents the results of furfural production with solid acid catalysts for furfural yield ≥50 mol%.

#### 2.2.2.1 Zeolites (Microporous Catalysts)

The application of zeolites to chemical reactions is based on shape selectivity where substrates, intermediates and products can be selected according to the pore size of the material. Zeolites are usually highly structured crystalline inorganic aluminosilicates with well-defined pores between 5 and 13 Å. These materials mostly present Brønsted acidity giving them the ability to transform substrates into products, for instance pentose dehydration into furfural. The most common zeolites used in the dehydration of pentoses into furfural are mordenites in their proton form (H-mordenite or H-M) [90, 111, 112]. Moreau et al. stated that although xylose conversion was low at 170 °C, H-M has high selectivity for furfural production (90–95%), due to the absence of cavities within catalyst structure that avoid formation of secondary products [111]. In contrast to the results obtained with zeolites, such as H-Ferrierite, H-ZSM and H-beta, higher furfural yield (39 mol%) and selectivity was obtained with H-M in the presence of the selective solvent DMSO. Reactions in aqueous medium showed lower xylose conversions in comparison to DMSO reaction system, due to the formation of side reaction products that caused blocking and poisoning of catalytic sites [112]. Nevertheless, H-ZSM-5 (a ZSM-5 variant), with a pore size of 12 Å, have been demonstrated to produce 46 mol% furfural yield at 200 °C in aqueous medium. Above this temperature, oligomerization of furfural to bi- and trimeric species was observed as a result of the intrinsic structure of H-ZSM-5. It was concluded that pore size of H-ZSM-5 allows longer residence time of furfural in the vicinity of the catalyst, thus rearrangements based on furfural ring breaking down may occur in parallel to oligomers' formation. Scanning electron microscopy showed partial coating of catalyst surfaces by oligomer deposits and narrow openings of catalyst pores [113]. Those moderate furfural yields were surpassed (to 98 mol%) when H-M-13 (a H-M variant) in water/toluene system at 260 °C under pressure of 55 atm in a plug-flow reactor were used (Table 2.6) [90]. Nevertheless, the very high temperature and pressure required for this system, makes it undesirable for industrial application. In other study, furfural was obtained at a much lower temperature (170 °C) using a specific approach in catalyst synthesis adopted by Lima et al. [114]. The researchers performed delamination of H-Nu-6 zeolite, increasing the specific surface area almost seven times. Subsequently, conversion rate was as twice high as that obtained from the original catalyst. After 4 h, higher furfural yields were reached with delaminated H-Nu-6 (47 mol% ) than those with H-M (34 mol%) or phosphate- based (SAPOs) zeolites (34–38 mol%) [114, 115]. Nonetheless, impressive improvements in furfural yields were accomplished with H-M zeolite through two different approaches [91, 92]. Among several zeolites examined, H-M demonstrated the highest efficiency by

**Table 2.6** Furfural production with zeolites, mesoporous acid-catalysts and metal oxides (furfural yield ≥50 mol%)

| Starting material | Catalyst | Solvent(s) | Conditions | | FUR (mol%) | Ref. |
|---|---|---|---|---|---|---|
| | | | T/°C | t/h | | |
| **Pentoses** | | | | | | |
| Xylose | H-M-13 | Water/Toluene | 260 | 3/60 | 98 | [90] |
| Xylose | H-beta | Water/Sulfolane | 175 | Up to 12 | >75 | [91] |
| Xylose | H-M | Water/GVL | 175 | 2 | 81 | [92] |
| Xylose | Sn-MMT | Water/DMSO | 180 | 0.5 | 77 | [93] |
| Xylose | MCM-41/$SO_3H$ | Water/Toluene | 140 | 24 | 76 | [94] |
| Xylose | MCM-41/ $MePrSO_3H$ | Water/Toluene | 155 | 2 | 93 | [95] |
| Xylose | MCM-41/PW | DMSO | 140 | 4 | 52 | [96] |
| Xylose | MCM-41/$ZrO_2$/ $SO_4^2$ | Water/Toluene | 160 | 4 | 50 | [97] |
| Xylose | SBA-15/$SO_3H$ | Water/Toluene | 160 | 4 | 68 | [98] |
| Xylose | SBA-arene$SO_3H$ | Water/Toluene | 160 | 20 | 86 | [99] |
| Xylose | SBA-propyl$SO_3H$ | Water/Toluene | 170 | 20 | 82 | [100] |
| Xylose | SBA-15/$ZrO_2$- $Al_2O_3$/$SO_4^2$ | Water/Toluene | 160 | 4 | 53 | [101] |
| Xylose | TUD-1/H-beta | Water/Toluene | 170 | 8 | 74 | [102] |
| Xylose | H-beta | Water/Toluene | 170 | 6 | 54 | [102] |
| Xylose | TUD-1/Al | Water/Toluene | 170 | 6 | 60 | [103] |
| Xylose | Nafion 117 | DMSO | 150 | 2 | ∽60 | [104] |
| Xylose | $HTiNbO_5$-MgO | Water/Toluene | 160 | 4 | 55 | [105] |
| Xylose | $SO_4^{2-}$/$ZrO_2$–$TiO_2$ | Water/Butanol | 170 | 3 | 54 | [106] |
| Xylose | $MgF_2$- Perfluorosulfonic | Water/Toluene | 180 | 20 | 78 | [107] |
| Xylose | Zr-W/Al | Water/Toluene | 170 | 6 | 51 | [108] |
| Xylose | $Nb_2O_5$ | Water | 175 | 3 | 78 | [109] |
| **Hemicellulose** | | | | | | |
| Hemicellulose-rich liquor from wood | H-beta | Water/Sulfolane | 175 | Up to 12 | 75 | [91] |
| Hemicellulose-rich liquor from corncob | Sn-MMT | Water/DMSO | 180 | 2 | 54 | [93] |
| **Lignocellulosic biomass** | | | | | | |
| Hemp shives | $Al_2(SO_4)_3 \cdot 18H_2O$ | Water/Steam distillation | 180 | 1.5 | 74 | [110] |
| Rice hull | $H_2SO_4$/$TiO_2$ | Water/Steam distillation | 125 | 0.5 | 53 | [18] |

converting 75 mol% xylose into furfural, but distillation of furfural was simultaneously performed during the process. This approach avoided consumption of furfural in undesirable side reactions [91]. On the other hand, using γ-valerolactone/water (9/1 v/v) as solvent system coupled with H-M zeolite allowed the production of 81 mol% furfural at 175 °C. The bio-based γ-valerolactone increased the selectivity of the reaction medium by inhibiting condensation reactions between furfural and pentose intermediates. In contrast, using H-M solely in water resulted in practically half of the furfural yield [92].

Another strategy to enhance furfural yield lies in the combination of Lewis acid zeolites with Brønsted acids [116]. Sn-beta zeolite was combined with Amberlyst 15 giving approximately 60 mol% conversion of xylose at 100 °C, which is at much lower temperature than that typically employed for this reaction. Sn-beta acted as a Lewis acid performing xylose isomerisation into intermediates (xylulose and lyxose), while the Brønsted acid Amberlyst 15 allowed dehydration of intermediates into furfural. Nevertheless, furfural yield was lower than 25 mol% [116]. In other work, a similar strategy was demonstrated to be more efficient in furfural production applying solely tin-montmorillonite (Sn-MMT), which possesses both Brønsted and Lewis active sites. In this case, ∽77 mol% furfural yield with ∽83% selectivity was obtained from xylose conversion shown in Table 2.6 [93].

An upgrade of more complex substrates, such as hemicellulose and/or raw biomass itself, has been showed [91, 93, 117–119]. Several zeolites including H-USY, H-beta and H-M were applied for simultaneous hydrolysis and conversion processes of wood-derived hemicellulose in aqueous medium at 170 °C. However only ∽12 mol% of furfural yield was obtained [117]. In the conversion of water-insoluble hemicellulose and water-soluble fraction from corncob processing with Sn-MMT, only ∽40 mol% and 54 mol% furfural yields were obtained, respectively [93]. The differences were justified by the polymerisation degree of carbohydrates present in each sample. Lower furfural yields were obtained for water-insoluble hemicellulose as this substrate is mostly in a polymeric form decreasing the efficiency of the catalyst [93]. Furthermore, condensation and polymerisation reactions between furfural and its intermediates with other compounds may proceed during the reaction, decreasing the final furfural yield. To avoid furfural polymerisation, Chen et al. added 4-methoxyphenol as inhibitor in the system containing steam explosion liquor of rice straw (hemicellulose-based stream) and H-ZSM-5 zeolite. This innovative approach allowed an increase of about 21% furfural yield in comparison to the control reaction (absence of 4-methoxyphenol) [119].

#### 2.2.2.2 Mesoporous Acid-Catalysts

The application of zeolites in catalytic processes is restricted due to the small pore size of their structure. Therefore, mesoporous materials constituted of amorphous silica wall that have pore diameters between 20–100 Å have been designed. The larger pore size of these materials offers the advantage of performing task of specific syntheses by introducing diverse functional groups into the material surface.

The main examples of mesoporous structures are MCM-41 and SBA-15 that have been used as supports for a diverse range of acids to catalyse xylose into furfural with high selectivity.

The hexagonally arranged MCM-41 is probably one of the most versatile mesoporous structures for acid catalysis functionalisation towards furfural production. Initially, $SO_3H$ groups were anchored to the silica structure of MCM-41 to perform xylose conversion to furfural [94]. The furfural yield increased especially when the catalyst was used in the toluene/water system, although the selectivity was reduced. It was suggested that the hydrophobic nature of the supporting material interferes in furfural loss reactions. Nevertheless, one study shows the contrary [95]. The addition of methyl propyl sulphonic acid (butane-2-sulphonic acid) to MCM-41 (MCM-41/MePr$SO_3H$) allowed >90 mol% of furfural yield from xylose with an elevated selectivity (93%) as shown in Table 2.6 [95]. Precisely, the attached methyl group repelled furfural from the reactive $SO_3H$ groups, thus avoiding it from further degradation [95]. Nevertheless, the reuse of the catalyst was inefficient, once the furfural yield was reduced to <50 mol%, resulting in carbon deposition that blocked the pores of the catalyst [94, 95]. In this case, thermal regeneration by heating the recovered catalyst at elevated temperatures could be an option for improving the efficiency of catalyst reuse, although the applied temperatures cannot compromise the stability of the catalyst. The addition of 12-tungstophosphoric acid to MCM-41 surface was used to produce furfural [96, 120]. It was demonstrated that MCM-41 supported acid-materials presented higher activity than non-supported acids. Increasing three parameters: i) 12-tungstophosphoric acid loading; ii) pore diameter of the parent silica support; and iii) reaction temperatures, lead to higher furfural yields. Furthermore, the catalysis in the presence of DMSO was more effective than in the toluene/water solvent system, since the acidic-supporting matrix demonstrated a higher stability and reusability in DMSO [96, 120]. Other studies performed functionalisations of MCM-41 with niobium silicate, sulphated and persulphated zirconia, however conversions of xylose into furfural were demonstrated to be moderate (yields ≤50 mol%) [97, 121].

As aforementioned, MCM-41 with $SO_3H$ groups was demonstrated to be high efficiency and selectivity in the production of furfural, but after reuse, furfural yield and selectivity drastically dropped. This was also confirmed by Jeong et al., thus they constructed a different support consisting of mesoporous silica shells and containing $SO_3H$ groups (MSHS/$SO_3H$) as an alternative to MCM-41/$SO_3H$. Although lower furfural yields were obtained with MSHS/$SO_3H$, the selectivity was maintained after each recycle run [122]. Another study demonstrated other alternatives based on the application of SBA-15 mesoporous matrix, which generally is easier to be synthesised and has a bigger pore size than MCM-41 [123]. Different $SO_3H$ sources, including propylsulphonic and arenesulphonic acids, were anchored to the surface of SBA-15, giving it the ability to convert xylose to furfural [98–100]. Furfural yields ranged from 68–86 mol% were obtained by using these $SO_3H$ functionalised SBA-15 mesoporous catalysts, thus presenting higher efficiencies than those composed of MCM-41 support. The co-condensation methodology in SBA-15/$SO_3H$ synthesis demonstrated homogenous distribution of acidic sites in the supporting materials, leading to high furfural yields (ca. 68 mol%) [98]. On the other hand, the high thermal stability of SBA-15 modified with arenesulphonic acids allowed conversion of xylose at 160 °C favouring the production of furfural (up to 86 mol% yield) [99]. Furthermore, it also allowed consecutive reaction runs without efficiency lost (∽75 mol% furfural yield after 3 runs) [99]. Another study demonstrated the functionalisation of SBA-15 with sulphated zirconia, but low furfural yields were obtained (53 mol%) [98, 101], similar to MCM-41 [97]. The reason lies in the deactivation of the catalyst's acid sites through the adsorption of by-products [98, 101].

Other less common mesoporous materials, such as TUD-1 and Nafion, have been applied for furfural production [102–104]. An interesting approach was the impregnation of the aforementioned zeolite, H-beta, in silicious TUD-1 [102] resulting in higher activity of supported zeolite than the zeolite itself. The furfural yield obtained for TUD-1/H-beta and H-beta was 74 mol% and 54 mol%, respectively. Furthermore, an advance was made by converting xylan into furfural solely with TUD-1/Al, constituted by both Lewis and Brønsted acids, but only 18 mol% of furfural yield was produced [103]. On the other hand, Nafion-based polymers demonstrated to be highly stable and robust catalysts for the production of furfural (∽60 mol% yield) within 15 repetitive runs without any regeneration step. It was also mentioned that the presence of DMSO as solvent reduces the carbonaceous compounds deposition on the catalyst surface [104].

The use of supported aluminosilicate mesoporous materials for the conversion process of xylose into furfural has demonstrated some potential. However, further improvements in catalytic performance must be achieved by changing the properties of the mesoporous materials. For instance, varying the Si/Al ratio and acid loading during synthesis is expected to change the total amount and the dispersity of acid sites as well as the catalyst surface polarity.

#### 2.2.2.3 Metal Oxides

Metal oxides are ordered crystalline materials that can have both Lewis and Brønsted acidity and are applied as catalysts. Metal oxides possess a rigid structure allowing high stability in chemical reactions and making further recovery and reuse more efficient. These features could facilitate dehydration of xylose into furfural as demonstrated in a wide range of studies, which extensively explored the activity of metal oxides in the transformation of xylose and also lignocellulosic biomass into furfural. Non-functionalised metal oxides, such as Zr–P, $SiO_2$–$Al_2O_3$, $WO_x$–$ZrO_2$ and $Al_2O_3$ were screened in water batch conditions to evaluate the influence of acidity type on furfural production from xylose [124]. The highest furfural yield was obtained by the catalysis of Zr-P which has the highest Brønsted acidity among the metal oxides examined. In contrast, $Al_2O_3$ demonstrated the highest xylose conversion for the higher Lewis acid sites' concentration. It was suggested that metal oxides containing high ratios of Brønsted to Lewis acid sites favoured furfural production [124]. In this context, Dias et al. performed exfoliation of $HTiNbO_5$-MgO aiming to increase the specific surface area of the material. This process allowed higher number of Brønsted acid sites and consequently higher catalytic activity than the original layered material. After 4 h reaction in water/toluene system at 160 °C, furfural yield reached 55 mol% [105]. In other studies, functionalisation of metal oxides with sulphate groups allowed an increased selectivity in xylose conversion [19, 20, 106, 125, 126]. Actually, furfural is barely produced without $SO_4^{2-}$, albeit high xylose conversion yields are obtained with $SnO_2$ metal oxide in water/toluene system. It has been suggested that the presence of both Lewis ($SnO_2$) and Brønsted ($SO_4^{2-}$) acid sites shift the reaction to xylulose route, thus lowering the reaction temperature that avoids side reactions [94]. A more efficient approach was demonstrated with the application of $SO_4^{2-}/ZrO_2–TiO_2$ within a biphasic system composed of water/butanol [106]. The furfural yield was 54.3 mol% at 170 °C, and the catalyst was reused multiple times without substantial change in furfural formation efficiency. Furthermore, $SO_4^{2-}/ZrO_2–TiO_2$ showed higher ability to produce furfural than zeolites, such as H-ZSM-5 [106]. In another study, the functionalisation of hydroxylated $MgF_2$ with fluorosulphonic acid allowed it to have more acidic groups on the metal oxide surface. Operating at 160 °C and assisted by $N_2$ stripping process, a 78 mol% of furfural yield was achieved making this process very effective [107] as presented in Table 2.6.

Another strategies are focused on the combination of metal oxides with other agents, such as materials with mesoporous structures [108], acidic resins [109] and organic acids [127] aiming at efficient reaction and selective furfural production. In the first case, the synthesised Zr-W in a surfactant template and doping with aluminium was applied for xylose conversion. Using solely Zr-W permitted 99% conversion, but only 35 mol% of furfural was obtained in water/toluene system at 170 °C. The mesoporosity of the material together with the incorporation of aluminium allowed high furfural selectivity and gave furfural yields up to 51 mol% [108]. In another work, $Nb_2O_5$ (possessing only Lewis acid sites) was mixed with resin Amberlyst 70 (possessing Brønsted acid sites) for furfural production in water/toluene at 175 °C [109]. In this case, only ∽35 mol% of furfural yield was achieved, however the reaction was assisted with $N_2$ stripping to remove furfural from the reaction system to give yields of ∽78 mol%. These results show that the *in-situ* extraction of furfural is crucial to increase the efficiency of the process. Besides $N_2$ stripping, furfural extraction can be also performed by supercritical $CO_2$ during the dehydration of xylose by metal oxides, which increases the furfural yield up to 60 mol% [128]. Another synergetic combination is the mixture of niobium hydroxide with acetic acid (weak Brønsted acid) [127], which besides Lewis and Brønsted acids that favour furfural formation have the presence of acetic acid in aqueous medium, which creates a suitable physico-chemical environment to enhance the interaction between the surface of the catalyst and xylose, thus improving furfural production [127].

The ability of metal oxides to effectively manufacture furfural from biomass has been addressed [18–21, 110]. In an early study, mixing Lewis $TiO_2$-type catalysts with $H_2SO_4$ in rice hulls processing was demonstrated to have very low efficiency in the production of furfural (< 10 mol%). It was assumed that the activation of the

$TiO_2$ surface with $H_2SO_4$ reduced its catalytic activity, due to the excess of sulphate groups. Nevertheless, a two-stage process using the action of each catalyst in consecutive steps improved furfural yield. Pre-hydrolysis of biomass with $H_2SO_4$ allowed pentoses release, which were subsequently used as the substrate for a steam distillation process catalysed by metal oxides. A maximum 53 mol% of furfural yield was then achieved [18]. Nowadays, efficient utilisation of lignocellulosic biomass aims at selective processing of each biomass fraction within the biorefinery. In the processing of corncob with $SO_4^{2-}/TiO_2$-$ZrO_2/La^{3+}$ catalyst in aqueous medium, the obtained solid residues are mainly constituted of lignin and cellulose indicating a selective action of the catalyst over hemicellulose [20]. This catalyst demonstrated high thermal stability and strong acid sites capable to promote hemicellulose hydrolysis and further dehydration of pentoses to furfural in one step. The highest furfural yield obtained was 6.18 g/100 g corncob [20]. The application of $Al_2(SO_4)_3 \cdot 18H_2O$ in the processing of hemp shives allowed an impressive 73.7 mol% of furfural yield at 180 °C for 90 min. Nevertheless, the authors established optimal conditions at 160 °C for 90 min reaction time, considering the benefit of preserving 95.8% of cellulose in the solid (slightly depolymerised) along with a 62.7 mol% of furfural yield [110].

## 2.3 Conclusions and Future Outlook

This chapter aiming to provide a background of present industrial developments in the furfural production and to describe novel, and more mass- and energy-efficient processes. One of the general conclusions is that the furfural yield is strongly dependent on the inherent properties of feedstock and on the catalytic system used in the process. The methods described provide a wide spectrum of catalytic systems among which, some of them might be a potential solution if scaled up. Among several novel approaches given in literature, one of the possible solutions which, at least can minimise the energy demands and reduce the reaction time is the use of microwave irradiation. Microwave irradiation has been shown to be a powerful tool to drastically reduce the time required to produce furfural within minutes or even seconds [38, 40]. Other aspects which are practically unexplored are the employment of solvents that readily dissolve biomass, *e.g.* 1-ethyl-3-methylimidazolium acetate, with catalysts (either another IL, homogeneous or heterogeneous catalysts), assisted by an immiscible extracting organic solvent to continuously remove furfural from the reaction medium. Contrary to ILs or other catalytic systems, $CO_2$ and $H_2O$ opens new possibilities to produce furfural in high yields, however the research in this area is still very scarce. These green and promising solvents for furfural production generate *in-situ* catalytic reversible environment and in combination with some solvent, *e.g.* THF can also protect furfural from excessive consumption in subsequent reactions. However, all the above limitations are related to achieve maximal furfural production from hemicellulose. From the material balance point of view, furfural yield from biomass can be improved significantly if the cellulose fraction in lignocelluloses can also be converted to furfural. However, only few literature reports show that furfural can be obtained from glucose or cellulose in GVL in the presence of heterogeneous acid catalysts [64, 92]. This can be one of the most relevant tendencies to intensify furfural production and hence to obtain a significant reduction in the investment and operational costs.

**Acknowledgements** This work was supported by the Fundação para a Ciência e a Tecnologia (FCT/MEC, Portugal) through SFRH/BD/90282/2012, SFRH/BD/94297/2013 and IF/00471/2015 grants and by BBRI – Biomass and Bioenergy Research Infrastructure (ROTEIRO/0189/2013) and the Associated Laboratory for Sustainable Chemistry–Clean Processes and Technologies–LAQV which is financed by national funds from FCT/MEC (UID/QUI/50006/2013) and co-financed by the ERDF under the PT2020 Partnership Agreement (POCI-01-0145-FEDER-007265).

## References

1. International Furan Chemicals B.V. – Historical overview and industrial development of furfural. 2016. http://www.furan.com/furfural_historical_overview.html Accessed 5 Aug 2016.
2. Zeitsch KJ. The chemistry and technology of furfural and its many by-products, vol 13. Elsevier; 2000.
3. Werpy T, Petersen GR. Top Value Added Chemicals From Biomass – Volume I: Results of Screening for Potential Candidates from Sugars and Synthesis Gas vol I. the Pacific Northwest National Laboratory (PNNL) and the National Renewable Energy Laboratory (NREL). 2004.
4. Kamm B, Gruber PR, Kamm M. Biorefineries–industrial processes and products. Status Quo and future directions. KGaA: Wiley-VCH Verlag GmbH & Co; 2006.
5. Magalhães da Silva SP, Morais ARC, Bogel-Lukasik R. The $CO_2$-assisted autohydrolysis pre-treatment of wheat straw. Green Chem. 2014;16(1):238–46.
6. Morais ARC, da Costa Lopes AM, Bogel-Lukasik R. Carbon dioxide in biomass processing: contributions to the green biorefinery concept. Chem Rev. 2015;115(1):3–27.
7. Garrote G, Dominguez H, Parajo JC. Kinetic modelling of corncob autohydrolysis. Process Biochem (Amsterdam, Neth). 2001;36(6):571–8.
8. Antal MJ, Leesomboon T, Mok WS, Richards GN. Mechanism of formation of 2-furaldehyde from D-xylose. Carbohydr Res. 1991;217:71–85.
9. Feather MS, Harris DW, Nichols SB. Routes of conversion of D-xylose, hexuronic acids, and L-ascorbic-acid to 2-furaldehyde. J Org Chem. 1972;37(10):1606–&.
10. Ahmad T, Kenne L, Olsson K, Theander O. The formation of 2-furaldehyde and formic-acid from pentoses in slightly acidic deuterium-oxide studied by H-1-Nmr spectroscopy. Carbohydr Res. 1995;276(2):309–20.
11. Bonner WA, Roth MR. The conversion of D-Xylose-1-$C^{14}$ into 2-Furaldehyde-α-$C^{14}$. J Am Chem Soc. 1959;81(20):5454–6.
12. Garrett ER, Dvorchik BH. Kinetics and mechanisms of the acid degradation of the aldopentoses to furfural. J Pharm Sci. 1969;58(7):813–20.
13. Nimlos MR, Qian X, Davis M, Himmel ME, Johnson DK. Energetics of xylose decomposition as determined using quantum mechanics modeling. J Phys Chem A. 2006;110(42):11824–38.
14. Wu J, Serianni AS, Vuorinen T. Furanose ring anomerization: kinetic and thermodynamic studies of the d-2-pentuloses by 13 Cn. mr spectroscopy. Carbohydr Res. 1990;206(1):1–12.
15. Harris DW, Feather MS. Evidence for a C-2→ C-1 intramolecular hydrogen-transfer during the acid-catalyzed isomerization of D-glucose to D-fructose ag. Carbohydr Res. 1973;30(2):359–65.

16. Hurd CD, Isenhour LL. Pentose reactions. I Furfural formation. J Am Chem Soc. 1932;54(1):317–30.
17. Binder JB, Blank JJ, Cefali AV, Raines RT. Synthesis of furfural from xylose and xylan. ChemSusChem. 2010;3(11):1268–72.
18. Mansilla HD, Baeza J, Urzúa S, Maturana G, Villaseñor J, Durán N. Acid-catalysed hydrolysis of rice hull: evaluation of furfural production. Bioresour Technol. 1998;66(3):189–93.
19. Chareonlimkun A, Champreda V, Shotipruk A, Laosiripojana N. Reactions of C-5 and C 6-sugars, cellulose, and lignocellulose under hot compressed water (HCW) in the presence of heterogeneous acid catalysts. Fuel. 2010;89(10):2873–80.
20. Li H, Deng A, Ren J, Liu C, Lu Q, Zhong L, Peng F, Sun R. Catalytic hydrothermal pretreatment of corncob into xylose and furfural via solid acid catalyst. Bioresour Technol. 2014;158:313–20.
21. You SJ, Park N, Park ED, Park M-J. Partial least squares modeling and analysis of furfural production from biomass-derived xylose over solid acid catalysts. J Ind Eng Chem. 2015;21:350–5.
22. Marcotullio G, De Jong W. Chloride ions enhance furfural formation from D-xylose in dilute aqueous acidic solutions. Green Chem. 2010;12(10):1739–46.
23. Morais ARC, Bogel-Lukasik R. Highly efficient and selective $CO_2$-adjunctive dehydration of xylose to furfural in aqueous media with THF. Green Chem. 2016;18(8):2331–4.
24. Peleteiro S, da Costa Lopes AM, Garrote G, Parajó JC, Bogel-Łukasik R. Simple and efficient furfural production from xylose in media containing 1-butyl-3-methylimidazolium hydrogen sulfate. Ind Eng Chem Res. 2015;54(33):8368–73.
25. Win DT. Furfural-gold from garbage. Au J Technol. 2005;8:185–90.
26. De Jong W, Marcotullio G. Overview of biorefineries based on co-production of furfural, existing concepts and novel developments. Int J Chem React Eng. 2010;8(1):A69.
27. Fitzpatrick SW. Lignocellulose degradation to furfural and levulinic acid. US4897497A. 1990.
28. Hayes DJ, Fitzpatrick S, Hayes MHB, Ross JRH. The biofine process – production of levulinic acid, furfural, and formic acid from lignocellulosic feedstock. In: Kamm B, Gruber PR, Kamm M, editors. Biorefineries – industrial processes and products, vol. 1. Winheim: Wiley-VCH Verlag GmbH & Co.; 2006. p. 139–64.
29. Ritter S. Biorefinery gets ready to deliver the goods. Sci Technol. 2006;84(34):47.
30. Gravitis J, Vedernikov N, Zandersons J, Kokorevics A. Furfural and levoglucosan production from deciduous wood and agricultural wastes. In: ACS symposium series, 2001. Washington, DC: American Chemical Society, 1974. p. 110–22.
31. Delmas M, Mlayah B. Organic pulping of cereal straw: from pilot plant to the first factory. In: 16th European Biomass Conference and Exhibition–from Research to Industry and Markets Valencia, Spain, ETA Florence, 2008. p. 1660–4.
32. CIMV. CIMV The biorefinery concept. 2015. http://www.cimv.fr/cimv-technology/cimv-technology/5-.html. Accessed 11 Feb 2015.
33. Arato C, Pye EK, Gjennestad G. The lignol approach to biorefining of woody biomass to produce ethanol and chemicals. Appl Biochem Biotechnol. 2005;123(1–3):871–82.
34. Zeitsch KJ. Process for the manufacture of furfural. PCT/ZA2000/000024. 2004.
35. Steiner P. CEO – International Furan Technology (Pty) Ltd, Personal communication. 2016.
36. DalinYebo. 2016. http://dalinyebo.com/project/large-scale-furfural-production-from-bagasse Accessed 5 Aug 2016.
37. Yu QA, Zhuang XS, Yuan ZH, Qi W, Wang QO, Tan XS. The effect of metal salts on the decomposition of sweet sorghum bagasse in flow-through liquid hot water. Bioresour Technol. 2011;102(3):3445–50.
38. Zhang LX, Yu HB, Wang P, Dong H, Peng XH. Conversion of xylan, D-xylose and lignocellulosic biomass into furfural using AlCl3 as catalyst in ionic liquid. Bioresour Technol. 2013;130:110–6.
39. vom Stein T, Grande PM, Leitner W, de Maria PD (2011) Iron-catalyzed furfural production in biobased biphasic systems: from pure sugars to direct use of crude xylose effluents as feedstock. ChemSusChem 4 (11):1592–1594

40. Zhang Z, Zhao ZK. Microwave-assisted conversion of lignocellulosic biomass into furans in ionic liquid. Bioresour Technol. 2010;101(3):1111–4.
41. Kamireddy SR, Li JB, Tucker M, Degenstein J, Ji Y. Effects and mechanism of metal chloride salts on pretreatment and enzymatic digestibility of corn stover. Ind Eng Chem Res. 2013;52(5):1775–82.
42. Roman-Leshkov Y, Davis ME. Activation of carbonyl-containing molecules with solid Lewis acids in aqueous media. ACS Catal. 2011;1(11):1566–80.
43. Grzybkowski W. Nature and properties of metal cations in aqueous solutions. Pol J Environ Stud. 2006;15(4):655–63.
44. Howell I, Neilson G. The coordination of Ni2+ in aqueous solution at elevated temperature and pressure. J Chem Phys. 1996;104(5):2036–42.
45. Cotton FA, Wilkinson G. Advanced inorganic chemistry, vol. 594. New York: Wiley; 1988.
46. Marcotullio G, Krisanti E, Giuntoli J, de Jong W. Selective production of hemicellulose-derived carbohydrates from wheat straw using dilute HCl or FeCl3 solutions under mild conditions. X-ray and thermo-gravimetric analysis of the solid residues. Bioresour Technol. 2011;102(10):5917–23.
47. Voigt B, Gobler A. Formation of pure hematite by hydrolysis of iron(Iii) salt-solutions under hydrothermal conditions. Cryst Res Technol. 1986;21(9):1177–83.
48. Danon B, Marcotullio G, de Jong W. Mechanistic and kinetic aspects of pentose dehydration towards furfural in aqueous media employing homogeneous catalysis. Green Chem. 2014;16(1):39–54.
49. Liu L, Sun JS, Cai CY, Wang SH, Pei HS, Zhang JS. Corn stover pretreatment by inorganic salts and its effects on hemicellulose and cellulose degradation. Bioresource Technol. 2009;100(23):5865–71.
50. Mao LY, Zhang L, Gao NB, Li AM. Seawater-based furfural production via corncob hydrolysis catalyzed by FeCl3 in acetic acid steam. Green Chem. 2013;15(3):727–37.
51. Marcotullio G, de Jong W. Furfural formation from D-xylose: the use of different halides in dilute aqueous acidic solutions allows for exceptionally high yields. Carbohydr Res. 2011;346(11):1291–3.
52. Enslow KR, Bell AT. The role of metal halides in enhancing the dehydration of xylose to furfural. ChemCatChem. 2015;7(3):479–89.
53. Yang Y, Hu CW, Abu-Omar MM. Synthesis of furfural from xylose, xylan and biomass using $AlCl_3{\cdot}6H_2O$ in biphasic media via xylose isomerization to xylulose. ChemSusChem. 2012;5(2):405–10.
54. Parr RG, Pearson RG. Absolute Hardness - Companion Parameter to Absolute Electronegativity. J Am Chem Soc. 1983;105(26):7512–6.
55. Liu CG, Wyman CE. The enhancement of xylose monomer and xylotriose degradation by inorganic salts in aqueous solutions at 180 degrees C. Carbohydr Res. 2006;341(15):2550–6.
56. Cai CM, Nagane N, Kumar R, Wyman CE. Coupling metal halides with a co-solvent to produce furfural and 5-HMF at high yields directly from lignocellulosic biomass as an integrated biofuels strategy. Green Chem. 2014;16(8):3819–29.
57. Wang W, Ren J, Li H, Deng A, Sun R. Direct transformation of xylan-type hemicelluloses to furfural via SnCl 4 catalysts in aqueous and biphasic systems. Bioresour Technol. 2015;183:188–94.
58. Wiberg N. Holleman-Wiberg's inorganic chemistry. New York: Academic; 2001.
59. Wang W, Li H, Ren J, Sun R, Zheng J, Sun G, Liu S. An efficient process for dehydration of xylose to furfural catalyzed by inorganic salts in water/dimethyl sulfoxide system. Chin J Catal. 2014;35(5):741–7.
60. Peleteiro S, Garrote G, Santos V, Parajó JC. Conversion of hexoses and pentoses into furans in an ionic liquid. Afinidad. 2014;71(567)
61. Peleteiro S, Garrote G, Santos V, Parajó JC. Furan manufacture from softwood hemicelluloses by aqueous fractionation and further reaction in a catalyzed ionic liquid: a biorefinery approach. J Clean Prod. 2014;76:200–3.

62. Sievers C, Musin I, Marzialetti T, Valenzuela Olarte MB, Agrawal PK, Jones CW. Acid-catalyzed conversion of sugars and furfurals in an ionic liquid phase. ChemSusChem. 2009;2(7):665–71.
63. Zhang LX, Yu HB. Conversion of xylan and xylose into furfural in biorenewable deep eutectic solvent with trivalent metal chloride added. Bioresources. 2013;8(4):6014–25.
64. Zhang LX, Yu HB, Wang P, Li Y. Production of furfural from xylose, xylan and corncob in γ-valerolactone using $FeCl_3{\cdot}6H_2O$ as catalyst. Bioresour Technol. 2014;151:355–60.
65. Dunlop A. Furfural formation and behavior. Ind Eng Chem. 1948;40(2):204–9.
66. Brunner G. Stofftrennung mit ueberkritischen gasen (gasextraktion). Chem Ing Tech. 1987;59(1):12–22.
67. Morais ARC, Matuchaki MDJ, Andreaus J, Bogel-Lukasik R. A green and efficient approach of selective conversion of xylose and biomass hemicellulose into furfural in aqueous media using high-pressure $CO_2$ as sustainable catalyst. Green Chem. 2016;18:2985–94.
68. Girio FM, Fonseca C, Carvalheiro F, Duarte LC, Marques S, Bogel-Lukasik R. Hemicelluloses for fuel ethanol: a review. Bioresour Technol. 2010;101(13):4775–800.
69. Brunner G. Supercritical fluids: technology and application to food processing. J Food Eng. 2005;67(1–2):21–33.
70. Clark JH, Deswarte FEI, Farmer TJ. The integration of green chemistry into future biorefineries. Biofuels Bioprod Biorefining. 2009;3(1):72–90.
71. Morais ARC, Mata AC, Bogel-Lukasik R. Integrated conversion of agroindustrial residue with high pressure $CO_2$ within the biorefinery concept. Green Chem. 2014;16(9):4312–22.
72. Relvas FM, Morais ARC, Bogel-Lukasik R. Selective hydrolysis of wheat straw hemicellulose using high-pressure $CO_2$ as catalyst. RSC Adv. 2015;5:73935–44.
73. Sako T, Sugeta T, Nakazawa N, Okubo T, Sato M, Taguchi T, Hiaki T. Kinetic-study of furfural formation accompanying supercritical carbon-dioxide extraction. J Chem Eng Jpn. 1992;25(4):372–7.
74. Sangarunlert W, Piumsomboon P, Ngamprasertsith S. Furfural production by acid hydrolysis and supercritical carbon dioxide extraction from rice husk. Korean J Chem Eng. 2007;24(6):936–41.
75. Gairola K, Smirnova I. Hydrothermal pentose to furfural conversion and simultaneous extraction with SC-CO2 – Kinetics and application to biomass hydrolysates. Bioresour Technol. 2012;123:592–8.
76. Jing Q, Lu XY. Kinetics of non-catalyzed decomposition of D-xylose in high temperature liquid water. Chin J Chem Eng. 2007;15(5):666–9.
77. Ryu J, Suh YW, Suh DJ, Ahn DJ. Hydrothermal preparation of carbon microspheres from mono-saccharides and phenolic compounds. Carbon. 2010;48(7):1990–8.
78. Pollet P, Davey EA, Urena-Benavides EE, Eckert CA, Liotta CL. Solvents for sustainable chemical processes. Green Chem. 2014;16(3):1034–55.
79. Cai CM, Zhang TY, Kumar R, Wyman CE. THF co-solvent enhances hydrocarbon fuel precursor yields from lignocellulosic biomass. Green Chem. 2013;15(11):3140–5.
80. Bogel-Lukasik R. Ionic liquids in the biorefinery concept. RSC Green Chemistry, vol 36. RSC, Cambridge; 2015.
81. Tao F, Song H, Chou L. Efficient process for the conversion of xylose to furfural with acidic ionic liquid. Can J Chem. 2011;89(1):83–7.
82. Serrano-Ruiz JC, Campelo JM, Francavilla M, Romero AA, Luque R, Menendez-Vazquez C, García AB, García-Suárez EJ. Efficient microwave-assisted production of furfural from C 5 sugars in aqueous media catalysed by Brönsted acidic ionic liquids. Cat Sci Technol. 2012;2(9):1828–32.
83. Lima S, Neves P, Antunes MM, Pillinger M, Ignatyev N, Valente AA. Conversion of mono/di/polysaccharides into furan compounds using 1-alkyl-3-methylimidazolium ionic liquids. Appl Catal A-Gen. 2009;363(1–2):93–9.
84. Xu H, Zhao H, Song H, Miao Z, Yang J, Zhao J, Liang N, Chou L. Functionalized ionic liquids supported on silica as mild and effective heterogeneous catalysts for dehydration of biomass to furan derivatives. J Mol Catal A Chem. 2015;410:235–41.

85. Peleteiro S, Santos V, Garrote G, Parajó JC. Furfural production from Eucalyptus wood using an acidic ionic liquid. Carbohydr Polym. 2016;146:20–5.
86. Peleteiro S, da Costa Lopes AM, Garrote G, Bogel-Łukasik R, Parajó JC. Manufacture of furfural in biphasic media made up of an ionic liquid and a co-solvent. Ind Crop Product. 2015;77:163–6.
87. Carvalho AV, da Costa Lopes AM, Bogel-Lukasik R. Relevance of the acidic 1-butyl-3-methylimidazolium hydrogen sulphate ionic liquid in the selective catalysis of biomass hemicellulose fraction. RSC Adv. 2015;5:47153–64.
88. Brandt A, Ray MJ, To TQ, Leak DJ, Murphy RJ, Welton T. Ionic liquid pretreatment of lignocellulosic biomass with ionic liquid–water mixtures. Green Chem. 2011;13(9):2489–99.
89. Rinaldi R, Schüth F. Design of solid catalysts for the conversion of biomass. Energy Environ Sci. 2009;2(6):610–26.
90. Lessard J, Morin J-F, Wehrung J-F, Magnin D, Chornet E. High yield conversion of residual pentoses into furfural via zeolite catalysis and catalytic hydrogenation of furfural to 2-methylfuran. Top Catal. 2010;53(15–18):1231–4.
91. Metkar PS, Till EJ, Corbin DR, Pereira CJ, Hutchenson KW, Sengupta SK. Reactive distillation process for the production of furfural using solid acid catalysts. Green Chem. 2015;17(3):1453–66.
92. Gürbüz EI, Gallo JMR, Alonso DM, Wettstein SG, Lim WY, Dumesic JA. Conversion of hemicellulose into furfural using solid acid catalysts in γ-valerolactone. Angew Chem Int Ed. 2013;52(4):1270–4.
93. Li H, Ren J, Zhong L, Sun R, Liang L. Production of furfural from xylose, water-insoluble hemicelluloses and water-soluble fraction of corncob via a tin-loaded montmorillonite solid acid catalyst. Bioresour Technol. 2015;176:242–8.
94. Dias AS, Pillinger M, Valente AA. Dehydration of xylose into furfural over micro-mesoporous sulfonic acid catalysts. J Catal. 2005;229(2):414–23.
95. Kaiprommarat S, Kongparakul S, Reubroycharoen P, Guan G, Samart C. Highly efficient sulfonic MCM-41 catalyst for furfural production: Furan-based biofuel agent. Fuel. 2016;174:189–96.
96. Dias AS, Pillinger M, Valente AA. Mesoporous silica-supported 12-tungstophosphoric acid catalysts for the liquid phase dehydration of D-xylose. Microporous Mesoporous Mater. 2006;94(1):214–25.
97. Dias AS, Lima S, Pillinger M, Valente AA. Modified versions of sulfated zirconia as catalysts for the conversion of xylose to furfural. Catal Lett. 2007;114(3–4):151–60.
98. Shi X, Wu Y, Yi H, Rui G, Li P, Yang M, Wang G. Selective preparation of furfural from xylose over sulfonic acid functionalized mesoporous Sba-15 materials. Energies. 2011;4(4):669–84.
99. Agirrezabal-Telleria I, Requies J, Güemez M, Arias P. Dehydration of d-xylose to furfural using selective and hydrothermally stable arenesulfonic SBA-15 catalysts. Appl Catal B Environ. 2014;145:34–42.
100. Agirrezabal-Telleria I, Requies J, Güemez M, Arias P. Pore size tuning of functionalized SBA-15 catalysts for the selective production of furfural from xylose. Appl Catal B Environ. 2012;115:169–78.
101. Shi X, Wu Y, Li P, Yi H, Yang M, Wang G. Catalytic conversion of xylose to furfural over the solid acid/ZrO 2–Al 2 O 3/SBA-15 catalysts. Carbohydr Res. 2011;346(4):480–7.
102. Lima S, Antunes MM, Fernandes A, Pillinger M, Ribeiro MF, Valente AA. Catalytic cyclodehydration of xylose to furfural in the presence of zeolite H-Beta and a micro/mesoporous Beta/TUD-1 composite material. Appl Catal A Gen. 2010;388(1):141–8.
103. Lima S, Antunes MM, Fernandes A, Pillinger M, Ribeiro MF, Valente AA. Acid-catalysed conversion of saccharides into furanic aldehydes in the presence of three-dimensional mesoporous Al-TUD-1. Molecules. 2010;15(6):3863–77.
104. Lam E, Majid E, Leung AC, Chong JH, Mahmoud KA, Luong JH. Synthesis of furfural from xylose by heterogeneous and reusable nafion catalysts. ChemSusChem. 2011;4(4):535–41.

105. Dias AS, Lima S, Carriazo D, Rives V, Pillinger M, Valente AA. Exfoliated titanate, niobate and titanoniobate nanosheets as solid acid catalysts for the liquid-phase dehydration of D-xylose into furfural. J Catal. 2006;244(2):230–7.
106. Zhang J, Lin L, Liu S. Efficient production of furan derivatives from a sugar mixture by catalytic process. Energ Fuel. 2012;26(7):4560–7.
107. Agirrezabal-Telleria I, Hemmann F, Jäger C, Arias P, Kemnitz E. Functionalized partially hydroxylated MgF 2 as catalysts for the dehydration of d-xylose to furfural. J Catal. 2013;305:81–91.
108. Antunes MM, Lima S, Fernandes A, Candeias J, Pillinger M, Rocha SM, Ribeiro MF, Valente AA. Catalytic dehydration of d-xylose to 2-furfuraldehyde in the presence of Zr-(W, Al) mixed oxides. Tracing by-products using two-dimensional gas chromatography-time-of-flight mass spectrometry. Catal Today. 2012;195(1):127–35.
109. Agirrezabal-Telleria I, Garcia-Sancho C, Maireles-Torres P, ARIAS PL. Dehydration of xylose to furfural using a Lewis or Brönsted acid catalyst and N 2 stripping. Chin J Catal. 2013;34(7):1402–6.
110. Brazdausks P, Paze A, Rizhikovs J, Puke M, Meile K, Vedernikovs N, Tupciauskas R, Andzs M. Effect of aluminium sulphate-catalysed hydrolysis process on furfural yield and cellulose degradation of *Cannabis sativa* L. shives. Biomass Bioenerg. 2016;89:98–104.
111. Moreau C, Durand R, Peyron D, Duhamet J, Rivalier P. Selective preparation of furfural from xylose over microporous solid acid catalysts. Ind Crop Product. 1998;7(2):95–9.
112. Kim SB, You SJ, Kim YT, Lee S, Lee H, Park K, Park ED. Dehydration of D-xylose into furfural over H-zeolites. Korean J Chem Eng. 2011;28(3):710–6.
113. O'Neill R, Ahmad MN, Vanoye L, Aiouache F. Kinetics of aqueous phase dehydration of xylose into furfural catalyzed by ZSM-5 zeolite. Ind Eng Chem Res. 2009;48(9):4300–6.
114. Lima S, Pillinger M, Valente AA. Dehydration of D-xylose into furfural catalysed by solid acids derived from the layered zeolite Nu-6 (1). Catal Commun. 2008;9(11):2144–8.
115. Lima S, Fernandes A, Antunes MM, Pillinger M, Ribeiro F, Valente AA. Dehydration of xylose into furfural in the presence of crystalline microporous silicoaluminophosphates. Catal Lett. 2010;135(1–2):41–7.
116. Choudhary V, Pinar AB, Sandler SI, Vlachos DG, Lobo RF. Xylose isomerization to xylulose and its dehydration to furfural in aqueous media. ACS Catal. 2011;1(12):1724–8.
117. Dhepe PL, Sahu R. A solid-acid-based process for the conversion of hemicellulose. Green Chem. 2010;12(12):2153–6.
118. Bruce SM, Zong Z, Chatzidimitriou A, Avci LE, Bond JQ, Carreon MA, Wettstein SG. Small pore zeolite catalysts for furfural synthesis from xylose and switchgrass in a γ-valerolactone/water solvent. J Mol Catal A Chem. 2016;422:18–22.
119. Chen H, Qin L, Yu B. Furfural production from steam explosion liquor of rice straw by solid acid catalysts (HZSM-5). Biomass Bioenerg. 2015;73:77–83.
120. Dias AS, Lima S, Pillinger M, Valente AA. Acidic cesium salts of 12-tungstophosphoric acid as catalysts for the dehydration of xylose into furfural. Carbohydr Res. 2006;341(18):2946–53.
121. Dias AS, Lima S, Brandão P, Pillinger M, Rocha J, Valente AA. Liquid-phase dehydration of D-xylose over microporous and mesoporous niobium silicates. Catal Lett. 2006;108(3–4):179–86.
122. Jeong GH, Kim EG, Kim SB, Park ED, Kim SW. Fabrication of sulfonic acid modified mesoporous silica shells and their catalytic performance with dehydration reaction of d-xylose into furfural. Microporous Mesoporous Mater. 2011;144(1):134–9.
123. Galarneau A, Cambon H, Martin T, De Ménorval L-C, Brunel D, Di Renzo F, Fajula F. SBA-15 versus MCM-41: are they the same materiais? Stud Surf Sci Catal. 2002;141:395–402.
124. Weingarten R, Tompsett GA, Conner WC, Huber GW. Design of solid acid catalysts for aqueous-phase dehydration of carbohydrates: the role of Lewis and Brønsted acid sites. J Catal. 2011;279(1):174–82.

125. Suzuki T, Yokoi T, Otomo R, Kondo JN, Tatsumi T. Dehydration of xylose over sulfated tin oxide catalyst: influences of the preparation conditions on the structural properties and catalytic performance. Appl Catal A Gen. 2011;408(1):117–24.
126. Li H, Deng A, Ren J, Liu C, Wang W, Peng F, Sun R. A modified biphasic system for the dehydration of d-xylose into furfural using SO 4 2−/TiO 2-ZrO 2/La 3+ as a solid catalyst. Catal Today. 2014;234:251–6.
127. Doiseau A-C, Rataboul F, Burel L, Essayem N. Synergy effect between solid acid catalysts and concentrated carboxylic acids solutions for efficient furfural production from xylose. Catal Today. 2014;226:176–84.
128. Kim Y-C, Lee HS. Selective synthesis of furfural from xylose with supercritical carbon dioxide and solid acid catalyst. J Ind Eng Chem. 2001;7(6):424–9.

# Chapter 3
# Catalytic Production of 5-Hydroxymethylfurfural from Biomass and Biomass-Derived Sugars

**Xinli Tong, Song Xue, and Jianli Hu**

**Abstract** Production of 5-hydroxymethylfurfural (5-HMF) from biomass has gained considerable attention in catalysis research because 5-HMF is a potential substitute of petroleum-based monomers of various polymers and it can be used as a starting material for biofuels and value-added chemicals. Generally, 5-HMF can be obtained *via* the catalytic transformation of biomass-derived carbohydrates such as fructose, glucose, sucrose and inulin, but direct transformation of cellulose and lignocelluloses into 5-HMF is of the highest interest. Catalysts used in production of 5-HMF include mineral acids, organic acids, solid acids and metal-containing compounds. Corresponding reaction media can be organic solvents, ionic liquids, water and biphasic systems. In this chapter, an overview of catalytic production methods of 5-HMF from biomass-derived sugars and lignocelluloses is given. Discussion is focused on the characteristics and superiority of different catalysts on the catalytic transformation of various feedstocks.

**Keywords** Catalysis • Sugar • 5-Hydroxymethylfurfural • Cellulose • Biomass conversion

---

X. Tong (✉) • S. Xue
Tianjin Key Laboratory of Organic Solar Cells and Photochemical Conversion, School of Chemistry and Chemical Engineering, Tianjin University of Technology, Tianjin 300384, People's Republic of China
e-mail: tongxinli@tju.edu.cn; tongxli@sohu.com

J. Hu (✉)
Department of Chemical and Biomedical Engineering, West Virginia University, Morgantown, WV 26506, USA
e-mail: jianlihu@hotmail.com



Z. Fang et al. (eds.), *Production of Platform Chemicals from Sustainable Resources*, Biofuels and Biorefineries 7, DOI 10.1007/978-981-10-4172-3_3

## 3.1 Introduction

Transformation of biomass to value-added chemicals and liquid fuels has attracted considerable attention and investment for applications in sustainable energy [1–8]. Due to concerns on global warming associated with the use of fossil fuels, catalytic biomass conversion has become a highly-active topical area. In the last decade, tremendous research effort has been devoted to developing efficient conversion methods for biomass and biomass-derived sugars [9–11].

Biomass feedstocks are renewable resources that can be defined as plant, urban waste and agricultural residues used to prepare fuels such as bioethanol, butanol, diesel, and liquid hydrocarbon fuels. Common examples of biomass feedstocks include woody plant, corn starch, sugarcane juice, crop residues and grass crops. In general, lignocellulosic biomass is often classified into virgin biomass such as trees, bushes and grass, waste biomass (corn stover, paper mill discards, sugarcane bagasse, straw, etc.) and energy crops (switch grass and Elephant grass). Worldwide production of plant-derived non-edible biomass feedstocks, including agricultural and forest residues is enormous and estimated to be around $1.8 \times 10^{12}$ t/a [12]. Lignocellulosic biomass is mainly made up of 45% cellulose, 25% hemicellulose, and 20% lignin with some other minor ingredients such as nutrients, proteins, and wax [13, 14]. In particular, cellulose is a homopolysaccharide derived from glucose units and made up of several $\beta$-D glucose units linked together *via* $\beta$-1, 4 glycosidic linkages. Hemicelluloses contain primarily pentose sugars, and occasionally small amounts of L-sugars as well and they are made up of shorter chains that contain 500 to 3000 sugar units.

Today, direct transformation of biomass feedstocks for the generation of bioethanol using fermentation produces carbon dioxide which remarkably cuts down on the carbon efficiency of the conversions and requires expensive enzymes [15]. A promising method for biomass conversion is fast pyrolysis technology [16–18]. However, bio-oils obtained from the pyrolysis technologies are generally of poor quality containing more than 300 different compounds that cannot be used for the direct replacement of gasoline and diesel fuels [19]. Consequently, catalytic transformation of lignocellulosic biomass to value-added chemicals through identified platform compounds is an attractive and promising strategy [20, 21].

It is well-known that the presence of oxygenated groups in lignocellulosic biomass could facilitate chemical reactions during the production of valuable chemicals. However, such characteristics impede their potentials of conversion into liquid fuels [22–24]. Nevertheless, direct production of fuel grade liquid alkanes from lignocellulosic biomass has many opportunities for meeting future regulations placed on transportation fuel [25]. The C5 and C6 sugars of holocellulose can be exploited for the production of fuel chemical intermediates by converting them into platform chemicals including 5-hydroxymethylfurfural (5-HMF), furfural and levulinic acid (LA) [26, 27]. Especially, 5-HMF derived from cellulose has emerged as

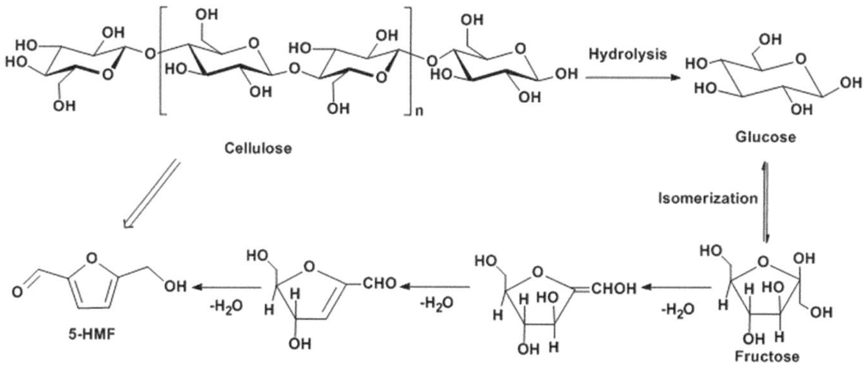


**Scheme 3.1** Reaction pathway for the transformation of cellulose into 5-hydroxymethyl furfural

an important platform chemical for meeting chemical and fuel needs of society [28, 29]. The impact of initiatives for using biorenewable sources are not limited to develop new energy platforms and $CO_2$ reduction, as they also create opportunities to secure the local supplies of energy and to support local agricultural economies [30, 31].

Hydrolysis of cellulose gives monosaccharides which in turn can be converted to 5-HMF by dehydration and removal of three water molecules ($-3H_2O$). Dehydration to yield furans from sugars is often considered as cyclodehydration or dehydrocyclization (Scheme 3.1). During the transformation of cellulose to 5-HMF, hexoses (glucose and fructose) are the main reaction intermediates, which are C6 sugars.

Six carbon sugars, e. g. hexoses, are the most plentiful monosaccharide which occurs in the nature. Fructose and glucose are commonly used as low cost feedstock in chemical syntheses [32–35]. Therefore, elucidation of reaction mechanisms that occur in the conversion of hexoses into furans is crucial for developing catalytic technology. In general, biomass transformation includes the following elementary reaction steps: dehydration, hydrolysis, isomerization, reforming, aldol condensation, hydrogenation and oxidation. Typically, products such as 5-HMF, 2, 5-diformylfuran (2,5-DFF), 2, 5-furandicarboxylic acid (2,5-FDCA), 2, 5-bis(hydroxymethyl)-furan (2,5-BHF) and 2, 5-dimethylfuran (2,5-DMF) are obtained. These compounds can be used as the feedstock or intermediates to replace petroleum-derived chemicals [20, 36–38].

In summary, production of chemicals and fuels from lignocellulosic biomass requires effective pretreatment and hydrolysis of cellulose to form the corresponding hexose sugar units, followed by catalytic dehydration of sugars to derive the corresponding 5-HMF products. Catalytic transformation of biomass-derived sugar to fine chemicals is very important for both scientific research and technological applications [32, 39].

## 3.2 Platform Chemical 5-Hydroxymethylfurfural

In the late 19th century, 5-HMF was separated from a reaction mixture of fructose and sucrose in the presence of oxalic acid and since then it has attracted significant research interest [40, 41]. Fenton and coworkers carried out extensive research in the field [42–44]. In 1909, the chemical structure of 5-HMF was identified [45]. After detailed investigation, Middendorp [46] reported experimental results regarding the physical and chemical properties of 5-HMF as well as its preparation [46]. In following years, the reaction chemistry relevant to 5-HMF was further developed by Reichstein and Zschokke [47, 48]. Significant progress was made by Haworth and Jones, who proposed a reaction mechanism for a synthetic route to 5-HMF [49]. The synthetic method was based on dehydration of fructose which is still used in the present industrial processes. In addition to acid-catalyzed dehydration, research showed that 5-HMF could be made through heating a 30% aqueous solution of sucrose under the conditions of 170 °C and hydrogen pressure, where a 22% yield of 5-HMF was obtained [50].

Research interests on 5-HMF were continued in 1980s when two different 5-HMF production processes based on aqueous and non-aqueous systems were reported by van Dam [51] and Cottier [52]. It is found that both processes could lead to around 37% yield of 5-HMF. The aqueous solution was found to promote degradation of 5-HMF, whereas polymerization of 5-HMF occurred in both aqueous and non-aqueous reaction media. Antal et al. [53] reported that, in the presence of acidic catalysts, 5-HMF was formed via the acid-catalyzed dehydration reaction where three water molecules were removed from hexoses. Scheme 3.2 illustrates the dehydration pathways of the hexoses and possible by-products that are formed. In aqueous solutions, formic acid and levulinic acid can be further generated *via* a consecutive hydration of 5-HMF. Although the hydrolysis of 5-HMF can be sup-

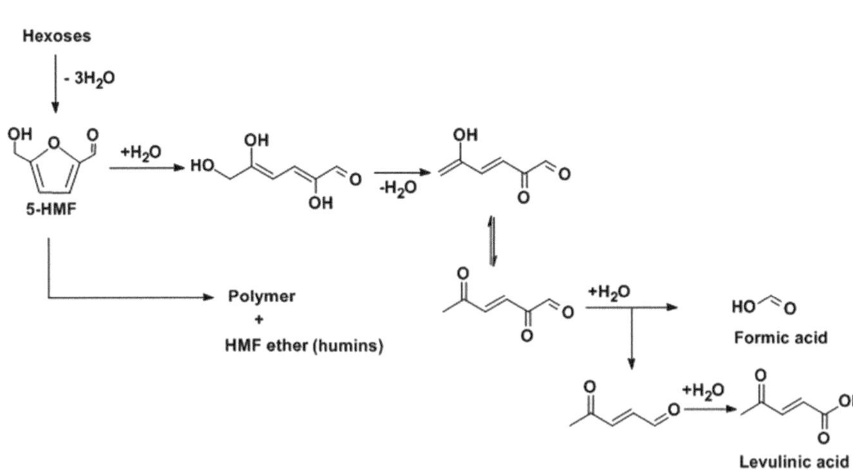


**Scheme 3.2** Transformation of hexoses to 5-hydroxymethyl furfural and other by-products

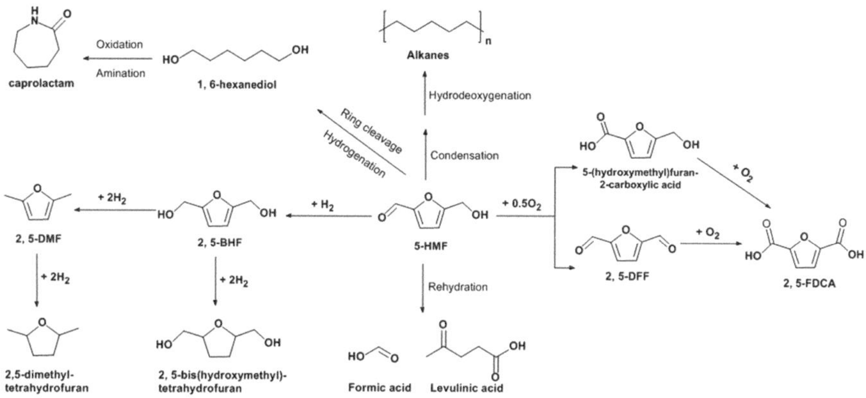


**Fig. 3.1** Application of 5-hydroxymethyl furfural as a platform compound for the production of value-added chemicals

pressed in non-aqueous solution, the cross-polymerization reactions take place leading to the formation of soluble polymers and insoluble brown humins [54]. To prevent side reactions and obtain a high 5-HMF yield, some of the adopted approaches are to design and synthesize catalysts that are tuned to promote formation of 5-HMF, but to inhibit the consecutive reactions. In addition, 5-HMF yield can be increased if it is continuously removed from the reaction mixture.

Catalytic transformation of 5-HMF into value-added chemicals has been explored by many researchers [55]. As illustrated in Fig. 3.1, compounds such as 2,5-FDCA, 2,5-DFF, 2,5-BHF and 2,5-DMF are derivatives of 5-HMF. For instance, 2,5-FDCA and 2,5-DFF can be produced from the complete or partial oxidation of 5-HMF and they can also be formed as co-products through simultaneous dehydration and oxidation of hexoses [38]. Many of these derivatives are presently used in chemical or pharmaceutical industries. For example, 2,5-FDCA can be used as a feedstock for the production of succinic acid that is widely used in industry. Thus, 2,5-FDCA has been identified by Department of Energy (USA) as one of the twelve platform chemicals that can be produced from sugars *via* biological or chemical processes [28]. Commercially, 2,5-FDCA has been found to be useful as a new fungicide, corrosion inhibitor and a melting product for the foundry sands and as an intermediate in pharmaceutical and lithographic fields [5, 56–58]. In the plastic industry, 2,5-FDCA has gained interest as a monomer for new polymeric materials in special applications [59–62]. It has been reported that 2, 5-FDCA is able to replace terephthalic acid, which is widely used in producing polyesters, such as polyethylene terephthalate (PET) and polybutylene terephthalate (PBT) [63, 64]. Similar to 2,5-FDCA, 2,5-DFF can be used to synthesize special polymers [65–67] or as an intermediate for pharmaceuticals [68] or as an antifungal agent [69]. Other applications of 2,5-FDCA include preparation of macrocyclic ligands [70, 71] and as a cross-linking agent for poly (vinyl alcohol) [72]. It is worthwhile to mention that 2,5-BHF and 2,5-DMF are derivatives from the partial or deep hydrogenation of 5-HMF. Research shows that 2, 5-DMF could be a promising liquid fuel, because it has a high energy density of 31.5 MJ/L, which is similar to that of gasoline (35.0 MJ/L) and 40% higher than that of ethanol (23.0 MJ/L) [73, 74]. Different from water-soluble oxygenated fuels, 2,5-DMF is immiscible with water and has similar characteristics to transportation fuels. The boiling point of 2,5-DMF (bp 92–94 °C) makes it less volatile than ethanol (bp 78 °C) thus it is more easily blended into gasoline within vapor pressure limitations than ethanol.

The other important derivatives from 5-HMF for bio-based fuels and chemicals include 2,5-furfuryldiamine, 2,5-furfuryldiisocyanate and di-5-hydroxymethyl furfural ether. Industry has special interest in using these derivatives as building blocks for the production of polymeric materials such as polyesters, polyamides and polyurethane [54, 56–57]. Furan-based polymers display excellent physicochemical properties such as strong resistance to thermal treatment. Interestingly, the Kevlar-like polyamides produced from furan diamine and the diacids exhibit liquid crystal behavior. Photoreactive polyesters have been used in the formulation of printing inks. Furthermore, furan-based polyconjugated polymers possess good electrical conductivity [54]. Although the length of the carbon chain in 5-HMF is relatively short, long-chain hydrocarbons can be produced from 5-HMF through condensation reactions [39, 75]. Catalytic selective hydrogenation of 5-HMF produces 1,6-hexanediol directly, which is an intermediate for the production of caprolactam (monomer of nylon-6) *via* further amination [76, 77]. Further rehydration of 5-HMF yields low-molecular-weight organic acids such as formic and levulinic acids, which are generally employed in the preparation of fine chemicals [78, 79]. Moreover, the selective etherification of 5-HMF can also generate 5-alkoxymethyl furfural as diesel additives, or further produce alkyl levulinates from ethanolysis reaction which are used as organic solvents and as additives [80, 81].

In 1990s and 2000s, Kuster [82] and Lewkowski [54] reported on the kinetics of catalytic dehydration reactions of hexoses for the formation of 5-HMF, respectively. After that, progress in the synthesis of 5-HMF through conversion of fructose, glucose, polysaccharides and biomass feedstocks has continued and is discussed in detail in the next section.

## 3.3 Catalytic Production of 5-Hydroxymethylfurfural (5-HMF) from Fructose

### *3.3.1 Mineral Acid and Organic Acid as Catalysts*

The dehydration of fructose for 5-HMF production is generally catalyzed by mineral acids as well as by organic acids. The catalysts should possess either a proton or be a Lewis acid [83, 84]. Since the discovery of oxalic acid-mediated production of 5-HMF, nearly one hundred inorganic and organic acidic compounds have been listed as potential catalysts for 5-HMF production. The most commonly used inexpensive mineral acids are $H_2SO_4$, $H_3PO_4$ and hydrochloric acid (HCl) [85–87].

Among the mineral acids, iodine-catalyzed dehydration of hexoses using hydroiodic acid (HI) has also been reported [88]. Although there are many choices of organic acids for 5-HMF dehydration, commonly used organic acid catalysts are oxalic, levulinic or *p*-toluenesulfonic acids [49, 89–92].

Early works almost 30 years ago reported on the synthesis of 5-HMF from the dehydration of fructose under supercritical and sub-critical water conditions [53, 93]. Yields of 5-HMF as high as 53% are obtained at 250 °C. Bicker et al. [32] investigated the synthesis of 5-HMF in the presence of $H_2SO_4$ when subcritical or supercritical acetone-water mixtures were employed as the reaction medium. Maximum yields of 5-HMF were about 78% at 180 °C. The advantage of using subcritical or supercritical aqueous solvents or solvent mixtures is that there is high carbon efficiency and low solid impurity formation.

A two-phase reactor that integrates 5-HMF production and separation into the reactor design was reported [33]. During the reaction, 5-HMF was continuously extracted into an organic methylisobutylketone (MIBK) phase containing 2-butanol that was used to enhance the partitioning from the reactive aqueous solution. This innovative approach led to 80% 5-HMF selectivity at a 90% conversion using a feedstock containing 10 wt% D-fructose. Other biphasic solvent systems were reported by Román-Leshkov and Dumesic [94] who investigated the solvent effect on the dehydration of fructose in the presence of inorganic salts, where tetrahydrofuran exhibited high extracting ability as a reaction media resulting in 83% 5-HMF selectivity. For the aqueous phase dehydration of D-fructose to 5-HMF using HCl catalyst, the addition of dimethylsulfoxide (DMSO) and poly(1-vinyl-2-pyrrolidinone) (PVP) suppressed the formation of byproducts. Improvement was made in HCl-catalyzed dehydration of fructose in aqueous solution using a continuous flow microreactor system [95]. The process was designed to rapidly elevate reaction temperatures from room temperature to 185 °C and 17 bar within 1 minute, leading to 54% 5-HMF yield at 71% D-fructose conversion.

A synthesis procedure based on microwave irradiation was explored for production of 5-HMF with inorganic acid catalysts. Hansen et al. [96] investigated microwave-assisted dehydration of highly concentrated aqueous fructose solutions (27 wt.%) to form 5-HMF in the presence of HCl. Compared with conventional heating, microwave irradiation gave 52% fructose conversion and 63% selectivity to 5-HMF within a short reaction time. Pawar et al. [97] found that the use of *p*-toluene sulphonic acid as catalyst for microwave (2.45 GHz) assisted dehydration of fructose and gave >98% fructose conversion with >90% 5-HMF yield in 90 s with isopropyl alcohol as solvent. Moreover, the dehydration of fructose in aqueous and organic media using three methods, i.e., conventional heating, ultrasonication and microwave irradiation was further investigated [98]. Results showed that the 5-HMF yield in organic reaction media was higher than in aqueous media, with the highest yields obtained from conventional, ultrasonicated and microwave-assisted reactions being 87%, 53%, and 38%, respectively. It was concluded that the microwave method was the fastest one, and the 5-HMF selectivity varied from 60 to 90% depending on the reaction media and assistance method.

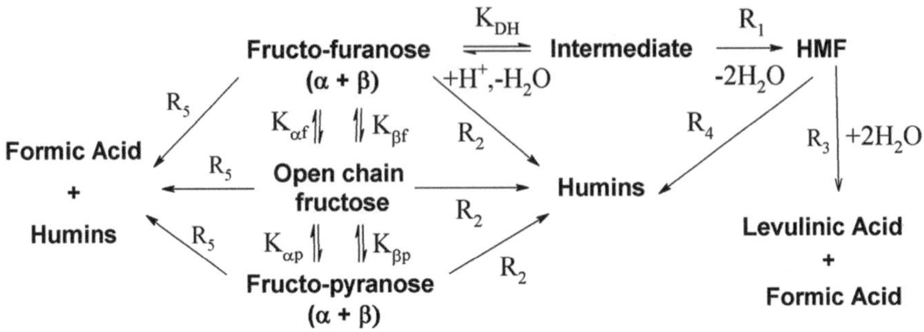


**Scheme 3.3** Acid-catalyzed reactions of fructose and 5-hydroxymethyl furfural in aqueous solution (Reprinted from Ref. [101] with permission of American Chemical Society Publishing Corporation. 2014 Copyright © American Chemical Society)

Caes et al. [99] found that high yields of 5-HMF from D-fructose could be achieved with HBr as catalyst in sulfolane, in which a 93% yield of 5-HMF was obtained with 5 wt.% HBr at 100 °C for 1 h. Qi et al. [100] found that using sulfuric acid as the catalyst and γ-valerolactone (GVL) as solvent, a 75% yield of 5-HMF could be obtained. Swift et al. [101] carried out a kinetic study on dehydration of fructose and rehydration of 5-HMF at different reaction conditions by employing HCl and KCl to adjust pH value [reaction temperature 70–150 °C, pH values 0.7–1.6, initial fructose concentration 5–20% (w/v), 5-HMF 2.5–10% (w/v)]. Scheme 3.3 shows that the reaction network has three important proportions: (1) fructose tautomerization, (2) fructose dehydration illuminated by the microkinetic model according to Quantum Mechanics/Molecular Mechanic (QM/MM) simulations and KIE experiments and (3) a direct route from the fructose to formic acid. The network includes five reactions R1-R5 which contain 4 directly detectable species (fructose, 5-HMF, formic acid, and levulinic acid) and 6 implicit species (five fructose tautomers and a reactive intermediate). The basic reaction mechanism includes two steps which are fast protonation and dehydration, followed by intramolecular hydride transfer as the rate-limiting step during reaction. These conclusions match the experimental kinetics and KIE experiments well. For dehydration of fructose, formic acid can be in stoichiometric excess relative to levulinic acid. Moreover, all reactions are found to be pseudo-first order for both catalyst and substrate.

Jiang et al. [102] studied the effects of mechanical agitation and addition of high-concentration fructose solutions on the dehydration of D-fructose to 5-HMF using HCl as catalyst. For a stirring rate of 100 rpm, a 5-HMF yield of 81.7% was obtained in the presence of 0.3 M HCl. Yields of 5-HMF all exceeded 55% in the experiments with the addition of concentrated fructose solution (54.6 wt.%). Pedersen et al. [103] performed simultaneous dehydration of fructose and glucose catalyzed by HCl in acetone-water mixtures, and it was found that a recirculating reactor configuration was more effective, where the equilibrium controlled by-products (anhydroglucose and glucose dimers) were recirculated into the dehydration reactor. The model predicted a selectivity of 5-HMF close to 70% in a recirculating reactor under

**Table 3.1** Fructose dehydration with selected solid acid catalysts

| Catalyst | Solvent | T (°C) | Time (min) | Conv. (%) | 5-HMF yield (%) | Ref. |
|---|---|---|---|---|---|---|
| H-form mordenite | $H_2O$-MIBK | 165 | 60 | 76.0 | 69.2 | [106] |
| vanadyl phosphate | $H_2O$ | 50 | 60 | 50.2 | 41.9 | [107] |
| $NbOPO_4$ | $H_2O$ | 100 | 120 | 61.4 | 21.6 | [108] |
| C-$ZrP_2O_7$ | $H_2O$ | 100 | 30 | 44.4 | 44.3 | [111] |
| Amberlyst-15 | [BMIM]$BF_4$ | 80 | 180 | – | 52 | [125] |
| Dowex 50wx8–100 | Acetone/$H_2O$ | 150 | 15 | 95.1 | 73.4 | [128] |
| Dowex 50wx8–100 | Acetone/DMSO | 150 | 10 | 96.4 | 82.1 | [129] |
| D001-cc resin | [BMIM]Cl | 75 | 20 | – | 93.0 | [130] |
| Anatase $TiO_2$ | $H_2O$ | 200 | 5 | 83.6 | 38.1 | [132] |
| $ZrO_2$ | $H_2O$ | 200 | 5 | 65 | 30.5 | [132] |
| $Ag_3PW_{12}O_{40}$ | $H_2O$-MIBK | 120 | 60 | 82.8 | 77.7 | [134] |
| Lignin-derived carbon | DMSO-[BMIM]Cl | 110 | 10 | 98 | 84 | [136] |
| $SO_4/TiO_2$-$SiO_2$ | DMSO | 110 | 180 | 77 | 68.5 | [141] |
| $SO_4^{2-}/SnO_2$-$ZrO_2$ | DMSO | 120 | 150 | – | 75 | [144] |
| KIT-6 | DMSO | 165 | 30 | 100 | 84.1 | [146] |

conditions where 5-HMF degradation was avoided. Brasholz et al. [104] reported that efficient conversion of D-fructose to 5-(chloromethyl) furfural (CMF) could be achieved using continuous flow reactor where ca. 81% yield was obtained in 32% HCl aqueous solution in only 60 s; then, the generated CMF was hydrolyzed to attain 5-HMF in a biphasic THF-$H_2O$ solvent and the isolated 5-HMF reached 74% under the suitable conditions.

### 3.3.2 *Solid Acids as Catalysts*

Heterogeneous catalysis is thought to be preferable to homogeneous catalysis due to ease of product separation. Heterogeneous catalysts possess several advantages over liquid acid catalysts: a) they facilitate reactor design and product separation, which allows recycle of unconverted feedstock; b) they can be utilized at high temperature to increase reaction kinetics therefore shortening the reaction time, increasing 5-HMF selectivity and eliminating side reactions; c) their surface acidity can be adjusted to improve the selectivity of 5-HMF. The surface characteristics of the heterogeneous catalysts greatly affect and are critical to the conversion of polysaccharides and biomass feedstock. In the transformation of sugars, the solid acid catalysts include H-form zeolites, ion-exchange resins, vanadyl phosphate and $ZrO_2$ as supports or components of the catalyst design. Typical results for D-fructose dehydration are summarized in Table 3.1.

Dehydration of D-fructose in the presence of the de-aluminated H-form mordenite was reported by Moreau and coworkers [105, 106]. The experiments were carried out at 165 °C in a solvent consisting of water and MIBK, where a conversion of 76% and a 5-HMF selectivity of 92% were achieved. The shape selective catalysis property of the solid acids was found to affect the performance of 5-HMF formation. The conversion of D-fructose and the selectivity towards 5-HMF are strongly influenced by catalyst micropore versus mesopore volume distribution. The Si/Al ratio of H-mordenite solid acids could affect the fructose transformation. The maximum reaction rate of D-fructose was obtained over H-mordenite catalyst with a Si/Al ratio of 11.

In solid acid-catalyzed dehydration of fructose, continuous product separation affects 5-HMF selectivity. A significant increase (*ca.* 10%) in 5-HMF selectivity was obtained by extracting 5-HMF into MIBK circulated in a countercurrent mode in a continuous heterogeneous pulsed column reactor [106]. Different from Si/Al based H-mordenite catalysts, vanadyl phosphate (VOP) was investigated by Carniti et al. [107] for the dehydration of D-fructose to 5-HMF in aqueous solutions. The results showed that using VOP as a catalyst, 40.2% and 32.9% 5-HMF yields were obtained when 6.0 wt.% and 30 wt.% aqueous fructose solutions were used as feedstock, respectively. In the literature, other VOP-based catalysts containing different trivalent metal ions ($Fe^{3+}$, $Cr^{3+}$, $Ga^{3+}$, $Mn^{3+}$ or $Al^{3+}$) were also reported. When Fe-containing VOP catalyst was employed and 40 wt.% fructose solution was used as feedstock, 87.3% 5-HMF selectivity at 50.4% yield was obtained.

Another category of phosphate catalysts used in the dehydration of fructose is Nb-based phosphate [108–110]. Niobium phosphate ($NbOPO_4$) and phosphoric acid treated niobium oxide exhibit high catalytic activity in dehydration of fructose. About 70–80% selectivity of 5-HMF at a D-fructose conversion of 30–50% was obtained at 100 °C without using any extraction solvent [108, 109]. The initial catalytic performance of $NbOPO_4$ catalyst was compared with niobate acid in the dehydration of fructose in aqueous phase. Results showed that $NbOPO_4$ catalyst was more active than niobate acid that was ascribed to the higher surface acidity of $NbOPO_4$ in the polar liquid [110]. Other metal-promoted phosphate systems reported for dehydration of fructose are based on Zr and Ti [111]. Cubic zirconium pyrophosphate ($ZrP_2O_4$) exhibited excellent performance under mild conditions. A 44.3% yield and a 99.8% selectivity for 5-HMF were obtained at 100 °C. Under similar reaction conditions, 35.3% yield of 5-HMF and 96.1% selectivity were obtained over γ-titanium phosphate catalyst.

There have been numerous reports on the application of ion-exchange resins for the synthesis of 5-HMF from sugars. Nakamura [112] examined a strong acidic ion exchange resin for D-fructose dehydration, where an 80% yield of 5-HMF was obtained. Using the Levatit® SPC-108 as a catalyst, Gaset and coworkers [113, 114] reported 39–80% 5-HMF yields. There are several categories of ion-exchange resins among which Diaion® PK-216 resin is found to be more effective for D-fructose dehydration with about 90% yield of 5-HMF being obtained [115]. Although an ion exchange resin has advantages over homogeneous mineral acids as catalyst, improvement in 5-HMF selectivity is still a challenge [116–118]. Performance improvement over Diaion® PK216 resin was made by Chheda and Dumesic [119] who showed that the yield of 5-HMF from dehydration of fructose was enhanced in water-MIBK biphasic systems which were generally modified by DMSO or NMP solvents.

Cottier et al. [120] reported that D-fructose could be effectively transformed with the ion-exchange resin in aqueous solution where the conversion was high, whereas, the yield of 5-HMF was low (*ca.* 28%) in water. It was further shown that the efficiency may be increased by dilution. The possible reason for low yields in water is that there exists the hydronium species in the macropore of resins, which can promote 5-HMF to be further converted into byproducts such as levulinic acid and formic acid [121]. Many experimental investigations show that the use of DMSO as the solvent improves the selectivity of 5-HMF from D-fructose under common reaction conditions [115, 122–124]. The primary merit of DMSO is due to its dipolar aprotic character which efficiently prevents the conversion of 5-HMF to generate levulinic acid or humins. However, there are some disadvantages when DMSO is used, for instance, separation of DMSO, 5-HMF and water are difficult and solvent decomposition can possibly occur during reaction that would release poisonous sulfur-containing compounds.

The D-fructose dehydration by Amberlyst-15 with ionic liquids as solvent was studied by Lansalot-Matras and Moreau [125]. Two kinds of usual ionic liquids, 1-butyl-3-methyl imidazolium tetrafluoroborate ([BMIM]$BF_4$) and 1-butyl-3-methyl imidazolium hexafluoro phosphate ([BMIM]$PF_6$), were used and the dehydration was carried out in a micro batch reactor. The hydrophilic [BMIM]$BF_4$ was found to be superior to hydrophobic ([BMIM]$PF_6$) as solvent. As a result, up to 50% yield of 5-HMF was obtained in [BMIM]$BF_4$ solvent at 80 °C for 3 h. Moreover, with DMSO as co-solvent, the catalytic reaction in both ionic liquids proceeded more rapidly than that in DMSO alone so that a yield of 5-HMF of nearly 80% could be obtained.

In the research of Ilgen et al. [126] yields of 5-HMF reached 40% in the presence of Amberlyst-15 catalyst for the mixture of choline chloride (ChCl) and fructose. Moreover, the influence of acetone, DMSO, ethanol, methanol, ethyl acetate, and supercritical $CO_2$ as co-solvents of ionic liquids for the dehydration of fructose with Amberlyst-15 catalyst at room temperature was studied in detail [127].

Qi et al. [128] investigated fructose dehydration in the mixed acetone-water solvent where a 73.4% yield of 5-HMF at 94.0% fructose conversion was obtained using cationic exchange resin Dowex 50wx8–100 as catalyst at 150 °C with microwave heating. Recycle experiments indicated that the activity and selectivity of the resin remained practically unchanged after being used for five times. Moreover, they found that the production of 5-HMF was improved when acetone was used as co-solvent of DMSO for the fructose dehydration process [129].

Li et al. [130] reported that acidic ion-exchange resin and ionic liquid influenced the dehydration of fructose to 5-HMF, in which the macroporous strong-acid resin displayed better performance than a gel strong-acid resin when ionic liquid [BMIM]Cl was used as the solvent. As a result, a 93.0% yield of 5-HMF was obtained at 75 °C for 20 min with D001-cc resin catalyst in [BMIM]Cl. The macroporous resin had large pore diameters, many apertures, a large surface area and rapid ion exchange characteristics. The large surface area of resin provided many active sites for the

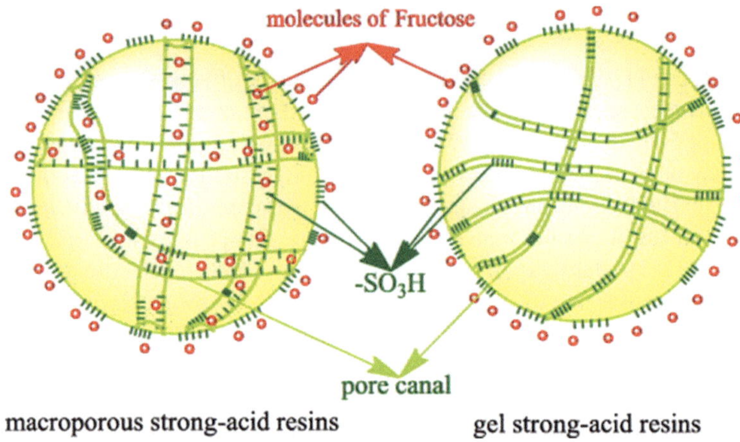


**Fig. 3.2** Schematic diagram of the interaction between the different resins and fructose (Reprinted from Ref. [130] with permission of Elsevier B. V. publishing corporation. 2014 Copyright © Elsevier)

dehydration of fructose; the large pore diameter and apertures increased diffusion of fructose into the inner of resins; the rapid ion exchange characteristics facilitated the interactions of fructose with acidic sites of resin. In contrast, the small surface area of gel resin reduced the total number of active sites. Moreover, the pore diameter of gel resin is relatively small which affects the diffusion of fructose as shown in Fig. 3.2.

Watanabe et al. [131] and Qi et al. [132] found that the synthesis of 5-HMF was favorable when the D-fructose dehydration was performed with microwave heating in the presence of $TiO_2$ and $ZrO_2$ catalyst. With $TiO_2$ as the catalyst, a 38.1% yield of 5-HMF with 83.6% conversion of D-fructose was achieved in 5 min at 200 °C while the yield of 5-HMF reached 30.5% at a 65% D-fructose conversion after 5 min with $ZrO_2$ as the catalyst. Shimizu et al. [133] studied the effect of water removal on the preparation of 5-HMF, and showed that 5-HMF yield could be improved in the dehydration with solid acids such as heteropoly acids, zeolite, or exchange resins when the reaction solution was handled through the evacuation at $0.97 \times 10^5$ Pa of pressure. In particular, the 5-HMF yield was close to 100% in the presence of about 0.15–0.053 mm size of Amberlyst-15 particle catalyst in DMSO solvent, which is attributed to the efficient removal of water removal from the small resin powder during reaction.

Fan et al. [134] reported on synthesis of 5-HMF from fructose using a solid heteropolyacid salt $Ag_3PW_{12}O_{40}$ as catalyst, on which fructose was selectively dehydrated to form 5-HMF with a yield of 77.7% and selectivity of 93.8% in 60 min at 120 °C. Phosphates of alkaline earth metals, $CaP_2O_6$ and a-$Sr(PO_3)_2$ were found to catalyze fructose dehydration in hot compressed water, in which 34–39% yields of 5-HMF were obtained [135]. Guo et al. [136] reported that lignin-derived solid acid catalyst efficiently catalyzed the dehydration of fructose, in which a 98% conversion of fructose and 84% yield of 5-HMF were obtained under microwave irradiation at 110 °C for 10 min. Lucas et al. [137] found that fructose could be efficiently dehydrated to produce 5-HMF with mesoporous AlSBA-15 catalysts under biphasic conditions. The catalyst with lower acid site density but medium to strong acid strength favors 5-HMF formation such that 5-HMF selectivity was as high as 88% at 59% conversion of fructose under the optimal conditions. Kruger et al. [138] studied the aqueous-phase dehydration of fructose to prepare 5-HMF with zeolite catalysts. About 1–2 wt.% of zeolite dissolved in the aqueous-phase, resulting in a solution of aluminosilicate species with Al being 1.2–26.2 mg/L and Si up to 226 mg/L. Active species dissolved from the zeolite are probably oligomeric aluminosilicate fragments in which the Al-O-Si and Si-O-Si connectivity is similar to that of amorphous aluminosilicates.

Alam et al. [139] investigated the catalytic activity of titanium hydrogen phosphate, a dual acid catalyst, for the conversion of fructose, in which a 55% yield of 5-HMF was obtained from fructose in a THF-$H_2O$ biphasic solvent. Xu [140] synthesized a series of ordered mesoporous zirconium oxophosphates with different P/Zr molar ratios *via* an evaporation-induced self-assembly method. The catalytic performance of zirconium oxophosphate with P/Zr of 0.75 was measured. Conversion (up to 97.4%) and selectivity (79.6%) of 5-HMF were obtained under relatively mild conditions in DMSO solvent. Different sulfated catalysts including $SO_4/TiO_2$-$SiO_2$, $SO_4$/Ti-SBA-15, $SO_4/ZrO_2$, $SO_4$/AC, and $SO_4/SiO_2$ were investigated in fructose dehydration to 5-HMF, in which 89% selectivity at 77% fructose conversion was obtained over the $SO_4/TiO_2$-$SiO_2$ catalyst owing to the highest amount of Brønsted sites on the catalyst [141]. Jain et al. [142] used mesoporous zirconium phosphate as catalyst for the conversion of fructose into 5-HMF, and *ca.* 80% yield of 5-HMF was obtained at 150 °C in 1 h using $H_2O$-diglyme mixture as solvent. Metal-containing silicoaluminophosphate molecular sieves (MeSAPO-5) with different Si/Al ratios were found to be effective in the dehydration of fructose. A 5-HMF yield of 73.9% was obtained over MeSAPO-5(0.25Si) catalyst at 170 °C in 2.0 h [143].

Our group [144] studied the production of 5-HMF from the fructose in the presence of heterogeneous Sn-based catalyst. A mixed $SnO_2$-$ZrO_2$ catalyst was prepared from zirconiumn-propoxide and different metal Sn precursors using a sol-gel method. The sulfated $SnO_2$-$ZrO_2$ ($SO_4^{2-}/SnO_2$-ZrO2) was obtained by the impregnation method using $H_2SO_4$ solution. When these materials were used in the dehydration of fructose, it was found that a suitable ratio of Sn/Zr was 0.5, and the catalytic activity of $SO_4^{2-}/SnO_2$-$ZrO_2$ was higher than that of $SnO_2$-$ZrO_2$ where more than 75.0% yield of 5-HMF was obtained in 2.5 h at 120 °C. Recycle experiments showed that the catalytic activity remained constant after being used five times. Furthermore, selective dehydration of fructose to 5-HMF using heterogeneous Ge (IV) catalysts was successful. Based on $GeO_2$ material, two catalyst systems are efficient for the dehydration process. In the presence of dual $GeO_2$-$Ge_3N_4$ catalyst, fructose is converted to 5-HMF in a 62% yield at 150 °C in 100 min. With sulfated $GeO_2$ ($SO_4^{2-}/GeO_2$) as catalyst, a 68% yield of 5-HMF from fructose is obtained in DMSO solvent [145]. The existence of $SO_4^{2-}$ contributes to the increase of more acid sites on the catalyst surface, leading to a high catalytic activity and high product yield.

Hafizi et al. [146] prepared well-ordered KIT-6 mesoporous silica *via* a sol-gel process and functionalized with sulfonic acid groups. Catalytic activity in the conversion of fructose to 5-HMF was measured. Results showed that 100% conversion and 84.1% yield of 5-HMF was obtained in DMSO at 165 °C in 30 min reaction time.

### 3.3.3 Metal-Containing Catalysts

Transition metal-promoted selective transformation of fructose has been studied since the 1960s [147]. Trapmann and Sethi [148] demonstrated the catalytic performance of thorium- and zirconium metals and showed that fructose is converted into 5-HMF with these transition metals in dilute solutions. Ishida and Seri [149] reported that glucose could be transformed to 5-HMF through promotion by lanthanide metals, and 5-HMF was further decomposed during the reaction. It was revealed that a special relationship exists between the activity and atomic weight of the lanthanide (III) ions, which are double arc-formed shapes with an extreme point of $Sm^{3+}$. This is quite useful to the development of novel catalyst. Further investigation showed that any of the lanthanide (III) ions ($La^{3+}$- $Lu^{3+}$) could accelerate effectively the conversion of various hexoses into 5-HMF in aqueous solutions at 140 °C in which no levulinic acid forms [150, 151]. Moreover, kinetic experiments revealed that generation of $La^{3+}$ complexes with hexose molecules is not relatively slow for dehydration. Therefore, it should not be the rate-determining step and is the subsequent process of a reactant-catalyst complex.

Metal-catalyzed transformation of hexoses to produce 5-HMF has had some exciting breakthroughs [152–154]. Zhao et al. [152] found that metal halides efficiently catalyze the dehydration of hexoses in the ionic liquid [EMIM]Cl where the $CrCl_2$ is highly active and *ca.* 70% yield of 5-HMF is obtained during the transformation of fructose and glucose. Yong et al. [153] revealed that *N*-heterocyclic carbene-Cr/ionic liquid has good catalytic performance for hexose dehydration. They found that about 96% yield of 5-HMF was obtained after extraction using diethyl ether. In addition, a considerable yield of 5-HMF could be achieved for high substrate loading and the catalyst was recyclable. Moreover, the same group further reported that D-fructose dehydration occurred with $WCl_6$ as catalyst in the [BMIM]Cl ionic liquid-THF biphasic solvent system. As a result, 5-HMF yield could reach 72% at 50 °C for 4 h reaction time [154].

Ilgen et al. [126] found that the D-fructose was effectively dehydrated to 5-HMF with $CrCl_3$ as catalyst in a mixture of choline chloride and 50 wt.% of D-fructose where *ca.* 60% yield could be obtained under suitable conditions. Our group [155] also investigated the synthesis of 5-HMF using different metal chlorides as catalysts in the THF-$H_2O$ biphasic system and found that antimony chloride ($SbCl_3$) was more effective than other metal chlorides for D-fructose dehydration in THF-$H_2O$

solvent, in which about 84.0% of yield of 5-HMF was obtained with $SbCl_3$-LiBr catalytic system at 120 °C in 1.5 h. The catalytic action of $SbCl_3$ is due to its suitable acidic character, and the addition of LiBr can improve the stability of 5-HMF during the entire reaction.

To achieve recycle of metal catalysts, polymer-supported NHC-metals have been prepared from chloromethyl polystyrene resin *via* two-step reaction. As a result, 1.6–16 mol % of metal components have been successfully loaded into total imidazolium, whereas the ionic liquid moiety in the imidazolium chloride salt remains almost unchanged [156]. When the polymer-supported NHC-metal catalysts were used for dehydration of fructose, about 70% yield of 5-HMF with high fructose conversion was attained in the presence of the polymer-supported NHC-$Fe^{III}$ catalyst.

### 3.3.4 Other Catalytic Systems

In the dehydration of D-fructose, some other catalytic systems have been employed to increase reaction efficiency. Mednick [89, 157] found that production of 5-HMF was achieved in the presence of ammonium phosphate, triethylamine phosphate or pyridinium phosphate catalyst. The yield of 5-HMF was as high as 44% when pyridinium phosphate was employed as the mediator. Fayet and Gelas [158] carried out the dehydration of D-fructose with pyridinium trifluoroacetate, hydrochloride, hydrobromide, perbromate or *p*-toluenesulfonate as catalyst, in which the 5-HMF yield could reach almost 70% at 120 °C after 30 min reaction time. Smith [159] found that ammonium sulfate or sulfite could be used as catalyst for the dehydration of D-fructose and reported that in the presence of $NH_4Al(SO_4)_2$ catalyst, a 5-HMF yield as high as 50% was obtained after 12 s at 170 °C [160]. Activated carbon as adsorbent showed a positive effect on the acid catalyzed fructose dehydration [161], in which both the yield and selectivity of 5-HMF could be increased during reaction.

Nowadays, research on catalytic processes is closely related to methods of green chemistry [162, 163]. On the stability, low vapor pressure and recyclability, ionic liquids have been considered to have many merits for their use as reaction solvents or as catalysts [164–166]. Ionic liquids have advantages as solvents in sugar dehydration with solid resins and metal chlorides [125, 153, 167]. The following discussion concentrates on the conversion of hexoses using ionic liquids as catalyst.

Catalytic activity of pyridinium-containing ionic liquid was first demonstrated for D-fructose dehydration [158]. Moreau et al. [168] found that D-fructose could be effectively and selectively dehydrated at 90 °C with 1-H-3-methyl imidazolium chloride ([HMIM]Cl) as solvent and catalyst, in which the yield of 5-HMF reached 92% for reaction times of 15 to 45 min. Bao et al. [169] investigated the catalytic performance of ionic liquids, 3-allyl-1-(4-sulfobutyl)imidazolium trifluoromethanesulfonate ([ASBI][Tf]), as well as its Lewis acid derivatives, 3-allyl-1-(4-sulfurylchloride butyl) imidazolium trifluoromethanesulfonate ([ASCBI][Tf]) on the dehydration of D-fructose and found that these type of ionic liquids have distinct effects on the dehydration process, namely, the Lewis acid ionic liquids are more efficient than Brønsted acid ionic liquids. Furthermore, the *N*-methyl-2-pyrrolidonium based ionic liquids were found to very active on the dehydration of D-fructose. Among the different ionic liquids, the *N*-methyl-2-pyrrolidonium methyl sulfonate ([NMP][$CH_3SO_3$]) and *N*-methyl-2-pyrrolidonium hydrogen sulfate ([NMP][$HSO_4$]) can efficiently catalyze the dehydration of D-fructose to produce 5-HMF with DMSO as the solvent under suitable conditions [170]. As a result, it is found that a 72.3% yield and 87.2% selectivity of 5-HMF was obtained in the presence of 7.5 mol% [NMP][$CH_3SO_3$] at 90 °C for 2 h reaction time.

The 1-methylimidazolium-based and N-methylmorpholinium-based ionic liquids have been employed as catalysts for dehydration of fructose. Yields of 74.8% and 47.5% of 5-HMF are obtained from D-fructose and sucrose, respectively, at 90 °C for 2 h under nitrogen atmosphere when N-methylmorpholinium methyl sulfonate ([NMM][$CH_3SO_3$]) is used as catalyst in an *N, N*-dimethylformamide-lithium bromide (DMF-LiBr) solvent system [171]. The 1-ethyl-3-methylimidazolium hydrogen sulfate ([EMIM][$HSO_4$]) is effective for converting fructose into 5-HMF. An 88% 5-HMF yield was obtained in 30 min using MIBK as a co-solvent [172]. In a biphasic system composed of choline chloride/citric acid ionic liquid and ethyl acetate, the yield of 5-HMF from fructose reached 91.4% with 93.6% 5-HMF selectivity at 80 °C in 1 h [173]. Kotadia et al. [174] further investigated the catalytic activities of symmetrical and unsymmetrical Brønsted acidic ionic liquids, and the results showed that the unsymmetrical [PSMBIM][$HSO_4$] exhibited high activity with 72.8% yield of 5-HMF in DMSO at 80 °C for 1 h.

Our group [175] reported that the selective conversion of fructose to 5-HMF was achieved with *N*-bromosuccinimide (NBS) as a promoter. In the presence of single NBS, a 64.2% yield of 5-HMF from fructose was obtained in *N*-methylpyrrolidone for 2 h. The 5-HMF yield could be enhanced to 79.6% and 82.3% when $FeCl_3$ and $SnCl_4$ were used as the additives, respectively.

Brønsted acidic ionic liquid 1-methyl-3-(3-sulfopropyl)-imidazolium hydrogen sulfate has been employed to catalyze the production of 5-HMF from highly concentrated solutions of D-fructose (40 wt.%), in which 73% yield of 5-HMF with 87% selectivity was obtained in 0.5 h at 150 °C [176]. Jackson et al. [177] synthesized heterogeneous sulfonic acid catalysts using an amine-catalyzed atomic layer deposition process where the performance of 3-(mercaptopropyl) trimethoxy-silane was tested in the dehydration of fructose. We further [178] reported that the selective conversion of D-fructose to 5-HMF was achieved in the presence of of graphite derivatives such as graphite oxide (GO), reduced graphite oxide (RGO) and sulfated reduced graphite oxide ($GSO_3H$). A 5-HMF yield of up to 60.8% was obtained in the presence of GO under suitable conditions. Mondal et al. [179] reported the dehydration of fructose using graphene oxide (GO) as acid catalyst and choline chloride (ChCl) as additive in ethyl lactate with a 76.3% molar yield. Zhang et al. [180] investigated the dehydration of D-fructose to 5-HMF in the presence of tin salt and hydrogen phosphate. It is found that the precipitate freshly formed from

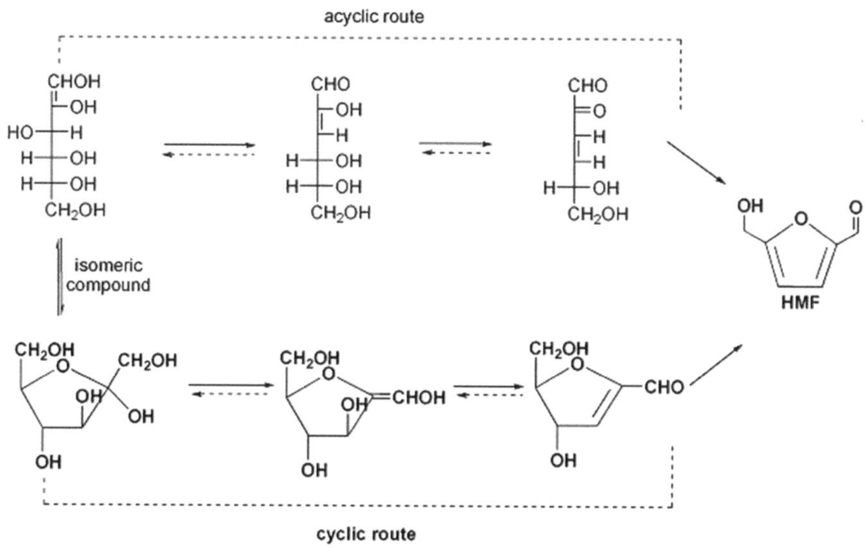


**Fig. 3.3** The possible mechanism for the dehydration of hexoses (Reprinted from Ref. [56] with permission of Elsevier B. V. publishing corporation. 2010 Copyright © Elsevier)

$SnCl_4$ and $(NH_4)_2HPO_4$ showed a good performance in a water-DMSO mixed solvent. The highest 5-HMF yield of 71% was achieved at 135 °C in 1 h. The tin valence number, the type of phosphate and the molar ratio of $Sn/PO_4$ all affect the 5-HMF yield in D-fructose dehydration.

Haworth and Jones [49] developed a reaction mechanism for the fructose dehydration to 5-HMF. Van Dam et al. [51], Antal et al. [53] and Kuster [55] supposed that the transformation of hexose was on the basis of two probable pathways: one depends on the conversion of ring compounds, while the another follows the acyclic molecules (Fig. 3.3). Generally, the possible reaction routes for the synthesis of 5-HMF from hexoses are made up of the isomerization, dehydration, fragmentation, reversion and condensation reactions. Some investigations confirmed that production of 5-HMF is *via* either open-chain 1, 2-enediol mechanism or based on fructofuranosyl compound [106, 181, 182]. However, Antal et al. [183] and Newth [184] suggested that the dehydration of fructose to form 5-HMF proceeded through intermediate ring structures. The proof given by them mainly consisted of: 1) simple transformation of 2, 5-anhydro-D-mannose (a formed molecular enol during cyclic mechanism) to 5-HMF; 2) easily available production of 5-HMF from fructose, but low production from glucose that is based on dehydration of sucrose; 3) the absence of carbon-deuterium occurrence in 5-HMF relying on the keto-enol tautomerism in open-chain mechanism if the reaction took place in $D_2O$ solvent. Amarasekara et al. [185] proposed two main intermediates as (4R, 5R)-4-hydroxy-5-hydroxymethyl-4, 5-dihydrofuran-2-carbaldehyde on the dehydration according to the information of $^{1}H$ and $^{13}C$ NMR measurements.

## 3.4 Catalytic Production of 5-Hydroxymethylfurfural (5-HMF) from Glucose

### 3.4.1 *Mineral Acids as Catalysts*

Comparing glucose with fructose, the former is cheaper and a more plentiful feedstock [55], but the lower dehydration rate and the poor selectivity to 5-HMF limit glucose as a starting material for 5-HMF production. The dehydration of glucose with $H_3PO_4$ catalyst only affords 15.5% yield of 5-HMF at 190 °C [186], which may be ascribed to the rigid tending aldose formation. A small quantity of open-chain substrates in this reaction system result in low enolization rate, which is the route for generating 5-HMF. Nevertheless, motivation for developing a process using glucose as feedstock to prepare high-value 5-HMF is very high.

The influences of solvents with mineral acids on the dehydration of glucose have been investigated [90]. With water as solvent, glucose dehydration had very low selectivity (ca. 6–11% for 5-HMF), while, if the aprotic polar DMSO was used as solvent, the 5-HMF yield was no more than 42% for a 3 wt.% concentration of substrate. Interestingly, specially designed biphasic solvent systems such as mixtures of DMSO and water, or a mixed system including MIBK and 2-butanol (70: 30) can efficiently elevate the yield of 5-HMF during the glucose dehydration. When 10 wt.% glucose was employed as substrate, the selectivity of 5-HMF was up to 53% in DMSO-water solvent mixtures, which is much higher than the selectivity of 11% in water [187].

Huang et al. [188] developed a novel and effective process for the synthesis of 5-HMF from glucose transformation. The used mediator included glucose isomerase and HCl and solvent in a mixture of water and butanol. It is found that 5-HMF yield reached 63.3% under mild conditions. Ståhlberg et al. [189] studied the boric acid-promoted dehydration of glucose to 5-HMF in imidazolium-based ionic liquids, where a yield of 42% was obtained. It was confirmed that the formation of 1:1 glucose-borate complexes facilitated the conversion pathway from glucose to fructose. Deuterium-labeling studies showed that the isomerization proceeded *via* an ene-diol reaction pathway, which is different from that of enzyme-mediated isomerization of glucose to fructose. Sulfuric acid has been used as catalyst for the transformation of glucose to produce 5-HMF in the γ-valerolactone solvent system; however, only 13% yield of 5-HMF was obtained at 130 °C in 1 h reaction time [100].

### 3.4.2 *Solid Acids as Catalysts*

Considering that solid acid catalyst probably accelerate glucose isomerization and fructose dehydration, Amberlyst-15 and Mg-Al hydrotalcite (HT) have been studied to promote the conversion of glucose and sucrose to 5-HMF [190]. Yields of 42–54%

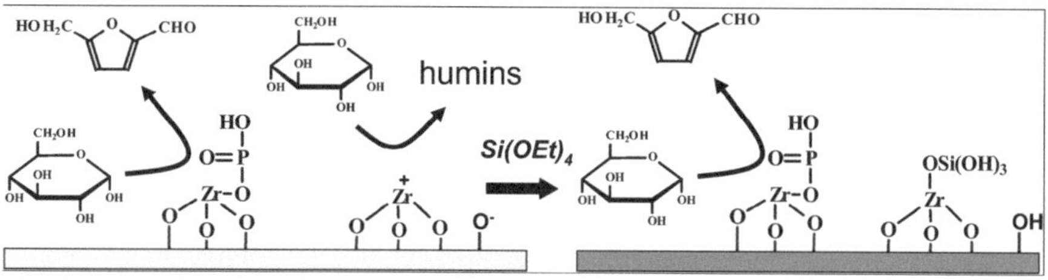


**Fig. 3.4** Glucose dehydration over parent ZrPO catalyst (*left*) and that after silylation (*right*) (Figure is adapted from Ref. [194] with permission of John Wiley and Sons Publishing Corporation. 2013 Copyright © WILEY-VCH Verlag GmbH & Co. KGaA, Weinheim)

of 5-HMF were obtained at 100 °C in DMF solvent. Furthermore, in the presence of $TiO_2$ and $ZrO_2$ catalysts, selective dehydration of glucose was also successfully carried out, where TPD detection showed that the quantity of basic sites plays an important role in glucose isomerization; however, the acidity density and basicity density are responsible for the production of 5-HMF during reaction [131]. The following investigations further showed molecular sieves such as HY zeolite, aluminum-pillared montmorillonite, MCM-20 or MCM-41 can also efficiently catalyze the transformation of glucose [191]; but, this route easily produces formic acid and 4-oxopentanoic acid. Hu's group [192] mentioned that the 5-HMF yield could reach 47.6% at 403 K for 4 h with $SO_4^{2-}/ZrO_2$-$Al_2O_3$ in a 1:1 Zr/Al molar ratio as the catalyst. They also showed that stronger acidity and appropriate basicity in the solid acid is more beneficial to catalytic transformation of glucose to 5-HMF under certain conditions.

Nikolla et al. [193] studied the combination of Sn-Beta with acid catalysts to accelerate the synthesis of 5-HMF from glucose in the biphasic $H_2O$/THF solvent system. A 79% conversion of glucose with 72% selectivity of 5-HMF was obtained under suitable conditions.

A foam-structured acid catalyst of a binderless zirconium phosphate (ZrPO) coating on aluminum foam was studied for dehydration of glucose to synthesize 5-HMF in a biphasic solvent system [194]. Based on extensive investigations of the ZrPO catalyst, it was shown that excessive Lewis acidity could lead to conversion of glucose to humins with a decrease of 5-HMF selectivity. Accordingly, 35% yield of 5-HMF at 70% conversion of glucose was obtained over silylated ZrP-f/Al(lc)-10 catalyst. The silylation procedure was found to be helpful in improving the 5-HMF selectivity over the foam catalyst, which was attributed to the deactivation of partial Lewis acid sites (Fig. 3.4).

Hara's group [195] found that the phosphate-immobilized anatase $TiO_2$ was efficient to catalyze 5-HMF production from glucose in THF/water (90/10 vol.%) solutions. An 81.2% yield of 5-HMF was obtained at 120 °C in 2 h reaction time, which was attributed to water-tolerant Lewis acid sites on phosphate/$TiO_2$ catalyst. They further explored the reaction mechanism for the formation of 5-HMF from glucose in water over $TiO_2$ and phosphate-immobilized $TiO_2$ using isotopically labeled molecules and $^{13}C$ NMR spectroscopy. They confirmed that stepwise dehydration of glucose forms 5-HMF over the $TiO_2$ catalysts, in which Lewis acid sites are effective for the dehydration of glucose and its derivatives [196]. Atanda et al. [197] investigated the catalytic performance of nanosized phosphated $TiO_2$ catalysts with different phosphate content on the conversion of glucose into 5-HMF, and discovered that 15 wt.% phosphate loading was desirable. Under optimal reaction conditions, a 97% glucose conversion and 81% 5-HMF yield was obtained in the water-butanol biphasic system. Jiménez-Morales et al. [198] employed aluminum doped MCM-41 silica as catalyst to effectively produce 5-HMF from glucose. By using a biphasic water/MIBK as reaction medium and a 30 wt.% of Al-MCM as catalyst, about 87% of glucose conversion and 36% of 5-HMF yield were obtained at 195 °C after 150 min. Furthermore, the conversion of glucose and yield of 5-HMF could be further enhanced to 98% and 63%, respectively, when a sodium chloride aqueous solution (20 wt.%) and MIBK were employed as reaction medium. It was also reported that for the conversion of glucose to 5-HMF on zirconium phosphate catalyst, a 63% yield of 5-HMF was obtained at 180 °C in 3 h in an $H_2O$-diglyme mixture [142]. Wang et al. [199] synthesized a series of cerous phosphates with different crystal structures and studied their catalytic performance on the dehydration of glucose to 5-HMF, in which a 97% conversion of glucose and 61% yield of 5-HMF were obtained using cerous phosphate with a hexagonal crystal structure as catalyst. It was shown that Lewis acidity played an important role in the dehydration of glucose to 5-HMF with cerous phosphate catalysts. Moreno-Recio et al. [200] investigated the catalytic dehydration of glucose to 5-HMF with ZSM-5 zeolites (H-, Fe- and Cu-ZSM-5). It is found that a glucose conversion of 80% and a 5-HMF yield of 42% were obtained using HZSM-5 as a catalyst at 195 °C after 30 min in a biphasic NaCl (20 wt.%) aqueous solution/MIBK system, which showed that the HZSM-5 catalyst was suitable Lewis/Brønsted molar ratio for conversion of glucose.

Kinetics of tandem glucose isomerization and fructose dehydration were determined by Swift et al. [201] who performed combined experimental and computational studies on catalytic production of 5-HMF from glucose in water. It is found that with a bifunctional zeolite HBEA-25, the octahedrally coordinated extra-framework Al site was effective to catalyze the isomerization of glucose to fructose during reaction. The activation energy of isomerization on H-BEA-25 is between that of Ti-BEA and Sn-BEA zeolite. The connection of 5-HMF formation rate versus ratio of Lewis to Brønsted acid sites has a volcano-shaped curve when the total number of acid sites over catalyst is fixed (e.g. aluminosilicate zeolites). On the other hand, when Lewis and Brønsted acid sites are varied independently (e.g., in Sn-BEA and HCl), the formation rate of 5-HMF first increased and then remained constant along with the elevation of Brønsted acid site density. These results are consistent with common features of tandem processes using multiple or multifunctional catalysts. Both the formation rate and yield of 5-HMF were at a maximum at an L/B ratio of 0.3, which is predicted to be close to 60% at 130 °C. It is considered that the degradation reaction of 5-HMF can be suppressed under certain conditions.

### 3.4.3 Metal-Containing Catalysts

Zhao et al. [152] studied the catalytic performance of metal halides for conversion of glucose in the ionic liquid [EMIM]Cl and found that $CrCl_2$ exhibited high efficiency for generation of 5-HMF, where nearly 70% yield of 5-HMF was obtained. Those authors proposed that the formed $CrCl_3^-$ anion was the active species in the reaction system that accelerated the process of glucose isomerization to produce fructose as reaction intermediate.

The *N*-heterocyclic carbene-Cr combined with an ionic liquid as catalyst for selective transformation of glucose to 5-HMF has been reported by Yong et al. [153]. The yield of 5-HMF was as high as 81% and its isolated yield is very high when the mixture is extracted by diethyl ether after reaction. Particularly, a large amount of glucose as substrate also exists in this system in which the catalyst is stable and recyclable for the catalytic dehydration reaction. Furthermore, it is found that $SnCl_4$ in the ionic liquid [EMIM]$BF_4$ could catalyze the conversion of glucose to 5-HMF under mild conditions [167]. Based on the investigation of the catalytic mechanism, the Sn atom forms a chelate with molecular glucose that creates a pentabasic cyclic intermediate that plays a crucial role in the efficient generation of 5-HMF during the reaction. Li et al. [202] showed that about 90% yield of 5-HMF from glucose could be achieved using $CrCl_3$ as catalyst in ionic liquids with the assistance of microwave.

Ståhlberg et al. [203] investigated the catalytic activities of $YbCl_3$ and $Yb(OTf)_3$ in the transformation of glucose. These two salts were very efficient for the synthesis of 5-HMF in alkylimidazolium chlorides and the promotion mechanism of lanthanides differs from that of the chromium compounds.

To realize the transformation at high concentrations of glucose, Ilgen et al. [126] employed choline chloride to dissolve more than 50 wt% of glucose and found that the 5-HMF yield still reached about 31% in the presence of $CrCl_3$ catalyst. Yuan et al. [204] studied the conversion of glucose to produce 5-HMF using $CrCl_2$ or $CrCl_3$ as the mediator in the DMSO and ionic liquids where some co-catalysts and solvents containing the halides were used. For these reaction systems, a 54.8% yield of 5-HMF was obtained in the presence of $CrCl_2$ and tetraethyl ammonium chloride at 120 °C for 1 h under a $N_2$ atmosphere. Bali et al. [205] proposed that Cr(III), instead of Cr(II), is the active species in the transformation of glucose to 5-HMF in ionic liquids, and the yield of 5-HMF can be high when a very weak ligand is possibly incorporated into the metallic Cr ion.

Our group [175] found that the conversion of glucose to 5-HMF occurred in the presence of NBS and $CrCl_3$, in which a 57.3% 5-HMF yield was obtained under optimal conditions. Moreover, we studied the conversion of glucose to 5-HMF using a combination of $SnCl_4$ and different quaternary ammonium salts [206] and found that tetrabutyl ammonium bromide was able to efficiently promote conversion of glucose to 5-HMF in the presence of $SnCl_4$. As a result, a 69.1% yield of 5-HMF was obtained with $SnCl_4$-tetrabutyl ammonium bromide ($SnCl_4$-TBAB) system in DMSO for 2 h reaction time at 100 °C in air.

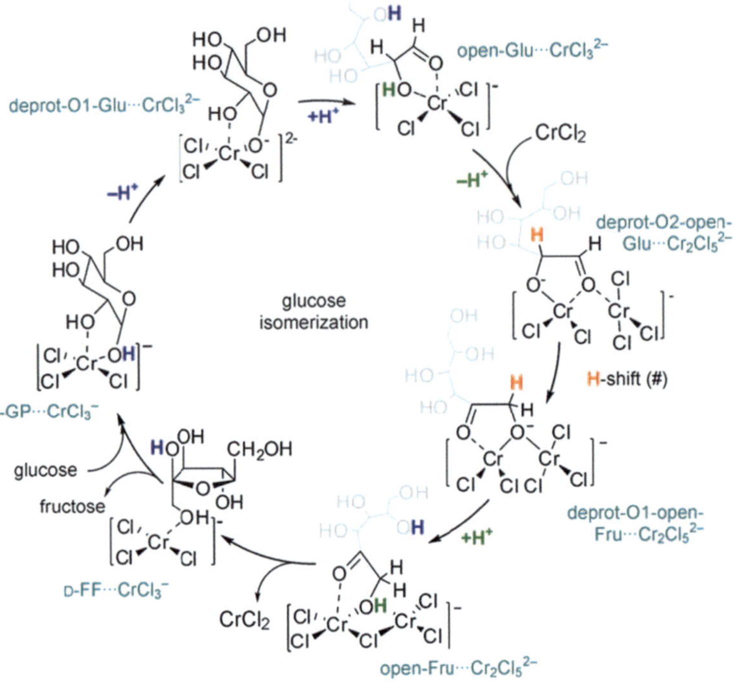


**Fig. 3.5** Catalytic cycle for isomerization of glucose in the presence of $CrCl_2$ in a model 1,3-dimethylimidazolium ionic-liquid medium. *d-FF* D-fructofuranose, *d-GP* D-glucopyranose (Adapted from Ref. [207] with permission of John Wiley and Sons Publishing Corporation. 2010 WILEY-VCH Verlag GmbH & Co. KGaA, Weinheim)

Extensive studies on the catalytic mechanism of the glucose dehydration in the presence of metal chlorides have been performed and show that mutarotation and isomerization reactions are essential to the transformation of glucose [152, 167]. The cyclic reaction line plays a key role in dehydration of glucose to 5-HMF based on molecular simulations [181]. Pidko et al. [207] reported that the coordination of a metallic Cr center happens in the opening and closure of the sugar ring when $CrCl_2$ is employed as catalyst. The rate-controlling H-shift reaction can be facilitated by transient self-organization of the Lewis acidic $Cr^{2+}$ centers into a binuclear complex with the open form of glucose, which is possible due to dynamic nature of the Cr complexes and to the presence of moderately basic functional groups in the ionic liquid. As shown in Fig. 3.5, the isomerization of glucose is involved with mono- and binuclear Cr complexes, in which the free-energy barrier is low for the generation of the open form of D-glucose during reaction. Moreover, the $O_2$-deprotonated glucose intermediate can be stabilized with the participation of a second Cr center.

Computational and experimental studies on the activation of glucose with Lewis acidic catalysts $CrCl_2$, $CuCl_2$ and $FeCl_2$ in dialkylimidazolium chloride ionic liquids have been performed [208]. The results show that there exists a coordination complex between the Lewis acidic $Cr^{2+}$ center and molecular glucose which directly affects transformation of glucose to fructose. It is further proven that formation of the intermediate compound is crucial to conversion of glucose. In case of $Cu^{2+}$, the direct coordination of glucose to the copper (II) chloride is difficult. However, molecular glucose can be deprotonated by the $Cl^-$ ligand of $CuCl_4^{2-}$ complex. During the period of the reaction, the $Cu^{2+}$ may be reduced to $Cu^+$ ions. On the other hand, both of these two pathways do not occur for the $FeCl_2$ catalyst.

### 3.4.4 Other Catalytic Systems

Based on a comprehensive study on dehydration catalysts, direct conversion of glucose to 5-HMF can be achieved by solely using acidic ionic liquid 1-butyl-3-methylimidazolium chloride [$C_4SO_3HMIM$]Cl as catalyst in a single-pot reaction [209]. The mechanism for conversion of glucose to 5-HMF catalyzed by the [$C_4SO_3HMIM$]Cl ionic liquid was studied using density functional theory (DFT) [210]. It was shown that the transformation can proceed *via* two feasible pathways where the cation of the ionic liquid plays a significant role, acting like a mobile proton carrier to accelerate the transformation. The chloride anions function with the reactant and acidic protons in the imidazolium ring *via* hydrogen-bonding, and supply a polar environment to stabilize the intermediate with assistance of the imidazolium cation. Furthermore, the overall barriers of the two potential pathways were calculated for catalytic transformation of glucose, and they are 32.9 kcal/mol and 31.0 kcal/mol, respectively.

Cao et al. [211] used an ammonium resin ($PBnNH_3Cl$) as a single catalyst for glucose dehydration to produce 5-HMF. Greater than >80% product selectivity was achieved in a DMSO or biphasic system. Further, they proposed the reaction pathway for glucose dehydration to 5-HMF by this ammonium resin catalyst (Scheme 3.4). They considered that an amine-catalyzed glucose isomerization *via* an enediol pathway and ammonium salt or acid (P-$BnNH_2$/HCl)-catalyzed fructose dehydration mechanism was feasible. The hydrogen bonds between N-H and O-H that exist in all intermediates and transition states should stabilize these compounds and facilitate the reaction. A sulfonic acid functionalized metal-organic framework MIL-101(Cr)-$SO_3H$ was used as the catalyst for the glucose dehydration to 5-HMF in γ-valerolactone : water (9 : 1) solvent [212]. Under optimal conditions, a 44.9% yield of 5-HMF with a 45.8% selectivity was obtained. Analysis of the reaction kinetics showed that the glucose isomerization in this catalytic system is second-order with an apparent activation energy of 100.9 kJ/mol. Moreover, the catalyst is highly stable and able to provide a steady 5-HMF yield in fixed-bed reactors.

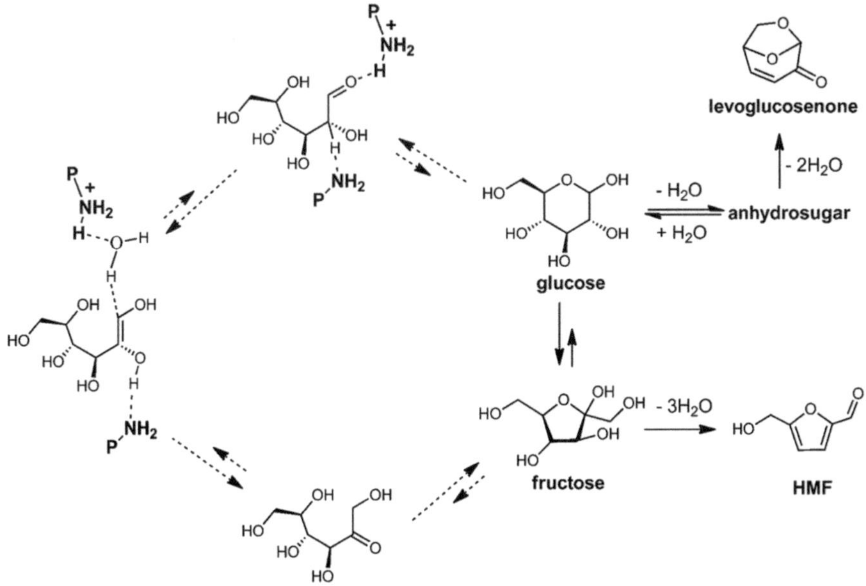


**Scheme 3.4** Reaction pathway for glucose dehydration in $PBnNH_3Cl$/DMSO catalytic system (Adapted from Ref. [211] with permission of Royal Society of Chemistry publishing corporation. 2015 Copyright © Royal Society of Chemistry)

## 3.5 Catalytic Production of 5-Hydroxymethylfurfural (5-HMF) from Polysaccharides

It is well-known that employing polysaccharides and cellulose as feedstocks for the direct production of 5-HMF is vital for sustainable processes. Chheda et al. [187] found that good selectivity for 5-HMF at high conversion from polysaccharides, such as sucrose, starch, cellobiose and xylan could be obtained using mineral acids (HCl, $H_2SO_4$ or $H_3PO_4$) as the catalyst, in a biphasic reaction phase consisting of water modified with DMSO and an organic extraction phase that was composed of a mixture of MIBK and 2-butanol (7:3, w/w).

Ilgen et al. [126] investigated the conversion of different polysaccharides in the choline chloride system with *p*-toluenesulfonic acid as the catalyst, where more than 50 wt% of substrate was used. Yields of 5-HMF reached 25% and 57% from the conversion of inulin and sucrose under the mild conditions. Moreover, the yields of 5-HMF from inulin and sucrose were 46% and 43%, respectively, when $CrCl_3$ catalyst was employed in the systems composed of choline chloride and more than 50% sugar. Yields of 27% and 54% of 5-HMF were obtained from sucrose and inulin using Amberlyst-15 solid acid catalyst in the mixture, respectively. Carlini et al. [108] reported that niobium phosphate had high activity in the transformation of sugars in aqueous medium, where the 5-HMF yields could reach 27% and 31% from solutions of 12.7% sucrose and 6.0% inulin, respectively. Chheda and Dumesic

[119] reported that a 69% or 43% yield of 5-HMF was obtained from inulin or sucrose with the Diaion® PK-216 resin as the catalyst in a biphasic system that was composed of water and NMP solvent (4: 6, w/w). For a continuous flow reactor, 61% product yield was obtained in the dehydration of sucrose to 5-(chloromethyl) furfural within 60 s in the biphasic system including HCl aqueous solution and organic solvent [104]. The 5-(chloromethyl) furfural can be further converted to 5-HMF through a simple hydrolysis process in the continuous flow reactor.

The conversion of disaccharides and polysaccharide to 5-HMF in imidazolium-based ionic liquids with boric acid as a promoter have been reported, in which 66% 5-HMF yield is obtained from sucrose and 33% 5-HMF yield is obtained from maltose or starch under suitable conditions [189]. Our group [175] found that the conversion of sucrose and inulin to 5-HMF occurred in the presence of NBS and $CrCl_3$, in which a 68.2% or 62.4% yield of 5-HMF from sucrose or inulin was achieved. Besides, we studied the selective production of 5-HMF from inulin with graphite oxide as the catalyst, in which a 58.2% of 5-HMF yield was obtained at 160 °C for 100 min in DMSO solvent [178]. A 61.2% of 5-HMF yield from inulin can be obtained in THF-$H_2O$ (3:1) mixture. Furthermore, 88.2% and 76.7% yields of 5-HMF from sucrose and maltose are possible using a combination of [bi-$C_3SO_3$HMIM][$CH_3SO_3$] and manganese chloride($MnCl_2$) as the catalyst in [BMIM] Cl, respectively [213]. About 38% yield of 5-HMF from sucrose was obtained in the presence of 5 mol/L $H_2SO_4$ and 10 mL $\gamma$-valerolactone at 130 °C [100].

Jain et al. [142] reported that in the conversion of glucose to produce 5-HMF with the zirconium phosphate catalyst, a 53% yield of 5-HMF was obtained at 180 °C for 2 h in the $H_2O$-diglyme mixture. Zhang et al. [180] found that a 45% yield of 5-HMF was attained from sucrose with $SnCl_4$ and $(NH_4)_2HPO_4$ as catalyst in water-DMSO mixture (35: 65, w/w) at 135 °C after 1 h.

## 3.6 Catalytic Production of 5-Hydroxymethylfurfural (5-HMF) from Biomass Feedstocks

Binder and Raines [214] reported that a combination of alkali metal halide and *N*, *N*-dimethylacetamide allows selective transformation of lignocellulosic biomass to 5-HMF. The obtained yields are comparable to those from the dehydration of fructose, glucose and cellulose with mineral acid catalysts. For example, 5-HMF yield reached 92% from lignocellulosic biomass in DMAc-KI reaction medium when a solution of 6.0 mol% $H_2SO_4$ was used as catalyst.

The selective transformation of lignocelluosic biomass to 5-HMF was studied by Zhang and Zhao [215], who found that 45–52% yields of 5-HMF are obtained from corn stalk, rice straw and pine wood under microwave in 3 min. In particular, the direct synthesis of 5-HMF from cellulose could be efficiently carried out with the catalytic system composed of $CuCl_2$ and $CrCl_2$ in [EMIM]Cl solvent in which 55.4 ± 4.0% yield of 5-HMF was obtained [216]. Zhang et al. [217] reported that the conversion reaches 89% with $CrCl_2$ using [EMIM]Cl and water as solvent. Kim et al. [218] carried out a one-pot transformation of cellulose into 5-HMF employing a combination of metal chlorides as catalysts with [EMIM]Cl solvent. Based on extensive screening of different metal chlorides, the combination of $CrCl_2$ and $RuCl_3$ was considered as the most effective catalyst. About 60% yield of 5-HMF was obtained from cellulose under certain conditions. Furthermore, this catalyst system was used to convert a lignocellulosic biomass feedstock Reed (*Phragmites communis*, Trin.) in 1-ethyl-3-methyl imidazolium chloride ([EMIM]Cl) solvent that gave yields of 5-HMF of *ca.* 41% using $CrCl_2/RuCl_3$ (4: 1) at 120 °C for 2 h in which the conversion of reed closely corresponded to the glucan content.

An efficient two-step process for converting microcrystalline cellulose into 5-HMF with the acidic resin and $CrCl_3$ in ionic liquids was demonstrated [219]. In the first step, high glucose yields of above 80% could be obtained from the hydrolysis of cellulose with the strong acidic cation exchange resin catalyst in ionic liquid [EMIM]Cl with gradual addition of water. In the second step, the used resin was separated from reaction mixture and $CrCl_3$ was added, which resulted in a 5-HMF yield of 73% based on cellulose substrate.

Wang' group [220] reported that the efficient transformation of microcrystalline cellulose to 5-HMF was achieved using the acidic ionic liquids and metal salts as the catalysts in the 1-ethyl-3-methylimidazolium acetate ([EMIM][Ac]) solvent. The results showed that 1-(4-sulfonic acid) butyl-3-methylimidazolium methyl sulfate ([$C_4SO_3$HMIM][$CH_3SO_3$]) and $CuCl_2$ were most efficient for transformation of cellulose, in which a 69.7% yield of 5-HMF was obtained. The same group [213] further studied the direct transformation of cellulose into 5-HMF in the [BMIM]Cl solvent using dual-core sulfonic acid ionic liquids as catalysts and metal salts as co-catalysts. The combination of [bi-$C_3SO_3$HMIM][$CH_3SO_3$] and manganese chloride ($MnCl_2$) was found to be the most effective catalyst in which 5-HMF was directly produced from cellulose in 66.5% yields. The catalyst could maintain its good performance for the cellulose conversion in this system even after being recycled for four runs. Moreover, in their further study on lignocellulosic biomass materials, the transformation of filter paper and reed gave a 40.28% and 32.62% yield of 5-HMF, respectively. The lowest yield was obtained from straw (without being activated) which was, nevertheless, as high as 28.54%.

Cai et al. [221] studied HY promoted hydrolysis of cellulose in ionic liquids and showed that 23.6% yield of 5-HMF could be obtained from cellulose in the presence of HY with the acid amount of 11.1 mol% by supplying water gradually during the period of reaction in [BMIM]Cl ionic liquid. Nandiwale et al. [222] systematically studied the catalytic synthesis of 5-HMF by hydrolysis of microcrystalline cellulose over bimodal-HZ-5 zeolite, in which bimodal-HZ-5 was prepared by post-synthesis modification of H-ZSM-5 with desilication. A 67% cellulose conversion and 46% yield of 5-HMF were obtained and the catalyst was found to be reusable for four cycles without loss of activity. Tan et al. [223] found a feasible way to operate chemical process using inexpensive chromium catalyst and low catalyst loading to convert cellulose to 5-HMF at mild temperatures (<120 °C) in [BMIM]Cl) solvent. Under optimal conditions, a 47.5% yield of 5-HMF was obtained in the ionic liquid/

zeolite/$CrCl_2$ system, where the zeolite with moderate acidity was used to promote cellulose hydrolysis and to slow down the decomposition process of 5-HMF in side-reactions.

Tao et al. [224] employed $CoSO_4$ as catalyst in 1-(4-sulfonic acid) butyl-3-methylimidazolium hydrogen sulfate ionic liquid for the hydrolysis of cellulose, affording an 84% conversion of cellulose at 150 °C after 300 min reaction time. In the presence of a catalytic amount of $CoSO_4$, the yields of 5-HMF and furfural were up to 24% and 17%, respectively. The $CoSO_4$-ionic liquid hydrolysis system exhibited favorable catalytic activity over five repeated runs in the ionic liquid. They further investigated the effects of ionic liquids and metal ions on the hydrolysis of microcrystalline cellulose [225] and found that 1-(4-sulfonic acid) butyl-3-methylimidazolium hydrogen sulfate showed the highest catalytic activity, and the promotion of $Cr^{3+}$, $Mn^{2+}$, $Fe^{3+}$, $Fe^{2+}$, $Co^{2+}$ were much better than that of other metal ions ($Zn^{2+}$, $Al^{3+}$, $Ni^{2+}$, $La^{3+}$, $Cu^{2+}$, $Ce^{3+}$, $Cs^{+}$, $Sn^{2+}$). Catalytic activities of metal nitrates were not as high as expected as those of chlorides and sulfates. Typically, using $MnCl_2$ as the co-catalyst, an 88.6% conversion of cellulose and 37.5% yield of 5-HMF was obtained at 150 °C and 2.5 MPa. Zhao et al. [226] reported on a novel method for the production of 5-HMF in high yields from cellulose using a Brønsted-Lewis-surfactant-combined heteropolyacid (HPA) $Cr[(DS)H_2PW_{12}O_{40}]_3$ as the catalyst. It is found that a 77.1% conversion and 52.7% yield of 5-HMF was obtained at 150 °C within 2 h, and that the micellar HPA catalyst could be recycled by a simple separation process. Abou-Yousef et al. [227] studied the conversion of cellulose to 5-HMF which was performed by using single or combined metal chloride catalysts in 1-ethyl-3-methylimidazolium chloride ([EMIM]Cl) ionic liquid. They found that $CrCl_3$ was the most effective single metal catalyst for selective conversion of cellulose into 5-HMF with 35.6% yield, and the $CrCl_3$/$CuCl_2$ was the most selective combination to convert cellulose into 5-HMF with a 39.9% yield. Dutta et al. [228] reported that the direct transformation of cellulose into 5-HMF occurred using single or combined metal chloride catalysts in the DMA-LiCl solvent system under microwave-assisted heating. Those authors showed that $Zr(O)Cl_2$/$CrCl_3$ combined catalyst was most effective as it enables about 57% yield of 5-HMF from cellulose fiber. Moreover, it is noteworthy that the combined $Zr(O)Cl_2$/$CrCl_3$ catalyst was also effective for transformation of sugarcane bagasse to 5-HMF, in which a 38% product yield was obtained under optimal conditions.

Sawdust has been directly converted into 5-HMF in 1,3-dimethyl-2-imidazolidinone (DMT) solvent using alkali halides with $Cr(NO_3)_3{\cdot}9H_2O$ as catalysts [229]. The alkali halides played an important role in the transformation of sawdust, in which a 28% or 25.5% yield of 5-HMF was obtained with the assistance of NaCl and KCl, respectively. Qu et al. [230] used ionic liquids as catalyst for the transformation of microcrystalline cellulose to 5-HMF under microwave irradiation in *N,N*-dimethylacetamide (DMAc) containing LiCl. They found that 1,1,3,3-tetramethylguanidine tetrafluoroborate ([TMG][$BF_4$]) showed high catalytic activity and a 5-HMF yield of 28.63% was achieved. Moreover, Lee et al. [231] provided a novel method to realize transformation of cellulose to 5-HMF, and a 46.1% yield was obtained by integrating a sequential enzyme cascade technique in an aqueous system with solid acid catalysis in an organic solvent system, in which solid acid was the sulfonic acid-functionalized mesoporous silica nanoparticles (MSN), and the used enzyme system contain cellulase-$Fe_3O_4$@MSN and isomerase-$Fe_3O_4$@MSN as catalysts.

The synthesis of 5-HMF from cellulose was investigated using phosphates as the catalyst, where a 31.2% product yield was obtained with subcritical water [232]. A 53% yield of 5-HMF from cellulose has been reported with $NaHSO_4$ and $ZnSO_4$ as the catalyst in a biphasic $H_2O$-THF solvent at 3.0 MPa pressure [233].

The acid ($-SO_3H$) impregnated low cost cellulose-derived carbonaceous catalyst (CCC), a new mesoporous material`, is highly effective for the production of 5-HMF from cellulose in which a 40.5% product yield has been obtained [234]. Numerous CCC mesoporous materials can be synthesized using reducing sugars as starting materials. Zhang et al. [235] studied the transformation of cellulose to 5-HMF catalyzed by the heteropolyacid $HOCH_2CH_2N(CH_3)_3)_xH_{3\text{-}x}PW_{12}O_{40}$ (abbreviated as $Ch_xH_{3-x}PW_{12}O_{40}$,x = 1, 2 and 3) catalysts in the methyl isobutyl ketone-$H_2O$ solvent. An outstanding 5-HMF yield of 75.0% was obtained using the $ChH_2PW_{12}O_{40}$ catalyst at 140 °C in 8 h reaction time, which can be attributed to the high Brønsted acidity and thermo-regulatable nature compared with the homogeneous $H_3PW_{12}O_{40}$ catalyst. Moreover, a 27.6% yield of 5-HMF was obtained from raw lignocellulosic biomass straw under similar conditions.

Based on above discussion, it is concluded that metal salts as the catalysts have favorable reactivity in ionic liquids which can be attributed to the cellulose unit being distorted during the dissolution stage. After the metal salts (as Lewis acid catalyst) interact with the dissolved cellulose, a higher surface area is formed that is helpful for depolymerization to glucose monomers. The following glucose isomerization should depend on the stability of the metal ion complex in dimers in ionic liquids. The performance of metal salts is related to the character of the Lewis acid as well as to the stabilization of the dimers in the complex stage [236]. Zhang et al. [237] reported on a new approach to synthesize 5-HMF through a tandem pathway in a molten hydrate solution which allows decrystallization, depolymerization and conversion of microcrystalline cellulose (MCC). When a 71.62 wt% $ZnCl_2$ aqueous solution was used as the reaction medium and methylisobutylketone (MIBK) as the extraction solvent, about 80.6% yield of 5-HMF was obtained with a 0.2 mol/L concentration of HCl as the catalyst under optimal conditions. The aqueous solution could be recycled without loss of catalytic activity.

An effective integrated process was developed for the transformation of cellulose, straw and barley husk into 5-HMF with organic salts as bi-functional organocatalysts in biphasic water/MIBK systems [238]. A 52% yield of 5-HMF at 83% conversion of cellulose was obtained with sulphanilic acid as catalyst at 150 °C within 60 minutes under metal-free conditions. As depicted in Scheme 3.5, the hydrolysis of cellulose to glucose and the dehydration of fructose to 5-HMF are both acid-catalyzed processes. The isomerization of glucose was preferable to be carried out in the presence of an amine-based catalyst, which is similar to the glucose isomerase enzyme in nature that uses an amino group. Thus, it was concluded that ammonium salts with an appropriate acidity/basicity balance could effectively

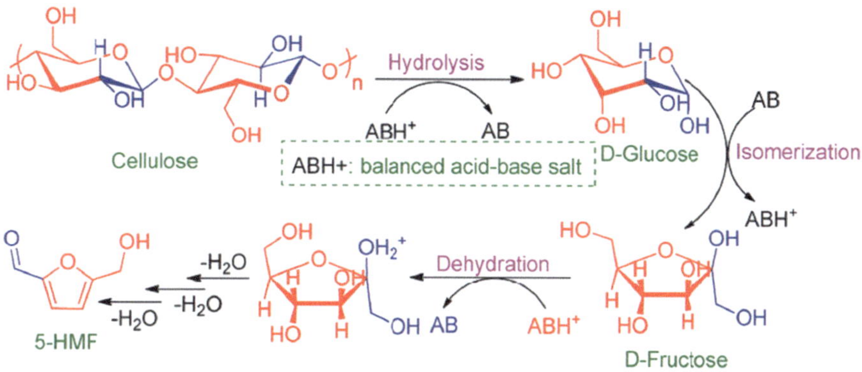


**Scheme 3.5** Proposed mechanism for cellulose conversion into 5-hydroxymethyl furfural using bifunctional acid-base salt catalyst (Adapted from Ref. [238] with permission of RSC publishing corporation. 2016 Copyright © Royal Society of Chemistry)

promote selective transformation of cellulose to generate 5-HMF in the presence of sulphanilic acid. The catalyst was further employed for the transformation of waste lignocellulosic biomass residues including the straw and barley husk. The experimental data showed that lignocellulosic materials were selectively converted to 5-HMF in approximately 41% yields. This catalyst and procedure provides a promising process for direct conversion of cellulose and raw biomass feedstocks to 5-HMF and furanic derivatives.

## 3.7 Conclusions and Future Outlook

In conclusion, 5-HMF is a useful and promising bio-based platform compound for the production of liquid fuels and highly valuable chemicals, and therefore can be regarded as a vital alternative to replace fossil-based energy resource in the future. Over the past decade, the synthesis of 5-HMF has primarily focused on the utilization of homogeneous mineral acids, metal halides and heterogonous solid acid catalysts in pure water or high boiling point solvents (DMSO, DMF, and ILs) alone or in modified high boiling point solvents mixed with aqueous media. Considerable improvement has been achieved in the conversion of fructose to 5-HMF, whereas the transformation of glucose, sucrose and cellulose are still relatively difficult. The main bottleneck for the conversion of glucose-based substrates to 5-HMF is the isomerization to fructose, which requires different conditions from the fructose dehydration reaction. Up until now, desirable methods rely on the two independent steps, in which several efficient catalysts composed of bases and acids or enzyme and acids have been developed and reported. To achieve large-scale industrial applications, further elevation of conversion and selectivity of 5-HMF for the catalytic transformation of biomass feedstocks and biomass-derived carbohydrates is essential. Thus, exploring the highly-active and selective catalysts and corresponding reactors to obtain high yields of 5-HMF from biomass feedstocks and biomass-derived sugars will be necessary.

Based on trends in green chemistry and sustainable production, the future orientation of catalytic technology will be focused on efficient, facile and environmentally benign preparation of 5-HMF. Especially, multifunctional heterogeneous catalysts that are synthesized by incorporation of solid acid or solid base catalysts with metal constituents needs to be developed. The application of the bifunctional metallic catalysts with both Brønsted acid and Lewis acid sites for the conversion of cellulose and hemicelluloses to furans *via* C6 and C5 sugars is very promising. Novel reactors for the application of the multifunctional catalysts need to be designed, which can be helpful to explore their economics. In addition, optimization of reaction conditions, use of mild condition for high substrate conversions that have high 5-HMF selectivity are desirable, in which these methods must be environmentally-friendly. The recyclability of heterogeneous catalysts and efficient separation and purity of the main products are also important topics to promote development in the catalytic field.

## References

1. Ragauskas AJ, Williams CK, Davison BH, Britovsek G, Cairney J, Eckert CA, Frederick WJ, Hallett JP, Leak DJ, Liotta CL, Mielenz JR, Murphy R, Templer R, Tschaplinski T. The path forward for biofuels and biomaterials. Science. 2006;311:484–9.
2. Huber GW, Iborra S, Corma A. Synthesis of transportation fuels from biomass: chemistry, catalysts, and engineering. Chem Rev. 2006;106:4044–98.
3. Luo C, Wang S, Liu H. Cellulose conversion into polyols catalyzed by reversibly formed acids and supported ruthenium clusters in hot water. Angew Chem Int Ed. 2007;46:7636–9.
4. Huber GW, Corma A. Synergies between bio- and oil refineries for the production of fuels from biomass. Angew Chem Int Ed. 2007;46:7184–201.
5. Christensen CH, Rass-Hansen J, Marsden CC, Taarning E, Egeblad K. The renewable chemicals industry. ChemSusChem. 2008;1:283–9.
6. Gallezot P. Catalytic conversion of biomass: challenges and issues. ChemSusChem. 2008;1:734–7.
7. Takagaki A, Tagusagawa C, Domen K. Glucose production from saccharides using layered transition metal oxide and exfoliated nanosheets as a water-tolerant solid acid catalyst. Chem Commun. 2008;42:5363–5.
8. Saha B, Abu-Omar MM. Advances in 5-hydroxymethylfurfural production from biomass in biphasic solvent. Green Chem. 2014;45:24–38.
9. Gilkey MJ, Xu B. Heterogeneous catalytic transfer hydrogenation as an effective pathway in biomass upgrading. ACS Catal. 2016;6(3):1420–36.
10. Pileidis FD, Titirici MM. Levulinic acid Biorefineries: new challenges for efficient utilization of biomass. ChemSusChem. 2016;47(21):562–82.
11. Lucia-Lucian A, Argyropoulos-Dimitris S, Adamopoulos L, Gaspar AR. Chemicals, materials, and energy from biomass: a review. ACS Symp. Ser. 2007;954:2–30.
12. Fan LT, Gharpuray MM, Lee YH. Cellulose hydrolysis. Berlin/Heidelberg: Springer; 1987. p. 198.
13. Kobayashi H, Fukuoka A. Synthesis and utilisation of sugar compounds derived from lignocellulosic biomass. Green Chem. 2013;15:1740–63.

14. Maki-Arvela P, Salmi T, Holmbom B, Willfor S, Murzin DY. Synthesis of sugars by hydrolysis of hemicelluloses-a review. Chem Rev. 2011;111(9):5638–66.
15. Mascal M, Nikitin EB. Direct, high-yield conversion of cellulose into biofuel. Angew Chem Int Ed. 2008;47:7924–6.
16. Agblevor FA, Mante O, Abdoulmoumine N, McClung R. Production of stable biomass pyrolysis oils using fractional catalytic pyrolysis. Energy Fuel. 2010;24(7):4087–9.
17. Carpenter D, Westover TL, Czernik S, Jablonski W. Biomass feedstocks for renewable fuel production: a review of the impacts of feedstock and pretreatment on the yield and product distribution of fast pyrolysis bio-oils and vapors. Green Chem. 2014;16(2):384–406.
18. Tong X, Chen H, Hu J, Bi Y, Sun Z, Fan W. The efficient and sustainable pyrolysis and gasification of biomass by catalytic processes. Chem Bio Eng Rev. 2015;2(3):157–74.
19. Mohan D, Singh P, Sarswat A, Steele PH, Pittman CU. Lead sorptive removal using magnetic and nonmagnetic fast pyrolysis energy cane biochars. J Colloid Interface Sci. 2015;448:238–50.
20. Zhou C, Xia X, Lin C, Tong D, Beltramini J. Catalytic conversion of lignocellulosic biomass to fine chemicals and fuels. Chem Soc Rev. 2011;40(11):5588–617.
21. Climent MJ, Corma A, Iborra S. Conversion of biomass platform molecules into fuel additives and liquid hydrocarbon fuels. Green Chem. 2014;16(2):516–47.
22. Moreau C, Belgasem MN, Gandini A. Recent catalytic advances in the chemistry of substituted furans from carbohydrate and in the ensuing polymers. Top Catal. 2004;27:11–30.
23. Rosatella AA, Simeonov SP, Frade RFM, Afonso CAM. 5-hydroxymethylfurfural(HMF) as a building block platform: biological properties, synthesis and synthetic applications. Green Chem. 2011;13(4):754–93.
24. Bhaumik P, Dhepe PL. Solid acid catalyzed synthesis of furans from carbohydrates. Catal Rev. 2015;58(1):36–112.
25. Bond JQ, Alonso DM, Wang D, West RM, Dumesic JA. Integrated catalytic conversion of gamma-valerolactone to liquid alkenes for transportation fuels. Science. 2010;327(5969):1110–4.
26. Chheda JN, Huber GW, Dumesic JA. Liquid-phase catalytic processing of biomass-derived oxygenated hydrocarbons to fuels and chemicals. Angew Chem Int Ed. 2007;38(50):7164–83.
27. Hayes DJ, Fitzpatrick S, Hayes MHB, Ross RH. Biorefineries-industrial processes and products. Weinheim Wiley-VCH. 2006;1:139–64.
28. Werpy T, Peterson G. Top value added chemicals from biomass. Pacific northwestern national laboratory, vol 1. 2004. available electronically at http://www.osti.gov/bridge
29. Wettstein SG, Alonso DM, Gürbüz EI, Dumesic JA. A roadmap for conversion of lignocellulosic biomass to chemicals and fuels. Chem Eng. 2012;1(3):218–24.
30. Lange JP. Catalysis for renewables: from feedstock to energy production. Wiley-VCH, Weinheim. 2007:21–51.
31. Lange JP. Lignocellulosic conversion: an introduction to chemistry, process and economics. Biofuels Bioprod Biorefin. 2007;1(1):39–48.
32. Bicker M, Hirth J, Vogel H. Dehydration of fructose to 5-hydroxymethylfurfural in sub- and supercritical acetone. Green Chem. 2003;5(2):280–4.
33. Roman-Leshkov Y, Chheda JN, Dumesic JA. Phase modifiers promote efficient production of hydroxymethylfurfural from fructose. Science. 2006;312(5782):1933–7.
34. Asghari FS, Yoshida H. Acid-catalyzed production of 5-hydroxymethyl furfural from d-fructose in subcritical water. Ind Eng Chem Res. 2005;45(45):2163–73.
35. Gallezot P. Catalytic routes from renewables to fine chemicals. Catal Today. 2007;121(121):76–91.
36. Gandini A, Belgacem MN. Furfural and furanic polymers. Lactualité Chimique. 2002;11:56–61.
37. Gandini A, Belgacem MN. Recent contributions to the preparation of polymers derived from renewable resources. J Polym Environ. 2002;10(3):105–14.

38. Vinke P, van Dam HE. New developments in selective oxidation. Elsevier. 1990;55:147–51.
39. Huber GW, Chheda JN, Barrett CJ, Dumesic JA. Production of liquid alkanes by aqueous-phase processing of biomass-derived carbohydrates. Science. 2005;308(5727):1446–50.
40. Dull G. Action of oxalic acid on inulin. Chem Zeit. 1895;19:216–7.
41. Kiermayer J. A derivative of furfuraldehyde from laevulose. Chemiker-Zeitung. 1895;19:1003–6.
42. Fenton HJH, Gostling M. The oxidation of polyhydric alcohols in presence of iron. J Chem Soc. 1899;75:1–11.
43. Fenton HJH, Gostling M. Derivatives of methylfurfural. J Chem Soc. 1901;79(79):807–16.
44. Fenton HJH, Robinson F. Homologues of furfuraldehyde. J Chem Soc. 1909;95:1334–40.
45. Van Enenstein WA, Blanksma JJ. Derivatives of furfural and of honey. Chem Weeklad. 1909;6:717.
46. Middendorp JA. Hydroxymethyl furfural. Rec trav chim. 1919;38(1):1–71.
47. Reichstein T. Notiz über 5-Oxymethyl-furfurol. Helv Chim Acta. 1926;9:1066–8.
48. Reichstein T, Zschokke H. Über 5-methyl-furfuryl-chlorid. Helv Chim Acta. 1932;15(1):249–53.
49. Haworth WN, Jones WGM. Some derivatives of glucosaccharic acid. J Chem Soc. 1944:65–7.
50. Montgomery R, Wiggins LF. The anhydrides of polyhydric alcohols; the constitution of dianhydro sorbitol. J Chem Soc. 1946;32(1–2):390–3.
51. Dam H, Kieboom A, Bekkum H. Alkaline degradation of monosaccharides part VIII. A $^{13}$C NMR spectroscopic study. Starch-Stärke. 1986;38:95–101.
52. Cottier L, Descotes G, Neyret C, Nigay H. Pyrolyse de sucres. Analyse des vapeurs de caramels industriels Industries Aliment Agricol. 1989:567–70.
53. Antal MJ, Mok WSL, Richards GN. Mechanism of formation of 5-(hydroxymethyl)-2-furaldehyde from D-fructose an sucrose. Carbohydr Res. 1990;199(1):91–109.
54. Lewkowski J. Synthesis, chemistry and applications of 5-hydroxymethyl-furfural and its derivatives. ARKIVOC. 2003;34(2):17–54.
55. Tong X, Ma Y, Li Y. Biomass into chemicals: conversion of sugars to furan derivatives by catalytic processes. Appl Catal A Gen. 2010;385(1–2):1–13.
56. Kröger M, Prube U, Vorlop KD. A new approach for the production of 2,5-furandicarboxylic acid by in situ oxidation of 5-hydroxymethylfurfural starting from fructose. Top in Catal. 2000;13(3):237–42.
57. Grushin V, Young RJ, Halliday GA. One-pot, two-step, practical catalytic synthesis of 2,5-diformylfuran from fructose. Org Let. 2003;5(11):2003–5.
58. Merat N, Verdeguer P, Rigal L, Gaset A. Process for the manufacture of furan-2,5-dicarboxylic acid. FR. 1992;2:669,634.
59. Costantin JM, Humphreys TW, Lange HB. Method for preventing the growth of fungi in leathers, paints, foods and fabrics. US3, 080, 279. 1963.
60. Hartzler JD, Morgan PW. Liquid crystalline solutions from polyhydrazides in aqueous organic bases. Contemporary Topics in Polymer Science. 1977:19–53.
61. Chundury D, Szmant HH. Preparation of polymeric building blocks from 5-hydroxymethyl- and 5-chloromethylfurfuraldehyde. Ind Eng Chem Prod Res Dev. 1981;20(1):158–63.
62. Hui Z, Gandini A. Polymeric schiff bases bearing furan moieties. Eur Polym J. 1992;28(12):1461–9.
63. Meàlares C, Gandini A. Polymeric schiff bases bearing furan moieties 2. Polyazines and polyazomethines. Polym Int. 1996;40(1):33–9.
64. Gandini A. Polymers from renewable resources: a challenge for the future of macromolecular materials. Macromolecules. 2008;41(24):37–59.
65. Gandini A, Belgacem NM. Recent advances in the elaboration of polymeric materials derived from biomass components. Polym Int. 1998;47(3):267–76.
66. Baumgarten M, Tyutyulkov N. Nonclassical conducting polymers: new approaches to organic metals? Chem Eur J. 1998;4(6):987–9.

67. Benahmed-Gasmi AS, Frere p, Jubault M, Gorgues J, Cousseau J, Garrigues B. 2,5-bis(1,4-dithiafulven-6-yl) substituted furans, thiophenes and N-methyl pyrroles as precursors for organic metals. Syn met. 1993;56(1):1751–1755.
68. Hopkins KT, Wilson WD, Bender BC, McCurdy DR, Hall JE, Tidwell RR, Kumar A, Bajia M, Boykin DW. Extended aromatic furan amidino derivatives as anti-pneumocystis carinii agents. J Med Chem. 1998;41(20):3872–8.
69. Del Poeta M, Schell WA, Dykstra CC, Jones SK, Tidwell RR, Kumar A, Boykin DW, Perfect JR. Structure-in vitro activity relationships of pentamidine analogues and dication-substituted bis-benzimidazoles as new antifungal agents. Antimicrob Agents Chemother. 1998;42(10):2495–502.
70. Richter DT, Lash TD. Oxidation with dilute aqueous ferric chloride solutions greatly improves yields in the '4+1' synthesis of sapphyrins. Tetrahedron Lett. 1999;40(40):6735–8.
71. Howarth OW, Morgan GG, McKee V, Nelson J. Conformational choice in disilver cryptates; an $^{1}H$ NMR and structural study. J Chem Soc Dalton Trans. 1999;12(12):2097–102.
72. Sheibley DW, Manzo MA, Gonzalez-Sanabria OD. Cross-linked polyvinyl alcohol films as alkaline battery separators. J Electrochem Soc. 1983;127(8):255–9.
73. Nisbet HB. The blending octane numbers of 2, 5-dimethylfuran. J Inst Petrol. 1946;32:162–6.
74. Roman-Leshkov Y, Barrett CJ, Liu ZY, Dumesic JA. Production of dimethylfuran for liquid fuels from biomass-derived carbohydrates. Nature. 2007;447(7147):982–5.
75. Dedsuksophon W, Faungnawakij K, Champreda V, Laosiripojana NN. Hydrolysis/dehydration/aldol-condensation/hydrogenation of lignocellulosic biomass and biomass-derived carbohydrates in the presence of $Pd/WO_3$-$ZrO_2$ in a single reactor. Bioresour Technol. 2011;102(2):2040–6.
76. Buntara T, Noel S, Phua PH, Melian-Cabrera I, Vries JGD, Heeres HJ. Caprolactam from renewable resources: catalytic conversion of 5-hydroxymethylfurfural into caprolactone. Angew Chem. 2011;50(31):7083–7.
77. Jaya T, Hemant C, Shun N, Kohki E. Direct synthesis of 1,6-hexanediol from HMF over a heterogeneous Pd/ZrP catalyst using formic acid as hydrogen source. ChemSusChem. 2014;7(1):96–100.
78. Ya'aini N, Amin NAS, Asmadi M. Optimization of levulinic acid from lignocellulosic biomass using a new hybrid catalyst. Bioresour Technol. 2012;116(4):58–65.
79. Chen H, Yu B, Jin S. Production of levulinic acid from steam exploded rice straw via solid superacid, $S_2O_8^{2-}/ZrO_2$-$SiO_2$-$Sm_2O_3$. Bioresour Technol. 2011;102(3):3568–70.
80. Balakrishnan M, Sacia ER, Bell AT. Etherification and reductive etherification of 5-(hydroxymethyl)furfural: 5-(alkoxymethyl)furfurals and 2,5-bis(alkoxymethyl)furans as potential biodiesel candidates. Green Chem. 2012;14(6):1626–34.
81. Demolis A, Essayem N, Rataboul F. Synthesis and applications of alkyl levulinates. ACS Sustain Chem Eng. 2014;2(6):1338–52.
82. Kuster BMF. 5-hydroxymethylfurfural (HMF). A review focussing on its manufacture. Starch/Stärke. 1990;42(8):314–21.
83. Moye CJ. 5-hydroxymethylfurfural. Rev Pure Appl Chem. 1964;14:161–70.
84. Feather MS, Harris JF. Dehydration reactions of carbohydrates. Adv Carbohydr Chem. 1973;28:161–224.
85. Harris DW, Feather MS. Intramolecular carbon-2.Far. Carbon-1 hydrogen transfer reactions during the conversion of aldoses to 2-furaldehydes. J Org Chem. 1974;39(5):724–5.
86. Kuster BFM, Baan HSVD. The influence of the initial and catalyst concentrations on the dehydration of D -fructose. Carbohydr Res. 1977;54(2):165–76.
87. Moye CJ, Goldsack RJ. Reaction of ketohexoses with acid in certain non-aqueous sugar solvents. J Appl Chem. 1966;16(7):206–8.
88. Moye CJ, Krzerninski ZS. The formation of 5-hydroxymethylfurfural from hexoses. Aust J Chem. 1963;16:258–69.

89. Mendnick ML. The acid-base-catalyzed conversion of aldohexose into 5-(hydroxymethyl)-2-furfural. J Org Chem. 1962;27:398–403.
90. Szmant HH, Chundury DD. The preparation of 5-hydroxymethylfurfuraldehyde from high fructose corn syrup and other carbohydrates. J Chem Technol Biotechnol. 1981;31(1):135–45.
91. Jow J, Rorrer GL, Hawley MC. Dehydration of d-fructose to levulinic acid over LZY zeolite catalyst. Biomass. 1987;14(3):185–94.
92. Chen J, Kuster BFM, Wiele KVD. Preparation of 5-hydroxymethylfurfural *via* fructose acetonides in ethylene glycol dimethyl ether. Biomass Bioenergy. 1991;1(4):217–23.
93. Antal MJ, Mok WS. A study of the acid catalyzed dehydration of fructose in near-critical water. Res Thermochem Biomass Conv. 1988;464–72.
94. Román-Leshkov Y, Dumesic JA. Solvent effects on fructose dehydration to 5-hydroxymethylfurfural in biphasic systems saturated with inorganic salts. Top Catal. 2009;52(3):297–303.
95. Tuercke T, Panic S, Loebbecke S. Microreactor process for the optimized synthesis of 5-hydroxymethylfurfural: a promising building block obtained by catalytic dehydration of fructose. Chem Eng Technol. 2009;32(11):1815–22.
96. Hansen TS, Woodley JM, Riisager A. Efficient microwave-assisted synthesis of 5-hydroxymethylfurfural from concentrated aqueous fructose. Carbohydr Res. 2009;344(18):2568–72.
97. Pawar H, Lali A. Microwave assisted organocatalytic synthesis of 5-hydroxymethyl-furfural in a monophasic green solvent system. RSC Adv. 2014;4(15):26714–20.
98. Esmaeili N, Zohuriaan-Mehr MJ, Bouhendi H, Baghlam-Marandi G. HMF synthesis in aqueous and organic media under ultrasonication, microwave irradiation and conventional heating. Korean J Chem Eng. 2016;33(6):1964–70.
99. Caes BR, Raines RT. Conversion of fructose into 5-(hydroxymethyl)furfural in sulfolane. ChemSusChem. 2011;4(3):353–6.
100. Qi L, Mui YF, Lo SW, Lo SW, Horváth HT. Catalytic conversion of fructose, glucose, and sucrose to 5-(hydroxymethyl)furfural and levulinic and formic acids in γ-valerolactone as a green solvent. ACS Catal. 2014;4(5):1470–7.
101. Swift TD, Bagia C, Choudhary V, Peklaris G, Nikolalds V, Vlachos DG, Nikolalds V. Kinetics of homogeneous Brønsted acid catalyzed fructose dehydration and 5-fydroxymethyl furfural rehydration: a combined experimental and computational study. ACS Catal. 2014;4:259–67.
102. Jiang N, Qi W, Huang R, Wang M, Su R, He Z. Production enhancement of 5-hydroxymethyl furfural from fructose via mechanical stirring control and high-fructose solution addition. Chem Technol Biotechnol. 2014;89:56–64.
103. Pedersen AT, Ringborg R, Grotkjar T, Pedersen S, Woodley JM. Synthesis of 5-hydroxymethylfurfural (HMF) by acid catalyzed dehydration of glucose-fructose mixtures. Chem Eng J. 2015;273:455–64.
104. Brasholz M, Von Känel K, Hornung CH, Saubern S, Tsanaktsidis J. Highly efficient dehydration of carbohydrates to 5-(chloromethyl)furfural (CMF), 5-(hydroxymethyl)furfural (HMF) and levulinic acid by biphasic continuous flow processing. Green Chem. 2011;13:1114–7.
105. Moreau C, Durand R, Pourcheron C, Razigade S. Preparation of 5-hydroxymethylfurfural from fructose and precursors over H-form zeolites. Ind Crop Prod. 1994;3:85–90.
106. Moreau C, Durand R, Razigade S, Duhamet J, Faugeras P, Rivalier P, Ros P, Avignon G. Dehydration of fructose to 5-hydroxymethylfurfural over H-mordenites. Appl Catal A Gen. 1996;145:211–24.
107. Carlini C, Patrono P, Galletti AMR, Sbrana G. Heterogeneous catalysts based on vanadyl phosphate for fructose dehydration to 5-hydroxymethyl-2-furaldehyde. Appl Catal A Gen. 2004;275:111–8.
108. Carlini C, Giuttari M, Galletti AMR, Busca G. Selective saccharides dehydration to 5-hydroxymethyl-2-furaldehyde by heterogeneous niobium catalysts. Appl Catal A Gen. 1999;183:295–302.

109. Armaroli T, Busca G, Carlini C, Giuttari M, Galletti AMR, Sbrana G. Acid sites characterization of niobium phosphate catalysts and their activity in fructose dehydration to 5-hydroxymethyl-2-furaldehyde. J Mol Catal A Chem. 2000;151:233–43.
110. Carniti P, Gervasini A, Biella S, Auroux A. Niobic acid and niobium phosphate as highly acidic viable catalysts in aqueous medium: fructose dehydration reaction. Catal Today. 2006;118:373–8.
111. Benvenuti F, Carlini C, Patrono P, Galli P. Heterogeneous zirconium and titanium catalysts for the selective synthesis of 5-hydroxymethyl-2-furaldehyde from carbohydrates. Appl Catal A Gen. 2000;193:147–53.
112. Nakamura Y. Preparation of 5-hydroxymethylfurfural. II-continuous dehydration of d-fructose Noguchi Kenkyusho Jiho. 1981;24:42–9.
113. Mercadier D, Rigal L, Gaset A, Gorrichon JP. Synthesis of 5-hydroxymethyl-2-furancarboxaldehyde catalysed by cationic exchange resins. Part 1. Choice of the catalyst and the characteristics of the reaction medium. J Chem Tech Biotechnol. 1981;31:489–96.
114. Rigal L, Gorrichon JP, Gaset A. Optimization of the conversion of d-fructose to 5-hydroxym ethyl-2-furancarboxaldehyde in a water-solvent-ion exchanger triphasic system- part I. Investigation of the main effects of the major parameters and of their interactions on the reaction. Biomass. 1985;7:27–45.
115. Nakamura Y, Morikawa S. The dehydration of D-fructose to 5-hydroxymethyl-2- furaldehyde. Bull Chem Soc Jpn. 1980;53:3705–6.
116. Mercadier D, Rigal L, Gaset A, Gorrichon JP. Synthesis of 5-hydroxymethyl-2-furancarboxaldehyde catalysed by cationic exchange resins. Part 3. Kinetic approach of the D-fructose dehydration. J Chem Technol Biotechnol. 1981;31:503–8.
117. Flèche G, Gaset A, Gorrichon JP, Truchot E, Sicard P. Process for manufacturing 5-hydroxymethylfurfural. FR. 1982;2(464):260.
118. El Hajj T, MasRoua A, Martin JC, Descotes G. Synthesis of 5-hydroxymethylfuran-2- carboxaldehyde and its derivatives by acidic treatment of sugars on ion-exchange resin. Bull Soc Chim Fr. 1987;5:855–60.
119. Chheda JN, Dumesic JA. An overview of dehydration, aldol-condensation and hydrogenation processes for production of liquid alkanes from biomass-derived carbohydrates. Catal Today. 2007;123:59–70.
120. Cottier L, Descotes G, Neyret C, Nigay H. FR. 1990;9:008,065.
121. Moreau C. Zeolites and related materials for the food and non food transformation of carbohydrates. Agro-Food-Industry Hi-Tech. 2002;13:17–26.
122. Gaset A, Rigal L, Paillassa G, Salome JP, Fleche G. Procede de fabrication du 5-hydroxymethylfurfural. FR. 1985;2(551):754.
123. Musau RM, Munavu RM. The preparation of 5-hydroxymethyl-2-furaldehyde (HMF) from d-fructose in the presence of DMSO. Biomass. 1987;13:67–74.
124. Bazoa CM, Franck R, Rigal L, Gaset A. Process for the manufacture of high purity hydroxymethylfurfural (HMF). FR. 1992;2(669):635.
125. Lansalot-Matras C, Moreau C. Dehydration of fructose into 5-hydroxymethylfurfural in the presence of ionic liquids. Catal Commun. 2003;4:517–20.
126. Ilgen F, Ott D, Kralisch D, Reil C, Palmberger A, Konig B. Conversion of carbohydrates into 5-hydroxymethylfurfural in highly concentrated low melting mixtures. Green Chem. 2009;11:1948–54.
127. Qi XH, Watanabe M, Aida TM, Smith RL. Efficient catalytic conversion of fructose into 5-hydroxymethylfurfural in ionic liquids at room temperature. ChemSusChem. 2009;2:944–6.
128. Qi X, Watanabe M, Aida TM, Smith RL. Catalytic dehydration of fructose into 5-hydroxymethylfurfural by ion-exchange resin in mixed-aqueous system by microwave heating. Green Chem. 2008;10:799–805.
129. Qi X, Watanabe M, Aida TM, Smith RL. Selective conversion of D-fructose to 5-hydroxymethylfurfural by ion-exchange resin in acetone/dimethyl sulfoxide solvent mixtures. Ind Eng Chem Res. 2008;47:9234–9.

130. Li Y, Liu H, Song C, Gu X, Li H, Zhu W, Yin S, Han C. The dehydration of fructose to 5-hydroxymethylfurfural efficiently catalyzed by acidic ion-exchange resin in ionic liquid. Bioresour Technol. 2013;133:347–53.
131. Watanabe M, Aizawa Y, Iida T, Inomata H. Catalytic glucose and fructose conversions with $TiO_2$ and $ZrO_2$ in water at 473 K: relationship between reactivity and acid–base property determined by TPD measurement. Appl Catal A Gen. 2005;295:150–6.
132. Qi X, Watanabe M, Aida TM, Smith RL. Catalytical conversion of fructose and glucose into 5-hydroxymethylfurfural in hot compressed water by microwave heating. Catal Commun. 2008;9:2244–9.
133. Shimizu K, Uozumi R, Satsuma A. Enhanced production of hydroxymethylfurfural from fructose with solid acid catalysts by simple water removal methods. A Catal Commun. 10:1849–53.
134. Fan C, Guan H, Zhang H, Wang X. Conversion of fructose and glucose into 5-hydroxymethylfurfural catalyzed by a solid heteropolyacid salt. Biomass Bioenergy. 2011;35:2659–65.
135. Daorattanachai P, Khemthong P, Viriya-empikul N, Faungnawakij K. Conversion of fructose, glucose, and cellulose to 5-hydroxymethylfurfural by alkaline earth phosphate catalysts in hot compressed water. Carbohydr Res. 2012;363:58–61.
136. Guo F, Fang Z, Zhou T. Conversion of fructose and glucose into 5-hydroxymethyl furfural with lignin-derived carbonaceous catalyst under microwave irradiation in dimethyl sulfoxide–ionic liquid mixtures. Bioresour Technol. 2012;112:313–8.
137. Lucas N, Kokate G, Nagpure A, Chilukuri S. Dehydration of fructose to 5-hydroxymethyl furfural over ordered AlSBA-15 catalysts. Microporous Mesoporous Mater. 2013;181:38–46.
138. Kruger JS, Nikolakis V, Vlachos DG. Aqueous-phase fructose dehydration using Brønsted acid zeolites: catalytic activity of dissolved aluminosilicate species. Applied Catalysis A General. 2014;469:116–23.
139. Alama MI, De S, Singh B, Saha B, Abu-Omar MM. Titanium hydrogenphosphate: an efficient dual acidic catalyst for 5-hydroxymethylfurfural (HMF) production. Appl Catal A General. 2014;486:42–8.
140. Xu H, Miao Z, Zhao H, Yang J, Zhao J, Song H, Liang N, Chou L. Dehydration of fructose into 5-hydroxymethylfurfural by high stable ordered mesoporous zirconium phosphate. Fuel. 2015;145:234–40.
141. Kılıç E, Yılmaz S. Fructose dehydration to 5-hydroxymethylfurfural over sulfated $TiO_2$-$SiO_2$, Ti-SBA-15, $ZrO_2$, $SiO_2$, and activated carbon catalysts. Ind Eng Chem Res. 2015;54:5220–5.
142. Jain A, Shore AM, Jonnalagadda SC, Ramanujachary KV, Mugweru A. Conversion of fructose, glucose and sucrose to 5-hydroxymethyl-2-furfural over mesoporous zirconium phosphate catalyst. Appl Catal A Gen. 2015;489:72–6.
143. Shao H, Chen J, Zhong J, Wang J. Development of MeSAPO-5 molecular sieves from attapulgite for dehydration of carbohydrates. Ind Eng Chem Res. 2015;54:1470–7.
144. Wang Y, Tong X, Yan Y, Xue S, Zhang Y. Efficient and selective conversion of hexose to 5-hydroxymethylfurfural with tin-zirconium-containing heterogeneous catalysts. Catal Commun. 2014;50:38–43.
145. Tong X, Wang Y, Nie G, Yan Y. Selective dehydration of fructose and sucrose to 5-hydroxymethyl-2-furfural with heterogeneous Ge (IV) catalysts. Environ Prog Sustain Energy. 2015;34:207–10.
146. Hafizi H, Chermahini AN, Saraji M. The catalytic conversion of fructose into 5-hydroxymethylfurfural over acid-functionalized KIT-6, an ordered mesoporous silica. Chem Eng J. 2016;294:380–8.
147. Jones RE, Lange HB. Conversion of invert molasses. US. 1962;3:066,150.
148. Trapmann H, Sethi VS. On the effect of thorium-and zirconium ions on aldoses. Arch Pharm (Weinheim, Ger.). 1966;299:657–2.

149. Ishida H, Seri K. Catalytic activity of lanthanoide(III) ions for dehydration of d-glucose to 5-(hydroxymethyl) furfural. J Mol Catal A Chem. 1996;112:L163–5.
150. Kei-Ichi S, Inoue Y, Ishida H. Catalytic activity of lanthanide(III) ions for the dehydration of hexose to 5-hydroxymethyl2-furaldehyde in water. Bull Chem Soc Jpn. 2001;74:1145–50.
151. Seri K, Sakaki T, Shibata M, Inoue Y, Ishida H. Lanthanum(III)-catalyzed degradation of cellulose at 250 °C. Bioresour Technol. 2002;85:257–60.
152. Zhao H, Holladay JE, Brown H, Zhang ZC. Metal chlorides in ionic liquid solvents convert sugars to 5-hydroxymethylfurfural. Science. 2007;316:1597–600.
153. Yong G, Zhang Y, Ying JY. Efficient catalytic system for the selective production of 5-hydroxymethylfurfural from glucose and fructose. Angew Chem. 2008;120:9485–8.
154. Young J, Chan G, Zhang Y. Selective conversion of fructose to 5-hydroxymethyl furfural catalyzed by tungsten salts at low temperatures. ChemSusChem. 2009;2:731–4.
155. Tong X, Yu L, Nie G, Li Z, Liu J, Xue S. Antimony-meditated efficient conversion of carbohydrates to 5-hydroxymethylfurfural in a simple THF-$H_2O$ binary solvent. Environ Prog Sustain Energy. 2015;34:1136–41.
156. Kim Y, Shin S, Yoon HJ, Lee YS. Polymer-supported N-heterocyclic carbene-iron(III) catalyst and its application to dehydration of fructose into 5-hydroxymethyl-2-furfural. Catal Commun. 2013;40:18–22.
157. Mednick ML. Chem Eng News. 1961;11:75.
158. Fayet C, Gelas J. Nouvelle méthode de préparation du 5-hydroxyméthyl-2-furaldéhyde par action de sels dàmmonium ou d'immonium sur les mono-, oligo-et poly-saccharides. Accès direct aux 5-halogénométhyl-2-furaldéhydes. Carbohydr Res. 1983;122:59–68.
159. Smith NH. Preparation of hydroxymethylfurfural. US 3, 118, 912, 21. 1964.
160. Garder JD, Jones RF. Method for producing 5-hydroxymethylfurfural. US. 1969;3:483,228.
161. Vinke P, Van Bekkum H. The dehydration of fructose towards 5-hydroxymethyl furfural using activated carbon as adsorbent. Starch/Stärke. 1992;44:90–6.
162. Anastas PT, Kirchhoff MK, Williamson TC. Catalysis as a foundational pillar of green chemistry. Appl Catal A Gen. 2001;221:3–13.
163. Centi G, Perathoner S. Catalysis and sustainable (green) chemistry. Catal Today. 2003;77:287–97.
164. Sheldon RA. Catalytic reactions in ionic liquids. Chem Commun. 2001:2399–407.
165. Olivier-Bourbigou H, Magna L. Ionic liquids: perspectives for organic and catalytic reactions. J Mol Catal A Chem. 2002;182-183:419–37.
166. Zhao D, Wu M, Kou Y. Ionic liquids: applications in catalysis. Catal Today. 2002;74:157–89.
167. Hu S, Zhang Z, Song J, Zhou Y, Han B. Efficient conversion of glucose into 5-hydroxymethyl furfural catalyzed by a common Lewis acid $SnCl_4$ in an ionic liquid. Green Chem. 2009;11:1746–9.
168. Moreau C, Finiels A, Vanoye L. Dehydration of fructose and sucrose into 5-hydroxymethylfurfural in the presence of 1-H-3-methyl imidazolium chloride acting both as solvent and catalyst. J Mol Catal A Chem. 2006;253:165–9.
169. Bao Q, Qiao K, Tomida D. Preparation of 5-hydroymethylfurfural by dehydration of fructose in the presence of acidic ionic liquid. Catal Commun. 2008;9:1383–8.
170. Tong X, Li Y. Efficient and selective dehydration of fructose to 5-hydroxymethyl furfural catalyzed by Brönsted-acidic ionicliquids. ChemSusChem. 2010;3:350–5.
171. Tong X, Ma Y, Li Y. An efficient catalytic dehydration of fructose and sucrose to 5-hydroxymethylfurfural with protic ionic liquids. Carbohydr Res. 2010;345:1698–701.
172. Lima S, Neves P, Antunes MM, Pillinger M, Valente AA. Conversion of mono/di/polysaccharides into furan compounds using 1-alkyl-3-methylimidazolium ionic liquids. Appl Catal A Gen. 2009;363:93–9.
173. Hu S, Zhang Z, Zhou Y, Han B, Fan H, Li W, Song J, Xie Y. Conversion of fructose to 5-hydroxymethylfurfural using ionic liquids prepared from renewable materials. Green Chem. 2008;10:1280–3.

174. Kotadia DA, Soni SS. Symmetrical and unsymmetrical Brønsted acidic ionic liquids for the effective conversion of fructose to 5-hydroxymethyl furfural. Cat Sci Technol. 2013;3:469–74.
175. Tian G, Tong X, Wang Y, Yan Y, Xue S. Highly efficient and N-bromosuccinimide-mediated conversion of carbohydrates to 5-hydroxymethylfurfural under mild conditions. Res Chem Intermed. 2013;39:3255–63.
176. Matsagar BM, Munshi MK, Kelkar AA. Conversion of concentrated sugar solutions into 5-hydroxymethyl furfural and furfural using Brönsted acidic ionic liquids. Cat Sci Technol. 2015;5:5086–90.
177. Jackson DHK, Wang D, Gallo JMR, Crisci AJ, Scott SL, Dumesic JA, Kuech TF. Amine catalyzed atomic layer deposition of (3-mercaptopropyl)trimethoxysilane for the production of heterogeneous sulfonic acid catalysts. Chem Mater. 2013;25:3844–51.
178. Nie G, Tong X, Zhang Y, Xue S. Efficient production of 5-hydroxymethylfurfural (HMF) from d-fructose and inulin with graphite derivatives as the catalysts. Catal Lett. 2014;144:1759–65.
179. Mondal D, Chaudhary JP, Sharmaab M, Prasad K. Simultaneous dehydration of biomass-derived sugars to 5-hydroxymethyl furfural (HMF) and reduction of graphene oxide in ethyl lactate: one pot dual chemistry. RSC Adv. 2014;4:29834–9.
180. Zhang M, Tong X, Ma R, Li Y. Catalytic transformation of carbohydrates into 5-hydroxymethyl furfural over tin phosphate in a water-containing system. Catal Today. 2016;264:131–5.
181. Antaljr MJ, Leesomboon T, Mok WS, Richard GN. Mechanism of formation of 2-furaldehyde from d-xylose. Carbohydr Res. 1991;217:71–85.
182. Qian X, Nimlos MR, Davis M, Johnsn DK, Himmel ME. Ab initio molecular dynamics simulations of beta-D-glucose and beta-D-xylose degradation mechanisms in acidic aqueous solution. Carbohydr Res. 2005;340:2319–27.
183. Antal MJ, Mok WSL, Richards GN. Four-carbon model compounds for the reactions of sugars in water at high temperature. Carbohydr Res. 1990;199:111–5.
184. Newth FH. The formation of furan compounds from hexoses. Adv Carbohydr Chem. 1951;6:83–106.
185. Amarasekara AS, Williams LD, Ebede CC. Mechanism of the dehydration of D-fructose to 5-hydroxymethylfurfural in dimethyl sulfoxide at 150 degrees C: an NMR study. Carbohydr Res. 2008;343:3021–4.
186. Stone JE, Blundell MJ. A micromethod for the determination of sugars. Can J Res. 1950;28:676–82.
187. Chheda JN, Roman-Leshkov Y, Dumesic JA. Production of 5-hydroxymethylfurfural and furfural by dehydration of biomass-derived mono- and poly-saccharides. Green Chem. 2007;9:342–50.
188. Huang R, Qi W, Su R, He Z. Integrating enzymatic and acid catalysis to convert glucose into 5-hydroxymethylfurfural. Chem Commun. 2010;46:1115–7.
189. Ståhlberg T, Rodriguez-Rodriguez S, Fristrup P, Riisager A. Metal-free dehydration of glucose to 5-(hydroxymethyl)furfural in ionic liquids with boric acid as a promoter. Chem Eur J. 2011;17:1456–64.
190. Takagaki A, Ohara M, Nishimura S, Ebitani K. A one-pot reaction for biorefinery: combination of solid acid and base catalysts for direct production of 5-hydroxymethylfurfural from saccharides. Chem Commun. 2009:6276–8.
191. Lourvanij K, Rorrer GL. Reaction rates for the partial dehydration of glucose to organic acids in solid-acid, molecular-sieving catalyst powders. J Chem Tech Biotechnol. 1997;69:35–44.
192. Yan HP, Yang Y, Tong DM, Xiang X, Hu CW. Catalytic conversion of glucose to 5-hydroxymethylfurfural over $SO_4^{2-}/ZrO_2$ and $SO_4^{2-}/ZrO_2$-$Al_2O_3$ solid acid catalysts. Catal Commun. 2009;10:1558–63.
193. Nikolla E, Román-Leshkov Y, Moliner M, Davis ME. "one-pot" synthesis of 5-(hydroxymethyl)furfural from carbohydrates using tin-beta zeolite. ACS Catal. 2011;1:408–10.

194. Ordomsky VV, Van der Schaaf J, Schouten JC. Glucose dehydration to 5-hydroxymethylfurfural in a biphasic system over solid acid foams. ChemSusChem. 2013;6:1697–707.
195. Nakajima K, Noma R, Kitano M, Hara M. Selective glucose transformation by titania as a heterogeneous Lewis acid catalyst. J Mol Catal A Chemical. 2014;388-389:100–5.
196. Noma R, Nakajima K, Kamata K, Kitano M, Hayashi S, Hara M. Formation of 5-(hydroxymethyl)furfural by stepwise dehydration over $TiO_2$ with water-tolerant Lewis acid sites. J Phys Chem C. 2015;119:17117–25.
197. Atanda L, Mukundan S, Shrotri A, Ma Q, Beltramini J. Catalytic conversion of glucose to 5-hydroxymethylfurfural with a phosphated $TiO_2$ catalyst. ChemCatChem. 2015;7:781–90.
198. Jiménez-Morales I, Moreno-Recio M, Santamaría-González J, Maireles-Torres P, Jimenéz-Lopez A. Production of 5-hydroxymethylfurfural from glucose using aluminium doped MCM-41 silica as acid catalyst. Appl Catal B Environ. 2015;164:70–6.
199. Wang L, Yuan F, Niu X, Kang C, Li P, Li Z, Zhu Y. Effect of cerous phosphates with different crystal structures on their acidity and catalytic activity for dehydration of glucose into 5-(hydroxymethyl)furfural. RSC Adv. 2016;6:40175–84.
200. Moreno-Recio M, Santamaría-González J, Maireles-Torres P. Brönsted and Lewis acid ZSM-5 zeolites for the catalytic dehydration of glucose into 5-hydroxymethylfurfural. Chem Eng J. 2016;303:22–30.
201. Swift TD, Nguyen H, Erdman Z, Vlachos DG. Tandem Lewis acid/Brønsted acid-catalyzed conversion of carbohydrates to 5-hydroxymethylfurfural using zeolite beta. J Catal. 2016;333:149–61.
202. Li C, Zhang Z, Zhao ZK. Direct conversion of glucose and cellulose to 5-hydroxymethylfurfural in ionic liquid under microwave irradiation. Tetrahedron Lett. 2009;50:5403–5.
203. Ståhlberg T, Søensen MG, Riisager A. Direct conversion of glucose to 5-(hydroxymethyl) furfural in ionic liquids with lanthanide catalysts. Green Chem. 2010;12:321–5.
204. Yuan Z, Xu C, Cheng S. Catalytic conversion of glucose to 5-hydroxymethyl furfural using inexpensive co-catalysts and solvents. Carbohydr Res. 2011;346:2019–23.
205. Bali S, Tofanelli MA, Ernst RD, Eyring EM. Chromium(III) catalysts in ionic liquids for the conversion of glucose to 5-(hydroxymethyl)furfural (HMF): insight into metal catalyst:ionic liquid mediated conversion of cellulosic biomass to biofuels and chemicals. Biomass Bioenerg. 2012;42:224–7.
206. Tian G, Tong X, Cheng Y, Xue S. Tin-catalyzed efficient conversion of carbohydrates for the production of 5-hydroxymethylfurfural in the presence of quaternary ammonium salts. Carbohydr Res. 2013;370:33–7.
207. Pidko EA, Degirmenci V, Van Santen RA, Hensen EM. Glucose activation by transient $Cr^{2+}$ dimers. Angew Chem Int Ed. 2010;49:2530–4.
208. Pidko EA, Degirmenci V, Hensen EJM. Cover picture: on the mechanism of Lewis acid catalyzed glucose transformations in ionic liquids. ChemCatChem. 2012;4:1263–71.
209. Jiang F, Zhua Q, Ma D, Liu X, Han X. Direct conversion and NMR observation of cellulose to glucose and 5-hydroxymethylfurfural (HMF) catalyzed by the acidic ionic liquids. J Mol Catal A Chemical. 2011;334:8–12.
210. Li J, Zhang D, Liu C. Theoretical elucidation of glucose dehydration to 5-hydroxymethylfurfural catalyzed by a $SO_3H$-functionalized ionic liquid. J Phys Chem B. 2015;119:13398–406.
211. Cao X, Teong SP, Wu D, Yi G, Su H, Zhang Y. An enzyme mimic ammonium polymer as a single catalyst for glucose dehydration to 5-hydroxymethylfurfural. Green Chem. 2015;17:2348–52.
212. Su Y, Chang G, Zhang Z, Xing H, Su B, Yang Q, Ren Q, Yang Y, Bao Z. Catalytic dehydration of glucose to 5-hydroxymethylfurfural with a bifunctional metal-organic framework. AICHE J. 2016;62(12):4403–17.
213. Shi J, Gao H, Xia Y, Li W, Wang H, Zheng C. Efficient process for the direct transformation of cellulose and carbohydrates to 5-(hydroxymenthyl)-furfural with dual-core sulfonic acid ionic liquids and co-catalysts. RSC Adv. 2013;3:7782–90.

214. Binder JB, Raines RT. Simple chemical transformation of lignocellulosic biomass into furans for fuels and chemicals. J Am Chem Soc. 2009;131:1979–85.
215. Zhang Z, Zhao ZK. Microwave-assisted conversion of lignocellulosic biomass into furans in ionic liquid. Bioresour Technol. 2010;101:1111–4.
216. Su Y, Brown HM, Huang X, Zhou X, Amonette JE, Zhang ZC. Single-step conversion of cellulose to 5-hydroxymethylfurfural (HMF), a versatile platform chemical. Appl Catal A Gen. 2009;361:117–22.
217. Zhang Y, Du H, Qian X, Chen EYX. Ionic liquid−water mixtures: enhanced $K_w$ for efficient cellulosic biomass conversion. Energ Fuel. 2010;24:2410–7.
218. Kim B, Jeong J, Lee D, Kim S, Yong HJ, Lee YS, Cho JK. Direct transformation of cellulose into 5-hydroxymethyl-2-furfural using a combination of metal chlorides in imidazolium ionic liquid. Green Chem. 2011;13:1503–6.
219. Qi X, Watanabe M, Aida TM, Smith RL. Catalytic conversion of cellulose into 5-hydroxymethylfurfural in high yields via a two-step process. Cellulose. 2011;18:1327–33.
220. Ding Z, Shi J, Xiao J, Gu W, Zheng C, Wang H. Catalytic conversion of cellulose to 5-hydroxymethyl furfural using acidic ionic liquids and co-catalyst. Carbohydr Polym. 2012;90:792–8.
221. Cai H, Li C, Wang A, Xu G, Zhang T. Zeolite-promoted hydrolysis of cellulose in ionic liquid, insight into the mutual behavior of zeolite, cellulose and ionic liquid. Appl Catal B Environ. 2012;123-124:333–8.
222. Nandiwale KY, Galande ND, Thakur P, Sawant SD, Zambre VP, Bokade W. One-pot synthesis of 5-hydroxymethylfurfural by cellulose hydrolysis over highly active bimodal micro/mesoporous H-ZSM-5 catalyst. ACS Sustain Chem Eng. 2014;2:1928–32.
223. Tan MX, Zhao L, Zhang Y. Production of 5-hydroxymethyl furfural from cellulose in $CrCl_2$/zeolite/BMIMCl system. Biomass Bioenerg. 2011;35:1367–70.
224. Tao F, Song H, Chou L. Catalytic conversion of cellulose to chemicals in ionic liquid. Carbohydr Res. 2011;346:58–63.
225. Tao F, Song H, Chou L. Efficient conversion of cellulose into furans catalyzed by metal ions in ionic liquids. J Mol Catal A Chemical. 2012;357:11–8.
226. Zhao S, Cheng M, Li J, Tian J, Wang X. One pot production of 5-hydroxymethylfurfural with high yield from cellulose by a Brønsted-Lewis-surfactant-combined heteropolyacid catalyst. Chem Commun. 2011;47:2176–8.
227. Abou-Yousef H, Hassan EB, Steele P. Rapid conversion of cellulose to 5-hydroxymethylfurfural using single and combined metal chloride catalysts in ionic liquid. J Fuel Chem Technol. 2013;41:214–22.
228. Dutta S, De S, Alam MI. Direct conversion of cellulose and lignocellulosic biomass into chemicals and biofuel with metal chloride catalysts. J Catal. 2012;288:8–15.
229. Li Y, Yuan Y, Wang K, Jia J, Qin X, Xu Y. Conversion of sawdust into 5-hydroxymethylfurfura by using 1,3-dimethyl-2-imidazolidinone as the solvent. Iran J Chem Eng. 2013;32:75–9.
230. Qu Y, Wei Q, Li H, Jian X. Microwave-assisted conversion of microcrystalline cellulose to 5-hydroxymethylfurfural catalyzed by ionic liquids. Bioresour Technol. 2014;162:358–64.
231. Lee YC, Dutta S, Wu KC. Integrated, cascading enzyme-/chemocatalytic cellulose conversion using catalysts based on mesoporous silica nanoparticles. ChemSusChem. 2014;7:3241–6.
232. Shi N, Liu Q, Wang T, Ma L, Zhang Q, Zhang Q. One-pot degradation of cellulose into furfural compounds in hot compressed steam with dihydric phosphates. ACS Sustain Chem Eng. 2014;2:637–42.
233. Shi N, Liu Q, Zhang Q, Wang T, Ma L. High yield production of 5-hydroxymethylfurfural from cellulose by high concentration of sulfates in biphasic system. Green Chem. 2013;15:1967–74.
234. Hu L, Zhao G, Tang X, Liu S. Catalytic conversion of carbohydrates into 5-hydroxymethylfurfural over cellulose-derived carbonaceous catalyst in ionic liquid. Bioresour Technol. 2013;148:501–7.

235. Zhang X, Zhang D, Sun Z, Jiang Z. Highly efficient preparation of HMF from cellulose using temperature-responsive heteropolyacid catalysts in cascade reaction. Appl Catal B Environ. 2016;196:50–6.
236. Rout PK, Nannaware AD, Prakash O, Rajasekharan L. Synthesis of hydroxymethylfurfural from cellulose using green processes: a promising biochemical and biofuel feedstock. Chem Eng Sci. 2015;142:318–46.
237. Zhang Y, Li N, Li M. Highly efficient conversion of microcrystalline cellulose to 5-hydroxymethyl furfural in a homogeneous reaction system. RSC Adv. 2016;6:21347–51.
238. Mirzaei HM, Karimi B. Sulphanilic acid as a recyclable bifunctional organocatalyst in the selective conversion of lignocellulosic biomass to 5-HMF. Green Chem. 2016;18:2282–6.

# Chapter 4
# 5-(Halomethyl)furfurals from Biomass and Biomass-Derived Sugars

**Mark Mascal**

**Abstract** 5-(Halomethyl)furfurals (XMFs) are biomass-derived platform chemicals that are gaining traction as practical alternatives to 5-(hydroxymethyl)furfural (HMF). This chapter provides an overview of the historical role of XMFs in the chemical investigation of carbohydrates and describes multiple approaches to their preparation, including recent breakthroughs by which XMFs, and in particular 5-(chloromethyl)furfural (CMF), are obtained in high yield directly from raw biomass. Halomethylfurfurals have two basic derivative manifolds: furanic and levulinic, and this chapter will highlight commercial markets that can be unlocked by synthetic manipulation of CMF and its immediate derivatives.

**Keywords** Biofuels • Biomass • Biorefinery • Chloromethylfurfural • Furans • Halomethylfurfurals • Hydroxymethylfurfural • Levulinic acid • Renewable polymers

## 4.1 Perspective on the 5-(Halomethyl)furfurals

There are three general approaches to biomass processing: biocatalytic, thermochemical, and chemocatalytic, and of these, the latter is currently the least practiced commercially, but considered to offer the greatest promise for future embodiments of the biorefinery [1]. Chemocatalytic methods have the advantage of being faster and generally cheaper than enzymatic/fermentative processes, and are much more selective than thermolysis.

The icon of the chemocatalytic method has long been 5-(hydroxymethyl)furfural **1** (Fig. 4.1), or HMF, and to date over 1500 literature references include HMF in their title, while a *Chemical Abstracts* search of the structure of HMF gives >10,000 hits. However, despite all this attention, the production of HMF has not been

M. Mascal (✉)
Department of Chemistry, University of California Davis,
1 Shields Ave, Davis, CA 95616, USA
e-mail: mjmascal@ucdavis.edu



Z. Fang et al. (eds.), *Production of Platform Chemicals from Sustainable Resources*, Biofuels and Biorefineries 7, DOI 10.1007/978-981-10-4172-3_4

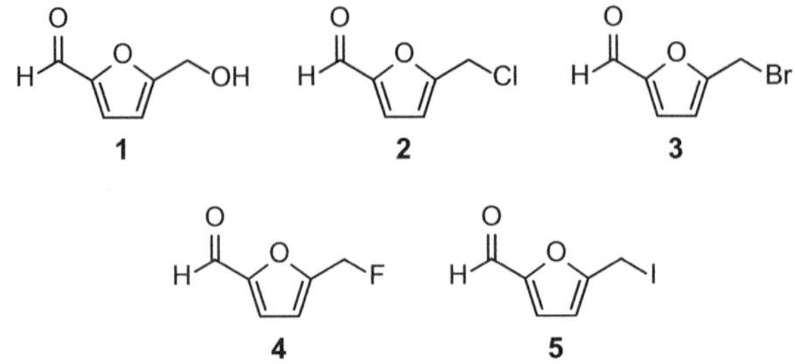


**Fig. 4.1** Structures of molecules associated with the development of XMF chemistry

commercialized, the reason being that the only high-yielding processes for obtaining HMF employ fructose as the feedstock. Beyond this, the isolation of HMF from the media in which it is produced (aqueous solution, highly polar solvents, or ionic liquids) is hampered by its high polarity and hydrophilicity. Finally, HMF is gradually decomposed into humins under the conditions of its formation (acid catalysis), and thus methods that involve long reaction times and/or high temperatures generally suffer from low selectivity.

A potential solution to all of the drawbacks of HMF presents itself in the form of the halomethylfurfurals (XMFs), and in particular 5-(chloromethyl)furfural **2** (CMF). CMF has the advantages of (1) being accessible in high yield directly from raw biomass; (2) being hydrophobic and thus easily extracted from aqueous media with common solvents; and (3) being comparatively stable in the presence of strong acid [2]. This chapter will describe the emergence of the halomethylfurfurals as platform chemicals with the disruptive potential to displace HMF as the go-to molecule of the chemocatalytic biorefinery movement.

## 4.2 Historical Reports of 5-(Halomethyl)furfural Preparation

The first report of the preparation of a halomethylfurfural was published as early as 1899 by Henry J. H. Fenton of Fenton's reagent fame. The procedure involved pouring an ethereal solution of hydrogen bromide onto solid fructose and allowing the mixture to stand for up to 24 h. The isolated product was 5-(bromomethyl)furfural (BMF) **3**, the structure of which was correctly assigned [3]. The method was later extended to various forms of cellulose, which gave BMF in yields around 30% [4]. Following the same procedure as used to produce **3** from cellulose, treatment with dry hydrogen chloride in chloroform gave the first sample of CMF **2**, albeit in only about 12% yield [5]. These works also describe multiple efforts to derivatize the halomethylfurfurals, including the preparation of HMF **1**, acyloxymethylfurfurals **6**, and 5-methylfurfural **7**, along with their hydrazone and oxime derivatives (see Scheme 4.2). In 1909, Fenton and Robinson described Friedel-Crafts reactions of CMF with benzene and toluene to give arylmethylfurfurals **8**, as well as the oxidation of CMF with $HNO_3$ to give 2,5-furandicarboxylic acid **9** [6]. They also used the chemistry of the halomethylfurfurals to derive the correct structure for HMF, which had been previously misassigned, which was confirmed in work by Erdmann [7] and van Ekstein [8] the following year. Further derivative chemistry was undertaken by Cooper and Nuttall, which led to the report of alkoxymethylfurfurals **10**, their corresponding furoic acids **11**, and 2,5-diformylfuran **12** [9, 10]. The celebrated carbohydrate chemist Emil Fischer undertook to develop a large-scale preparation of CMF and reported a yield of about 25% starting from 1 kg of sucrose [11]. The first report of an alkoxymethylfurfural acetal **13** also appears here. In 1923, XMF chemistry was employed to shed light on the disputed nature of the cellulose polymer itself. Thus it had been suggested that the negligible yield of BMF from glucose, versus its relatively good yield from fructose, was an indication that cellulose was composed of ketose monomers. It was shown however by Hibbert and Hill that glucose also could be converted into BMF, undermining the support for this theory [12]. CMF next appears as an intermediate in a 1934 *Organic Syntheses* procedure for the preparation of 5-methylfurfural **7** [13], essentially using the $SnCl_2$-based method of Fenton and Gostling [5]. Yet another Nobel Prize winning sugar chemist, Norman Haworth, studied the preparation of CMF employing for the first time a two-phase system, although the yield from sucrose was only 21%. Hydrolysis of CMF in boiling water was also shown to give HMF **1** in 90% yield [14]. A series of patents later described the use of a two-phase system to produce CMF from simple sugars in good yield, particularly in the case of fructose [15–17], using ionic modifiers and surface-active agents, and this was followed up by papers [18–20]. At around the same time, Szmant and Chundry published a detailed parameter study of the production of CMF from fructose and reported an optimized yield of 95%. Glucose and starch performed less well, giving 45 and 21% yields under the same conditions, respectively [21]. Finally, the production of XMFs by the straightforward reaction of HMF with halogenating agents has also been described [22, 23].

## 4.3 Modern Approaches to 5-(Halomethyl)furfural Preparation

Up until about the year 2000, multiple studies had been concluded describing the high-yielding production of halomethylfurfurals from fructose, but with less satisfactory outcomes using other sugars or cellulose, as described above. In 2008, Mascal and Nikitin reported the conversion of either glucose, sucrose, or cellulose to a mixture of CMF **2** (71-76%), 2-(hydroxyacetyl)furan **14** (6-8%), HMF **1** (4-8%), and levulinic acid (LA) **15** (1-5%) (Scheme 4.1) [24]. All three feedstocks gave similar product yields and showed similar kinetics, suggesting that hydrolysis of cellulose was not the rate-determining step of this reaction. The reactor setup involved a two-phase system of 35% hydrochloric acid and 1,2-dichloroethane (DCE) solvent, which extracted the products by continuous recirculation through

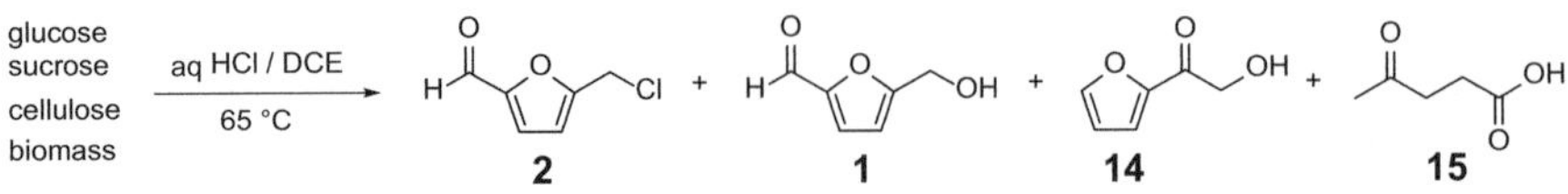


**Scheme 4.1** Process for conversion of sugars, cellulose, or cellulosic biomass into a mixture of CMF **2**, HMF **1**, 2-hydroxyacetylfuran **14**, and levulinic acid **15** in a biphasic reactor system

the reactor. At a reaction temperature of 65 °C, complete conversion was observed within 30 h. A key finding in this study was that cellulose performed nearly identically to the monomeric sugars, and indeed the following year the process was applied with equal effect to sources of raw biomass (cotton, newspaper, birch sawdust, corn stover, and straw). Based on a full carbohydrate analysis of one of these feedstocks (corn stover), the yield of CMF and minor co-products **1**, **14**, and **15** was shown to be essentially identical to when simple sugars or pure cellulose were used [25, 26]. A dramatic decrease in reaction time was later achieved by increasing the temperature of the system to 100 °C and operating in a closed reactor with periodic extractions in place of a continuous solvent loop. Under these conditions, CMF was the only product found in the organic phase and was isolated in yields of 70-90% depending on feedstock and loading, with a further 5–9% LA **15** in the aqueous phase [27]. Finally, a 2010 study extended the range of feedstock to include oil seeds, wherein the carbohydrate content (starch, sugars, fiber) was converted into CMF while the lipids were simply extracted into the organic phase. Thermolysis of the CMF-lipid mixture in ethanol gave a biofuel cocktail of ethyl levulinate and biodiesel ethyl ester [28].

In related work in the area of halomethylfurfural production, Bols and co-workers obtained up to 80% yields of BMF **3** from cellulose using essentially the method of Mascal and Nikitin [24] but substituting HBr for HCl [29]. Another study extended this same production method to various species of wood and also looked into the effects of varying the solvent, temperature, and presence of an LiBr additive [30]. A highly useful advancement in the carbohydrate to CMF conversion process was the adaptation of the reaction to continuous flow by Brasholz and co-workers [31]. Inputting 5 mL $min^{-1}$ of a 10% solution of fructose in 32% HCl along with the same volume of $CH_2Cl_2$ and flowing through a coil at 100 °C with a residence time of about 2 min, 300 mg $min^{-1}$ of CMF could be produced with a yield >80%. Another innovation in CMF production technology was the use of microwave radiation as the heat source, which heats the aqueous phase while avoiding excessive heating of the relatively microwave-transparent solvent layer. This led to a substantial acceleration of the reaction in batch mode while maintaining yields up to 85% [32, 33]. Although this study found DCE to be the most effective solvent for this process, cyclohexane was also shown to perform well. The same authors also demonstrated that the reduction of cellulose crystallinity by ball-milling pretreatment likewise increased CMF yield. Gao and co-workers studied the effect of using a mixed HCl-$H_3PO_4$ acid system in the CMF process at only 45 °C, although yields of CMF from fructose were no better than 50% after 20 h [34, 35]. Another alternative medium involves the use of a choline chloride-based ionic liquid for fructose to CMF conversion in the presence of an immiscible organic solvent layer. Although the yield is relatively low (~50%), the formal need for an aqueous acid phase is removed [36]. In this case, it must however also be noted that a similar study in the 1980s likewise showed that ammonium salts were effective for the conversion of monomeric sugars into HMF and in some cases halomethylfurfurals [37]. Most recently, the effects of mass transfer, reaction temperature, Hansen solvent parameters, solvent fraction, and feedstock loading on yields of CMF from glucose have been studied in detail, and multiple lines of evidence suggest that the yield of CMF is mainly limited by the extraction of the HMF intermediate into the solvent phase [38]. This finding suggests that this partitioning effect increases CMF yields either by lowering the aqueous layer concentration of HMF and thereby shielding it from decomposition, or by providing a medium conducive to the conversion of HMF into CMF.

The production of halomethylfurfurals is also described in the recent patent literature, variously involving the use of Lewis acids, gaseous HCl, ionic modifiers, assorted feedstocks, and the effect of varying $H^+$ and $Cl^-$ concentrations, temperature, and solvent [39–43]. However, these disclosures are largely redundant with previously published work.

Finally, the direct production of 5-(fluoromethyl)furfural (FMF) **4** from sugars or other carbohydrates is not known. The only reported routes to date involve the reaction of BMF **3** with either AgF [44] or KF in the presence of a crown ether [45]. Although 5-(iodomethyl)furfural (IMF) **5** has been invoked as an intermediate in certain biomass transformation processes [46, 47], no report of its isolation and characterization has been published.

## 4.4 Halomethylfurfural Derivative Chemistry – Furanic Manifold

Some of the most basic derivatization work involving the halomethylfurfurals was initially reported alongside the early synthetic approaches. Thus, as noted above, XMF hydrolysis, alcoholysis, acetolysis, and hydrogenolysis, alongside Friedel-Crafts reactions and varying extents of oxidation, had all been described prior to 1920. These simple derivatizations of CMF are summed up in Scheme 4.2. Later work has reproduced and in some cases improved access to these derivatives. For example, a 1982 patent describes the catalytic hydrogenation of CMF for the production of 5-methylfurfural **7** [48], while in a 1985 paper the same reduction is performed electrochemically [49], both of which are improvements over the *Organic Syntheses* method that requires $SnCl_2$ [13]. Acyloxymethylfurfurals **6** were originally prepared by reaction of **3** with silver salts of carboxylic acids [5]. Recent work has promoted 5-(acetoxymethyl)furfural (AMF) **6** ($R=CH_3$) as a more hydrophobic and stable version of HMF and simplifies the synthesis by reaction of CMF with tetraalkylammonium acetate salts [50]. Acylation of CMF has also been carried out using a palladium-catalyzed carboxylative coupling reaction with allyltributylstannane [51]. Related work involves the preparation of an $\eta^3$ $Pd(PPh_3)_2$ complex **16**

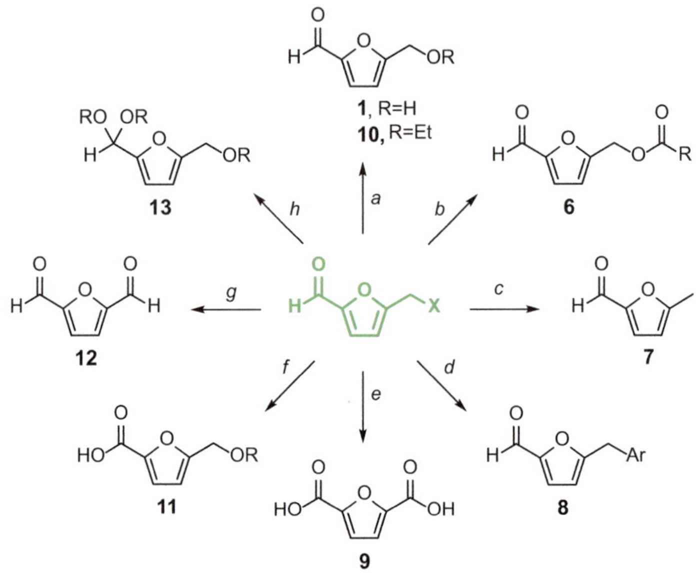

**Scheme 4.2** Historic derivatizations of XMFs (X = Br, Cl). *Reagents and conditions*: *a*. for **1**: aq. $AgNO_3$ or $H_2O$, $BaCO_3$, Δ; for **10**: ROH, $AgNO_3$ or ROH, $CaCO_3$, Δ; *b*. $RCO_2^-$ $Ag^+$; *c*. $SnCl_2$, aq. HCl; *d*. $AlCl_3$, ArH; *e*. $HNO_3$, Δ; *f*. (from **10**) $Ag_2O$, Δ; *g*. $HNO_3$; *h*. (from **10**) EtOH, $(EtO)_3CH$, $NH_4Cl$ (R=Et). See text for updated versions of some of these reactions

from CMF (Scheme 4.3) [52]. A study by Zhou and Rauchfuss in 2013 [53] involving the Friedel-Crafts reactions of CMF essentially reproduced results obtained more than a century earlier [6]. DFF **12** has been another recent XMF-derived target of interest, and while early work relied on gentle nitric acid oxidation to obtain modest yields of **12** from CMF [10], reaction under Kornblum oxidation conditions (heating in DMSO at 150 °C for 18 h) provides DFF in 81% yield [54]. More forcing nitric acid oxidation of CMF gives 2,5-furandicarboxylic acid (FDCA) **9** [6], a reaction that was later repeated and found to proceed in 59% yield [31]. An alternative two-step process reported by Mascal and co-workers gave the bis(acid chloride) of FDCA, *i.e.* 2,5-furandicarbonyl chloride **17** [55] in high yield. Compound **17** is a platform for 2,5-furandicarboxylic esters **18** and can also undergo twofold Friedel-Crafts acylations to give products such as **19**. In the same paper, reaction of CMF itself with *t*-BuOCl gave 5-(chloromethyl)furan-2-carbonyl chloride **20**, which reacts with alcohols under mild conditions to give 5-(chloromethyl)furan-2-carboxylic esters **21**, which themselves react with alcohols under more forcing conditions to give ester derivatives of 5-(alkoxymethyl)furan-2-carboxylic acids **11**, i.e. **22** [55]. 2,5-Dimethylfuran **24** is a high-octane fuel additive [56] and a precursor to *p*-xylene [57], the utility of which is described later. A high-yielding route to **24** via a CMF acetal **23** has been described [58]. Another method for producing novel CMF derivatives was reported in a 2012 patent whereby the combination of an

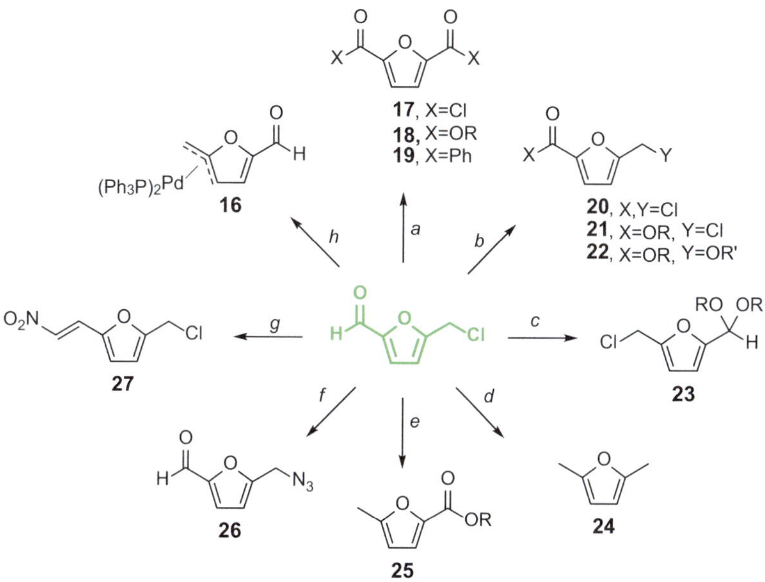


**Scheme 4.3** Recent derivatizations of CMF. *Reagents and conditions*: *a*. for **17**: 1) DMSO, Δ, 2) *t*-BuOCl; for **18** (from **17**) ROH; for **19** (from **17**) $AlCl_3$, PhH; *b*. for **20**: *t*-BuOCl; for **21** (from **20**) ROH; for **22** (from **21**) R'OH, Δ; *c*. BuOH, HCl (R = Bu); *d*. (from **23**) $H_2$, Pd/C; *e*. ROH, NHC cat.; *f*. $NaN_3$; *g*. $MeNO_2$, poly(vinylpyridine); *h*. $Pd(PPh_3)_4$, $AgBF_4$

alcohol and N-heterocyclic carbene (NHC) catalyst leads directly to 5-methylfuroate esters **25** [59]. These compounds are being commercialized as renewable motor fuel blendstocks. The selective introduction of nitrogen and carbon nucleophiles to CMF is complicated by the presence of two different types of electrophile (aldehyde and primary alkyl halide). Reaction of CMF with cyanide fails to give a nitrile but succeeds with azide to give azidomethylfurfural **26** [60]. Protection of the carbonyl group of CMF as an acetal **23** enables reactions at the C-Cl bond with simple carbon nucleophiles such as cyanide and acetylenide as well as organometallics, as will be discussed later. A selective carbon-carbon bond forming reaction at the carbonyl group, on the other hand, has been accomplished by Masuno and coworkers by treatment of CMF with nitromethane to give **27** [61].

## 4.5 Halomethylfurfural Derivative Chemistry – Levulinic Manifold

A key reaction of CMF is its hydrolysis or alcoholysis under forcing conditions to give levulinic acid **15** or levulinic esters **28** respectively, along with the corresponding formate co-product **29** (Scheme 4.4) [62]. This same chemistry has been observed stemming from HMF **1** since the late 19th century [63], and the conversion of CMF to LA likely proceeds via **1**. Since CMF can be produced from raw biomass

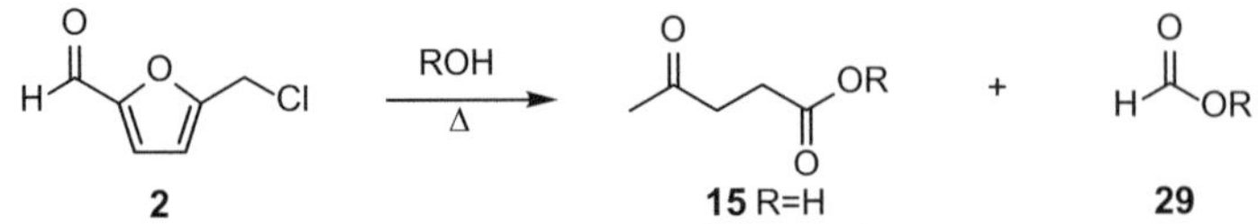


**Scheme 4.4** Conversion of CMF to levulinic and formic derivatives

in yields around 80% [27], and the CMF to LA reaction proceeds in at least 90% yield [62], the overall yield of LA from biomass is >70%, which compares favorably with the only industrially competitive technology for the production of LA (Biofine) [64]. Like CMF, LA is a platform chemical in its own right and appears in the NREL top 12 value added chemicals from biomass [65]. A number of high-level reviews of LA chemistry have been published and Chap. 6 of this volume is also dedicated to this molecule, therefore this contribution will not consider this topic further. Halomethylfurfurals have been employed as intermediates in a large number of synthetic studies across medicinal, polymer, macrocycle, biofuel, and value-added product chemistries. Of these, selected highlights are covered below.

## 4.6 Halomethylfurfural Derivative Chemistry – Advanced Targets

### *4.6.1 Medicinal Chemistry*

CMF can be used as the starting material for an efficient synthesis of the first $1 billion selling drug ranitidine **33**, which is used in the management of gastroesophageal reflux disease (GERD) and the treatment of gastric and duodenal ulcers. First introduced in 1981, it is currently sold over the counter under the trade name Zantac. Mascal and Dutta published a novel, 4-step synthesis of **33** in 2011 involving an initial substitution of an N-protected 2-aminoethanethiol on CMF followed reductive amination with dimethylamine to give intermediate **31** (Scheme 4.5). Deprotection of the primary amino group and subsequent condensation with inexpensive, commercial N-methyl-1-(methylthio)-2-nitroethenamine gave ranitidine **33** in 68% overall yield [66].

An example which highlights the application of acetal formation to mask the reactive aldehyde of CMF is embodied in the synthesis of a furan fatty acid (FFA) **36**. FFAs are naturally-occurring dietary antioxidants which are proposed to have potent anti-atherosclerotic properties. They are found in high levels in fish and may be responsible in part for the cardioprotective effects of diets high in seafood. The synthesis of an FFA starting from CMF is shown in Scheme 4.6. Here, protection of the aldehyde function as an acetal is the key to implementing the required chain elongation at the methylene group. Thus CMF is converted to the dibutyl acetal **23**

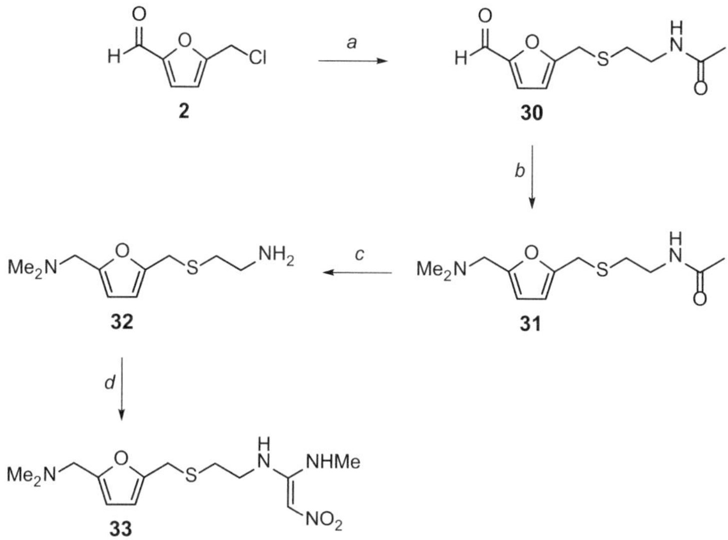


**Scheme 4.5** Synthesis of ranitidine from CMF. *Reagents and conditions*: *a*. $HSCH_2CH_2NHAc$, NaH, THF, 91%; *b*. $Me_2NH$, $NaBH_4$, MeOH, 90%; *c*. 2 M KOH, 94%; *d*. (HS)(MeNH)C=$CHNO_2$, $H_2O$, 88%

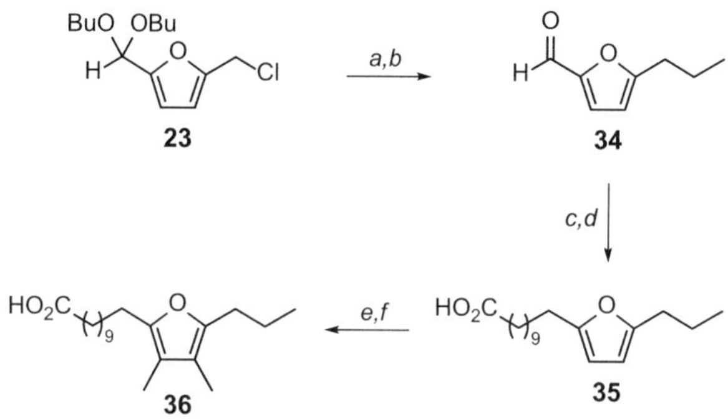


**Scheme 4.6** Synthesis of a furan fatty acid from CMF. *Reagents and conditions*: *a*. EtMgCl, $Ni(acac)_2$, DAE, THF; *b*. HCl/$H_2O$, 83% over 2 steps; *c*. (9-carboxynonyl)triphenylphosphoniumiodide, LiHMDS, THF/DMSO; *d*. $H_2$, Pd/C, THF, 92% over 2 steps; *e*. $(CH_2O)_n$, HBr, AcOH; *f*. $H_2$, Pd/C, THF/$H_2O$, 80% over 2 steps

in nearly quantitative yield in n-butanol solution containing a few drops of aq HCl. Nickel-catalyzed $sp^3$-$sp^3$ coupling converts the chloromethyl function to a propyl group, and Wittig olefination of the deprotected aldehyde gives desmethyl intermediate **35**. Methylation of the furan ring as shown completes the synthesis of FFA **36**. The complete process from CMF to **36** involves seven steps and a 60% overall isolated yield [67].

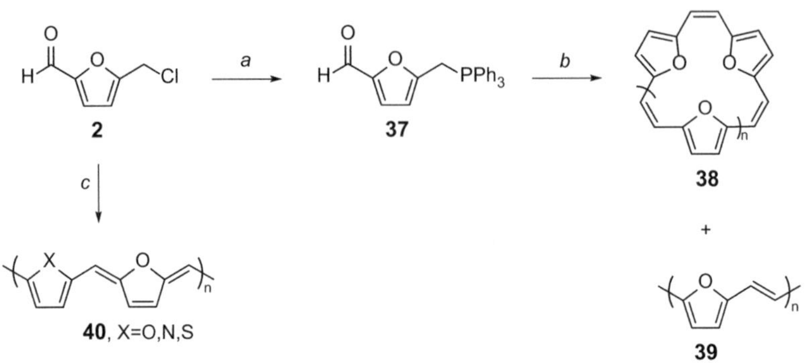


**Scheme 4.7** Synthesis of macrocycles and polymers from CMF. *Reagents and conditions*: *a*. $PPh_3$,PhH, reflux; *b*. LiOEt, DMF-EtOH, Δ; *c*. $BF_3OEt$, $CHCl_3$

An additional 27 research papers and patents to date describe molecules that incorporate CMF-derived 2,5-furandiyl units, which have been synthesized as potential drug targets for a diverse range of medicinal applications, including anti-microbial, anti-fungal, anti-viral, anti-tumor, anti-muscarinic, anti-angiogenic, anti-diabetic, anti-osteoporotic, and LDL-lowering activity. The range of structures is too diverse to include here, but it is clear that the furan moiety constitutes a useful pharmacophore.

### *4.6.2 Macrocyclic and Polymer Chemistry*

While Chundry and Szmant published a detailed paper describing the preparation of several novel monomers incorporating the $-CH_2$(2,5-furandiyl)CHO group [68], only a few macrocycles and polymers have actually been produced directly from CMF. The substitution of chlorine of CMF with triphenylphosphine leads to a phosphonium salt **37** that can be deprotonated and undergo self-condensation to give either macrocycles **38** or poly(2,5-furanylvinylene) **39** [69]. A different approach was taken by Jira and Bräunling, who reacted CMF with either furan, pyrrole, or thiophene to obtain conjugated polymers of the formula **40** (Scheme 4.7) which, when doped, were shown to possess conductivities up to $10^{-2}$ S/m [70]. Cram and co-workers used CMF in the synthesis of crown ethers incorporating furan rings [71, 72].

In terms of current and potential commercial monomer markets, CMF has been employed to supply a renewable source of 2,5-dimethylfuran **24** [58], which undergoes cycloaddition with ethylene to give *para*-xylene (PX) **41** (Scheme 4.8) [57]. This “bio-PX” can then be oxidized to terephthalic acid, which is used to produce the high volume polyester polyethylene terephthalate (PET) [73].

Yet another high-profile innovation in polymer chemistry that may be approached via CMF is the production of 2,5-furandicarboxylic acid (FDCA) **9**. The furanic

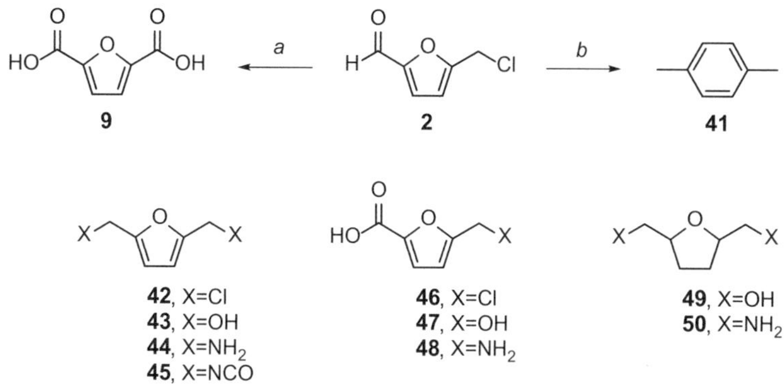


**Scheme 4.8** Synthesis of bio-PX **41** and FDCA **9** from CMF. *Reagents and conditions*: *a*. $HNO_3$, Δ or (1) DMSO, Δ; (2) t-BuOCl; (3) $H_2O$; *b*. (1) $H_2$, cat.; (2) $H_2C{=}CH_2$, cat

equivalent of PET is polyethylene furoate (PEF), a novel polymer currently being commercialized not only as a green substitute for PET, but as a material possessing superior attributes (better gas barrier properties, higher $T_g$ and lower $T_m$ than PET) [74]. While FDCA is currently produced from fructose via HMF, a biomass-based approach via CMF is available that delivers the bis(acid chloride) of FDCA **17**, from which polymers can be directly made. Intermediate **17** is also easier to work with than the highly insoluble FDCA itself. Other polymers of FDCA involving a variety of diols and diamines have been described and are under active investigation [75]. In fact, a Chemical Abstract search of FDCA based polymers shows >50 hits for binary copolymers and >200 that involve FDCA as a component of a higher-order blend.

Given its bifunctional nature, a range of novel monomers can be envisaged stemming from CMF, many of which (**42-50**) have already been described. With their increased availability via CMF, these derivatives may help define a future polymer market based exclusively on renewables.

### 4.6.3 *Biofuels*

Like furfural and HMF [76], CMF has been recruited as a platform for carbon chain extension via aldol and related reactions for the purpose of making biofuels. Since carbohydrate-based substrates are $C_6$ or less, condensations are necessary to achieve the hydrocarbon volatility range required for automotive fuels. In most cases poly-condensation products are catalytically hydrodeoxygenated with $H_2$ to give linear alkanes which are suitable as diesel and jet fuels. The high degree of branching necessary for gasoline is on the other hand more difficult to achieve via biomass.

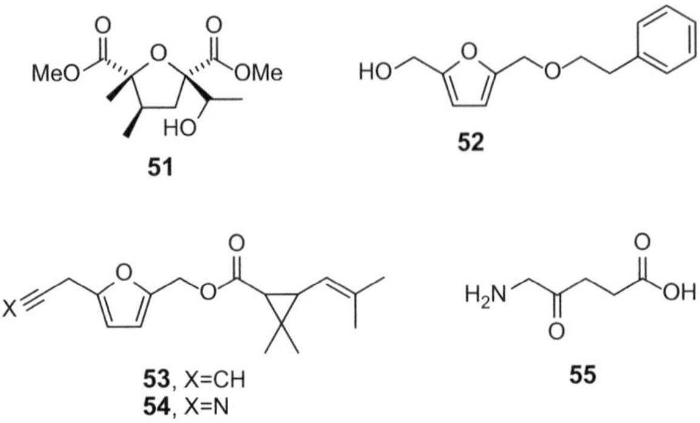


**Fig. 4.2** Value-added products derived from CMF

A number of groups have used CMF either as an intermediate or a direct reactant in condensation reactions with simple aldehydes or ketones to give >$C_6$ products. Thus, Corma describes the use of CMF as a precursor to 5-methyl-furfural **7**, which undergoes various couplings to give a $C_9$-$C_{16}$ diesel grade biofuel [77]. Seck likewise hydrotreats mixtures of CMF-derived 5-methylfurfural and levulinate ester condensates to arrive at diesel and jet fuel formulations [78]. Another group converted CMF into 5-(ethoxymethyl)furfural before condensation with acetone and mixing with fatty acid esters to give aviation fuel after hydrogenation [79]. Finally, a team from Los Alamos National Lab describes the direct condensation of CMF with ketones and hydroxyketones catalyzed by proline metal chelates and/or substituted benzimidazole catalysts to give $C_8$-$C_{15}$ hydrocarbon precursors [80].

### *4.6.4 Miscellaneous Value-Added Products*

Finally, CMF has also proved useful in the synthesis of chemicals with varied applications, some examples of which are given in Fig. 4.2. The first use of CMF in complex natural product synthesis was in an approach to dimethyl jaconate **51**, a metabolite of the complex pyrrolizidine alkaloid jacobine. (±)-**51** was produced in nine steps from CMF [81]. A simpler natural product, pichiafuran C **52**, which is a metabolite of a marine fungus, could be synthesized in two steps from CMF [82]. Prothrin **53**, a synthetic pyrethroid insecticide, was prepared in six steps and overall 65% yield from CMF using a tactic which, like that of FFA **36**, involved the di-n-butyl acetal **23** in order to substitute the chloromethyl group without interference from the aldehyde [83]. A cyano analogue **54** was likewise produced but found to be less active than prothrin. Finally, CMF was used as the starting point for the most efficient synthesis to date of 5-aminolevulinic acid **55**, a compound used in photodynamic therapy but also of strong interest as a natural herbicide. The approach took advantage of the ability to selectively introduce nitrogen at the chloromethyl group of CMF in the form of 5-(azidomethyl)furfural **26**. Compound **26** is only two steps away from target molecule **55**, which was obtained in an overall yield of 68% from CMF [84].

## 4.7 Conclusions and Future Outlook

The modern concept of the biorefinery is characterized by the production of fuels, commodity chemicals, and value-added products from non-petroleum based carbon sources [85]. Biorefinery technologies will only be competitive when they have the following characteristics: (1) operate on raw biomass; (2) involve short reaction sequences running at high performance (high selectivity at high conversion); (3) involve no major waste streams; (4) operate under mild conditions; (5) produce multifunctional platform chemicals with the potential to unlock renewable fuel, materials, and bulk chemicals markets; and (6) outperform other biocatalytic, thermochemical or chemocatalytic methods to the same products. As this chapter demonstrates, the technology for producing XMFs from biomass has been proven by multiple studies to conform to the above six characteristics. This is in stark contrast to the processes leading to HMF **1** which, as discussed in Section 1.1, have performance issues including low yield from raw biomass sources and difficult product isolation and catalyst recovery from the media in which it is produced. CMF **2** is the functional equivalent of HMF and can undergo virtually any reaction in which HMF can participate. Further, since CMF can be converted into HMF in high yield, all the derivative chemistry of HMF applies by proxy to CMF. CMF further benefits from greater stability, higher solubility in nonpolar solvents, and better reactivity in direct substitutions at the methylene group than HMF.

The only legitimate practical detraction from CMF commercialization involves the necessity for chloride management, both in the handling of HCl and its recovery after CMF derivatization. Here, it can only be recognized that the use of HCl has a long history in the chemical industry and multiple materials have been developed to accommodate it and other corrosives, including a range of metal alloys, glass- or ceramic-lined steel, and various plastics and elastomers for piping [86]. Multiple technologies for the recovery of HCl from solution are also practiced, including membrane distillation [87], pervaporation [88], acid-base couple extraction [89, 90], solvent extraction [91–93], diffusion dialysis [94, 95], and electrodialysis [96]. Thus, although the use of HCl is a CAPEX issue, it is not a prohibitive one.

In conclusion, no single renewable technology will provide the solution to the economic, environmental, and political issues raised by the unabated worldwide use of petroleum. The integrated biorefinery represents a broad composite of processes which itself merges with non-biomass based alternative energy technologies to create a better future world economy. Chemocatalytic technologies have a large and increasingly recognized role to play in this context, and the halomethylfurfurals, and in particular CMF, have the potential to become the dominant platform chemical representing the chemocatalytic approach to biomass valorization.

## References

1. Farmer TJ, Mascal M. Platform molecules. In: Clark J, Deswarte F, editors. Introduction to chemicals from biomass. 2nd ed. Chichester: Wiley; 2015. doi:10.1002/9781118714478.ch4.
2. Mascal M. 5-(Chloromethyl)furfural is the new HMF: functionally equivalent but more practical in terms of its production from biomass. ChemSusChem. 2015;8:3391–5.
3. Fenton HJH, Gostling M. Bromomethylfurfuraldehye. J Chem Soc Trans. 1899;75:423–33.
4. Fenton HJH, Gostling M. The action of hydrogen bromide on car bohydrates. J Chem Soc Trans. 1901;79:361–5.
5. Fenton HJH, Gostling M. Derivatives of methylfurfural. J Chem Soc Trans. 1901;79:807–16.
6. Fenton HJH, Robinson F. Homologues of furfuraldehyle. J Chem Soc Trans. 1909;95:1334–40.
7. Erdmann E. ω-Hydroxy-sym-methylfurfuraldehyde and its relation ship to cellulose. Berichte. 1910;43:2391–8.
8. van Ekstein WA, Blanksma JJ. ω-Hydroxymethylfurfuraldehyde as the cause of certain color reactions of the hexoses. Berichte. 1910;43:2355–61.
9. Cooper WF, Nuttall WH. Some Reactions of ω-bromomethylfurfur aldehyde. J Chem Soc Trans. 1911;99:1193–200.
10. Cooper WF, Nuttall WH. Furan-2:5-dialdehyde. J Chem Soc Trans. 1912;101:1074–81.
11. Fischer E, von Neyman H. ω-Chloromethyl- and ethoxymethyl-fur fural. Berichte. 1914;47:973–7.
12. Hibbert H, Hill HS. Studies on cellulose chemistry II. The action of dry hydrogen bromide on carbohydrates and polysaccharides. J Am Chem Soc. 1923;45:176–82.
13. Rinkes IJ. 5-Methylfurfural. Org Synth. 1934;14:62–3.
14. Haworth WN, Jones WGM. The conversion of sucrose into furan compounds. Part 1. 5-Hydroxymethylfurfuraldehyde and some derivatives. J Chem Soc, 1944;667–70.
15. Hamada K, Suzukamo G, Nagase T. Furaldehydes. Ger Offen DE. 1978;2745743.
16. Hamada K, Suzukamo G. 5-Chloromethylfurfural. Jpn. Kokai Tok kyo Koho JP. 1978;53105472.
17. Hamada K, Yoshihara H, Suzukamo G 5-Halomethylfurfural. Eur Pat Appl EP. 1983;79206.
18. Hamada K, Yoshihara H, Suzukamo G. An improved method for the conversion of saccharides into furfural derivatives. Chem Lett. 1982;5:617–8.
19. Hamada K, Yoshihara H, Suzukamo G. Surface active agent-cata lyzed conversion of saccharides to furfural derivatives. J Oleo Sci. 2001;50:207–9.
20. Hamada K, Yoshihara H, Suzukamo G. Novel synthetic route to 2,5-disubstituted furan derivatives through surface active agent-catalyzed dehy dration of D-(-)-fructose. J Oleo Sci. 2001;50:533–6.
21. Szmant HH, Chundury DD. The preparation of 5-Chloromethyl furfuraldehyde from High Fructose Corn Syrup and other carbohydrates. J Chem Tech Biotechnol. 1981;31:205–12.
22. Sanda K, Rigal L, Gaset A. Synthèse du 5-Bromométhyl- et du 5-Chlorométhyl-2-furannecarboxaldéhyde. Carbohydrate Res. 1989;187:15–23.
23. Sanda K, Rigal L, Gaset A. Optimisation of the Synthesis of 5-Chloro methyl-2-furancarboxaldehyde from D-Fructose Dehydration and in-situ Chlorination of 5-Hydroxymethyl-2-furancarboxaldehyde. J Chem Tech Bi otechnol. 1992;55:139–45.
24. Mascal M, Nikitin EB. Direct, high-yield conversion of cellulose into biofuel. Angew Chem Int Ed. 2008;47:7924–6.
25. Mascal M, Nikitin EB. Towards the efficient, total glycan utilization of biomass. ChemSusChem. 2009;2:423–6.
26. Mascal M. High-yield conversion of cellulosic biomass into furanic biofuels and value-added products. US. 2010;7:829,732.
27. Mascal M, Nikitin EB. Dramatic advancements in the saccharide to 5-(chloromethyl)furfural conversion reaction. ChemSusChem. 2009;2:859–61.
28. Mascal M, Nikitin EB. Co-processing of carbohydrates and lipids in oil crops to produce a hybrid biodiesel. Energy Fuels. 2010;24:2170–1.

29. Kumari N, Olesen JK, Pedersen CM, Bols M. Eur J Org Chem. 2011;1266–70.
30. Bredihhin A, Mäeorg U, Vares L. Evaluation of carbohydrates and lignocellulosic biomass from different wood species as raw material for the synthesis of 5-bromomethyfurfural. Carbohydrate Res. 2013;375:63–7.
31. Brasholz M, von Känel K, Hornung CH, Saubern S, Tsanaktsidis J. Green Chem. 2011;13:1114–7.
32. Breeden SW, Clark JH, Farmer TJ, Macquarrie DJ, Meimoun JS, Nonne Y, Reid JESJ. Microwave heating for rapid conversion of sugars and polysaccharides to 5-chloromethyl furfural. Green Chem. 2013;15:72–5.
33. Budarin VL, Shuttleworth PS, De bruyn M, Farmer TJ, Gronnow MJ, Pfaltzgraff L, Macquarrie DJ, Clark JH. The potential of microwave technology for the recovery, synthesis and manufacturing of chemicals from bio-wastes. Catalysis Today. 2015;239:80–9.
34. Gao W, Li Y, Xiang Z, Chen K, Yang R, Argyropoulos DS. Efficient one-pot synthesis of 5-chloromethylfurfural (CMF) from carbohydrates in mild biphasic systems. Molecules. 2013;18:7675–85.
35. Mascal M. Comment on Gao, W., et al. "Efficient one-pot synthesis of 5-chloromethylfurfural (CMF) from carbohydrates in mild biphasic sys tems.". Molecules. 2014;18:7675–85.
36. Miao Z, Li Z, Jiang Y, Tang X, Zeng X, Sun Y, Lin L. Green catalytic conversion of bio-based sugars to 5-chloromethyl furfural in deep eutectic solvent, catalyzed by metal chlorides. RSC Adv. 2016;6:27004–7.
37. Fayet C, Gelas J. A new method for the preparation of 5-hydroxyme thyl-2-furaldehyde by the reaction of ammonium or immonium salts with mono-, oligo- and polysaccharides. Direct access to 5-halomethyl-2-fural dehydes. Carbohydrate Res. 1983;122:59–68.
38. Lane DR, Mascal M, Stroeve P. Experimental studies towards opti mization of the production of 5-(chloromethyl)furfural (CMF) from glucose in a two-phase reactor. Renewable Energy. 2016;85:994–1001.
39. Masuno, MN, Bissell J, Smith RL, Higgins B, Wood AB, Foster M. Utilizing a multiphase reactor for the conversion of biomass to produce sub stituted furans in the presence of gaseous acid, proton donor, and solvent. 2012;WO 2012170520.
40. Cho J-K, Kim S-Y, Lee D-H, Kim BR, Jung J-W. Method for prepar ing 5-chloromethyl-2-furfural using galactan derived from seaweed in two-component phase. 2012;WO 2012111988.
41. Browning SM, Bissell J, Smith RL, Masuno MN, Nicholson BF, Wood AB. Methods for producing 5-(halomethyl)furfural from feedstock com prising six-carbon sugars. 2014;WO 2014066746.
42. Mikochik P, Cahana A, Nikitin E,; Standiford J, Ellis K, Zhang L, George T. Efficient, high-yield conversion of saccharides in a pure or crude form to 5-(chloromethyl)-2-furaldehyde. 2014;US 20140187802.
43. Wood AB, Masuno MN; Smith RL, Bissell J; Hirsch-Weil DA.; Araiza RJ, Henton DR, Plonka JH. Methods for producing 5-(halomethyl) furfu ral from renewable biomass resources. 2015;WO 2015042407.
44. Chernyak MY, Tarabanko VE, Sokolenko VA, Sharypov VI, Morozov AA, Suchkova EO. Interaction of 5-bromomethylfurfural with silver flu oride in methanol and toluene. J Siberian Federal University, Chemistry. 2011;4(2):191–8.
45. Chernyak MY, Tarabanko VE, Sokolenko WA, Morozov AA. Syn thesis of 5-fluoromethylfurfural from 5-bromomethylfurfural in presence of dibenzo-24-crown-8. Khimiya Rastitel'nogo Syr'ya. 2012;3:223–4.
46. Yang W, Sen A. Direct catalytic synthesis of 5-methylfurfural from biomass-derived carbohydrates. ChemSusChem. 2011;4:349–52.
47. Yang W, Grochowski MR, Sen A. Selective reduction of biomass by hydriodic acid and its in situ regeneration from iodine by metal/hydrogen. ChemSusChem. 2012;5(7):1218–22.
48. Hamada K, Suzukamo G, Fujisawa K. 5-Methylfurfural. Eur Pat Appl EP. 1982;44186
49. Lund T, Lund H. Electrochemical reduction of furan derivatives de rived from biomass. Acta Chem Scand. 1985;B39:429–35.

50. Kang E-S, Hong Y-W, Chae DW, Kim B, Kim B, Kim YJ, Cho JK, Kim YG. From lignocellulosic biomass to furans via 5-acetoxymethylfurfural as an alternative to 5-hydroxymethylfurfural. ChemSusChem. 2015;8:1179–88.
51. Sun J, Bao M, Feng X, Yu X, YamamotoY AAI, Arumugam N, Kumar RS. Carboxylative coupling reaction of five-membered (chloromethyl) heteroarenes with allyltributylstannane catalyzed by palla dium nanoparticles. Tetrahedron Lett. 2015;56:6747–50.
52. Shi Y, Brenner P, Bertsch S, Radacki K, Dewhurst RD. $\eta^3$-Furfuryl and $\eta^3$-thienyl complexes of palladium and platinum of relevance to the functionalization of biomass-derived furans. Organometallics. 2012;31:5599–605.
53. Zhou X, Rauchfuss TB. Production of hybrid diesel fuel precursors from carbohydrates and petrochemicals using formic acid as a reactive sol vent. ChemSusChem. 2013;6:383–8.
54. Laugel C, Estrine B, Le Bras J, Hoffmann N, Marinkovic S, Muzart J. NaBr/DMSO-Induced synthesis of 2,5-diformylfuran from fructose or 5-(hydroxymethyl)furfural. ChemCatChem. 2014;6:1195–8.
55. Dutta S, Wu L, Mascal M. Production of 5-(chloromethyl)furan-2-carbonyl chloride and furan-2,5-dicarbonyl chloride from biomass-derived 5-(chloromethyl)furfural (CMF). Green Chem. 2015;17:3737–9.
56. Chen G, Shen Y, Zhang Q, Yao M, Zheng Z, Liu H. Experimental study on combustion and emission characteristics of a diesel engine fueled with 2,5-dimethylfuran-diesel, n-butanol-diesel and gasoline-diesel blends. Energy. 2013;54:333–42.
57. (a) Brandvold TA. Carbohydrate route to para-xylene and terepthalic acid. 2010;US 0331568 A1.; (b) Masuno MN, Bissell J, Smith R, Higgins B, Wood AB, Foster M. Utilizing a multiphase reactor for the conversion of biomass to produce substituted furans. 2012;WO 170521; (c) Shiramizu M, Toste FD. On the diels–alder approach to solely biomass-derived polyethylene terephthalate (PET): conversion of 2,5-dimethylfuran and acrolein into p-xylene. Chem Eur J. 2011;17:12452–7.; (d) Williams CL, Chang C-C, Do P, Nikbin N, Caratzoulas S, Vlachos DG, Lobo RF, Fan W, Dauenhauer PJ. Cycloaddition of biomass-derived furans for catalytic production of renewable p-xylene. ACS Catal. 2012;2:935–9.
58. Dutta S, Mascal M. Novel Pathways to 2,5-dimethylfuran via biomass-derived 5-(chloromethyl)furfural. ChemSusChem. 2014;7:3028–30.
59. Mikochik P, Cahana A. Conversion of 5-(chloromethyl)-2- furaldehyde into 5-methyl-2-furoic acid and derivatives thereof. PCT Int Appl. 2012;WO 2012024353
60. Mascal M, Dutta S. Synthesis of the natural herbicide d-aminolevulinic acid from cellulose-derived 5-(chloromethyl)furfural. Green Chem. 2011;13:40–41.
61. Wood AB, Masuno MN, Smith RL, Bissell JA, Araiza RJ, Hirsch-Weil DA. Methods for production of aryldiamine compounds, aryldinitro compounds and other compounds. PCT Int Appl. 2016;2016033348
62. Mascal M, Nikitin EB. High-yield conversion of plant biomass into the key value-added feedstocks 5-(hydroxymethyl)furfural, levulinic acid, and levulinic esters via 5-(chloromethyl) furfural. Green Chem. 2010;12:370–3.
63. Kiermayer J. A derivative of furfuraldehyde from laevulose. Chemiker-Zeitung. 1895;19:1003–5.
64. Hayes DJ, Fitzpatrick S, Hayes MHB, Ross JRH. The biofine process – production of levulinic acid, furfural, and formic acid from lignocellulosic feedstock. In: Kamm B, Gruber PR, Kamm M, editors. Biorefineries – industrial processes and products, vol 1. Weinheim: Wiley-VCH Verlag GmbH & Co. 2006. p. 139–164
65. Top value added chemicals from biomass. Volume I: results of screening for potential candidates from sugars and synthesis gas. Technical report identifier PNNL-14804, Pacific Northwest National Laboratory and the Na tional Renewable Energy Laboratory, 2004. http://www1.eere.energy.gov/biomass/pdfs/35523.pdf.
66. Mascal M, Dutta S. Synthesis of ranitidine (Zantac) from cellulose-derived 5-(chloromethyl) furfural. Green Chem. 2011;13:3101–2.

67. Chang F, Hsu W-H, Mascal M. Synthesis of anti-inflammatory furan fatty acids from biomass-derived 5-(chloromethyl)furfural. Sustainable Chem Pharm. 2015;1:14–8.
68. Chundury D, Szmant HH. Preparation of polymeric buildlng blocks from 5-hydroxymethyl- and 5-chloromethylfurfuraldehyde. Ind Eng Chem Prod Res Dev. 1981;20:158–63.
69. Elix JA. Synthesis and properties of annulene polyoxides. Aus J Chem. 1969;22:1951–62.
70. Jira R, Bräunling H. Synthesis of polyarenemethines - a new class of conducting polymers. Synthetic Metals. 1987;17:691–6.
71. Timko JM, Cram DJ. Furanyl unit in host compounds. J Am Chem Soc. 1974;96:7159–60.
72. Timko JM, Moore SS, Walba DM, Hiberty PC, Cram DJ. Host-guest complexation. 2. Structural units that control association constants between polyethers and tert-butylammonium salts. J Am Chem Soc. 1977;99:4207–19.
73. Smith PB. Bio-based sources for terephthalic acid. ACS Symposium Series 1192 (Green Polymer Chemistry: Biobased Materials and Biocataly sis). 2015. p. 453–69.
74. Burgess SK, Leisen JE, Kraftschik BE, Mubarak CR, Kriegel RM, Koros WJ. Chain mobility, thermal, and mechanical properties of poly(ethylene furanoate) compared to poly(ethylene terephthalate). Macro molecules. 2014;47:1383–91.
75. Sousa AF, Vilela C, Fonseca AC, Matos M, Freire CSR, Gruter G-J M, JFJ C, AJD S. Biobased polyesters and other polymers from 2,5-furandicarboxylic acid: a tribute to furan excellency. Polymer Chem. 2015;6:5961–83.
76. Wu L, Moteki T, Gokhale AA, Flaherty DW, Toste FD. Production of fuels and chemicals from biomass: condensation reactions and beyond. Chem. 2016;1:32–58.
77. Corma A, de la Torre O, Renz M. High-quality diesel from hexose- and pentose-derived biomass platform molecules. ChemSusChem. 2011;4:1574–7.
78. Seck KA. Biorefinery for conversion of carbohydrates and lignocel lulosics via primary hydrolysate CMF to liquid fuels. PCT Int Appl WO. 2013;2013122686
79. Stepan E, Velea S, Oancea F, Bombos M, Vasilievici G, Parvulescu V, Blajan O, Crucean A. Process for obtaining aviation biofuel from microalgal biomass, by processing its components in an integrated system. PCT Int Appl. 2015;WO 2015076687.
80. Silks LA, Gordon JC, Wu R, Hanson SK. Process for preparation of furan derivatives by carbon chain extension through aldol reaction. PCT Int. Appl. WO. 2011;2011022041
81. Klein LL, Shanklin MS. Total synthesis of dimethyl jaconate. J Org Chem. 1988;53:5202–9.
82. Florentino HQ, Hernandez-Benitez RI, Avina JA, Burgueno-Tapia E, Tamariz J. Total synthesis of naturally occurring furan compounds 5-{[(4-hydroxybenzyl)oxy]methyl}-2-furaldehyde and pichiafuran C. Synthesis. 2011;1106–12.
83. Chang F, Dutta S, Becnel JJ, Estep AS, Mascal M. Synthesis of the insecticide prothrin and its analogues from biomass-derived 5-(chlorome thyl)furfural. J Agric Food Chem. 2014;62:476–80.
84. Mascal M, Dutta S. Synthesis of the natural herbicide δ-ami nolevulinic acid from cellulose-derived 5-(chloromethyl)furfural. Green Chem. 2011;13:40–1.
85. The biorefinery concept: an integrated approach. In: Clark J, Deswarte F, editors. Introduction to chemicals from biomass, 2nd ed. Chichester: Wiley; 2015.
86. Hisham MWM, Bommaraju TV. Kirk-Othmer Encyclopedia of Chemical Technology. 2005;13:808–37.
87. Tomaszewska M, Gryta M, Morawski AW. Recovery of hydrochloric acid from metal pickling solutions by membrane distillation. Sep Purif Tech nol. 2001;22-23:591–600.
88. Schuchardt U, Joekes I, Duarte HC. Hydrolysis of sugar cane bagasse with hydrochloric acid: separation of the acid by pervaporation. Evaluation of the Bergius process. J Chem Technol Biotechnol. 1988;41:51–60.
89. Baniel A, Eyal A. A process for the recovery of HCl from a dilute solution thereof and extractant composition for use therein. WO. 2009;2009125400
90. Sarangi K, Padhan E, Sarma PVRB, Park KH, Das RP. Removal/re covery of hydrochloric acid using alamine 336, aliquat 336, TBP and cyanex 923. Hydrometallurgy. 2006;84:125–9.

91. Gaddy JL, Clausen EC. Recovery of concentrated hydrochloric acid from a product comprising sugars and hydrochloric acid from acid hydroly sis of biomass. US. 1987:4645658.
92. Crittenden ED, Hixson AN. Extraction of hydrogen chloride from aqueous solutions. Ind Eng Chem. 1954;46:265–74.
93. Forster AV, Martz LE, Leng DE. Recovering concentrated hydrochlo ric acid from the crude product obtained from acid hydrolysis of cellulose. 1980;EP 18621.
94. Zhang X, Li C, Wang X, Wang Y, Xu T. Recovery of hydrochloric acid from simulated chemosynthesis aluminum foils wastewater: an inte gration of diffusion dialysis and conventional electrodialysis. J Membr Sci. 2012;409–410:257–63.
95. Xu J, Lu S, Fu D. Recovery of hydrochloric acid from the waste acid solution by diffusion dialysis. J Hazard Mater. 2009;165:832–7.
96. Rohman FS, Aziz N. Desalination. 2011;275:37.

# Part III
# Production of Acids

# Chapter 5
# Levulinic Acid from Biomass: Synthesis and Applications

**Buana Girisuta and Hero Jan Heeres**

**Abstract** Levulinic acid (LA) is a promising platform chemical that can be obtained from biomass. The potential to obtain useful chemical derivatives from levulinic acid is high due to the presence of both a ketone group and a carboxylic acid group. The synthesis of LA on the laboratory scale has been investigated extensively using homogeneous or heterogeneous catalysts. The highest reported yields of LA from monosaccharides, polysaccharides and lignocellulosic biomass and their reaction conditions are summarized in this chapter. In addition, an overview is given on process technology studies including kinetic models and the status of large scale production of LA from biomass. Levulinic acid derivatives and their application will be presented along with future prospects of LA synthesis in biorefineries.

**Keywords** Biomass hydrolysis • Levulinic acid • Kinetic models • 5-hydroxymethylfurfural • Catalysis

## 5.1 Introduction

The world consumption of fossil resources has increased rapidly the last 3 decades (Fig. 5.1). Total consumption of fossil resources in 2012 was almost twice the consumption in 1980. Several studies have projected that the world consumption of fossil resources will increase to 640 quadrillion BTU in 2040. These fossil resources are consumed to fulfill our energy needs, to produce transportation fuels and to manufacture a wide range of modern products like polymers, resins, textiles, lubricants and fertilizers.

B. Girisuta (✉)
Institute of Chemical and Engineering Sciences,
1 Pesek Road, Jurong Island 627833, Singapore
e-mail: buana_girisuta@ices.a-star.edu.sg

H.J. Heeres
Department of Chemical Engineering, University of Groningen,
Nijenborgh 4, Groningen 9747, AG, The Netherlands
e-mail: h.j.heeres@rug.nl



Z. Fang et al. (eds.), *Production of Platform Chemicals from Sustainable Resources*, Biofuels and Biorefineries 7, DOI 10.1007/978-981-10-4172-3_5

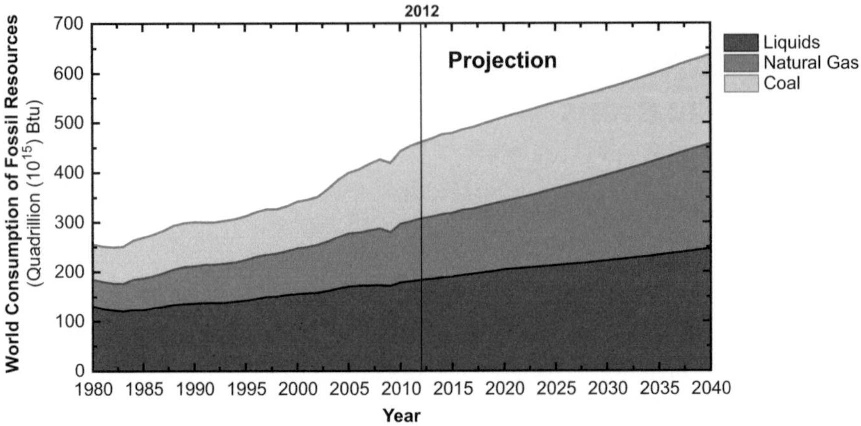


**Fig. 5.1** The world consumption of fossil resources 1980–2012 and its projection until 2040 (Source: U.S. Energy Information Administration (EIA), May 2016, acknowledgement is given to EIA) [1]

Fossil resources are not-renewable and their availability is irrevocably decreasing. Biomass-based technologies can provide alternative sustainable routes for the production of liquid transportation fuels and platform chemicals [2–5]. One of the platform chemicals that can be produced from sustainable resources is levulinic acid (LA). LA is accessible through the hydrolysis of lignocellulosic biomass at elevated temperatures (100–250 °C) using acid catalysts. The applications of LA and its derivatives have been widely reviewed [6–10].

Lignocellulosic biomass consists of three main biopolymers: cellulose, hemicellulose, and lignin. Depolymerization of the cellulose and hemicellulose fractions is an important step for many conversion routes. On a molecular level, hydrolysis of lignocellulosic biomass is complex, involving many reactions and intermediates (Fig. 5.2). In the presence of an acid catalyst and at elevated temperatures, the cellulose fraction is depolymerized to glucose, which is subsequently converted into 5-hydroxymethylfurfural (HMF) and finally to LA and formic acid. The hemicellulose fraction contains various sugar polymers such as xylan, arabinan, mannan, or galacto-glucomannan. During the hydrolysis reaction, these polymeric sugars are depolymerized to pentoses, hexoses, 4-O-methyl glucuronic acid, and acetic acid. All hexoses, like glucose, are converted to LA and formic acid as the final products. The pentoses are converted to furfural, which is known to be rather reactive under the prevailing conditions and may react further to form formic acid and other undesirable decomposition products. The lignin fraction in lignocellulosic biomass can be partly solubilized (acid-soluble lignin) in water.

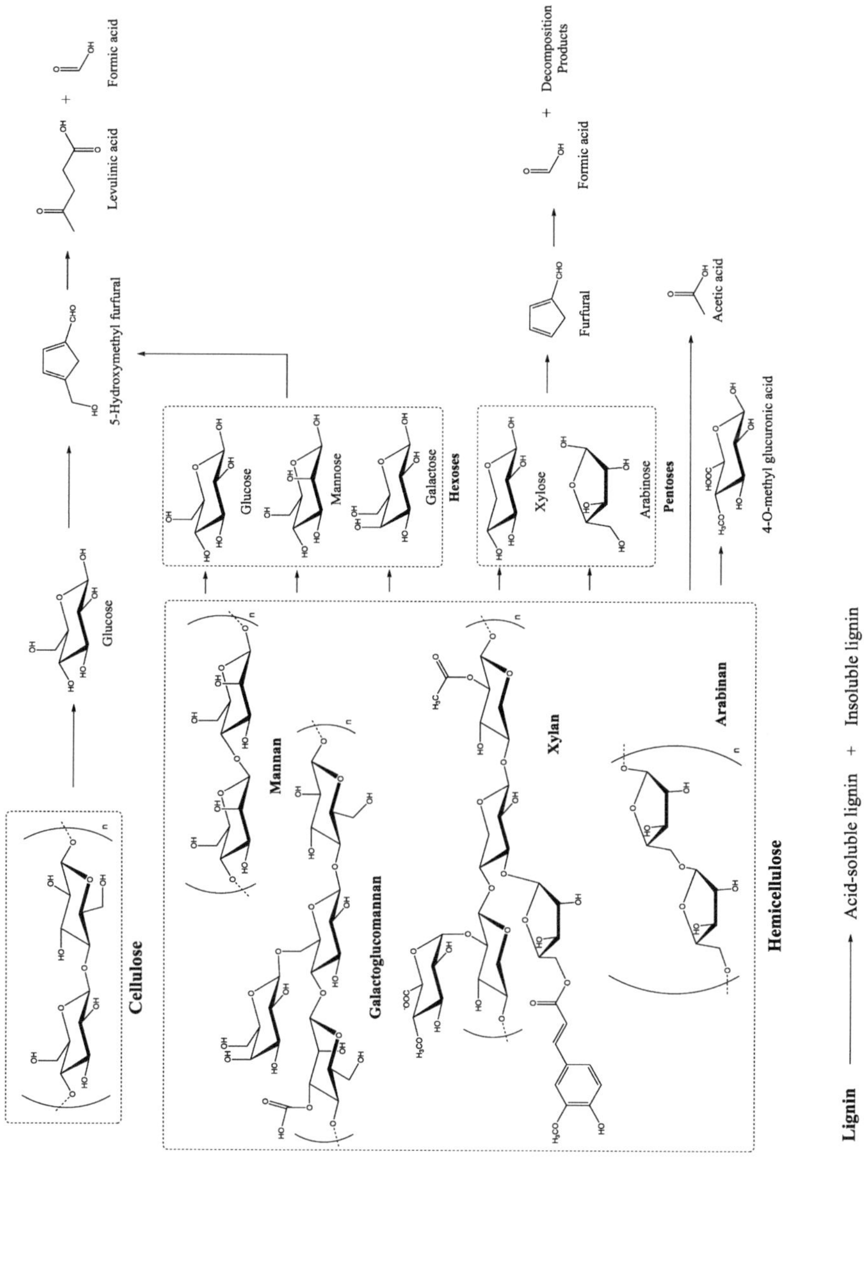


**Fig. 5.2** Reaction pathways of acid-catalyzed hydrolysis of lignocellulosic biomass

## 5.2 Chemistry and Catalysis Towards the Formation of LA

### 5.2.1 Reaction Mechanism of LA Formation from Sugars

Several reaction mechanisms for the formation of LA have been reported in the literature, see for instance, the review of van Putten et al. [11]. The first step in the formation of LA is the acid-catalyzed dehydration of the monosaccharides to form HMF. The mechanistic pathway for this dehydration reaction involves an acyclic or cyclic pathway (Fig. 5.3). The acyclic pathway assumes that the intermediate 1,2-enediol is formed from the monosaccharides by a Lobry de Bruyn-Alberda van Ekenstein mechanism. This intermediate is converted to HMF through two consecutive $\beta$-dehydration reactions, followed by a ring closure with water elimination. The cyclic pathway starts from the cyclic ketofuranose, which is dehydrated at C2 to form a tertiary carbenium cation. Subsequently, two consecutive $\beta$-dehydrations in the ring lead to the formation of HMF. In van Putten et al., it was concluded that fructose is more readily converted to HMF than glucose, and as a result the cyclic pathway for fructose is much more likely to occur than the acyclic pathway [11].

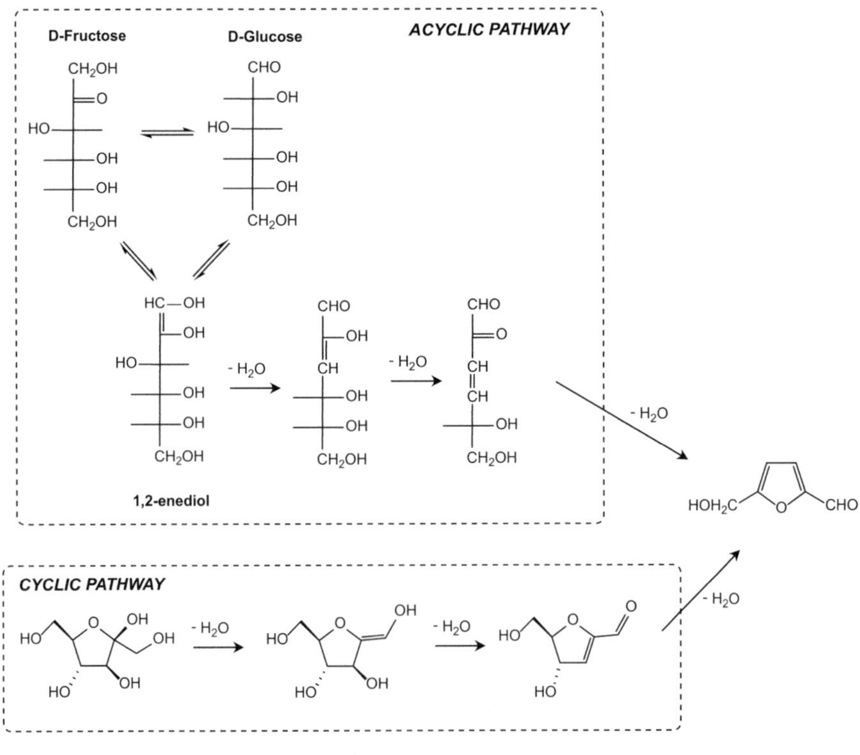


**Fig. 5.3** Reaction mechanism pathways of dehydration reactions of sugars to 5-hydroxymethylfurfural

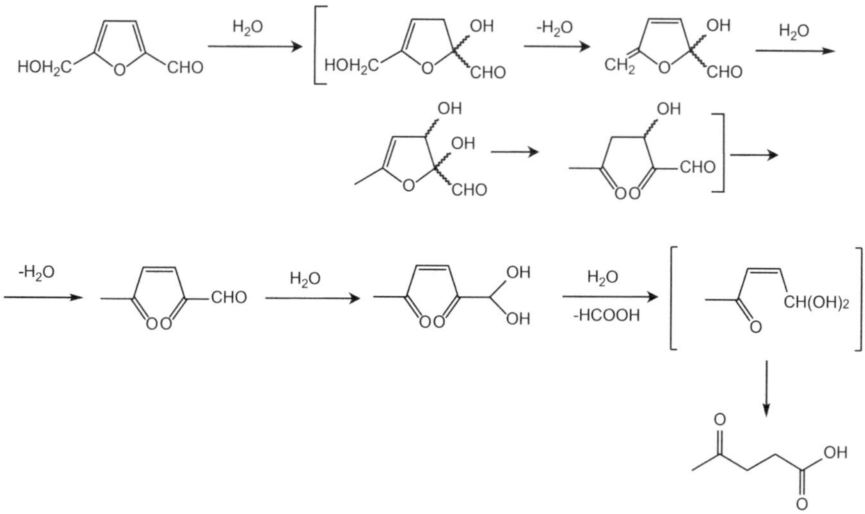


**Fig. 5.4** Reaction mechanism of 5-hydroxymethylfurfural hydration to levulinic acid

Studies by Horvat et al. using nuclear magnetic resonance (NMR) have shown that the conversion of HMF to LA involves the addition of water to the C2 – C3 bond of the furan [12] (Fig. 5.4). This mechanism was confirmed by a recent study using $^{13}C$–labeled fructose on the C1 and C6 position. It revealed that the carbonyl carbon of HMF is incorporated in formic acid, whereas the hydroxymethyl carbon forms the methyl carbon (C5) in LA [13].

The main byproducts of LA production from sugars are soluble and insoluble polymeric materials (mostly known as humins), which are formed by condensation of HMF with sugars through 2,6-anhydro-*β*-D-fructofuranose as the intermediate product [14]. These humins not only cause blocking of pipes and other process equipment, but also lower the carbon efficiency of the conversions considerably [15, 16]. In some cases, humins yields as high as 40% based on feed intake have been reported [17, 18].

A number of investigations have been reported on conversion methods of solid humins to higher value-added products. For instance, Hoang et al. studied the valorization of humins by steam reforming with alkali-metal-based catalysts (900–1200 °C) [19, 20] and found that the highest activity was obtained when using $Na_2CO_3$ as the catalyst. The $H_2$ to CO ratio of the produced syngas was about 2. However, substantial loss of carbon was observed during the heating up stage (up to 45%-wt. on intake). Humins derived from glucose can be depolymerized using a hydrotreatment with Ru/C in an isopropanol/formic acid mixture [21]. The reaction was carried out at 400 °C and humins conversions up to 69% were achieved [21]. Liquid products from that work were shown to consist of both mono- and oligomeric compounds, and the major GC detectable compounds were alkylphenolics, aromatics (mono and with multiple fused rings) and cyclic alkanes, though their yields based on humins intake was relatively low (< 10%). It was shown that molecular hydrogen is formed in situ by the catalytic conversion of formic acid and that this in situ produced hydrogen can be used for hydrogenolysis and hydro(deoxygenation) reactions involving the humins structure. In addition, IPA was shown to be reactive under these conditions and was (partly) converted into acetone and hydrogen. The latter finding shows that formic acid is possibly not required for the liquefaction reaction which would considerably simplify the process.

## 5.2.2 *Synthesis of LA*

The first-study on the preparation of LA was reported in the 1840s by the Dutch professor G. J. Mulder [22]. Numerous studies on the synthesis of LA from various feedstocks under different catalytic conditions have been reported in the last decades. Even though various solvents have been applied, water is still the most commonly used solvent for the synthesis of LA from monosaccharides, polysaccharides, and lignocellulosic biomass.

### 5.2.2.1 Homogeneous Catalysts

Table 5.1 shows an overview of LA yields obtained from various biomass-related feedstocks using homogeneous catalysts. Mineral acids, such as HCl and $H_2SO_4$, are the most commonly used homogeneous catalysts. Depending on reaction conditions, the maximum reported LA yields from fructose are 73% and 81%-mol when using $H_2SO_4$ and HCl as the catalyst, respectively. Lower yields of LA are obtained when using glucose as the starting material and HCl/$H_2SO_4$ as catalysts. Other types of homogeneous catalysts (e.g. $InCl_3$ and methanesulfonic acid) have been tested for the synthesis of LA from glucose, and a significant improvement in LA yield (64%-mol) is obtained when methanesulfonic acid is used as the catalyst. Unfortunately, isolated yields are hardly reported, and most are based on GC, HPLC or NMR analysis of the liquid phase. In addition, mass balances closures are in most cases also not provided, mainly because the soluble humins are difficult to quantify.

The cyclic pathway shown in Fig. 5.3 requires glucose to isomerize to fructose prior to the dehydration to HMF. The addition of Lewis acid catalysts (e.g. $CrCl_3$) enhances the rate of glucose isomerization to fructose, and a combination of Lewis and Brønsted acid catalysts is beneficial to obtain high yields of LA from glucose [23]. However, the reported yields are still lower than LA yields from fructose. Further research should focus on the screening and selection of novel Lewis acids and the optimization of reaction conditions.

LA yields from polysaccharides are close to those obtained from monosaccharides [7]. However, when using polysaccharides, more severe reaction conditions seem to be required, i.e. longer reaction time or higher amounts of acid catalysts to

**Table 5.1** Levulinic acid (LA) yields from various biomass-related feedstocks using homogeneous catalysts

| Feedstock | Catalyst | T (°C) | Reaction time | LA yield[a] (%-wt) | LA yield[a] (%-mol) | Ref. |
|---|---|---|---|---|---|---|
| Fructose | $H_2SO_4$ 0.1 M | 120 | 24 h | 40 | 62 | [24] |
| | $H_2SO_4$ 1 M | 140 | 30 min | 47 | 73 | [25] |
| | $H_2SO_4$ 2 M | 170 | 30 min | 28 | 43[b] | [26] |
| | HCl 2 M | 100 | 24 h | 52 | 81 | [27] |
| | HCl 2 M | 170 | 30 min | 32 | 49[b] | [26] |
| | Trifluoroacetic acid 0.5 M | 180 | 1 h | 37 | 57 | [28] |
| Glucose | $H_2SO_4$ 1 M | 140 | 2 h | 38 | 59 | [16] |
| | $H_2SO_4$ 2 M | 170 | 30 min | 26 | 41[b] | [26] |
| | HCl 0.1 M | 160 | 2.5 h | 35 | 54 | [29] |
| | HCl 2 M | 170 | 30 min | 31 | 49[b] | [26] |
| | $InCl_3$ 0.01 M | 180 | 1 h | 37 | 57 | [30] |
| | Methanesulfonic acid 0.5 M | 180 | 15 min | 41 | 64 | [31] |
| | HCl 0.1 M + $CrCl_3$ 0.02 M | 140 | 6 h | 30 | 47 | [23] |
| | $H_3PO_4$ 0.02 M + $CrCl_3$ 0.02 M | 170 | 4.5 h | 32 | 50 | [32] |
| Sucrose | HCl 0.2 M | 150 | 3 h | 24 | 71[c] | [33] |
| Extruded starch | $H_2SO_4$ 4%-wt | 200 | 40 min | 48 | 67 | [34] |
| Cellobiose | $H_2SO_4$ 2 M | 170 | 30 min | 28 | 41[b] | [26] |
| | HCl 2 M | 170 | 30 min | 30 | 44[b] | [26] |
| Cellulose | $H_2SO_4$ 1 M | 150 | 2 h | 43 | 60 | [35] |
| | $H_2SO_4$ 2 M | 170 | 50 min | 23 | 34[b] | [26] |
| | HCl 0.927 M | 180 | 20 min | 44 | 60 | [36] |
| | HCl 2 M | 170 | 50 min | 31 | 46[b] | [26] |
| | $CrCl_3$ 0.02 M | 200 | 3 h | 48 | 67 | [37] |
| | $[C_3SO_3Hmim]HSO_4$ | 160 | 30 min | 32 | 45[b] | [38] |
| | $[BSMim]HSO_4$ | 120 | 2 h | 40 | 56 | [39] |
| Water hyacinth (26%-wt glucan) | $H_2SO_4$ 1 M | 175 | 30 min | 9 | 53 | [40] |
| Pretreated hybrid poplar (92%-wt glucan) | $H_2SO_4$ 5%-wt | 190 | 50 min | 18 | 60 | [41] |
| Poplar sawdust (58%-wt glucan) | HCl 11.5 meq | 200 | 1 h | 21 | 37 | [42] |
| Giant reed (35%-wt glucan) | HCl 1.68%-wt | 190 | 20 min | 22 | 30 | [43] |
| Corncob residue (62%-wt glucan) | NaCl 6.8 M + $AlCl_3$ 0.08 M | 180 | 2 h | 33 | 47 | [44] |

[a]LA yields measured using HPLC or GC analysis and are based on the amount of C6 sugars in the feedstock. [b]LA yields measured using NMR analysis with internal standard. [c]Each mol of sucrose contains 1 mol of fructose and 1 mol of glucose

achieve similar LA yields as for monosaccharides. Several studies on the synthesis of LA from cellulose using a Lewis acid ($CrCl_3$) as catalyst [37] or acidic ionic liquids have been reported. In the latter case, the ionic liquid acts as the catalyst and reaction medium [38, 39]. The reported LA yields from these studies are more or less comparable with LA yields obtained using mineral acids as catalyst.

A large number of studies have been carried out on the production of LA from various types of actual biomass. Apart from the reaction conditions and catalysts used, LA yields from biomass strongly depend on the biomass composition, especially the amount of hexose sugars present in each type of biomass (Table 5.1). Experimental results given in Table 5.1 are not exhaustive and additional data are reported in reviews [6, 7]. The presence of lignin in the lignocellulosic biomass may have a negative effect on the effectiveness of the hydrolysis of cellulose and hemicellulose fractions and as such, also affects the yield of LA. For instance, Li et al. has published a review on correlations between lignin structure and biomass recalcitrance [45]. In that review, the role of lignin on the pretreatment process and enzymatic hydrolysis are comprehensively reviewed. The presence of lignin leads to a reduction in the enzymatic hydrolysis rates through physical barriers and non-productive binding to enzymes. As such, the presence of lignin in the early stage of enzymatic hydrolysis inhibits the conversion of cellulose to sugars [46]. One study has reported the effect of lignin on the rate of chemical hydrolysis of cellulose using an acidic ionic liquid, [BMIM]Cl as catalyst [47]. With an increase in the concentration of lignin, the glucose yield on cellulose intake decreased from 23 to 7%-wt., which indicates that lignin may also significantly inhibit the chemical hydrolysis of cellulose.

A consortium led by the University of Limerick completed a research project funded by the European Union entitled "Development of Integrated Biomass Approaches NETwork (DIBANET)". The main objective of that project was to develop technologies for the synthesis of ethyl-levulinate from organic wastes and residues. As part of that project, Dussan carried out an extensive study to determine the compositions of 15 different lignocellulosic biomass sources and to evaluate compositional differences on the LA yields [48]. The biomass composition indeed played a major role and high hexose content was shown to have a positive effect on LA yield. In general, lower LA yields were obtained at higher temperatures, which is in line with previous studies where high temperatures have shown to decrease the LA selectivity. The study also revealed that the amount of inorganic materials in the lignocellulosic biomass plays an important role by neutralizing acid catalysts. As a result, lignocellulosic biomass with a higher content of base (for instance, potassium hydroxide) such as in banana stalk, give lower yields of LA.

#### 5.2.2.2 Heterogeneous Catalysts

Heterogeneous catalysts have received high interest as catalysts for LA synthesis (Table 5.2). Heterogeneous catalysts offer several advantages compared with homogeneous ones, such as few issues with corrosion and ease in separation. Reported

**Table 5.2** Levulinic acid (LA) yields from biomass-related feedstocks using heterogeneous catalysts

| Feedstock | Catalyst | T (°C) | Reaction time | LA yield[a] (%-wt) | LA yield[a] (%-mol) | Ref. |
|---|---|---|---|---|---|---|
| Fructose | Amberlyst-15 | 120 | 24 h | 34 | 52 | [24] |
| | LZY-zeolite | 140 | 15 h | 43 | 66 | [50] |
| | 5-Cl-SHPAO [b] | 165 | 1 h | 43 | 66 | [51] |
| Glucose | Amberlyst-70 | 180 | 25 h | 32 | 50 | [52] |
| | Fe/HY zeolite | 180 | 4 h | 43 | 66 | [53] |
| | 5-Cl-SHPAO[b] | 165 | 5 h | 33 | 51 | [51] |
| | Sulfonated graphene oxide | 200 | 2 h | 50 | 78 | [49] |
| Sucrose | Amberlyst-36 | 150 | 3 h | 18 | 53 | [33] |
| | 5-Cl-SHPAO [b] | 165 | 5 h | 19 | 55 | [51] |
| Starch | 20%Ru-ZSM-5 | 300 | 1 h | 21 | 32 | [54] |
| | 5-Cl-SHPAO[b] | 165 | 7 h | 36 | 50 | [51] |
| Cellulose | $ZrO_2$ | 180 | 3 h | 39 | 54 | [55] |
| | Al–$NbOPO_4$ | 180 | 24 h | 38 | 53 | [56] |
| Nordic pulp (91%-wt glucan) | Amberlyst-70 | 180 | 60 h | 37 | 57 | [52] |
| Dried sugarcane bagasse (51%-wt glucan) | Acid-activated bentonite | 200 | 60 min | 16 | 45 | [57] |
| Empty fruit bunch (41%-wt glucan) | $CrCl_3$ + HY zeolite hybrid | 145 | 147 min | 16 | 59 | [58] |

[a]LA yields measured using HPLC or GC analysis. [b]5-chloro-sulfonated hyper branched poly-(arylene oxindole)

LA yields from low molecular weight sugars (such as fructose, glucose and sucrose) using heterogeneous catalysts (e.g. zeolites or Amberlyst resins) are typically >50%-mol and comparable with yields obtained using homogenous catalysts. Remarkably high LA yields can be obtained from glucose (78%-mol) when using graphene oxide with sulfonic acid groups as catalyst [49]. However, when considering available data, this value seems to be overestimated, possibly due to experimental/analytical issues. The main disadvantage of the use of heterogeneous catalysts is deactivation due to the deposition of humins leading to blocking of the catalyst surface or active sites.

Many studies have been carried out on the synthesis of LA from polysaccharides using heterogeneous catalysts (Table 5.2). For instance, LA yields of up to 54%-mol from cellulose using $ZrO_2$ as catalyst have been reported [55]. Heterogeneous catalysts have been used for the synthesis of LA from lignocelluosic biomass, such as sugar cane bagasse [57] and empty fruit bunch [58] with reported LA yields of 45 and 59%-mol, respectively. Even though heterogeneous catalysts are interesting catalysts for LA synthesis, the main challenge will be on the development of methodology to prevent catalyst deactivation caused by excessive humin deposition.

#### 5.2.2.3 Biphasic Systems

Biphasic liquid-liquid systems are applicable to LA synthesis, for example through the use of an aqueous and organic solvent. This approach was introduced to improve LA yields and to facilitate catalyst recycling concepts. For instance, conversion of cellulose to LA in a biphasic system (water - γ-valerolactone) was examined using aqueous HCl saturated with 35%-wt. NaCl (1:1 aqueous to organic ratio) [59]. The highest LA yield was 72%-mol when using 1.25 M HCl (155 °C, 1.5 h). Synthesis of LA from rice straw using $H_2SO_4$ was examined in a biphasic system with acetone and water at 180 °C [60]. The highest LA yield was 10%-wt. (on straw) after 5 h. Bamboo fibers used as a source for LA synthesis in a biphasic system consisting of $H_2O$/THF using sulfamic acid as the catalyst [53] allowed LA yields of up to 11%-mol with microwave heating (500 W) for a $H_2O$/THF phase ratio of 3 to 1.

## 5.3 Process Technology

### *5.3.1 Kinetic Studies on LA Synthesis*

Reliable kinetic models are essential in the design of optimum processes for production of LA. Kinetic studies of LA synthesis from various lignocellulosic feedstocks have been reported (Fig. 5.5). Most of the studies use the approach proposed by Saeman for Douglas fir [61]. In that study, two consecutive first-order reactions are used to model the hydrolysis reaction of biomass to glucose as the main product (Eq. 5.1).

$$\text{Cellulosic biomass} \xrightarrow{k_1} \text{Glucose} \xrightarrow{k_2} \text{Decomposition products} \tag{5.1}$$

The effects of temperature ($T$) and acid concentration ([*acid*]) are included in the reaction rate constants ($k_1$ and $k_2$) as follows:

$$k_{\mathrm{i}} = A_{\mathrm{o,i}} \left[acid\right]^{m_{\mathrm{i}}} \exp^{\frac{-E_{\mathrm{i}}}{RT}} \quad \mathrm{i} = 1,2 \tag{5.2}$$

Here, $A_{o,i}$ is the frequency factor, $m_i$ is the reaction order in acid, $R$ is the ideal gas constant and $E_i$ is the activation energy. An overview of kinetic studies is given in Girisuta et al. [62] and the reported activation energies for the formation and decomposition reactions of glucose from various feedstocks is shown in Fig. 5.5.

The activation energy of glucose formation from different types of lignocellulosic biomass or pure cellulose are between 165 and 190 kJ/mol [62]. These values show that hydrolysis of the cellulose fraction to glucose strongly depends on the type of feedstock. The lowest activation energy for glucose formation is observed when using cellobiose, which is apparently relatively easily hydrolyzed. Activation energies for glucose decomposition are relatively close and are between 130 and 140 kJ/mol [62], indicating that the activation energy for the decomposition reaction of

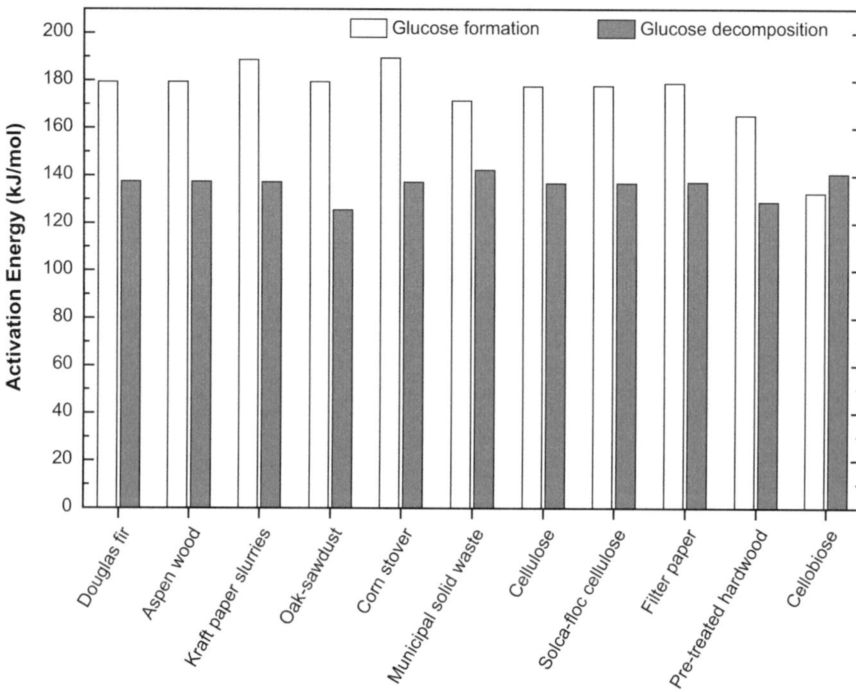


**Fig. 5.5** Activation energies of the formation and decomposition reactions of glucose

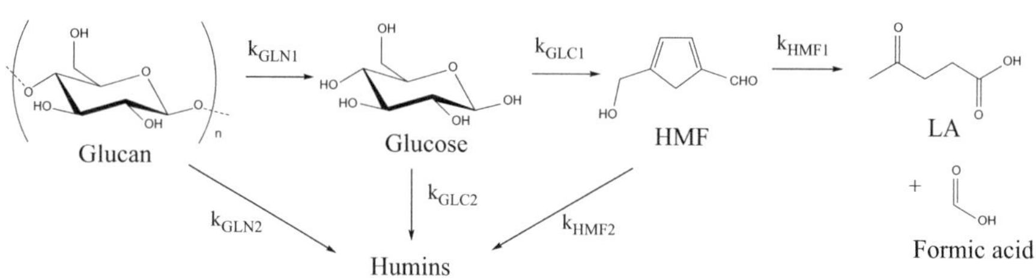


**Fig. 5.6** Reaction scheme of the acid-catalyzed hydrolysis of the glucan fraction in biomass to levulinic acid (LA)

glucose is relatively independent of the feedstock type. The data also reveal that the activation energies of glucose formation are higher than the activation energies of glucose decomposition. As a result, it is beneficial to conduct the hydrolysis reaction of lignocellulosic biomass at higher temperature to obtain higher glucose yields.

A generalized reaction scheme used for kinetic modelling studies on biomass conversion to LA is given in Fig. 5.6. Most of these kinetic studies assume that the consecutive reactions are first-order reactions in the substrates.

For homogeneous acid catalysts, the literature shows that the activation energies to the desired products (glucose, HMF and LA) are lower than those for the formation of undesirable humin byproducts [35, 62–64]. This observation indicates that lower temperatures are favored to reduce the rate of formation of humins. Due to the contrary effects of temperature on glucose and LA formation, it is preferable to perform LA synthesis from biomass in two stages at different temperatures. The first-stage is operated at higher temperatures to optimize the glucose yield from the cellulose fraction; the second-stage is preferably operated at a lower temperature to optimize the yield of LA from glucose.

### *5.3.2 Product Separation and Isolation*

One of the main challenges in the production of LA when using homogeneous catalysts is separation of the product from the aqueous phase containing the mineral acids [65]. A commonly used method to isolate LA from the product stream is by organic solvent extraction, for example with methyl isobutyl ketone (MIBK) [66]. The extraction solvent is separated from the LA in an evaporator and can be recycled back to the extraction unit. Further concentration and purification of LA is carried out in a fractionation unit (e.g. vacuum distillation). In a 2014 DSM patent [67], a method was disclosed to isolate LA involving nanofiltration. The first-step is a solvent extraction to yield an organic phase with LA. The organic phase is subjected to a nanofiltration step to remove soluble humins. Finally, LA is isolated after distillation and the solvent is recycled.

### *5.3.3 Commercial Status of LA Production*

Initial attempts to commercialize the production of LA were based on Biofine technology [68], which is schematically shown in Fig. 5.7. In this process concept, a carbohydrate-rich feedstock and sulfuric acid are mixed and fed into a reactor operated at 210–220 °C at a mean residence time of 12 s to effectively hydrolyze the polysaccharides into soluble sugar monomers and oligomers. The outlet stream of the first reactor is fed into a continuously stirred tank reactor operated at a lower temperature (190–200 °C) but having a longer mean residence time of 20 min than the first reactor. LA is removed as the liquid product from the second reactor, while formic acid and furfural are recovered from the vapor stream. Solid by-products are separated from the aqueous LA solution with a filter-press. Based on this technology, a 1 ton per day pilot plant was built in New York, USA [69]. The current status of the Biofine process is unknown, but it is probable that a commercial scale plant is not in operation yet at the time of this writing.

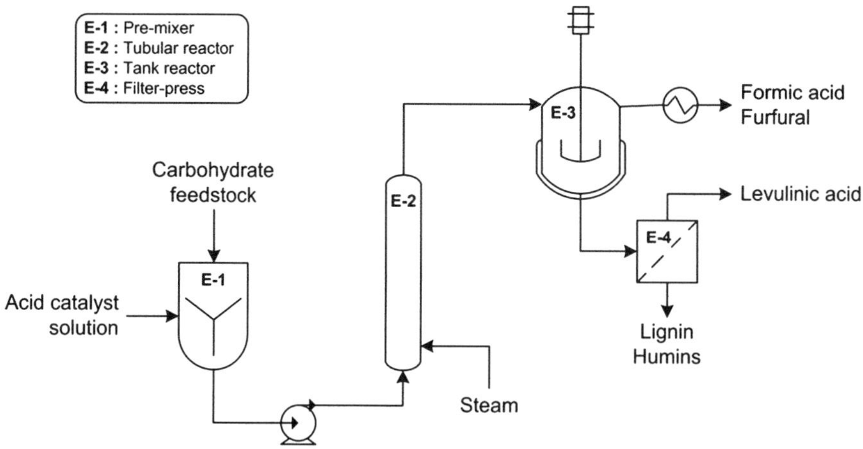


**Fig. 5.7** Biofine process for the production of levulinic acid

Another pilot plant for producing LA was built in 2013 by Segetis in Golden Valley, Minnesota [70]. The pilot plant was designed for a capacity of 80 metric tons of LA per year and uses corn sugar as the main feedstock, although the design allows the possibility of using a wide range of biomass feedstocks. Segetis intended to use the LA for the synthesis of levulinic ketals based on a highly selective process for esters of LA with alcohols derived from vegetable oils, e.g. glycerol. In 2016, Segetis was acquired by GFBiochemicals [71]. The latter company started LA production in 2009 in a pilot plant facility with a capacity of 2000 metric tons per year in Caserta, Italy. In 2015, GFBiochemicals announced to start the commercial scale production of LA with the objective to produce up to 10,000 metric tons of LA per year by 2017.

## 5.4 Potential Applications of LA and Its Derivatives

An overview of potentially interesting LA derivatives is presented in Fig. 5.8.

### *5.4.1 Diphenolic Acid*

Diphenolic acid (DPA) has been identified as a potential replacement for bisphenol A, which is one of the monomers for epoxy resins and polycarbonates. DPA can be made by the condensation reaction of LA with phenol in the presence of acid catalysts (Scheme 5.1).

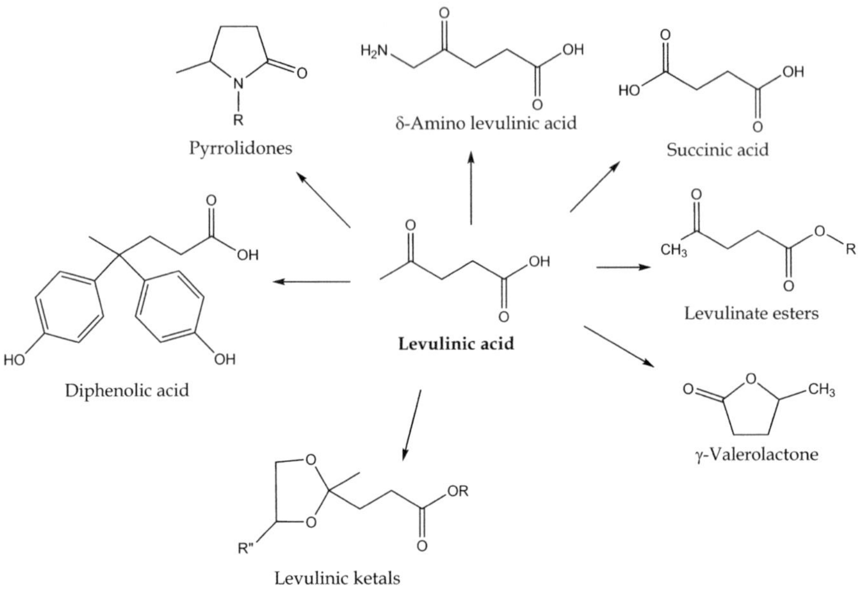


**Fig. 5.8** Potentially interesting levulinic acid derivatives

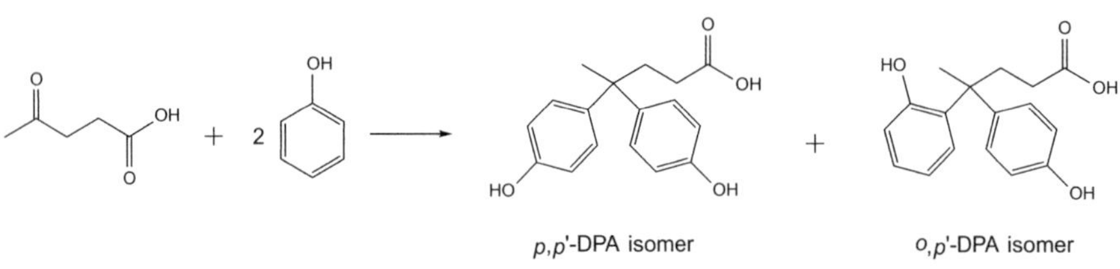


**Scheme 5.1** Synthesis of diphenolic acid from levulinic acid

Table 5.3 gives DPA yields for selected homogeneous and heterogeneous catalysts and reaction conditions. Some studies have reported DPA yields exceeding 50%, although the yield of the reaction is reduced by the formation of the undesirable *o,p′*-DPA isomer byproduct. The molar ratio of both isomers has a significant effect on important polycarbonate properties, such as color stability and crystallinity. One study suggests that the ratio of both isomers is a function of the temperature and that high reaction temperatures tend to lower the selectivity to the desired *p,p'*-DPA isomers [72].

**Table 5.3** Diphenolic acid (DPA) yields for selected catalyst types and reaction conditions

| Catalyst | T (°C) | Reaction time(h) | Phenol:LA mol ratio | DPA yield (%)[a] | Selectivity (%)[b] | Ref. |
|---|---|---|---|---|---|---|
| $H_2SO_4$ (300%-wt of LA) | 25 | 20 | 2:1 | 60 | N.A. | [73] |
| HCl (40%-mol of LA) | 100 | 6 | 3:1 | 28 | 69 | [74] |
| HCl (50%-mol of LA) | 60 | 24 | 4:1 | 62 | 67 | [75] |
| [BSMim]$HSO_4$ (50%-mol of LA) | 60 | 30 | 4.5:1 | 80 | 100 | [75] |
| $Cs_{1.5}H_{4.5}$–$P_2W_{18}O_{62}$ | 150 | 24 | 4:1 | 71 | 88 | [74] |
| $Cs_{2.5}H_{0.5}$–$PW_{12}O_{40}$ | 150 | 24 | 9:1 | 45 | 83 | [74] |
| $H_3PW_{12}O_{40}$ | 100 | 16 | 3:1 | 34 | 90 | [76] |
| SHPAO[c] + Ethanethiol | 100 | 16 | 3:1 | 53 | 95 | [76] |

[a]DPA yields based on molar amount of LA. [b]Selectivity of p,p′-DPA isomer. [c]Sulfonated hyperbranched poly-(arylene oxindole)

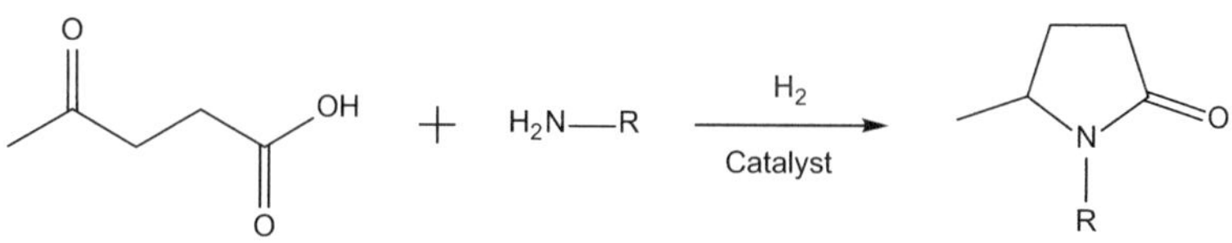


**Scheme 5.2** Synthesis of pyrrolidones from levulinic acid

## 5.4.2 *Pyrrolidones*

Reductive amination of LA is an effective route to synthesize N-(methyl, aryl, or cycloalkyl)-5-methyl-2-pyrrolidones (Scheme 5.2). These pyrrolidones may be used as solvents, surfactants, complexing agents and are important ingredients in topical formulations, such as ointments, creams, lotions, pastes and gels [77].

Manzer and co-workers have patented heterogenous catalysts for the reductive amination of LA in combination with molecular hydrogen. Examples are transition-metals (e.g. Ni, Cu, Rh, Ru, Ir, or Pt) supported on silica, alumina or carbon [77, 78]. The use of formic acid, which is the main byproduct of LA synthesis, instead of hydrogen, has gained high interest. Wang and co-workers have reported the use of homogeneous cyclometalated iridium complexes as catalyst for the reductive amination of LA using formic acid as the hydrogen source [79]. Shimizu's group reported the use of Pt and MoOx on $TiO_2$ as catalysts for the reductive amination of LA under solvent-free conditions [80]. Esposito's group reported the use of a continuous reactor for the reductive amination of LA using inexpensive FeNi catalysts [81]. Catalyst stability was good and the conversion and selectivity were about constant for runtimes up to 52.5 h. Sakai et al. demonstrated that $In(OAc)_3$ is a good catalyst for the reductive amination of LA using $PhSiH_3$ as the reducing agent [82].

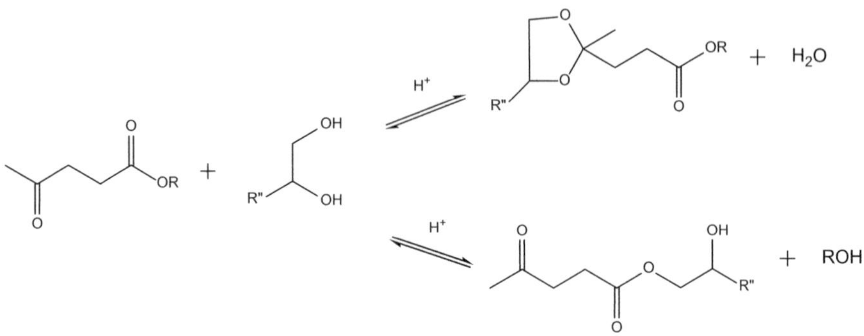


**Scheme 5.3** Synthesis of levulinic ketals from levulinic acid

**Table 5.4** Production of levulinic ketals from ethyl levulinate and glycerol catalyzed by homogeneous and heterogeneous catalysts [84]

| Ethyl levulinate (mol) | Glycerol (mol) | Catalyst | T (°C) | Reaction time (min) | Selectivity (%) |
|---|---|---|---|---|---|
| 2.95 | 0.56 | $H_2SO_4$ (3 mmol) | 80 | 20 | 96.8 |
| 2.95 | 0.56 | $H_2SO_4$ (0.03 mmol) | 80 | 20 | 100 |
| 0.63 | 0.14 | HCl (0.025 mmol) | 110 | 30 | 100 |
| 0.63 | 0.14 | $NH_3^+SO_3^-$ (0.025 mmol) | 110 | 30 | 100 |
| 0.28 | 0.14 | Amberlyst 15(0.024 mmol) | 110 | 135 | 98.7 |

## 5.4.3 *Levulinic Ketals*

Levulinic ketals are made by reacting LA or levulinate esters with a mono-alcohol or a diol in the presence of acidic catalysts (Scheme 5.3). The levulinic ketals may be used as building blocks in the production of plasticizers for PVC, or in polyols for polyurethane materials and in polymer synthesis [83].

Segetis has reported catalysts and reaction conditions to synthesize levulinic ketals from ethyl levulinate and glycerol (Table 5.4) [84]. The selectivity to ketal formation was usually >98% with almost full conversion. To shift the equilibrium reaction to the desired product, the reaction was carried out by continuously removing water under vacuum. Table 5.4 also shows that low acid loading is sufficient to achieve very high product selectivities.

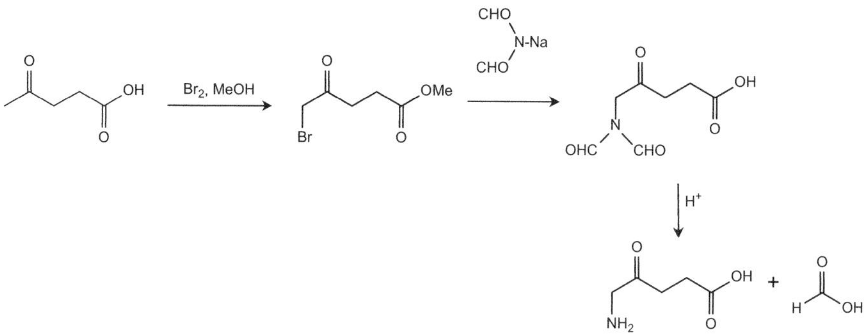


**Scheme 5.4** Synthesis of δ-aminolevulinic acid from levulinic acid

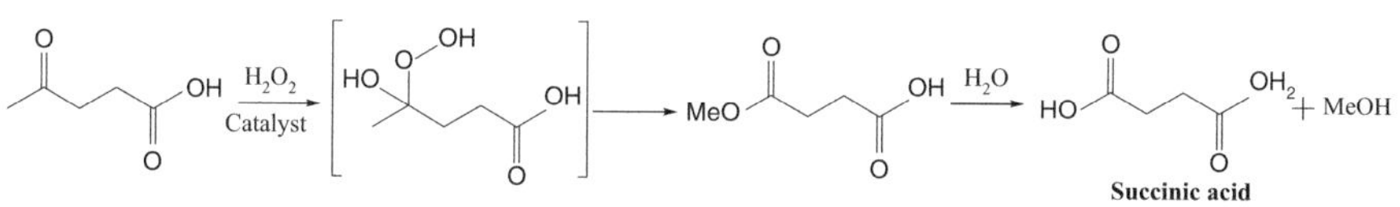


**Scheme 5.5** Synthesis of succinic acid from levulinic acid

## 5.4.4 *δ-Aminolevulinic Acid*

NREL showed that δ-Aminolevulinic acid (DALA) can be prepared from LA using a two-step procedure (Scheme 5.4). In the first step, LA is brominated in methanol to give methyl 5-bromolevulinate. This ester is reacted with a nitrogen-containing nucleophile, such as sodium diformylamide, to give DALA in high yields (>80 mol %) and purity (>90%) [15]. A biotechnological approach for producing DALA is by using certain bacteria [85].

DALA can be used as a biodegradable herbicide [86] as well as a growth promoter by increasing the chlorophyll content of cells. These effects are observed in potatoes, garlic, radishes and also blue-green algae (*Spirulina platensis*) [87]. In the medical field, DALA can be used to detect heavy-metal poisoning. Advanced medical applications of DALA involve its use in photodynamic therapies for cancer patients, for instance to treat brain tumors [88].

## 5.4.5 *Succinic Acid*

Succinic acid may be obtained by the catalytic oxidation of LA with oxidants such as oxygen and peroxides (Scheme 5.5).

Podolean and co-workers investigated the utilization of Ru(III)/silica-coated magnetic nanoparticles for the oxidation of LA to succinic acid under mild conditions and without the use of a base [89]. They reported 54%-mol conversion of LA with excellent selectivity (98–99%) to succinic acid. Dutta et al. showed that the reaction of LA with hydrogen peroxide with trifluoroacetic acid as the catalyst at 90 °C and 2 h reaction time gives succinic acid in good yields (62%-mol) [90]. Those authors proposed a work-up procedure to isolate the product from the volatile catalyst and byproducts (acetic acid, formic acid, and 3-hydroxypropanoic acid). Under basic conditions, those authors successfully converted LA to 3-hydroxypropanoic acid, which is a precursor to acrylic acid, in high yields [91].

### 5.4.6 *γ-Valerolactone*

LA can be catalytically reduced to γ-valerolactone (GVL). A large number of studies have been conducted on the catalytic hydrogenation of LA to GVL, and several reviews are available [92–95]. GVL has been identified as a platform chemical to produce valuable chemicals such as α-methylene-γ-valerolactone which is an acrylic monomer that has structural similarities with methyl methacrylate [96]. GVL has also been used as starting material for adipic acid, a precursor for nylon production [97]. Further details on the synthesis of GVL and its applications are provided in another chapter of this book.

### 5.4.7 *Levulinate Esters*

Levulinate esters are produced from the esterification of LA with alcohols using acid catalysts (Scheme 5.6).

Homogeneous catalysts, such as $H_2SO_4$ and HCl, are commonly used for the synthesis of levulinate esters from LA. Because these catalysts are difficult to separate and recycle, heterogeneous catalysts have been developed and tested for the synthesis of levulinate esters. Examples are heteropolyacids [98–100], zeolites [101, 102], sulfonic resins [103] and sulfated oxides [104, 105]. Levulinate ester yields of up to 95%-mol have been reported using heterogeneous catalysts.

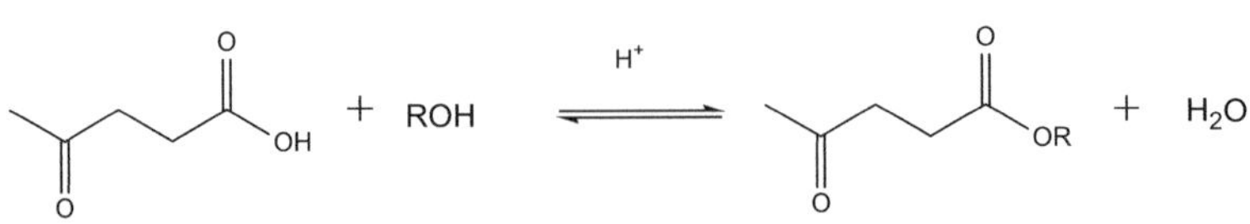


**Scheme 5.6** Synthesis of levulinate esters from levulinic acid

The synthesis of levulinates esters directly from monosaccharides, polysaccharides and lignocellulosic biomass has gained a lot of interest. This method involves contacting the biomass source with an alcohol in the presence of a homogeneous or heterogeneous catalyst. The yields of levulinate esters from these feedstocks are lower than from LA, due to the formation of by-products, such as formate ester, HMF ethers, HMF and humins. The presence of water, either from the biomass source or formed during the various intermediate reactions, may lead to incomplete esterification that is confirmed by the presence of LA in the reaction mixtures. In addition, under acidic conditions, the alcohol may be converted to the corresponding dialkylethers. It has been reported that humins formation is suppressed when using alcohols as the solvent [106]. The yields of levulinate esters from various feedstocks and using different types of catalysts are given in Table 5.5. Mostly, the yields are based on GC-FID measurements, and only a few papers report the isolated yields [107].

When using $H_2SO_4$ as the catalyst, the yields of levulinate esters from fructose are higher than those for other feedstocks. Application of 30%-wt of $H_2SO_4$ as catalyst allows operation at a lower reaction temperature and leads to higher yields of levulinate ester from cellulose [108]. However, conducting the reaction at higher concentrations of $H_2SO_4$ and elevated temperatures has some drawbacks such as corrosion and the decomposition of the alcohol to the corresponding dialkyl ether. Therefore, significant effort has been dedicated to the development of heterogeneous catalysts as evident by a review [109]. An example on the use of a heterogeneous catalysts ($TiO_2$ and sulfated $TiO_2$) is given in Fig. 5.9. It is clear that the yields of levulinate esters from fructose (and polysaccharides containing fructose moieties) are higher than those from glucose (and polysaccharides containing glucose moieties). Direct conversion of biomass to levulinates esters using 2-naphthalene sulfonic acid as catalyst has been reported with yields up to 97% [110]. However, a latter study failed to reproduce these high yields and obtained ethyl levulinate in only ca. 20% yield [111].

Even though heterogeneous catalysts give promisingly high yields of levulinate esters, catalyst stability is a key factor when considering implementation on larger scales. So far, most studies on catalyst stability involve experiments in batch set-ups using multiple recycle experiments. For further scale up, studies in continuous units for prolonged runtimes will be required to properly assess catalyst stability.

The use of ionic liquids has been investigated for the synthesis of ethyl levulinate from fructose (70%), glucose (13%) and sucrose (43%) [107]. In the case of glucose, alkyl-D-glucopyranoside with yield up to 50% was present in the reaction mixture.

Levulinate esters have potential applications in the synthesis of polymers, perfumes and flavoring formulations, processes for degreasing metallic surfaces, and latex coating compositions. Levulinate esters have been proposed as additives for diesel, gasoline, and biodiesel. However, a study on the performance of ethyl- and butyl-levulinate as diesel blending components shows that the levulinate esters have a very low cetane number and poor solubility in diesel at low temperatures [118]. Another disadvantage of the use of levulinate ester as a fuel blend is that a mixture of 5% ethyl levulinate in gasoline leads to higher levels of elastomer swell and water pickup [119]. A promising application of levulinate esters is the use as a precursor for the synthesis of levulinic ketals (see Sect. 5.4.3).

**Table 5.5** Selected studies on the synthesis of levulinate esters from various substrates

| Substrate | Alcohol | Catalyst | t (h) | T (°C) | Conversion (%-mol)[a] | Ester yield (%-mol)[a] | Ref. |
|---|---|---|---|---|---|---|---|
| Fructose | Ethanol | $H_2SO_4$ (10%-mol) | 30 | 120 | >99 | 56 | [112] |
| Glucose | Methanol | $H_2SO_4$ 0.02 M | 5 | 180 | N.A.[b] | 44 | [113] |
| Cellulose | Methanol | $H_2SO_4$ 0.02 M | 5 | 180 | N.A. | 43 | [113] |
| Cellulose | n-Butanol | $H_2SO_4$ 30%-wt | 5 | 130 | >99 | 60 | [108] |
| Cellulose | Methanol | $H_2SO_4$ 0.2 M | 2 | 170 | N.A. | 46 | [114] |
| Eucalyptus wood (49%-wt glucan) | Methanol | $H_2SO_4$ 0.2 M | 2 | 170 | N.A. | 44 | [114] |
| Levoglucosan | Methanol | Amberlyst-70 | 3 | 170 | >99 | 90 | [106] |
| Fructose | n-Propanol | CNT-PSSA[c] | 24 | 120 | >99 | 86 | [115] |
| Fructose | n-Propanol | Amberlyst-15 | 24 | 120 | >99 | 80 | [115] |
| Fructose | Methanol | $SO_4^{2-}$ / $TiO_2$ | 2 | 200 | N.A. | 59 (60)[d] | [116] |
| Glucose | Methanol | $SO_4^{2-}$ / $TiO_2$ | 2 | 200 | N.A. | 33 (35)[d] | [116] |
| Sucrose | Methanol | $SO_4^{2-}$ / $TiO_2$ | 2 | 200 | N.A. | 43 (41)[d] | [116] |
| Cellulose | Methanol | $SO_4^{2-}$ / $TiO_2$ | 2 | 200 | N.A. | 10 (8)[d] | [116] |
| Fructose | Methanol | $TiO_2$ nanoparticles | 1 | 175 | >99 | 80 | [117] |
| Glucose | Methanol | $TiO_2$ nanoparticles | 1 | 175 | >99 | 61 | [117] |
| Sucrose | Methanol | $TiO_2$ nanoparticles | 3 | 175 | >99 | 65 | [117] |
| Cellulose | Methanol | $TiO_2$ nanoparticles | 20 | 175 | 72 | 42 | [117] |
| Fructose | Ethanol | [BPyr-$SO_3$H][$HSO_4$][e] | 24 | 140 | >99 | 70 | [107] |
| Glucose | Ethanol | [BPyr-$SO_3$H][$HSO_4$][e] | 24 | 140 | >99 | 13 (50)[f] | [107] |
| Sucrose | Ethanol | [BPyr-$SO_3$H][$HSO_4$][e] | 24 | 140 | >99 | 43 (27)[f] | [107] |

[a]Conversion of feedstocks measured by HPLC. Yields of levulinate esters measured by GC-FID-MS. [b]Not available. [c]poly(p-styrenesulfonic acid)-grafted carbon nanotubes. [d]Values in parentheses are isolated yields. [e]1-(4-sulfobutyl)pyridinium hydrogensulfate. [f]Values in parentheses are yields of ethyl-D-glucopyranoside

## 5.5 Conclusions and Future Outlook

A considerable amount of research has been carried out to develop efficient, low cost, and environmentally benign processes for the production of levulinic acid from lignocellulosic biomass. Various studies have been reported dealing with the reaction mechanism of levulinic acid formation from various feedstocks on a molecular level. Better understanding of the formation of levulinic acid on a

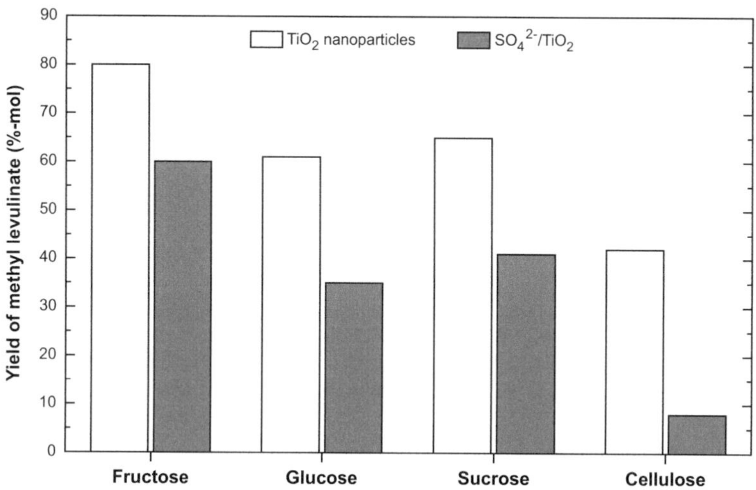


**Fig. 5.9** Yields of methyl levulinate for several biomass-related feedstocks and type of heterogeneous catalyst

molecular level is of high importance to develop strategies for improving levulinic acid yields and to suppress the formation of undesired humins. A wide range of different biomass feedstocks and catalysts have been tested for levulinic acid synthesis, and reaction conditions have been optimized to maximise the levulinic acid yields. Monosaccharides feedstocks give the highest levulinic acid yields; however, the use of cheap feedstocks such as lignocellulosic biomass is preferred for the commercial-scale production of levulinic acid. Homogeneous catalysts are widely used in laboratory scale syntheses. However, the separation and recycle of the homogeneous catalysts is a major issue and will require substantial research attention. In this respect, the use of biphasic reaction systems may be advantageous. Heterogeneous catalysts have been applied as well and are easier to recover and recycle than homogeneous catalysts. However, levulinic acid yields from heterogeneous catalysts tend to be slightly lower than those obtained for homogenous catalysts. In addition, humins deposition may have a negative effect on heterogeneous catalyst stability and advanced catalyst activation protocols may be required. Another issue is the possibility for loss in acidity of the solid acid catalyst by exchange of cations present in the mineral/ash fraction in most biomass sources. Kinetic studies and reactor development for the production of levulinic acid from biomass have also gained more and more interest in the last decade. A commercial scale levulinic acid production unit has been announced by GFBiochemicals and according to a recent press release, production started in 2016. This milestone will pave the way for the production of levulinic acid derivatives, which has so far been not possible due to the low market availability of levulinic acid.

## References

1. U.S. Energy Information Administration, International Energy Outlook. 2016, May 2016 (Access date: 14 September 2016). Retrieved from http://www.eia.gov/forecasts/ieo/
2. Climent MJ, Corma A, Iborra S. Conversion of biomass platform molecules into fuel additives and liquid hydrocarbon fuels. Green Chem. 2014;16:516–47.
3. de Jong E, Higson A, Walsh P, Wellisch M. Product developments in the bio-based chemicals arena. Biofuels Bioprod Biorefin-Biofpr. 2012;6:606–24.
4. Deuss PJ, Barta K, de Vries JG. Homogeneous catalysis for the conversion of biomass and biomass-derived platform chemicals. Cat Sci Technol. 2014;4:1174–96.
5. Sheldon RA. Green and sustainable manufacture of chemicals from biomass: state of the art. Green Chem. 2014;16:950–63.
6. Morone A, Apte M, Pandey RA. Levulinic acid production from renewable waste resources: bottlenecks, potential remedies, advancements and applications. Renew Sustain Energy Rev. 2015;51:548–65.
7. Mukherjee A, Dumont M-J, Raghauan V. Sustainable production of hydroxymethylfurfural and levulinic acid: challenges and opportunities. Biomass Bioenergy. 2015;72:143–83.
8. Pileidis FD, Titirici M-M. Levulinic acid biorefineries: new challenges for efficient utilization of biomass. ChemSusChem. 2016;9:562–82.
9. Yan K, Jarvis C, Gu J, Yan Y. Production and catalytic transformation of levulinic acid: a platform for speciality chemicals and fuels. Renew Sustain Energy Rev. 2015;51:986–97.
10. Rackemann DW, Doherty WOS. The conversion of lignocellulosics to levulinic acid. Biofuels Bioprod Biorefin. 2011;5:198–214.
11. van Putten R-J, van der Waal JC, de Jong E, Rasrendra CB, Heeres HJ, de Vries JG. Hydroxymethylfurfural, a versatile platform chemical made from renewable resources. Chem Rev. 2013;113:1499–597.
12. Horvat J, Klaic B, Metelko B, Sunjic V. Mechanism of levulinic acid formation. Tetrahedron Lett. 1985;26(17):2111–4.
13. Zhang J, Weitz E. An in situ NMR study of the mechanism for the catalytic conversion of fructose to 5-hydroxymethylfurfural and then to levulinic acid using C-13 labeled D-fructose. ACS Catal. 2012;2:1211–8.
14. Akien GR, Qi L, Horvath IT. Molecular mapping of the acid catalysed dehydration of fructose. Chem Commun. 2012;48:5850–2.
15. Bozell JJ, Moens L, Elliott DC, Wang Y, Neuenscwander GG, Fitzpatrick SW, Bilski RJ, Jarnefeld JL. Production of levulinic acid and use as a platform chemical for derived products. Resour Conserv Recycl. 2000;28:227–39.
16. Girisuta B, Janssen LPBM, Heeres HJ. A kinetic study on the conversion of glucose to levulinic acid. Chem Eng Res Des. 2006;84:339–49.
17. Funke A, Ziegler F. Hydrothermal carbonization of biomass: a summary and discussion of chemical mechanisms for process engineering. Biofuels Bioprod Biorefin. 2010;4:160–77.
18. van Zandvoort I, Wang Y, Rasrendra CB, van Eck ERH, Bruijnincx PCA, Heeres HJ, Weckhuysen BM. Formation, molecular structure, and morphology of humins in biomass conversion: influence of feedstock and processing conditions. ChemSusChem. 2013;6:1745–58.
19. Hoang TMC, Lefferts L, Seshan K. Valorization of humin-based byproducts from biomass processing-a route to sustainable hydrogen. ChemSusChem. 2013;6:1651–8.
20. Hoang TMC, van Eck ERH, Bula WP, Gardeniers JGE, Lefferts L, Seshan K. Humin based by-products from biomass processing as a potential carbonaceous source for synthesis gas production. Green Chem. 2015;17:959–72.
21. Wang Y, Agarwal S, Kloekhorst A, Heeres HJ. Catalytic hydrotreatment of humins in mixtures of formic acid/2-propanol with supported ruthenium catalysts. ChemSusChem. 2016;9:951–61.
22. Mulder GJ. Untersuchungen über die Humussubstanzen. J Prakt Chem. 1840;21:203–40.

23. Choudhary V, Mushrif SH, Ho C, Anderko A, Nikolakis V, Marinkovic NS, Frenkel AI, Sandler SI, Vlachos DG. Insights into the Interplay of Lewis and Bronsted acid catalysts in glucose and fructose conversion to 5-(Hydroxymethyl)furfural and levulinic acid in aqueous media. J Am Chem Soc. 2013;135:3997–4006.
24. Son PA, Nishimura S, Ebitani K. Synthesis of levulinic acid from fructose using Amberlyst-15 as a solid acid catalyst. React Kinet Mech Catal. 2012;106:185–92.
25. Fachri BA, Abdilla RM, van de Bovenkamp HH, Rasrendra CB, Heeres HJ. Experimental and kinetic modeling studies on the sulfuric acid catalyzed conversion of D-fructose to 5-hydroxymethylfurfural and levulinic acid in water. ACS Sustain Chem Eng. 2015;3:3024–34.
26. Szabolcs A, Molnar M, Dibo G, Mika LT. Microwave-assisted conversion of carbohydrates to levulinic acid: an essential step in biomass conversion. Green Chem. 2013;15:439–45.
27. Kuster BFM, Vanderbaan HS. Dehydration of D-fructose (Formation of 5-hydroxymethyl-2-furaldehyde and levulinic acid). 2. Influence of initial and catalyst concentrations on dehydration of D-fructose. Carbohydr Res. 1977;54:165–76.
28 Heeres H, Handana R, Chunai D, Rasrendra CB, Girisuta B, Heeres HJ. Combined dehydration/(transfer)-hydrogenation of C6-sugars (D-glucose and D-fructose) to [gamma]-valerolactone using ruthenium catalysts. Green Chem. 2009;11:1247–55.
29. Weingarten R, Cho J, Xing R, Conner Jr WC, Huber GW. Kinetics and reaction engineering of levulinic acid production from aqueous glucose solutions. ChemSusChem. 2012;5:1280–90.
30. Shen Y, Sun J, Yi Y, Wang B, Xu F, Sun R. 5-Hydroxymethylfurfural and levulinic acid derived from monosaccharides dehydration promoted by InCl3 in aqueous medium. J Mol Catal a-Chem. 2014;394:114–20.
31. Rackemann DW, Bartley JP, Doherty WOS. Methanesulfonic acid-catalyzed conversion of glucose and xylose mixtures to levulinic acid and furfural. Ind Crop Prod. 2014;52:46–57.
32. Yang F, Fu J, Mo J, Lu X. Synergy of Lewis and Bronsted acids on catalytic hydrothermal decomposition of hexose to levulinic acid. Energy Fuel. 2013;27:6973–8.
33. Tang P, Yu J. kinetic analysis on deactivation of a solid Bronsted acid catalyst in conversion of sucrose to levulinic acid. Ind Eng Chem Res. 2014;53:11629–37.
34. Cha JY, Hanna MA. Levulinic acid production based on extrusion and pressurized batch reaction. Ind Crop Prod. 2002;16:109–18.
35. Girisuta B, Janssen LPBM, Heeres HJ. Kinetic study on the acid-catalyzed hydrolysis of cellulose to levulinic acid. Ind Eng Chem Res. 2007;46:1696–708.
36. Shen J, Wyman CE. Hydrochloric acid-catalyzed levulinic acid formation from cellulose: data and kinetic model to maximize yields. AICHE J. 2012;58:236–46.
37. Peng L, Lin L, Zhang J, Zhuang J, Zhang B, Gong Y. Catalytic conversion of cellulose to levulinic acid by metal chlorides. Molecules. 2010;15:5258–72.
38. Ren H, Zhou Y, Liu L. Selective conversion of cellulose to levulinic acid via microwave-assisted synthesis in ionic liquids. Bioresour Technol. 2013;129:616–9.
39. Shen Y, Sun J-K, Yi Y-X, Wang B, Xu F, Sun R-C. One-pot synthesis of levulinic acid from cellulose in ionic liquids. Bioresour Technol. 2015;192:812–6.
40. Girisuta B, Danon B, Manurung R, Janssen LPBM, Heeres HJ. Experimental and kinetic modelling studies on the acid-catalysed hydrolysis of the water hyacinth plant to levulinic acid. Bioresour Technol. 2008;99:8367–75.
41. Runge T, Zhang C. Two-stage acid-catalyzed conversion of carbohydrates into levulinic acid. Ind Eng Chem Res. 2012;51:3265–70.
42. Galletti AMR, Antonetti C, De Luise V, Licursi D, Di Nasso NNO. Levulinic acid production from waste biomass. Bioresources. 2012;7:1824–35.
43. Antonetti C, Bonari E, Licursi D, Di Nasso NNO, Galletti AMR. Hydrothermal conversion of giant reed to furfural and levulinic acid: optimization of the process under microwave irradiation and investigation of distinctive agronomic parameters. Molecules. 2015;20:21232–53.
44. Li J, Jiang Z, Hu L, Hu C. Selective conversion of cellulose in corncob residue to levulinic acid in an aluminum trichloride-sodium chloride system. ChemSusChem. 2014;7:2482–8.

45. Li M, Pu YQ, Ragauskas AJ. Current understanding of the correlation of lignin structure with biomass recalcitrance. Front Chem. 2016;4:1–8.
46. Li H, Pu Y, Kumar R, Ragauskas AJ, Wyman CE. Investigation of lignin deposition on cellulose during hydrothermal pretreatment, its effect on cellulose hydrolysis, and underlying mechanisms. Biotechnol Bioeng. 2014;111:485–92.
47. Lee H-J, Sanyoto B, Choi J-W, Ha J-M, Suh DJ, Lee K-Y. Effects of lignin on the ionic-liquid assisted catalytic hydrolysis of cellulose: chemical inhibition by lignin. Cellulose. 2013;20:2349–58.
48. Dussan, K. Primary conversion of lignocellulosic biomass for the production of furfural and levulinic acid (PhD Thesis) University of Limerick. 2014.
49. Upare PP, Yoon J-W, Kim MY, Kang H-Y, Hwang DW, Hwang YK, Kung HH, Chang J-S. Chemical conversion of biomass-derived hexose sugars to levulinic acid over sulfonic acid-functionalized graphene oxide catalysts. Green Chem. 2013;15:2935–43.
50. Jow J, Rorrer GL, Hawley MC. Dehydration of D-fructose to levulinic acid over LZY zeolite catalyst. Biomass. 1987;14:185–94.
51. Yu F, Thomas J, Smet M, Dehaen W, Sels BF. Molecular design of sulfonated hyperbranched poly(arylene oxindole)s for efficient cellulose conversion to levulinic acid. Green Chem. 2016;18:1694–705.
52. Ahlkvist J, Ajaikumar S, Larsson W, Mikkola J-P. One-pot catalytic conversion of Nordic pulp media into green platform chemicals. Appl Catal a-General. 2013;454:21–9.
53. Ramli NAS, Amin NAS. Kinetic study of glucose conversion to levulinic acid over Fe/HY zeolite catalyst. Chem Eng J. 2016;283:150–9.
54. Suacharoen S, Tungasmita DN. Hydrothermolysis of carbohydrates to levulinic acid using metal supported on porous aluminosilicate. J Chem Technol Biotechnol. 2013;88:1538–44.
55. Joshi SS, Zodge AD, Pandare KV, Kulkarni BD. Efficient conversion of cellulose to levulinic acid by hydrothermal treatment using zirconium dioxide as a recyclable solid acid catalyst. Ind Eng Chem Res. 2014;53:18796–805.
56. Ding D, Wang J, Xi J, Liu X, Lu G, Wang Y. High-yield production of levulinic acid from cellulose and its upgrading to gamma-valerolactone. Green Chem. 2014;16:3846–53.
57. Putro JN, Kurniawan A, Soetaredjo FE, Lin S-Y, Ju Y-H, Ismadji S. Production of gamma-valerolactone from sugarcane bagasse over $TiO_2$-supported platinum and acid-activated bentonite as a co-catalyst. RSC Adv. 2015;5:41285–99.
58. Ya'aini N, Amin NAS, Asmadi M. Optimization of levulinic acid from lignocellulosic biomass using a new hybrid catalyst. Bioresour Technol. 2012;116:58–65.
59. Wettstein SG, Alonso DM, Chong Y, Dumesic JA. Production of levulinic acid and gamma-valerolactone (GVL) from cellulose using GVL as a solvent in biphasic systems. Energy Environ Sci. 2012;5:8199–203.
60. Amiri H, Karimi K, Roodpeyma S. Production of furans from rice straw by single-phase and biphasic systems. Carbohydr Res. 2010;345:2133–8.
61. Saeman JF. Kinetics of wood saccharification – hydrolysis of cellulose and decomposition of sugars in dilute acid at high temperature. Indust Eng Chem. 1945;37:43.
62. Girisuta B, Dussan K, Haverty D, Leahy JJ, Hayes MHB. A kinetic study of acid catalysed hydrolysis of sugar cane bagasse to levulinic acid. Chem Eng J. 2013;217:61–70.
63. Dussan K, Girisuta B, Haverty D, Leahy JJ, Hayes MHB. Kinetics of levulinic acid and furfural production from Miscanthus x giganteus. Bioresour Technol. 2013;149:216–24.
64. Jing Q, Lu XY. Kinetics of non-catalyzed decomposition of glucose in high-temperature liquid water. Chin J Chem Eng. 2008;16:890–4.
65. Serrano-Ruiz JC, Braden DJ, West RM, Dumesic JA. Conversion of cellulose to hydrocarbon fuels by progressive removal of oxygen. Appl Catal B Environ. 2010;100:184–9.
66. Dunlop AP, Wells JPA. Process for producing levulinic acid US. 1957:2813900.
67. Hoving HD, Rijke DEA, Wagemans GMC, Parton RFMJ, Babic K. Process for the separation of levulinic acid from biomass WO. 2014:2014037560.
68. Fitzpatrick SW. Production of levulinic acid from carbohydrate-containing materials. US. 1997;5:608,105.

69. Hayes DJ, Fitzpatrick SW, Hayes MHB, Ross JRH. The biofine process – production of levulinic acid, furfural, and formic acid from lignocellulosic feedstocks. In: Kamm B, Gruber PR, Kamm M, editors. Biorefineries – industrial processes and products: status Quo and future directions. vol 1. Weinheim: Wiley-VCH; 2006.
70. de Guzman D. Segetis starts levulinic acid pilot production (Access date: 16 December 2016). Retrieved from http://greenchemicalsblog.com/2013/10/10/segetis-starts-levulinic-acid-pilot-production/
71. Lane J. GFBiochemicals acquires Segetis, enters the US market (Access date: 16 December 2016). Retrieved from http://www.biofuelsdigest.com/bdigest/2016/02/19/gfbiochemicals-acquires-segetis-enters-the-us-market/
72. Guo YH, Li KX, Yu XD, Clark JH. Mesoporous H3PW12O40-silica composite: Efficient and reusable solid acid catalyst for the synthesis of diphenolic acid from levulinic acid. Appl Catal B-Environ. 2008;81:182–91.
73. Bader AR, Kontowicz AD. γ,γ-Bis-(p-hydroxyphenyl)-valeric Acid. J Am Chem Soc. 1954;76(17):4465–6.
74. Yu XD, Guo YH, Li KX, Yang X, Xu LL, Guo YN, Hu JL. Catalytic synthesis of diphenolic acid from levulinic acid over cesium partly substituted Wells-Dawson type heteropolyacid. J Mol Catal a-Chem. 2008;290:44–53.
75. Shen Y, Sun JK, Wang B, Xu F, Sun RC. Catalytic synthesis of diphenolic acid from levulinic acid over bronsted acidic ionic liquids. Bioresources. 2014;9:3264–75.
76. Van de Vyver S, Geboers J, Helsen S, Yu F, Thomas J, Smet M, Dehaen W, Sels BF. Thiol-promoted catalytic synthesis of diphenolic acid with sulfonated hyperbranched poly(arylene oxindole)s. Chem Commun. 2012;48:3497–9.
77. Manzer LE. Production of 5-methyl-N-(methyl aryl)-2-pyrrolidone, 5-methyl-N-(methyl cycloalkyl)-2-pyrrolidone and 5-methyl-N-alkyl-2-pyrrolidone by reductive amination of levulinic acid with cyano compounds. 2005; US 6841520.
78. Manzer L, Herkes F. Production of 5-methyl-1-hydrocarbyl-2-pyrrolidone by reductive amination of levulinic acid. 2004; US 2004192933.
79. Wei Y, Wang C, Jiang X, Xue D, Li J, Xiao J. Highly efficient transformation of levulinic acid into pyrrolidinones by iridium catalysed transfer hydrogenation. Chem Commun. 2013;49:5408–10.
80. Touchy AS, Siddiki SMAH, Kon K, Shimizu K-i. Heterogeneous Pt catalysts for reductive amination of levulinic acid to pyrrolidones. ACS Catal. 2014;4:3045–50.
81. Chieffi G, Braun M, Esposito D. Continuous reductive amination of biomass-derived molecules over carbonized filter paper-supported FeNi alloy. ChemSusChem. 2015;8:3590–4.
82. Ogiwara Y, Uchiyama T, Sakai N. Reductive amination/cyclization of keto acids using a hydrosilane for selective production of lactams versus cyclic amines by switching of the indium catalyst. Angewandte Chemie-International Edition. 2016;55:1864–7.
83. Leibig C, Mullen B, Mullen T, Rieth L, Badarinarayana V. Cellulosic-derived levulinic ketal esters: a new building block, in renewable and sustainable polymers. In: Payne GF, Smith PB, editors. American Chemical Society. 2011; p. 111–116.
84. Selifonov S, Rothstein SD, Mullen BD. Method of Making Ketals and Acetals US. 2010:20100292491.
85. Liu S, Zhang G, Li X, Zhang J. Microbial production and applications of 5-aminolevulinic acid. Appl Microbiol Biotechnol. 2014;98:7349–57.
86. Rebeiz CA, Montazer-Zouhoor A, Hopen HJ, Wu SM. Photodynamic herbicides: 1. Concept and phenomenology. Enzym Microb Technol. 1984;6:390–6.
87. Sasaki K, Watanabe M, Tanaka T. Biosynthesis, biotechnological production and applications of 5-aminolevulinic acid. Appl Microbiol Biotechnol. 2002;58:23–9.
88. Colditz MJ, Jeffree RL. Aminolevulinic acid (ALA)-protoporphyrin IX fluorescence guided tumour resection. Part 1: clinical, radiological and pathological studies. J Clin Neurosci. 2012;19:1471–4.
89. Podolean L, Kuncser V, Gheorghe N, Macovei D, Parvulescu VI, Coman SM. Ru-based magnetic nanoparticles (MNP) for succinic acid synthesis from levulinic acid. Green Chem. 2013;15:3077–82.

90. Dutta S, Wu L, Mascal M. Efficient, metal-free production of succinic acid by oxidation of biomass-derived levulinic acid with hydrogen peroxide. Green Chem. 2015;17:2335–8.
91. Wu L, Dutta S, Mascal M. Efficient, chemical-catalytic approach to the production of 3-hydroxypropanoic acid by oxidation of biomass-derived levulinic acid with hydrogen peroxide. ChemSusChem. 2015;8:1167–9.
92. Alonso DM, Wettstein SG, Dumesic JA. Gamma-valerolactone, a sustainable platform molecule derived from lignocellulosic biomass. Green Chem. 2013;15:584–95.
93. Horvath IT. Green or sustainable chemistry or both? Chem Tod. 2014;32:76–9.
94. Yan K, Yang Y, Chai J, Lu Y. Catalytic reactions of gamma-valerolactone: a platform to fuels and value-added chemicals. Appl Catal B-Environ. 2015;179:292–304.
95. Zhang Z. Synthesis of gamma-valerolactone from carbohydrates and its applications. ChemSusChem. 2016;9:156–71.
96. Manzer LE. Catalytic synthesis of α-methylene-γ-valerolactone: a biomass-derived acrylic monomer. Appl Catal A Gen. 2004;272:249–56.
97. Wong PK, Li C, Stubbs L, van Meurs M, Anak Kumbang DG, Lim SCY, Drent E. Synth Diacids. 2012:WO 2012134397.
98. Nandiwale KY, Sonar SK, Niphadkar PS, Joshi PN, Deshpande SS, Patil VS, Bokade VV. Catalytic upgrading of renewable levulinic acid to ethyl levulinate biodiesel using dodecatungstophosphoric acid supported on desilicated H-ZSM-5 as catalyst. Appl Catal a-General. 2013;460:90–8.
99. Pasquale G, Vazquez P, Romanelli G, Baronetti G. Catalytic upgrading of levulinic acid to ethyl levulinate using reusable silica-included Wells-Dawson heteropolyacid as catalyst. Catal Commun. 2012;18:115–20.
100. Su F, Ma L, Song D, Zhang X, Guo Y. Design of a highly ordered mesoporous H3PW12O40/ZrO2-Si(Ph)Si hybrid catalyst for methyl levulinate synthesis. Green Chem. 2013;15:885–90.
101. Melero JA, Morales G, Iglesias J, Paniagua M, Hernandez B, Penedo S. Efficient conversion of levulinic acid into alkyl levulinates catalyzed by sulfonic mesostructured silicas. Appl Catal a-General. 2013;466:116–22.
102. Patil CR, Niphadkar PS, Bokade VV, Joshi PN. Esterification of levulinic acid to ethyl levulinate over bimodal micro-mesoporous H/BEA zeolite derivatives. Catal Commun. 2014;43:188–91.
103. Fernandes DR, Rocha AS, Mai EF, Mota CJA, Teixeira da Silva V. Levulinic acid esterification with ethanol to ethyl levulinate production over solid acid catalysts. Appl Catal a-General. 2012;425:199–204.
104. Kuwahara Y, Fujitani T, Yamashita H. Esterification of levulinic acid with ethanol over sulfated mesoporous zirconosilicates: influences of the preparation conditions on the structural properties and catalytic performances. Catal Today. 2014;237:18–28.
105. Li Z, Wnetrzak R, Kwapinski W, Leahy JJ. Synthesis and characterization of sulfated $TiO_2$ nanorods and $ZrO_2/TiO_2$ nanocomposites for the esterification of biobased organic acid. ACS Appl Mater Interfaces. 2012;4:4499–505.
106. Hu X, Li C-Z. Levulinic esters from the acid-catalysed reactions of sugars and alcohols as part of a bio-refinery. Green Chem. 2011;13:1676–9.
107. Saravanamurugan S, Nguyen Van Buu O, Riisager A. Conversion of mono- and disaccharides to ethyl levulinate and ethyl pyranoside with sulfonic acid-functionalized ionic liquids. ChemSusChem. 2011;4:723–6.
108. Hishikawa Y, Yamaguchi M, Kubo S, Yamada T. Direct preparation of butyl levulinate by a single solvolysis process of cellulose. J Wood Sci. 2013;59:179–82.
109. Demolis A, Essayem N, Rataboul F. Synthesis and applications of alkyl levulinates. ACS Sustain Chem Eng. 2014;2:1338–52.
110. Bianchi, D., Romano AM. Process for the production of esters of levulinic acid from biomasses. 2008;WO2009-156842A1.
111. Mascal M, Nikitin EB. Comment on processes for the direct conversion of cellulose or cellulosic biomass into levulinate esters. ChemSusChem. 2010;3:1349–51.

112. Balakrishnan M, Sacia ER, Bell AT. Etherification and reductive etherification of 5-(hydroxymethyl)furfural: 5-(alkoxymethyl)furfurals and 2,5-bis(alkoxymethyl)furans as potential biodiesel candidates. Green Chem. 2012;14:1626–34.
113. Wu X, Fu J, Lu X. One-pot preparation of methyl levulinate from catalytic alcoholysis of cellulose in near-critical methanol. Carbohydr Res. 2012;358:37–9.
114. Kang S, Yu J. Effect of methanol on formation of levulinates from cellulosic biomass. Ind Eng Chem Res. 2015;54:11552–9.
115. Liu R, Chen J, Huang X, Chen L, Ma L, Li X. Conversion of fructose into 5-hydroxymethylfurfural and alkyl levulinates catalyzed by sulfonic acid-functionalized carbon materials. Green Chem. 2013;15:2895–903.
116. Peng L, Lin L, Li H, Yang Q. Conversion of carbohydrates biomass into levulinate esters using heterogeneous catalysts. Appl Energy. 2011;88:4590–6.
117. Kuo C-H, Poyraz AS, Jin L, Meng Y, Pahalagedara L, Chen S-Y, Kriz DA, Guild C, Gudz A, Suib SL. Heterogeneous acidic TiO2 nanoparticles for efficient conversion of biomass derived carbohydrates. Green Chem. 2014;16:785–91.
118. Christensen E, Williams A, Paul S, Burton S, McCormick RL. Properties and performance of levulinate esters as diesel blend components. Energy Fuel. 2011;25:5422–8.
119. Lange J-P, Price R, Ayoub PM, Louis J, Petrus L, Clarke L, Gosselink H. Valeric biofuels: a platform of cellulosic transportation fuels. Angew Chem Int Ed. 2010;49:4479–83.

# Chapter 6
# Catalytic Aerobic Oxidation of 5-Hydroxymethylfurfural (HMF) into 2,5-Furandicarboxylic Acid and Its Derivatives

**Zehui Zhang and Peng Zhou**

**Abstract** Catalytic synthesis of value-added chemicals from biomass is important for reducing current dependence on fossil-fuel resources. The bifunctional compound, 2,5-furandicarboxylic acid (FDCA) has wide application in many fields, particularly as a substitute for petrochemical-derived terephthalic acid in the synthesis of polymers. Therefore, much effort has been devoted to the catalytic synthesis of FDCA. In this chapter, a concise overview of up-to-date methods for the synthesis of FDCA from 5-hydroxymethylfurfural (HMF) or directly from carbohydrates by one-pot reaction is provided with special attention being given to catalytic systems, mechanistic insight, reaction pathway and catalyst stability. In addition, the one-pot oxidative conversion of carbohydrates into FDCA, and the one-pot synthesis of FDCA derivatives are discussed. It is anticipated that the chemistry detailed in this review will guide researchers to develop effective catalysts for economical and environmental-friendly synthesis of FDCA on a large-scale.

**Keywords** Biomass • Carbohydrates • 2,5-furandicarboxylic acid • 5-hydroxymethylfurfural • Catalytic oxidation

## 6.1 Introduction

The worldwide consumption of petrochemical products and fossil fuels is increasing rapidly [1]. Therefore, there is an urgent need to produce fuels and chemicals from renewable resources [2]. Biomass is one of the most abundant renewable resources with annual production of 170 billion metric tons, and it has been considered as a unique and promising candidate [3] Biomass plays a crucial role in the earth's

Z. Zhang (✉) • P. Zhou
Key Laboratory of Catalysis and Materials Sciences of the State Ethnic Affairs Commission & Ministry of Education, College of Chemistry and Material Science, South-Central University for Nationalities, Wuhan 430074, China
e-mail: zehuizh@mail.ustc.edu.cn



Z. Fang et al. (eds.), *Production of Platform Chemicals from Sustainable Resources*, Biofuels and Biorefineries 7, DOI 10.1007/978-981-10-4172-3_6

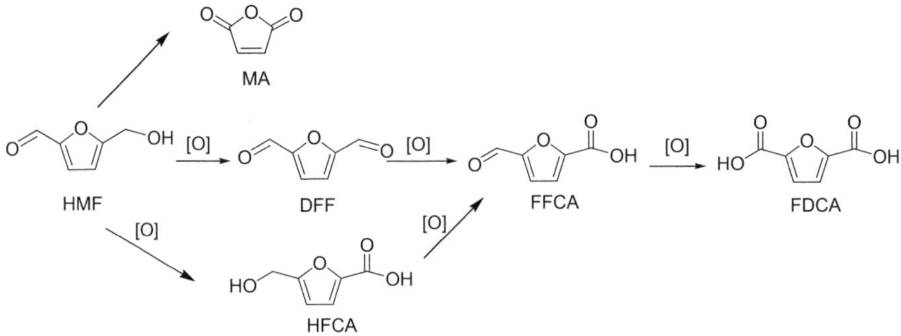


**Fig. 6.1** Schematic illustration of the potential oxidation products from HMF

carbon cycle, as the generated $CO_2$ output can be offset by $CO_2$ fixation through photosynthesis of plant growth [4]. Conversion of biomass into fuels and chemicals, generally called "biorefinery technology", has existed for centuries. Much effort is being devoted to develop new methods to convert biomass into valuable chemicals and fuels using various methods [5–8]. Catalytic oxidation of biomass or biomass-based chemicals have received much attention for the production of value-added chemicals in recent years [9–11]. The platform chemical, 5-Hydroxymethylfurfural (HMF), which is the dehydration product of C6 carbohydrates, has been considered as key for the production of a wide variety of commodity chemicals [12]. For example, as shown in Fig. 6.1, selective oxidation of HMF can generate several kinds of important furanic compounds such as maleic anhydride (MA), 2,5-diformylfuran (DFF), 5-hydroxymethyl-2-furancarboxylic acid (HFCA) and 2,5-furandicarboxylic acid (FDCA) [13–15].

Among the platform chemicals, FDCA is listed as one of the top-12 value-added chemicals from biomass by the US Department of energy [15]. FDCA is very stable, it has a high melting point of 342 °C [16] and it is insoluble in most common solvents. FDCA has been found to be useful in many fields. The most important application of FDCA is that it can serve as a polymer building block for the production of biobased polymers such as polyamides, polyesters, and polyurethanes [17–19]. The most attractive way that FDCA can be used is to replace petrochemical-derived terephthalic acid in the synthesis of biobased polyesters. Terephthalic acid has been used as a monomer in polyethylene terephthalate (PET) plastics for a long time [20]. PET is usually used for production of films, fibers, and in particular, for bottles in the packaging of soft drinks, water and fruit juices. One promising biobased polymer is polyethylene furanoate (PEF), which is the esterification product of ethane-1,2-diol and FDCA [21–23] PEF has similar properties as the petroleum-based PET. The Coca-Cola company has collaborated with Avantium, Danone, and ALPLA to develop and commercialize PEF bottles. Their research has shown that PEF bottles outperform PET bottles in many areas. Besides the main application as monomer for the production of biobased polymer, FDCA has been found to be useful in organic synthesis, pharmacology, and metal-organic frameworks materials [24–26].

In view of its wide application, catalytic synthesis of FDCA has been extensively studied. The focus of this review is to summarize the most recent findings with a critical discussion of the catalytic oxidation of HMF into FDCA. Research on the conversion of carbohydrates into FDCA as well as on FDCA derivatives is also discussed.

## 6.2 FDCA Production Using Different Methods in the Past

FDCA was first produced from the dehydration of mucic acid in the presence of strong acid catalyst (Fig. 6.2). In 1876, Fittig and Heinzelman first reported that the dehydration of mucic acid could produce FDCA using 48% aqueous HBr as the catalyst and the solvent [27]. Later on, other methods that changed dehydrating agents were reported for the synthesis of FDCA. However, the dehydration of mucic acid to FDCA requires severe conditions (highly concentrated acids, high reaction temperature 120 °C, long reaction times >20 h) and all the methods were non-selective with yields less than 50% [25]. More importantly, mucic acid itself is a rare organic acid and its price is high. Therefore, this method has not been studied for the synthesis of FDCA in modern times.

Another method for the synthesis of FDCA is via the oxidation of furfural or HMF by using inorganic oxidants. As shown in Fig. 6.3, several steps are required to achieve FDCA using furfural as the starting material [28]. Furfural is first oxidized to 2-furoic acid with nitric acid and the latter was then converted to its methyl ester. The ester undergoes chloromethylation at position C5 to give methyl 5-chloromethylfuroate. The oxidation of 5-chloromethylfuroate with nitric acid affords dimethyl 2,5-furandicarboxylate, followed by hydrolysis to give FDCA.

The synthesis of FDCA from furfural is complex, and the total FDCA yield is generally low after many reaction steps. A simple method for the synthesis of FDCA is via the direct oxidation of HMF. In the past decades, the oxidation of HMF into FDCA has been performed using stoichiometric oxidants such as $KMnO_4$ [29]. These methods, however, have some distinct drawbacks, such as low selectivity, high cost of oxidant, and generation of highly toxic waste to the environment.

## 6.3 Current Methods for the Oxidation of HMF into FDCA

As mentioned as in Sect. 6.2, many older methods for the synthesis of FDCA are against the principle of green chemistry and practical significance. Thus, many environmental-friendly and economical methods are currently being developed for the oxidation of HMF into FDCA.

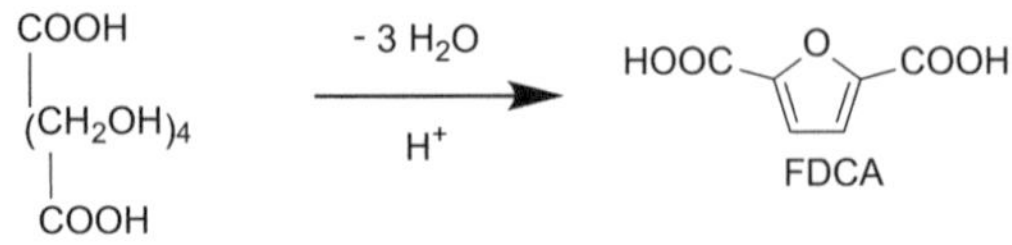


**Fig. 6.2** Schematic illustration of the dehydration of mucic acid into FDAC

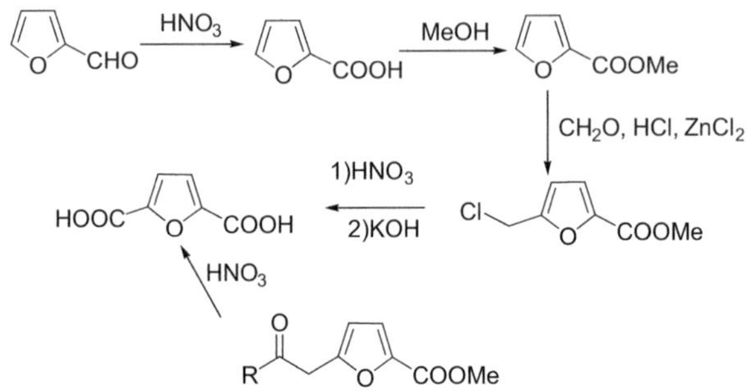


**Fig. 6.3** Synthesis of FDCA from furfural via multiple reaction steps [28]

## *6.3.1 Electrocatalytic Synthesis of FDCA from HMF*

Electrochemical oxidation is driven by electrochemical potential of an electrode and realizes electron transfer, without the use of $O_2$ or other chemical oxidants. Thus, electrochemical oxidation has been considered as a clean synthetic method and has received great interest [30, 31]. Electrochemical oxidation can also offer the advantage of controlling the oxidation potential and the faradaic current, which can be used to monitor the thermodynamic driving force, the selectivity of the surface reaction and the reaction rate.

In 1995, the electrochemical oxidation of HMF into FDCA was first reported in a H-shaped cell [32]. HMF electrochemical oxidation occurred near the anode (nickel oxide/hydroxide as the anode material), affording FDCA in a yield of 71% after 4 h in 1 M NaOH solution at a current density of 0.016 A/cm$^2$. However, the electrochemical oxidation of HMF has been scarcely explored after that work, possibly because the importance of FDCA has not been well recognized by researchers. However, the electrochemical oxidation of HMF has received fresh attention. Strasser and co-workers studied the electrochemical oxidation of HMF using a Pt electrode at pH 10 [33]. They found that a fraction of HMF was oxidized into DFF at a current density of 0.44 mA/cm$^2$. However, FDCA was obtained with a negligible yield (less than 1%). The authors claimed that water oxidation was the major competing reaction and probably limited the faradaic efficiency for HMF oxidation. Later, Li and co-workers studied the electrocatalytic oxidation of HMF in alkaline solution over carbon-black supported noble metal catalysts.[34] They found that the reaction was affected by the potential and the electrocatalyst (Fig. 6.4). The oxidation of the aldehyde group in HMF was much easier than the oxidation of the

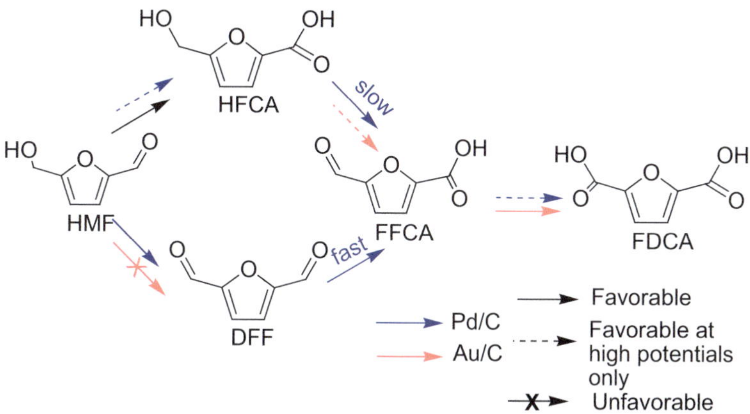


**Fig. 6.4** Proposed reaction pathways of HMF oxidation on Pd/C and Au/C electrocatalysts in alkaline media (Reprinted with permission from [34], Copyright © 2014 Royal Society of Chemistry)

alcohol group over Au/C catalyst, affording HFCA as the intermediate, but high electrode potential was required for further oxidation of the alcohol group in HFCA to FDCA. In the oxidation of HMF over Pd/C catalyst, two competitive routes were observed for the oxidation of HMF into the intermediate FFCA, which depended on the electrode potential. Oxidation of aldehyde groups occurred much slower on Pd/C than on Au/C at low potentials, but was greatly enhanced at increased potentials. It was found that Pd–Au bimetallic catalysts achieved highly oxidized products (FFCA and FDCA) at lower potentials than monometallic catalysts and the product distribution depended on the electrode potential and surface alloy composition. Bimetallic $PdAu_2/C$ catalyst significantly enhanced the efficiency of the electrochemical oxidation of HMF, affording full HMF conversion and a FDCA yield of 83% at a potential of 0.9 V, much higher than the monometallic catalyst, which was due to alloy effect [34]. However, FDCA was obtained with other oxidation intermediates, mainly HFCA, making it difficult to purify the main product from the liquid solution.

Choi and co-workers reported on the electrochemical oxidation of HMF reaction system that had high efficiency [35]. As reported by Strasser and co-workers [33], the low efficiency of the electrochemical oxidation of HMF is the competitive oxidation of water to $O_2$. Strasser and coworkers found that the necessary overpotential to initiate HMF oxidation was significantly reduced using 2,2,6,6-tetramethylpiperidine 1-oxyl (TEMPO) as the mediator, which inhibited the oxidation of water. The reaction mechanism is illustrated in Fig. 6.5, and the electrochemical cell is shown in Fig. 6.6a. TEMPO is oxidized to its oxidation state of $TEMPO^+$ in the vicinity of the Au electrode surface, which serves as a mediator and catalyst for HMF oxidation. High FDCA yield (≥ 99%) and faradaic efficiency (≥ 93%) were obtained in a pH 9.2 aqueous medium. Kinetic study and cyclic voltammetry indicated that DFF was the reaction intermediate.

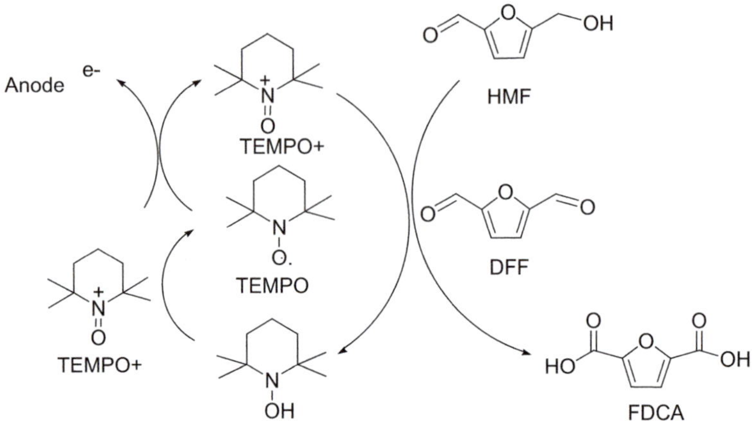


**Fig. 6.5** Scheme of electrocatalytic oxidation of HMF into FDCA with TEMPO as the mediator [35]

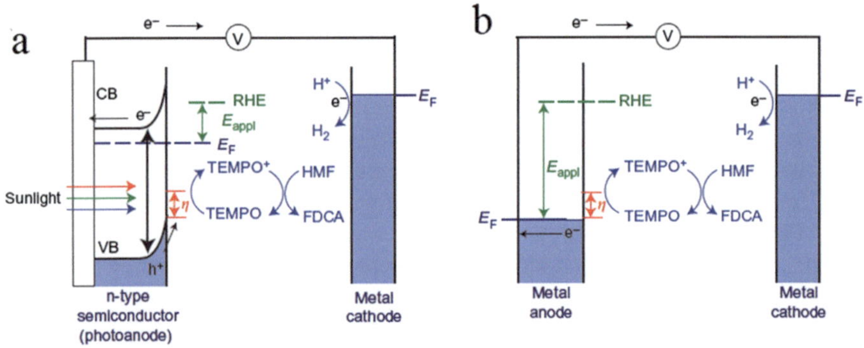


**Fig. 6.6** Schematic comparison of the photoelectrochemical and electrochemical cells. (**a**) Photoelectrochemical TEMPO-mediated HMF oxidation and (**b**) electrochemical TEMPO-mediated HMF oxidation. *CB* conduction band, $E_F$ Fermi energy (Reprinted with permission from [35], Copyright © 2015 Nature Publishing Group)

Furthermore, Strasser and coworkers constructed a photoelectrochemical cell (PEC) (Fig. 6.6b) that used TEMPO-mediated photoelectrochemical oxidation of HMF as the anode reaction. In this cell, an n-type nanoporous $BiVO_4$ electrode was used as a photo-anode that absorbed photons to generate and separate electron–hole pairs. After separation, electrons were transferred to the Pt counter electrode to reduce water to $H_2$ (Eq. 6.1), whereas the holes that reached the surface of $BiVO_4$ were used for TEMPO-mediated HMF oxidation (Eq. 6.2). The overall reaction achieved by this PEC is shown in Eq. (6.3). The photoelectrochemical method also generated high FDCA yields (≥99%) and faradaic efficiency (≥93%). This method did not require adjustment of pH during the reaction process as HMF oxidation tended to offset the pH change at the cathode. This method not only afforded a high FDCA yield, but it also simultaneously produced $H_2$ as a clean energy source.

$$\text{Cathode reaction}: 2\text{H}_2\text{O} + 6e^- \rightarrow 3\text{H}_2 + 6OH^- \quad (6.1)$$

$$\text{Anode reaction}: HMF + 6OH^- \rightarrow \text{FDCA} + 4\text{H}_2\text{O} + 6e^- \quad (6.2)$$

$$\text{Overall}: HMF + 2\text{H}_2\text{O} \rightarrow \text{FDCA} + 3\text{H}_2 \quad (6.3)$$

Although favorable results on the synthesis of FDCA have been obtained with the electrocatalytic method, there are still some issues in its practical application. High FDCA yield I obtained at the expense of the addition of high amounts of TEMPO (1.5 equiv. TEMPO) [35], thus the cost of the synthesis of FDCA is high [35]. Further, it is difficult to separate FDCA from TEMPO or the byproducts and the electrolyte. In addition, the initial HMF concentration in the reported methods was low in order to achieve high FDCA yield. Therefore, it will be very crucial to design robust electro-catalysts that will promote the oxidation of HMF with full conversion at high concentrations and selectivities.

### 6.3.2 *Biocatalyst Method for the Synthesis of FDCA from HMF*

Chemical oxidation reactions are typically performed at high temperature and high pressure. In contrast to chemical processes, biocatalytic transformations are typically carried out under relatively mild conditions and usually require fewer and less toxic chemicals [36]. Despite of these apparent advantages, biocatalytic approaches to FDCA production are less well-established.

Mara and co-workers reported that DFF could be oxidized to FDCA by in-situ produced peracids, which were formed in the presence of lipases as biocatalysts [37]. Using lipases as biocatalysts, alkyl esters as acyl donors, and aqueous solutions of hydrogen peroxide (30% v/v) added stepwise, peracids were formed in situ, which subsequently oxidized DFF to afford FDCA with high yield (>99%) and excellent selectivity (100%) (Fig. 6.7). However, this method was inactive for HMF. The use of DFF as a feedstock for the synthesis of FDCA requires an additional step for the oxidation of HMF into DFF. Thus, this method is a high cost way to produce FDCA. A chloroperoxidase from Caldariomyces fumago was found to have the biocatalytic activity towards the oxidation of HMF into FDCA [38]. But, this method could not provide a complete HMF oxidation, affording FDCA yields of 60–75% as well as 25–40% yields of HFCA. This property renders the *C. fumago* chloroperoxidase a poor biocatalyst for FDCA production, especially when FDCA of very high purity is required for specific applications such as for polymer manufacture. Later, a fermentative process using Pseudomonas putida S12 to host the oxidoreductase from Cupriavidusbasilensis HMF was studied for the oxidation of

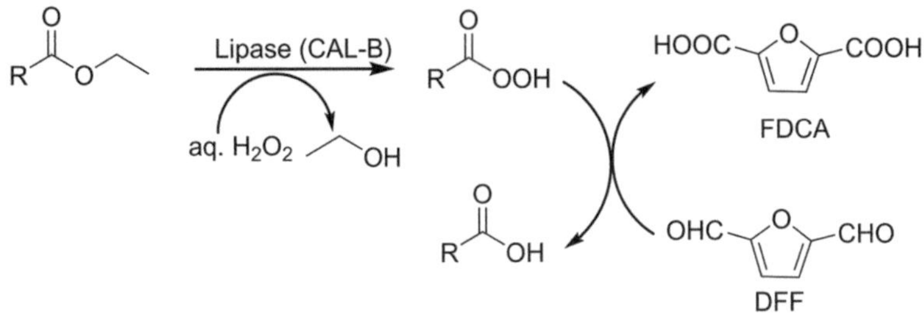


**Fig. 6.7** Envisaged lipase-catalyzed peracid formation in the chemo-enzymatic oxidation of DFF into FDCA (Reprinted with permission from [35], Copyright © 2013 John Wiley & Sons)

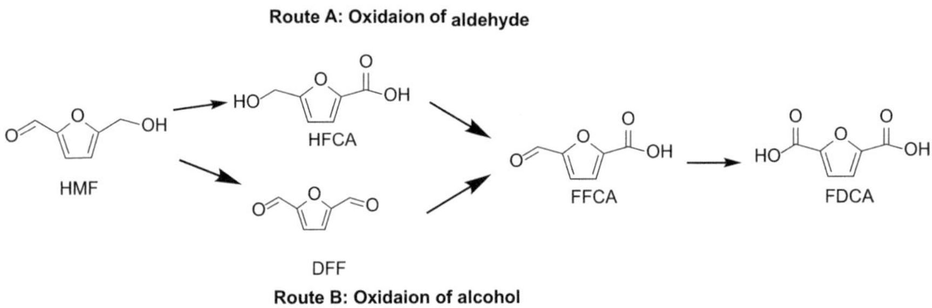


**Fig. 6.8** Reaction routes for enzymatic oxidation of HMF into FDCA by HMF oxidase (Reprinted with permission from [39], Copyright © 2010 Elsevier)

HMF into FDCA [39]. In fed-batch experiments using glycerol as the carbon source, FDCA was produced from HMF with a yield of 97%. FDCA was recovered from the culture broth as a 99.4% powder, at 76% recovery after acid precipitation and tetrahydrofuran extraction. This process relies on the activities of both oxidases and dehydrogenases. Most enzymes are restricted to either alcohol or aldehyde oxidations, while the full oxidation of HMF to FDCA requires the enzyme to act on both alcohol and aldehyde groups. Fraaije and co-workers identified an FAD-dependent oxidase of the glucose-methanol-choline oxidoreductase (GMC) family, named HMF oxidase (HMFO), which showed high catalytic activity toward the HMF oxidation [40, 41]. FDCA yields up to 95% with full HMF conversion were achieved at ambient pressure and temperature, but required a relatively long reaction time (24 h) at a low HMF concentrations (2 mM). Experiments confirmed that the oxidation of HMF by this method underwent two routes (Fig. 6.8). The reaction rate was controlled by the final step of the oxidation of FFCA to FDCA.

The conditions of biocatalytic oxidation are mild (room temperature and ambient atmosphere), but a long reaction time is required that use low concentration of HMF. Constructing enzymes by genetic engineering with high catalytic activity and stability that can oxidize HMF fast at high concentrations will make the biocatalytic method much more competitive in large-scale production of FDCA. It would also be exciting to construct a microorganism with multiple enzymatic activities that can use carbohydrates directly to produce FDCA, which will greatly decrease the production cost of FDCA.

### 6.3.3 *Chemical Synthesis of FDCA from HMF by Homogeneous Catalyst*

Although methods such as electrochemical oxidation and biocatalytic oxidation have been reported for the synthesis of FDCA, synthesis of FDCA is mainly carried out via chemical catalytic methods using homogeneous catalysts or heterogeneous catalysts. Compared with heterogeneous catalysts, there are fewer reports on the synthesis of FDCA using homogeneous catalysts. In 2001, Partenheimer and Grushin studied the aerobic oxidation of HMF to FDCA at 125 °C under 70 bar air pressure in acetic acid solvent using $Co(OAc)_2$, $Mn(OAc)_2$ and HBr as the catalysts, commonly known as the Amoco Mid-Century (MC) catalyst [42]. Similar to para-xylene oxidation into terephthalic acid over MC catalyst, the oxidation of HMF with the MC catalyst also proceeded via the formation of peroxyl radical in the chain propagation step. The peroxyl radicals were formed through the abstraction of H-atom of HMF by the bromide radical, generated in the catalytic cycle by the oxidation of HBr with Co(III) or Mn(III), followed by reaction of aryl alkyl radical with $O_2$. Although both hydroxymethyl and aldehyde groups of HMF could be simultaneously oxidized, the authors proposed that hydroxymethyl group might be possibly preferentially oxidized first. This method afforded FDCA with a yield of 60.9% together with other byproducts. Similar homogeneous catalysts of $Co(OAc)_2$/$Zn(OAc)_2$/NaBr were later used for the aerobic oxidation of HMF into FDCA [43]. DFF was observed as the sole oxidation product without an acid additive, but FDCA was obtained in a yield of 60% with trifluoroacetic acid as an additive.

Besides the use of oxygen as the oxidant, *t*-BuOOH was also used as the oxidant for the oxidation of HMF into FDCA. Riisager and co-workers studied the oxidation of HMF into FDCA in acetonitrile with *t*-BuOOH as the oxidant and copper salts as catalyst [44]. The use of CuCl together with LiBr as the additive afforded FDCA in a low yield of 43% after 48 h at room temperature, and that was 45% using $CuCl_2$ as the catalyst without additive. Homogeneous reaction systems suffered from two distinct drawbacks in practical applications. First, FDCA yield tends to be relatively low accompanied with the formation of some byproducts in the reaction solution. Second, it is difficult to recycle the homogeneous catalysts, and also separate the FDCA from the metal salts. The use of heterogeneous catalysts can possibly overcome the drawbacks of homogeneous catalysts, as heterogeneous catalysts can be easily separated from the reaction solution.

## 6.4 Catalytic Synthesis of FDCA from HMF by Supported Noble Metal Catalysts

Catalytic aerobic oxidation of HMF into FDCA has been extensively studied over heterogeneous catalysts. The use of molecular oxygen as the oxidant is cheap and environmental-friendly as water is the only reduction product. The heterogeneous catalysts are easily recycled and reused. Thus, the use of oxygen and heterogeneous catalysts is in accordance with the concept of "green and sustainable chemistry", which is the main philosophy for the synthesis of FDCA from HMF. As oxygen is not easy to be activated, supported Pt, Pd, Au and Ru catalysts with high activity are the main heterogeneous catalysts for the oxidation of HMF into FDCA.

### *6.4.1 Synthesis of FDCA from HMF Over Supported Pt Catalysts*

Compared with Pd, Au and Pd catalysts, supported Pt catalysts were found to be active towards the aerobic oxidation of HMF into FDCA (Table 6.1). In 1983, Verdeguer and co-workers studied the oxidation of HMF over Pt/C catalyst [45]. Those authors found that the addition of Pb greatly enhanced catalytic activity. Under reaction conditions (1.25 M NaOH solution, 25 °C, with $O_2$ flow rate at 2.5 mL/s), FDCA was obtained in a high yield of 99% within 2 h over Pt-Pb/C catalyst, while Pt/C catalyst only produced FDCA with a yield of 81% at HMF conversion of 100% (Table 6.1, Entries 1 and 2). HFCA was the reaction intermediate, suggesting that the oxidation of the formyl group was much easier than that of the hydroxymethyl group over Pt-Pb/C catalyst. Besides the additive Pb, Bi was also observed to have a positive effect on the catalytic performance of Pt/C catalyst [46]. The Pt-Bi/C catalyst with a Pt-Bi molar ratio of ca. 0.2 afforded FDCA in a high yield of 98% after 6 h at 100 °C, while it was 69% for Pt/C catalyst (Table 6.1, Entries 3 and 4). Observed initial intermediates were HFCA and DFF, which were rapidly oxidized to FFCA, and the oxidation of FFCA was the rate-determining step. The Pt-Bi/C catalyst demonstrated much higher stability than that of the Pt/C catalyst, as the addition of Bi increased the resistance to oxygen poisoning and prevented Pt leaching. The same group also observed superior catalytic activity and stability of Pt-Bi/$TiO_2$ to Pt/$TiO_2$ catalyst (Table 6.1, Entries 5 & 6) [47]. Besides active carbon, reduced graphene oxide (RGO) has been deemed as an excellent support due to its abundant surface functional groups to anchor metal nanoparticles. Tsubaki and co-workers (Table 6.1, Entry 7) studied RGO supported metal nanoparticles for the oxidation of HMF at 25 °C.[48] Fora reaction time of 6 h, both Pt/RGO and Pd/RGO afforded 100% HMF conversion, and produced FDCA with yields of 40.6% and 30.5%, respectively, while Ru/RGO and Rh/RGO could not afford FDCA. This means that the catalytic activity of Pt catalysts was higher than that of the Pd, Ru and Rh catalysts. Prolonging the reaction time to 24 h, 84% FDCA yield was

**Table 6.1** Oxidation of HMF into FDCA over supported Pt catalysts

| Entry | Catalyst | Oxidant | Base | T (°C) | Time (h) | HMF Conv. (%) | FDCA Yield (%) | Ref. |
|---|---|---|---|---|---|---|---|---|
| 1 | Pt-Pb/C | 1 bar $O_2$ | 1.25 M NaOH | 25 | 2 | 100 | 99 | [45] |
| 2 | Pt/C | 1 bar $O_2$ | 1.25 M NaOH | 25 | 2 | 100 | 81 | [45] |
| 3 | Pt-Bi/C | 40 bar air | 2 equiv. $Na_2CO_3$ | 100 | 6 | 100 | >99 | [46] |
| 4 | Pt/C | 40 bar air | 2 equiv. $Na_2CO_3$ | 100 | 6 | 99 | 69 | [46] |
| 5 | Pt-Bi/$TiO_2$ | 40 bar air | 2 equiv. $Na_2CO_3$ | 100 | 6 | >99 | 99 | [47] |
| 6 | Pt/$TiO_2$ | 40 bar air | 2 equiv. $Na_2CO_3$ | 100 | 6 | 90 | 84 | [47] |
| 7 | Pt/RGO | 1 bar $O_2$ | 5 equiv. NaOH | 25 | 24 | 100 | 84 | [48] |
| 8 | Pt/C | 0.69 bar $O_2$ | 2 equiv. NaOH | 22 | 6 | 100 | 79 | [49] |
| 9 | Pt/γ-$Al_2O_3$ | 0.2 bar partial $O_2$ | pH = 9 | 60 | 6 | 100 | 99 | [50] |
| 10 | Pt/γ-$Al_2O_3$ | 1 bar $O_2$ | 1 equiv. $Na_2CO_3$ | 75<br>140 | 12<br>12 | 96 | 96 | [52] |
| 11 | Pt/$ZrO_2$ | 1 bar $O_2$ | 1 equiv. $Na_2CO_3$ | 75<br>140 | 12<br>12 | 100 | 94 | [52] |
| 12 | Pt/C | 1 bar $O_2$ | 1 equiv. $Na_2CO_3$ | 75<br>140 | 12<br>12 | 100 | 89 | [52] |
| 13 | Pt/$TiO_2$ | 1 bar $O_2$ | 1 equiv. $Na_2CO_3$ | 75<br>140 | 12<br>12 | 96 | 2 | [52] |
| 14 | Pt/$CeO_2$ | 1 bar $O_2$ | 1 equiv. $Na_2CO_3$ | 75<br>140 | 12<br>12 | 100 | 8 | [52] |
| 15 | Pt/$Ce_{0.8}Bi_{0.2}O_{2-\delta}$ | 10 bar $O_2$ | 4 equiv. NaOH | 23 | 0.5 | 100 | 98 | [53] |
| 16 | Pt/$CeO_2$ | 10 bar $O_2$ | 4 equiv. NaOH | 23 | 0.5 | 100 | 20 | [53] |
| 17 | Pt/PVP | 1 bar $O_2$ | No base | 80 | 24 | 100 | 94 | [55] |

obtained over Pt/RGO catalyst (Table 6.1, Entry 7). HFCA was observed as the only intermediate. The Pt/RGO catalyst could be reused with full HMF conversion in each run, but with a slight decrease of FDCA yield and a slight increase of HFCA yield. Similar to the results reported by Tsubaki and co-workers [48]. Davis and co-workers [49] observed that carbon supported Pt catalyst (Pt/C) showed slightly higher catalytic activity than carbon supported Pd catalyst (Pd/C) [49]. Under the same reaction conditions (0.69 bar $O_2$, 2 equiv. NaOH, 22 °C), HMF was completely converted over the two catalysts, but Pt/C catalyst produced higher FDCA yield (79%) after 6 h (Table 6.1, Entry 8) than that of Pd/C catalyst (71%).

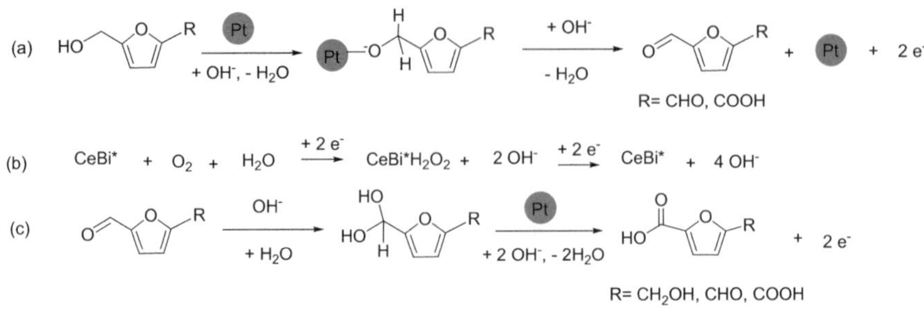


**Fig. 6.9** Proposed reaction mechanism for oxidation of HMF in alkaline aqueous solutions. CeBi* represents the oxygen vacancy accompanied with bismuth (Reprinted with permission from Ref. [53], Copyright © 2015 Royal Society of Chemistry)

Besides carbon material supported metal catalysts, metal oxide supported Pt catalysts have been studied for the oxidation of HMF into FDCA. In 1990, Vinke et al. found that Pt/$\gamma$-$Al_2O_3$ catalyst showed high catalytic activity towards oxidation of HMF into FDCA [50, 51]. Near quantitative FDCA yields (>99%) were obtained (Table 6.1, Entry 9). However, the Pt/$Al_2O_3$ catalyst became deactivated by oxygen chemisorption due to the high activity of Pt nanoparticles. Sahu and Dhepe compared the catalytic performance of metal oxide-supported Pt catalysts [52]. Under the same reaction conditions, Pt/$\gamma$-$Al_2O_3$, Pt/$ZrO_2$ and Pt/C showed high catalytic activity with FDCA yields being 96%, 94% and 89%, respectively, while Pt/$TiO_2$ and Pt/$CeO_2$ catalyst produced very poor FDCA yields (2 and 8%) even with HMF conversion of ca. 100%(Table 6.1, Entries 10 ~ 14). These results are evidence that non-reducible oxide ($\gamma$-$Al_2O_3$, $ZrO_2$ and C) supported Pt catalysts have higher catalytic performance than reducible oxide ($TiO_2$ and $CeO_2$) supported Pt catalysts. The authors claimed that the main reason was attributed to the different oxygen storage capacity (OSC) of each support. It is known that the OSC of catalysts like Pt/$\gamma$-$Al_2O_3$, Pt/$ZrO_2$ and Pt/C is quite low, which keeps the active sites in active form (not covered by oxygen), whereas, $TiO_2$ and $CeO_2$ have high OSC because of the presence of $Ce^{4+}/Ce^{3+}$or $Ti^{4+}/Ti^{3+}$ redox couple. Although high OSC is good for reactions such as CO oxidation, it is detrimental to HMF oxidation. Similar to the addition of Bi to Pt for Pt/C catalyst [46]. Yang and co-workers found that the addition of Bi to $CeO_2$ support could greatly improve catalytic performance of Pt/$CeO_2$ [53]. The Pt/$Ce_{0.8}Bi_{0.2}O_{2-\delta}$ afforded 100% HMF conversion and 98% yield of FDCA, while HMF conversion was less than 20% over Pt/$CeO_2$ catalyst (Table 6.1, Entries 15–16). As shown in Fig. 6.9, Pt nanoparticles react with the hydroxyl group in HMF to form the Pt–alkoxide intermediate, followed by $\beta$-H elimination with the help of hydroxide ions. Bi-containing ceria accelerates the oxygen reduction process because of the presence of a large amount of oxygen vacancies and the cleavage of the peroxide intermediate promoted by bismuth. Thus, surface electrons are consumed to reduce oxygen and the catalytic cycle can be smoothly continued. The Pt/$Ce_{0.8}Bi_{0.2}O_{2-\delta}$ catalyst could be reused for five runs without the loss of its catalytic performance (FDCA yield 98% in the first run *vs* 97% in the fifth run). To produce FDCA on a large scale, Lilga et al. performed oxidation of HMF into FDCA over Pt/C and Pt/$Al_2O_3$ catalysts in a continuous reactor and obtained nearly quantitative yields of FDCA over two catalysts for a feed of 1 wt% HMF at 100 °C under 7 bar air (LHSV =4.5 $h^{-1}$, GHSV =600 $h^{-1}$) using stoichiometric aqueous $Na_2CO_3$ [54].

As shown in Table 6.1, aerobic oxidation of HMF over Pt catalysts is generally carried out in the presence of excess base. The disadvantages of basic feeds are that product solutions must be neutralized, and inorganic salts must be separated. Yan and co-workers found that PVP stabilized Pt nanoparticles (Pt/PVP) could promote base-free aerobic oxidation of HMF into FDCA in water and obtained 100% conversion of HMF and 95% yield of FDCA at 80 °C after 24 h under 1 bar $O_2$. It was observed by these authors that PVP/Pt catalyst showed a slight decrease of its catalytic activity in the recycle runs. Although this base-free method is environmental-friendly, a high catalyst loading of 5 mol%, a low content of the feedstock (0.29 mmol in each run) and a long reaction time of 24 h were needed to achieve 84% yield of FDCA, which also makes the cost of producing FDCA high.

### 6.4.2 *Synthesis of FDCA from HMF Over Supported Pd Catalysts*

Supported Pd catalysts show excellent catalytic performance towards the aerobic oxidation of HMF into FDCA. As described above, Davis and co-workers found the catalytic activity of Pd/C catalyst was comparable to Pt/C catalyst [49]. Under the given reaction conditions (Entry 1, Table 6.2), Pd/C catalyst afforded HMF conversion of 100% and FDCA yield of 71% after 6 h, and Pt/C catalyst gave HMF conversion of 100% and FDCA yield of 79% (Table 6.2, Entry 1 and Table 6.1, Entry 4). DaSiyo and co-workers studied the aerobic oxidation of HMF over PVP stabilized Pd nanoparticles (Pd/PVP) [56] Pd/PVP were prepared in ethylene glycol with the addition of NaOH, and the particle sizes could be controlled by the amount of NaOH. It was found that Pd/PVP catalyst with smaller Pd nanoparticle size afforded FDCA with higher yields. Under the given reaction conditions (Table 6.2, Entry 2), maximum FDCA yield of 90% with full HMF conversion was obtained after 6 h with Pd diameter of 1.8 nm (Table 6.2, Entry 2). FDCA yield decreased to 81% when the Pd diameter was 2.0 nm. The higher catalytic activity of Pd/PVP catalyst with smaller particle size should be due to a higher number of surface atoms and a higher amount of coordinately unsaturated metal sites. Interestingly, the catalyst activity was also found to be dependent on the oxygen flow rate. If the oxygen flow rate was far away from the optimum, Pd nanoparticles deactivated quickly, probably through blocking of the active surface sites by byproducts (oxygen flow rate too low) or by interaction of the Pd surface with oxygen (oxygen flow rate too high). The optimal oxygen flow rate was 35 mL/min in their catalytic system. Similar phenomenon of the effect of oxygen flow rate on the Pt/$Al_2O_3$ catalyst was also

**Table 6.2** Oxidation of HMF into FDCA over supported Pd catalysts

| Entry | Catalyst | $O_2$ partial pressure (bar) | Base | T (°C) | Time (h) | HMF Conv. (%) | FDCA Yield (%) | Ref. |
|---|---|---|---|---|---|---|---|---|
| 1 | Pd/C | 6.9 | 2 equiv. NaOH | 23 | 6 | 100 | 71 | [49] |
| 2 | Pd/PVP | 1 | 1.25 equiv. NaOH | 90 | 6 | >99 | 90 | [56] |
| 3 | $Pd/ZrO_2/La_2O_3$ | 1 | 1.25 equiv. NaOH | 90 | 6 | >99 | 90 | [57] |
| 4 | $Pd/Al_2O_3$ | 1 | 1.25 equiv. NaOH | 90 | 6 | >99 | 78 | [57] |
| 5 | $Pd/Ti_2O_3$ | 1 | 1.25 equiv. NaOH | 90 | 6 | >99 | 53 | [57] |
| 6 | $\gamma$-$Fe_2O_3$@HAP-Pd | 1 | 0.5 equiv. $K_2CO_3$ | 100 | 6 | 97 | 92.9 | [58] |
| 7 | C-$Fe_3O_4$-Pd | 1 | 0.5 equiv. $K_2CO_3$ | 80 | 4 | 98.1 | 91.8 | [59] |
| 8 | Pd/C@$Fe_3O_4$ | 1 | 0.5 equiv. $K_2CO_3$ | 80 | 6 | 98.4 | 86.7 | [60] |

observed by Sahu and Dhepe [52]. The reaction pathway of the oxidation of HMF over Pd/PVP catalyst was affected by the reaction temperature. At low reaction temperature below 70 °C, the rate of oxidation of HFCA into FFCA was slower than the rate of the oxidation of FFCA to FDCA, and the main product was HFCA. The rate of the oxidation of HFCA into FFCA was close to the rate of the oxidation of FFCA to FDCA at the reaction temperature of 90 °C. The stability of PVP/Pd in alkaline solution decreased during the reaction, and PVP/Pd was difficult to be recycled and reused. To improve the stability of Pd nanoparticles and facilitate the catalyst recycle, DaSiyo and co-workers further deposited Pd/PVP onto different metal oxides ($TiO_2$, $\gamma$-$Al_2O_3$, $KF/Al_2O_3$, and $ZrO_2/La_2O_3$) and studied their catalytic performance toward HMF oxidation. $Pd/ZrO_2/La_2O_3$ showed the highest catalytic activity with the highest FDCA yield up to 90% and a relatively stable catalytic performance than other supported Pd catalysts (Table 6.2, Entries 4–6). TEM images indicated that there was no obvious aggregation of Pd nanoparticles in the spent $Pd/ZrO_2/La_2O_3$ catalyst, while others catalysts showed serious aggregation. XPS confirmed that most of the Pd was in its zero-valent form and that the electronic structure of the Pd nanoparticles was not changed in the spent $Pd/ZrO_2/La_2O_3$ catalyst. Nevertheless, the procedure of recycling the heterogeneous catalysts required a tedious recovery procedure *via* filtration or centrifugation and the inevitable loss of solid catalysts occurred during the separation process, which limits its application. Some improvements on the aerobic oxidation of HMF into FDCA using several kinds of magnetic Pd catalysts [58–60]. Zhang et al. (Table 6.2 Entry 6) prepared the magnetic γ-$Fe_2O_3$@HAP supported Pd catalysts (γ-$Fe_2O_3$@HAP-Pd) for the aerobic oxidation of HMF into FDCA [58]. The magnetic core γ-$Fe_2O_3$ was coated with a layer of HAP (HAP = hydroxyapatite, $Ca_{10}(PO_4)_6(OH)_2$), and the $Ca^{2+}$ in the HAP layer can be changed with $Pd^{2+}$, followed by reduction of the $Pd^{2+}$ to Pd(0) nanoparticles. Catalytic oxidation of HMF over γ-$Fe_2O_3$@HAP-Pd catalyst afforded 97% (Table 6.2, Entry 6). The γ-$Fe_2O_3$@HAP-Pd catalyst could be easily separated from the reaction solution by an external magnet and reused without loss of catalytic activity. TEM images indicated that particle size of Pd nanoparticles did not change in the spent catalyst.

Graphene oxide and carbon have been widely used as supports for the immobilization of metal nanoparticles due to its high surface area and abundant oxygen functional groups. However, the recycling of carbon supported catalysts is difficult due to its small size. Magnetically separable graphene oxide supported Pd catalyst (C-$Fe_3O_4$-Pd), in which $Fe_3O_4$ nanoparticles and Pd nanoparticles were simultaneously deposited on graphene oxide by a one-step solvothermal route was prepared by Zhang et al. [59]. The C-$Fe_3O_4$-Pd catalyst showed excellent catalytic performance in the aerobic oxidation of HMF into FDCA, giving high HMF conversion (98.1%) (Table 6.2, Entry 7). The C-$Fe_3O_4$-Pd catalyst could be easily recovered by an external magnet and reused without loss of catalytic activity. With the aim to achieve sustainability, the magnetic C@$Fe_3O_4$ supported Pd nanoparticles (Pd/C@$Fe_3O_4$)were prepared for aerobic oxidation of HMF into FDCA under mild conditions [60]. The core–shell structure C@$Fe_3O_4$ support was prepared by in situ carbonization of glucose on the surface of the $Fe_3O_4$ microspheres. HMF conversion of 100% and FDCA yield of 87.8% were obtained over Pd/C@$Fe_3O_4$ catalyst after 6 h at 80 °C (Table 6.2, Entry8). The Pd/C@$Fe_3O_4$ catalyst showed good stability in subsequent recycling experiments and XPS measurements confirmed that zero-valent Pd(0) remained in the spent catalyst. All catalytic systems for oxidation of HMF into FDCA over magnetic Pd catalysts showed common advantages: (a) they did not require a large amount of base; (b) they could be performed under atmospheric oxygen pressure; (c) the catalysts could be easily separated by an external magnet and reused without the loss of catalytic activity.

### *6.4.3 Synthesis of FDCA from HMF Over Supported Au Catalysts*

Compared with Pt and Pd catalysts, Au catalysts were once considered to be inactive in chemical reactions. The discovery in the 1980s that Au could behave as a catalyst has been one of the most stunning breakthroughs [61, 62] and opened up a new field of research that led to the discovery of very active catalysts for many

**Table 6.3** Oxidation of HMF over supported Au catalyst

| Entry | Catalyst | Oxidant | Base | T (°C) | Time (h) | HMF Con. (%) | FDCA Yield (%) | Ref. |
|---|---|---|---|---|---|---|---|---|
| 1 | $Au/CeO_2$ | 10 bar Air | 4 equiv. NaOH | 130 | 5 | 100 | 96 | [65] |
| 2 | $Au/TiO_2$ | 10 bar Air | 4 equiv. NaOH | 130 | 8 | 100 | 84 | [65] |
| 3 | $Au/Ce_{0.9}Bi_{0.1}O_{2-\delta}$ | 1 bar $O_2$ | 4 equiv. NaOH | 65 | 2 | 100 | >99 | [67] |
| 4 | Au/HY | 0.3 bar $O_2$ | 5 equiv. NaOH | 60 | 6 | >99 | >99 | [68] |
| 5 | $Au/CeO_2$ | 0.3 bar $O_2$ | 5 equiv. NaOH | 60 | 6 | >99 | 73 | [68] |
| 6 | $Au/TiO_2$ | 0.3 bar $O_2$ | 5 equiv. NaOH | 60 | 6 | >99 | 85 | [68] |
| 7 | $Au/Mg(OH)_2$ | 0.3 bar $O_2$ | 5 equiv. NaOH | 60 | 6 | >99 | 76 | [68] |
| 8 | Au/H-MOR | 0.3 bar $O_2$ | 5 equiv. NaOH | 60 | 6 | 96 | 15 | [68] |
| 9 | Au/Na-ZSM-5-25 | 0.3 bar $O_2$ | 5 equiv. NaOH | 60 | 6 | 92 | 1 | [68] |
| 10 | $Au/TiO_2$ | 20 bar $O_2$ | 20 equiv. NaOH | 30 | 18 | 100 | 71 | [69] |
| 11 | $Au-Cu/TiO_2$ | 10 bar $O_2$ | 4 equiv. NaOH | 95 | 4 | 100 | 99 | [72] |
| 12 | $Au_8-Pd_2/C$ | 30 bar $O_2$ | 2 equiv. NaOH | 60 | 2 | >99 | >99 | [74] |
| 13 | Au/HT | 1 bar $O_2$ | none | 95 | 7 | >99 | >99 | [75] |
| 14 | Au-Pd/CNT | 5 bar $O_2$ | none | 100 | 12 | 100 | 94 | [77] |
| 15 | Au-Pd/CNT | 10 bar air | none | 100 | 12 | 100 | 96 | [77] |

potential applications [63, 64]. Supported Au catalyst shave shown encouraging performance for aerobic oxidation of HMF to FDCA in water.

As discussed above, the catalytic performance of Pt and Pd catalysts towards HMF oxidation is greatly affected by its support. Corma and co-workers observed that the support has a large effect on the catalytic activity of Au catalysts [65]. $Au/CeO_2$ and $Au/TiO_2$ catalysts afforded quantitative FDCA yields (>99%) after 8 h (Table 6.3, Entries 5 and 6), while Au/C and $Au/Fe_2O_3$ catalysts produced FDCA yields of 44% and 15%, respectively. Further experiments confirmed that $Au/CeO_2$ showed higher catalytic activity and selectivity over $Au/TiO_2$, which was also observed by Albonetti et al. [66] High FDCA yields of 96% were obtained after 5 h over $Au/CeO_2$ catalyst, while others were 84% after 8 h for $Au/TiO_2$ catalyst (Table 6.3, Entries 1 and 2). For each case, HMF conversion was 100%. HFCA was determined to be the only intermediate. As shown in Fig. 6.10, the first step (very fast) in conversion of HMF to FDCA is the oxidation of HMF into HFCA via the

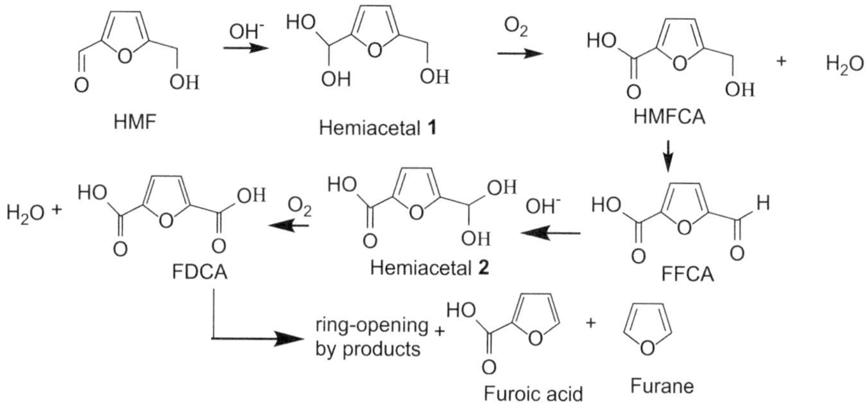


**Fig. 6.10** Proposed reaction pathway for aqueous HMF aerobic oxidation over $Au/CeO_2$ catalyst (Reprinted with permission from [65], Copyright © 2009 John Wiley & Sons)

formation of the intermediate Hemiacetal **1**. As no FFCA was observed in the reaction process, it is probable that FFCA formed from the oxidation of HFCA was rapidly converted into FDCA through a second hemiacetal intermediate (Hemiacetal **2**). Substrate degradation was strongly diminished with increase in catalyst life by performing the reaction in two steps at different reaction temperatures, first the oxidation of HMF into FDCA at are action temperature of 25 °C, and subsequent oxidation of HFCA into FDCA at a temperature of 130 °C. Reductive pretreatment of the $Au/CeO_2$ catalyst was shown to increase the catalytic activity because it increased the amount of $Ce^{3+}$ and oxygen vacancies. The increased $Ce^{3+}$ and oxygen vacancies were shown to have great influence in transferring hydride and activating $O_2$ during catalytic oxidation of alcohols in a former report [65]. It was also proposed that $Ce^{3+}$ centers (Lewis acid sites and stoichiometric oxidation sites of $CeO_2$) and $Au^+$ species of $Au/CeO_2$ could readily accept a hydride from the C–H bond in alcohol or in the corresponding alkoxide to form Ce–H and Au–H, with the formation of a carbonyl compound at the same time. The oxygen vacancies of ceria could activate $O_2$ and form cerium-coordinated superoxide(Ce–OO˙) species which subsequently evolved to cerium hydroperoxide by hydrogen abstraction from Au–H. The cerium hydroperoxide then interacted with Ce–H, producing $H_2O$ and recovering the $Ce^{3+}$ centers. Au–H donated H and changed back to the initial $Au^+$ species. To further improve the catalytic activity of $Au/CeO_2$, Yang and co-workers found that the doping $Bi^{3+}$ into the nano-$CeO_2$ support could improve the $O_2$ activation and hydride transfer of nano-$CeO_2$, which was enhanced by the lone electron pair of $Bi^{3+}$ [67]. Under the same reaction conditions (65 °C, 10 bar, HMF/Au = 150) as described by Corma and co-workers [65], Both $Au/CeO_2$ and $Au/Ce_{1-x}Bi_xO_{2-\delta}$ ($0.08 \leq x \leq 0.5$) gave 100% HMF conversion after 1 h, but FDCA yield largely increased from 39% over $Au/CeO_2$ catalyst to about 75% over $Au/Ce_{1-x}Bi_xO_{2-\delta}$ catalysts. After 2 h, FDCA was obtained in quantitative yield (>99%) over $Au/Ce_{0.9}Bi_{0.1}O_{2-\delta}$ catalyst (Table 6.3,

Entry 3). More importantly, the Au/$Ce_{0.9}Bi_{0.1}O_{2-\delta}$ catalyst showed much higher stability than the Au/$CeO_2$ catalyst, albeit a slight decrease of FDCA yield was still observed in the recycling experiments. Xu and co-workers also found that the catalytic performance of Au catalysts was greatly affected by the support [68]. HY zeolite supported Au catalyst (Au/HY) afforded quantitative FDCA yield (>99%) yield after 6 h under mild conditions, which was much higher than that of Au supported on $TiO_2$, $CeO_2$, and $Mg(OH)_2$ and channel-type zeolites (ZSM-5 and H-MOR) (Table 6.3, Entries 4~9). Detailed characterization revealed that Au nanoclusters were well encapsulated in the HY zeolite supercage, which was considered to restrict and avoid further growth of the Au nanoclusters into large particles. The acidic hydroxyl groups of the supercage were proven to be responsible for the formation and stabilization of the gold nanoclusters. TEM results indicated that the particle size was 1 nm for Au/HT catalyst, while it was (3–20) nm for other catalysts. Moreover, the electronic modification of the Au nanoparticles, which is caused by the interaction between hydroxyl groups in the supercage and Au nanoclusters was supposed to contribute to the high efficiency in the catalytic oxidation of HMF to FDCA. The catalyst could be recycled, but a slight decrease in its catalytic activity was observed.

Besides the support, the amount of base also had a large effect on the oxidation of HMF over Au catalysts. Riisager and co-worker studied the effect of NaOH amount on the catalytic performance of Au/$TiO_2$ catalyst toward HMF oxidation [69]. With the use of 20 equivalents of NaOH, 1 wt.% Au/$TiO_2$ catalyst was found to oxidize HMF into FDCA in 71% yield at 30 °C after 18 h with 20 bar $O_2$ (Table 6.3, Entry 10). Low concentrations of base (i.e., corresponding to less than five equivalents) afforded relatively more of the intermediate HFCA compared to FDCA. The Au/$TiO_2$ catalyst may become deactivated by the initially formed acids, as the conversion of HMF dropped to 13% under base-free conditions. In addition, FDCA precipitating on the catalyst surface due to low solubility may have also hampered the reaction without base. Similar results on the effect of base concentration on selectivity of FDCA were observed by Davis, in which Au/$TiO_2$ catalyst was used for the oxidation of HMF [49]. Thus, an excess amount of base is generally required to obtain high FDCA yields over supported Au catalysts.

Au containing bimetallic catalysts have been used for the oxidation of HMF into FDCA. Their chemical and physical properties of bimetallic catalysts may be easily tuned by varying the size, composition, and degree of mixing, thus the catalytic activity of Au containing bimetallic catalysts was often observed to be higher than that of the monometallic Au catalysts [70, 71]. Pasini and co-workers demonstrated that Au-Cu/$TiO_2$ showed higher catalytic activity and stability over Au/$TiO_2$ towards HMF oxidation into FDCA [72]. All of the bimetallic Au-Cu/$TiO_2$ catalysts with different Au/Cu mol ratio prepared *via* a colloidal route displayed improved activity, by at least a factor of 2 with respect to their corresponding monometallic Au catalysts. The post-deposition of a PVP-stabilized gold sol onto Cu/$TiO_2$ resulted in a less active catalyst as compared with the bimetallic sample, where a preformed Au-Cu sol was utilized. The post-deposited catalyst gave both lower yield and selectivity of FDCA. Thus, it was concluded that homogeneous Au site isolation effect benefited from AuCu alloying as being the main reason for the excellent catalytic activity of the Au-Cu/$TiO_2$ catalysts prepared *via* a colloidal route towards HMF oxidation into FDCA. Under optimal reaction conditions, HMF conversion of 100% and FDCA yield of 99% were attained after 4 h (Table 6.3, Entry 11). A strong synergistic effect was evident in terms of the catalyst stability and resistance to poisoning. The Au-Cu/$TiO_2$ catalyst could be easily recovered and reused without significant leaching and agglomeration of the metal nanoparticles. A strong synergistic effect of Au-Cu alloy was also observed by Albonetti et al. [67, 73], in which they used Au-Cu/$TiO_2$ and Au-Cu/$CeO_2$ catalysts for the oxidation of HMF into FDCA, respectively. Besides the strong synergistic effect of Au-Cu alloy, Villa and co-workers observed a synergistic effect of Au-Pd alloy during the aerobic oxidation of HMF over Au-Pd/AC catalyst [74]. Under reaction conditions (30 bar $O_2$, 2 equiv. NaOH and 60 °C), Au/AC or Pd/AC catalysts afforded full HMF conversion, but with the major product of the intermediate HFCA after 6 h, while the $Au_8$-$Pd_2$/AC catalyst produced quantitative FDCA yield (>99%) after 2 h (Table 6.3, Entry 12). In addition, the $Au_8$-$Pd_2$/AC catalyst showed higher stability than the monometallic Au/AC catalyst. The Au/AC catalyst showed good product selectivity, but it underwent deactivation, losing 20% of conversion efficiency after five runs. No Au leaching from the catalyst was detected, thus this deactivation was possibly due to irreversible adsorption of intermediates or Au particle agglomeration. However, its stability was increased extraordinarily by alloying Au with Pd. FDCA yield of 99% was still obtained even after five runs over $Au_8$-$Pd_2$/AC catalyst. Generally speaking, the monometallic Au catalysts easily suffers from deactivation due to byproducts or reaction intermediates. The alloying of another metal (e.g., Pd and Cu) with Au to form bimetallic alloy catalysts combines the advantages of different components in the atomic level and enhances the activity and stability.

The aerobic oxidation of HMF over Au catalysts is mainly carried out in the presence of excessive base, which needs additional acids to neutralize the base after reaction. Thus, base-free oxidation of HMF into FDCA should be an economical and environmentally benign way. There have been some reports on the oxidation of HMF into FDCA over Au catalysts without base. Ebitani and co-workers developed a base-free process for the oxidation of HMF into FDCA using hydrotalcite supported Au catalyst (Au/HT) [75]. FDCA was achieved with an excellent yield of 99% after 7 h at 95 °C under 1 bar $O_2$ in water (Table 6.3, Entry 13). Au deposited onto neutral supports ($Al_2O_3$, C) showed rare activity and acidic $SiO_2$ with no activity, suggesting that the importance of the basicity of HT in the oxidation of HMF into FDCA. Although MgO is a strong base, Au/MgO showed a much poorer catalytic activity with FDCA yield of 21% than Au/HT catalyst. TEM images indicated that the Au/HT catalyst had a particle size of 3.2 nm, while the particle size of Au/MgO was larger (>10 nm). The large size and low dispersion of Au nanoparticles on MgO is probably the reason of the lower catalytic activity of Au/MgO. Thus, both the basicity of solid support and the metal active sites played important roles in HMF oxidation. HFCA was the main intermediate, which was the same for Au/$TiO_2$ and Au/$CeO_2$ catalysts [65]. Au/HT catalyst could be reused, but with a slight

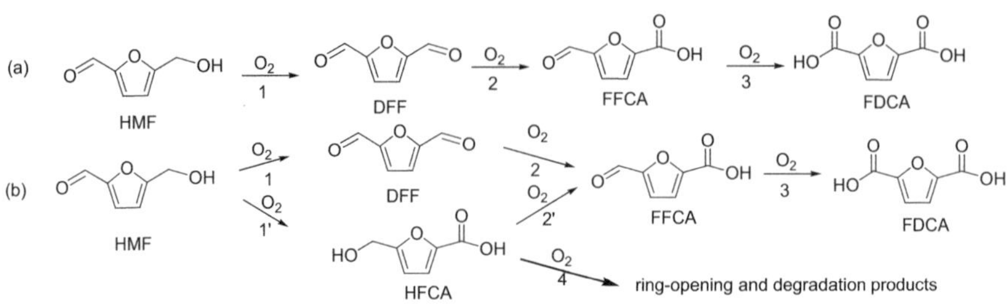


**Fig. 6.11** Reaction pathways for the aerobic oxidation of HMF: (**a**) Pd/CNT and Au − Pd/CNT Catalysts, (**b**) Au/CNT Catalyst (Reprinted with permission from [77], Copyright © 2014 American Chemical Society)

decrease in its catalytic activity, as complete conversion of HMF for all cycles was achieved with FDCA yields over 90% for the first 3 cycles. Although Ebitani and co-workers reported that Au/HT catalyst could be reused, Davis and co-workers observed that extensive leaching of $Mg^{2+}$ from HT occurred inevitably for the oxidation of HMF over Au/$TiO_2$ catalyst and HT as solid base, which was due to the chemical interactions between the basic HT and the formed FDCA [76]. Wang and co-workers developed an active and stable carbon nanotube (CNT)-supported Au − Pd catalyst for the base-free oxidation of HMF [77]. HMF conversion and FDCA selectivity over the Au − Pd/CNT catalyst could reach 100% and 94% after 12 h at 373 K under 5 bar $O_2$, respectively (Table 6.3, Entry 14). The Au − Pd/CNT catalysts showed high catalytic activity using air as the oxidant, affording HMF conversions of 100% and FDCA selectivities of 96% after 12 h at 373 K under 10 bar air (Table 6.3, Entry 15). The authors identified that the carbonyl/quinone and phenol (especially the former) groups on CNT surfaces facilitated the adsorption of HMF and the intermediate (DFF) but not FDCA, which was believed to be significant for the high catalytic activity of the Au − Pd/CNT catalyst. In addition to the support-enhanced adsorption effect, significant synergistic effect was present between Au and Pd in the alloy for the oxidation of HMF to FDCA. The catalytic performance of the Au-Pd bimetallic catalysts was much better than the monometallic Au or Pd catalysts. DFF and FFCA were identified as the reaction intermediates (Fig. 6.11a), suggesting that the oxidation of hydroxyl group in HMF was faster than the oxidation of aldehyde group over Au-Pd/CNT catalysts. The Au/CNT catalyst preferentially catalyzed the oxidation of the aldehyde group in HMF, forming HFCA, which was the same as the reaction pathway of other supported Au catalysts such as Au/$CeO_2$ and Au/$TiO_2$ in the presence of base [65]. But HFCA mainly underwent ring-opening and degradation reactions to form byproducts under base-free conditions (Fig. 6.11b). For the Pd/CNT catalyst, the same reaction pathway was observed as the Au-Pd/CNT catalyst. The addition of Pd to the Au/CNT catalyst changed the main route from HFCA formation to DFF formation by accelerating the oxidation of the hydroxyl group in HMF. In addition, the incorporation of Pd further enhanced the oxidation of FFCA to FDCA, which is a difficult elementary step over monometallic Au catalyst under base-free conditions. More importantly, the Au – Pd/CNT showed high stability. Although the selectivity of FDCA decreased slightly in the initial three recycles, both the HMF conversion and FDCA selectivity were sustained in the further recycling uses.

### 6.4.4 *Synthesis of FDCA from HMF Over Supported Ru Catalysts*

The noble metal Ru is the same group of Pt and it also shows the catalytic activity towards HMF oxidation. In the oxidation of HMF, Ru catalysts are mainly used for the oxidation of HMF into DFF in organic solvents [12, 78–80], but a few examples have been reported for the oxidation of HMF into FDCA in water. Riisager and co-workers deposited catalytically active $Ru(OH)_x$ species onto a series of metal oxides and studied their catalytic performance towards HMF oxidation in water without base [81, 82]. Under the reaction conditions (2.5 bar $O_2$, 140 °C, 6 h), all catalysts were found to have catalytic activity toward HMF oxidation with FDCA yields ranging from 20% to 100%, but $[Ru(OH)_x]$ deposited onto basic carrier materials such as MgO, $La_2O_3$, and HT gave excellent catalytic performance with FDCA yields above 95%. The pH measurement of the post-reaction mixture indicated that the reaction solution with basic carrier supported Ru catalysts was higher than those with none-basic carriers supported Ru catalysts and $Mg^{2+}$ was determined in the former reaction solution suggesting that basic supports act to promote the oxidation of HMF into FDCA. Through a series of control experiments, the authors found that DFF and HFCA were reaction intermediates at low temperatures and low pressures, but HFCA was the only intermediate at high temperatures and high oxygen partial pressures. These results indicate that the oxidation of aldehyde group is competitive with the oxidation of the alcohol group at low temperatures and low oxygen partial pressures and that the oxidation of the aldehyde group is much faster than the oxidation of the alcohol group at high temperatures and high oxygen partial pressures. Besides the reaction in $H_2O$, the same research groups studied the oxidation of HMF into FDCA over supported $[Ru(OH)_x]$ catalysts in ionic liquids [83]. However, this method showed no significance in the practical application. Firstly, the catalytic results were not satisfactory, with the maximum FDCA yield being around 48%. Secondly, the cost of ionic liquids is very high and it is also difficult to separate FDCA from ionic liquids.

According to the results of the aerobic oxidation of HMF over supported metal catalysts, much progress has already been achieved for the synthesis of FDCA in recent years. Au catalysts are more stable and selective for the aerobic oxidation of HMF into FDCA in water than Pt, Ru and Pd based catalysts, because the Au catalysts can offer better resistance to water and $O_2$. However, Au catalysts suffer from deactivation in some cases by the deposition of byproducts or intermediates on their active sites. To promote oxidation of HMF, an excess of base is required to accelerate the reaction and to maintain the FDCA formed in aqueous solution as the salt of the dicarboxylic acid, thus strong adsorption of the carboxylic acids on the catalyst is avoided. Selecting an appropriate support and the use of an alloy might improve of catalyst performance to be highly active and stable for base-free oxidation of HMF into FDCA.

### 6.4.5 *Mechanism of the Oxidation of HMF into FDCA Over Supported Metal Catalysts*

Although many researchers have proposed reaction pathways for oxidation of HMF into FDCA over different supported metal catalysts based on the concentration change of HMF and the detectable intermediates such as DFF, HFCA, and FFCA and FDCA, no single unambiguous reaction mechanism for HMF oxidation exists so far. Attempts have been made to understand the reaction mechanism using isotope labeling technology [84].

Davis and co-workers studied the mechanism of the oxidation of HMF into FDCA over supported Au and Pt catalysts by the isotope labeling technology [84, 85]. The oxidation of HMF catalyzed by Au/$TiO_2$ under 3.45 bar $O_2$ at 22 °C with 2 equivalents of NaOH gave mostly HFCA (≥98% selectivity after 6 h. To understand the role of $O_2$, oxidation of HMF over Au/$TiO_2$ catalyst was carried out using $^{18}O_2$ as the oxidant, but the product HFCA showed no incorporation of $^{18}O$ atoms. Thus, the oxygen of HFCA most likely comes from the solvent $H_2O$. To verify this, HMF oxidation was carried out in labeled water $H_2{}^{18}O$. As expected, mass spectrum analysis revealed two $^{18}O$ atoms were incorporated into HFCA and the Na-adduct of HFCA. Thus, $O_2$ was not essential for the oxidation of the aldehyde group in HMF to produce HFCA. The aldehyde side chain is believed to undergo rapid reversible hydration to ageminal diol *via* nucleophilic addition of a hydroxide ion to the carbonyl and subsequent proton transfer from water to the alkoxy ion intermediate (Fig. 6.12, step 1). This step accounts for the incorporation of two $^{18}O$ atoms in HFCA when the reaction is performed in $H_2{}^{18}O$. The second step is the dehydrogenation of the geminal diol intermediate, facilitated by the hydroxide ions adsorbed on the metal surface to produce the carboxylic acid (Fig. 6.12, step 2). Production of the desired FDCA requires further oxidation of the alcohol side-chain of HFCA. Without base, no FDCA was formed even in the presence of Au and Pt catalysts, indicating the base played a crucial role in the transformation of HFCA into FDCA. The base is thought to deprotonate the alcohol side-chain to form an alkoxy intermediate, which is a step that may occur primarily in the solution [86]. Hydroxide ions on the catalyst surface then facilitate the activation of the C–H bond in the alcohol side-chain to form the aldehyde intermediate (FFCA) (Fig. 6.12, step 3). The next two steps (Fig. 6.12, steps 4 and 5) oxidize the aldehyde side-chain of FFCA to form FDCA. These two steps are expected to proceed analogously to steps 1 and 2 for oxidation of HMF to HFCA. The reversible hydration of the aldehyde group in step 4 to a geminal diol accounts for two more $^{18}O$ atoms incorporated in

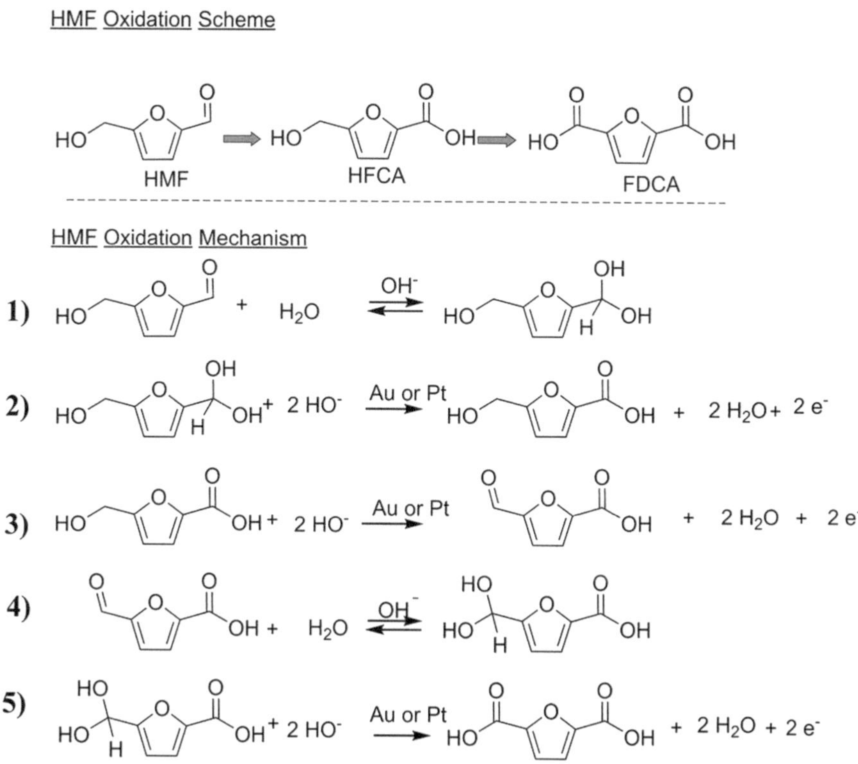


**Fig. 6.12** Proposed mechanism for the oxidation of HMF in aqueous solution in the presence of excess base ($OH^-$) and either Pt or Au [84, 85] (Reprinted with permission from [84], Copyright © 2011 Royal Society of Chemistry)

FDCA when the oxidation is performed in $H_2^{18}O$. Thus, the sequence in Fig. 6.12 explains the incorporation of all four $^{18}O$ atoms in FDCA when the reaction is performed in $H_2^{18}O$. Molecular $O_2$ is essential for the production of FDCA and plays an indirect role during oxidation by removing electrons deposited onto the supported metal particles.

The above mechanism involves the participation of base ($OH^-$) in the oxidation of HMF into FDCA, thus the reaction mechanism of the base-free reaction should be different from it. Yan and co-workers used isotope labeling technology to study the mechanism of base-free oxidation of HMF into FDCA over PVP or poly[bvbim]Cl stabilized Pd nanoparticles [55]. DFF and FFCA were found to be the reaction intermediates and HFCA was not determined during the reaction process. Other researchers have observed the same reaction pathway when the oxidation of HMF was performed without base or under low pH conditions [77, 85]. As shown in Fig. 6.13, the release of two $H^+$ from the hydroxyl group in HMF generates DFF. Then, they studied the oxidation of HMF over PVP/Pt catalyst in $H_2^{18}O$ for a reaction time of 4 h. Mass spectrometric analysis of the reaction mixture revealed

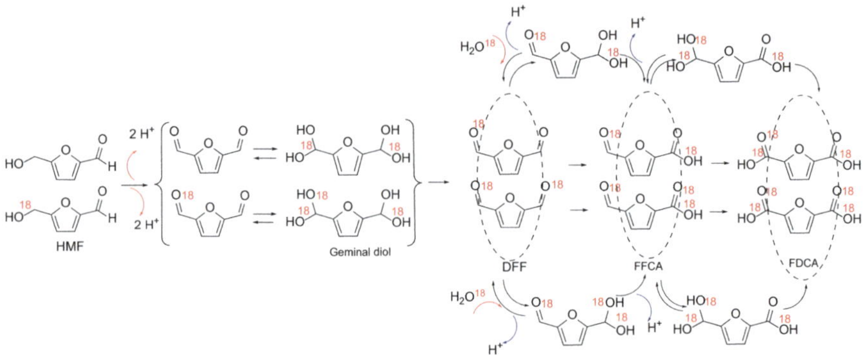


**Fig. 6.13** Incorporation of $^{18}O$ in the reaction steps: $^{18}O$ (*red*) and observed units in *dashed ellipses* (Reprinted with permission from Ref. [55], Copyright © 2014 Elsevier)

that the oxidation products (FDCA and FFCA) incorporated $^{18}O$ atoms. Peaks with m/z 163 and 161 correspond to four and three $^{18}O$ atoms incorporated in FDCA and peaks with m/z 145 and 143 corresponding to three and two $^{18}O$ atoms incorporated into FFCA. The aldehyde group is thought to undergo rapid reversible hydration to ageminal diol *via* nucleophilic addition of water to the carbonyl and subsequent proton transfer metal nanoparticles, which is similar to Fig. 6.12, step 1 in the above mechanism. Finally, the transfer of two $H^+$ to the surface of the metal generate the carboxylic groups. Molecular oxygen reacts with the surface hydride to release $H_2O$.

Although some information regarding the mechanism of HMF oxidation of HMF was obtained by the use of isotope labeling technology, much more effort should be devoted to obtain a deep understanding of the intrinsic kinetics and mechanisms by other methods. Understanding of the reaction mechanism will provide insight into the design of efficient and stable catalysts on the atomic level for the oxidation of HMF into FDCA.

### *6.4.6 Catalytic Synthesis of FDCA Over Non-Noble Metal Heterogeneous Catalysts*

Although noble metal catalysts generally show high catalytic performance towards the oxidation of HMF into FDCA, the industrially large-scale production of FDCA is limited by the use of noble metal catalysts due to their high cost. Inexpensive transition metal catalysts have been found to be active towards the oxidation of HMF into FDCA. Saha and co-workers prepared Fe catalyst by the incorporation of $Fe^{3+}$ in the porphyrin ring center of porphyrin-based porous organic polymer (Fe–POP) and studied the catalytic performance of the Fe–POP catalyst towards aerobic oxidation of HMF in water [87]. HMF conversion of 100% and FDCA yield of 79%

were attained after 10 h at 100 °C under 10 bar air. This catalytic system did not require the use of base, and is thought to proceed via a radical chain mechanism with the formation of peroxyl radical in the catalytic cycle through EPR spectra analysis. Zhang et al. found that Merrifield resin supported Co(II)-meso-tetra (4-pyridyl)-porphyrin (abbreviated as Merrifield resin-Co-Py) showed high catalytic activity towards the oxidation of HMF into FDCA with the use of *t*-BuOOH as the oxidant [88]. High HMF conversion of 95.6% and FDCA yield of 90.4% were attained in acetonitrile at 100 °C after 24 h. DFF was the intermediate of the oxidation of HMF into FDCA over Merrifield resin-Co-Py catalyst. In addition, the catalyst could be reused without significant loss of its catalytic activity. Later, Mugweru and co-workers performed the oxidation of HMF into FDCA in acetic acid by use of spinel mixed metal oxide catalyst($Li_2CoMn_3O_8$) and NaBr as the additive [89]. Full HMF conversion and 80% isolated yield of FDCA were obtained at 150 °C after 8 h under 55 bar air. The use of acetic acid and NaBr as the additive as well as the high reaction temperature and high pressure resulted in this catalytic system being less competitive for large-scale production of FDCA. To facilitate the recycling of the catalysts, our group has reported a new method for the oxidation of HMF into FDCA over magnetic Nano-$Fe_3O_4 - CoO_x$ catalyst [90]. High HMF conversion of 97.2% was obtained after 12 h at 80 °C, but FDCA yield was obtained at 68.6%. The catalyst could be recovered by an external magnet and reused with little mass loss.

Compared with noble metal catalysts, the use of inexpensive transition metal catalysts is promising for practical synthesis of FDCA, due to their low cost. However, current reported methods do not show selective production of FDCA, and the reactions are sometimes carried out in organic solvents, such as acetic acid or the use of t-BuOOH as the oxidant, which are less desirable for the "green chemistry". Therefore, further efforts should be devoted to developing new methods based on non-precious transition metal catalysts that can promote the oxidation of HMF into FDCA in water with high catalytic activity and selectivity with $O_2$ as the oxidant.

## 6.5 Catalytic Synthesis of FDCA from Carbohydrates

Although FDCA can be produced from HMF in nearly quantitative yields, the cost of HMF is high. Carbohydrates such as fructose, glucose and cellulose are much cheaper and more abundant than HMF. Therefore, it is more attractive to carry out the oxidative conversion of carbohydrates into FDCA by one-pot reaction over multiple functional catalysts combining acidic and metal sites.

The major problem of the direct conversion of carbohydrates into FDCA is the risk of simultaneous oxidation of carbohydrates. In 2000, Kröger and co-workers realized the one-pot conversion of fructose to FDCA using the strategy of two-phase system water/methyl isobutyl ketone (MIBK) [91]. As shown in Fig. 6.14. the reaction was carried out in a membrane reactor divided with a polytetrafluoroethylene

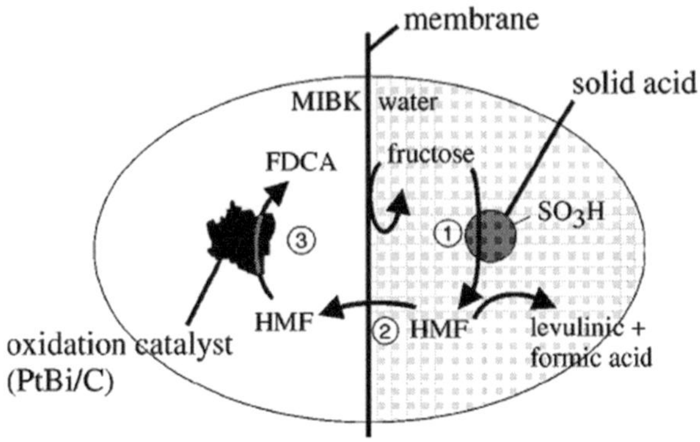


**Fig. 6.14** Scheme of the processes in membrane reactor: *1* – HMF formation in water phase, *2* – diffusion of HMF in MIBK phase and *3* – HMF oxidation (Reprinted with permission from Ref. [91], Copyright © 2000 Springer)

membrane to prevent fructose from oxidation. Fructose firstly dehydrated into HMF in water with a Lewatit SPC (trade name) 108 as the solid acid catalyst. Then HMF was extracted into MIBK due to the higher solubility of HMF in MIBK, followed by oxidation into FDCA over metal catalysts, while fructose is insoluble in MIBK, avoiding the oxidation of fructose. This process yielded a maximum 25% of FDCA. Interestingly, the authors observed that in a pure MIBK phase, DFF was the major product in HMF oxidation with this catalyst. It was also claimed that water was needed as a co-substrate for the formation of FDCA as final product. The overall reaction rate was related to the diffusion through the membrane. In addition, levulinic acid (25% yield) was formed as the by-product. This method produced low FDCA yield, and it was difficult to purify FDCA from the byproducts. Later, Ribeiro and Schuchardt prepared a bifunctional catalyst by the encapsulation of $Co(acac)_3$ in sol-gel silica, combining the acidic and redox ability, and studied the one-pot conversion of fructose into FDCA, affording fructose conversion of 72% and 99% selectivity of FDCA [92]. Compared with the method reported by Kröger and co-workers, both FDCA yield and selectivity was improved by a factor of several times. However, the reaction was carried out at 165 °C and at a high pressure of 20 bar air, which is difficult to be applied in practical applications.

Zhang and co-workers reported a two-step method to produce FDCA from fructose [93]. Firstly, fructose was dehydrated to HMF in *iso*-propanol with HCl as catalyst. Then, *iso*-propanol was collected for the next run by evaporation. Finally, HMF was purified with water-extraction and oxidized to FDCA over Au/HT catalyst. An overall FDCA yield of 83% was achieved from fructose. The water-extraction step was very important for the whole process. Under the same conditions, the oxidation of HMF without extraction gave only 51% yield of FDCA even after a long reaction time of 20 h, while it was 98% in 7 h after water extraction. It was claimed that byproducts such as humins caused deactivation of the Au/HT catalyst. This method also afforded an overall FDCA yield of 52% when Jerusalem artichoke tuber (major component of fructose unit) was used as the feedstock. The same group later reported a similar routine, in which poly-benzylic ammonium chloride resins were used as a solid catalyst for the dehydration of fructose in the first step, affording FDCA with a yield of 72% from fructose [94].

Zhang and co-workers demonstrated a one-pot process for the production of FDCA from sugars using a triphasic reactor [95]. This triphasic system is composed of tetraethylammoniumbromide (TEAB) or water (phase I)—methyl isobutyl ketone (MIBK) (phase II)—water (phase III).In the designed triphasic setup, sugars (fructose or glucose) were first dehydrated to HMF in Phase I. HMF was then extracted, purified, and transferred to Phase III via a bridge (Phase II). Finally, HMF was oxidized to FDCA over Au/HT catalyst in Phase III. Overall FDCA yields of 78% and 50% were achieved with fructose and glucose feedstock, respectively. Phase II plays multiple roles: as a bridge for HMF extraction, transport and purification. To facilitate the recycling of the catalysts and reduce the cost of the catalyst, Zhang et al. developed the one-pot conversion of fructose into FDCA via a two-step method by the combination of two paramagnetic catalysts [90]. Nano-$Fe_3O_4 - CoO_x$ catalyst showed high catalytic activity towards the oxidation of HMF into FDCA using *t*-BuOOH as the oxidant. A two-step strategy was applied for the synthesis of FDCA from fructose. HMF was first produced from the dehydration of fructose over the $Fe_3O_4@SiO_2 - SO_3H$ acid catalyst in DMSO. The magnetic $Fe_3O_4@SiO_2 - SO_3H$ could be easily separated from the reaction system with an external permanent magnet, and HMF in the remaining reaction solution was then oxidized into FDCA with *t*-BuOOH over nano-$Fe_3O_4 - CoO_x$ catalyst. FDCA was obtained in a yield of 59.8% after 15 h based on the starting fructose. The developed method shows the following two distinct advantages: (1) the use of paramagnetic catalyst facilitates catalyst recycle; (2) the use of transition metal catalyst makes this method much more economical for the practical synthesis of FDCA from renewable carbohydrates.

The direct conversion of carbohydrates into FDCA is much more attractive, but the current results are not satisfactory. The carbohydrates for FDCA production are focused on fructose. Much more work should be paid to the design novel catalysts with multiple catalytic sites for the conversion of other carbohydrates such as glucose even cellulose into FDCA. To avoid side reactions, the isolation of multiple catalytic sites is essential to realize the one-pot conversion of carbohydrates to FDCA. The acidic sites located in a hydrophilic environment would benefit the adsorption of the carbohydrates and promote the dehydration of carbohydrates into HMF, and the release of HMF into the reaction solution. The oxidative sites located in a hydrophobic environment enhance the absorption of the intermediate HMF and promote its oxidation to FDCA, and simultaneously release the high polar product FDCA in the reaction solution.

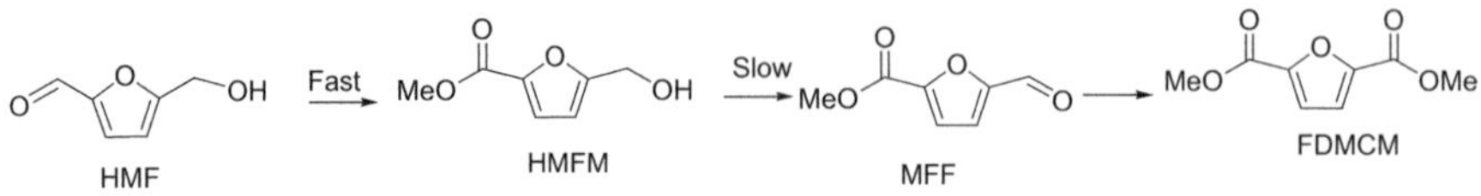


**Fig. 6.15** Proposed oxidation pathway of HMF to FDCDM (Reprinted with permission from Ref. [96], Copyright © 2007 John Wiley & Sons)

## 6.6 Catalytic Synthesis of FDCA Derivatives

Catalytic synthesis of FDCA from HMF can be obtained with high yields, but there are few reports on the purification of FDCA. As the oxidation of HMF into FDCA in most cases is carried out in the presence of excessive base, FDCA is in the form of its salts together with byproducts and the salts, making it difficult to purify FDCA. Hence, regarding FDCA purification, there is still a dearth of straightforward and eco-friendly methods. A possible way to overcome these issues is to produce the corresponding ester, 2,5-furandicarboxylicacid dimethyl ester (FDCDM), which can be easily purified through vacuum distillation and transformed into FDCA through a simple hydrolysis reaction. Moreover, instead of FDCA, FDCDM can also be used directly to synthesize polymers through transesterification reaction.

In 2008, Christensen and co-workers reported an effective way for the one-pot oxidative esterification of HMF into FDCDM over Au/$TiO_2$ catalyst. FDCDM could be achieved in an excellent yield of 98% at 130 °C under 4 bar $O_2$ in MeOH in the presence of 8% MeONa [96]. The addition of MeONa accelerated this reaction remarkably, because the reaction was incomplete without MeONa even when run for a long duration. Kinetic studies revealed that 5-hydroxymethyl methylfuroate (HMMF) was the reaction intermediate by the fast oxidative esterification of HMF (Fig. 6.15), which was then slowly converted to FDCDM, suggesting that the oxidation of the aldehyde moiety of HMF was faster than the hydroxymethyl group, which was similar as the oxidation of HMF into FDCA in water over Au catalysts [65]. In contrast to FDCA, FDCDM can be easily purified by sublimation at standard pressure and 160 °C to afford colorless crystals. Later, Corma and co-worker improved the oxidative esterification of HMF into FDCDM over Au/$CeO_2$ catalyst [97]. The Au/$CeO_2$ catalyst showed high catalytic activity without base, affording high FDCDM yield of 99% after 5 h at 130 °C under 10 bar $O_2$, whereas 8% MeONa was required to promote the reaction smoothly over Au/$TiO_2$ catalyst. Therefore, Au/$CeO_2$ showed higher catalytic activity than Au/$TiO_2$ catalyst, which was due to the formation of peroxo-and superoxo-species in monoelectronic defects of nanocrystalline ceria. Kinetic studies indicated that the oxidation of alcohol group was the rate-limiting step of the conversion of HMF into FDCDM, which was similar to the results reported by Christensen and co-workers using Au/$TiO_2$ catalyst [96]. Au/$CeO_2$ could be easily recovered and reused with a little loss of activity while maintaining high selectivity towards FDCDM. Although high FDMCA yields were obtained over $Au/CeO_2$ and $Au/TiO_2$ catalysts, the use of Au catalysts is high cost. Fu and co-workers developed an economical way for the oxidative esterification of HMF into FDCDM using cheap cobalt-based catalysts ($Co_xO_y$–N@C) [98]. However, the $Co_xO_y$–N@C catalyst is not as active as the Au catalysts. Under 10 bar $O_2$ partial pressure at 100 °C for 12 h, 95% conversion of HMF was obtained, but the FDCDM yield was only 38%, with other products mainly including 2,5-furandicarboxylic acid monomethyl ester (FDCMM) in a yield of 44% and 5-hydroxymethyl-2-furoic acid methyl ester (HMFM) in a yield of 12%. To improve the oxidation of alcohol group to aldehyde group, the strong oxidant K-OMS-2 was used to promote functional group change. Although HMFM yield decreased from 12% to 1%, FDCDM was still co-produced with FDCMM, in yields of 53% and 41%, respectively.

Taking the above described methods into consideration, the oxidative esterification of HMF into FDMC shows several advantages over the oxidation of HMF into FDCA, such as the facile product purification and the use of low catalytic amounts of base or even base-free conditions. Oxidative esterification of HMF to 2,5-furandimethylcarboxylate should be a direction to effectively and economically transform carbohydrates into platform molecules and polymers. These described methods using noble metal Au catalysts are effective, while the inexpensive Co catalysts are much less active. Therefore, developing none-noble metal catalysts with high catalytic activity and selectively should be given much more attention than their noble metal counterpart.

## 6.7 Conclusions and Future Outlook

### *6.7.1 Conclusions*

FDCA is a promising biomass-derived chemical and has wide application in many fields. More importantly, it can serve as a replacement for petrochemical derived terephthalic acid in production of bio-based polymers. Thus, effective catalytic methods for synthesis of FDCA will have huge economic benefits in the future, and will play a vital role in the development of sustainable economies. Currently, there are several methods for catalytic synthesis of FDCA from HMF or directly from carbohydrates by one-pot reaction, including biocatalytic, electrochemical and chemical catalytic approaches.

The chemical catalytic method is the main driving force for the synthesis of FDCA as it shows high potential for large-scale synthesis of FDCA by the use of molecular $O_2$ in an economical way. Much effort has been devoted to the design of chemical catalytic systems for aerobic oxidation of HMF into FDCA mainly using supported noble metal catalysts (Au, Pt and Pd). The activity of the catalyst and the reaction pathways are affected by the catalyst itself (such as the active phase, support, particle size) and reaction conditions (such as oxygen pressure, oxygen flow rate, pH and temperature). Au catalysts are more stable and selective than Pt and Pd catalysts because the Au catalysts can offer better resistance to $O_2$. However, Au catalysts are deactivated by byproducts or by intermediates in some cases. The alloying of another metal (e.g., Pd, Cu, Pt) with Au to form bimetallic alloy catalysts generally show high catalytic activity and stability as compared with the monometallic Au catalysts, due to alloying effect. HFCA is generally detected as the reaction intermediate in alkaline reaction solution especially for Au catalyst, while DFF is the reaction intermediate of base-free aerobic oxidation of HMF, suggesting that the oxidation of the aldehyde group in HMF is faster than the oxidation of alcohol group at low pH. Regardless of the reaction pathway, isotope labeling shows that the reaction mechanisms are similar for the oxidation of aldehyde group and alcohol group. The oxygen in FDCA is from water and molecular oxygen plays a key role in the catalytic recycle by removing electrons deposited into the supported metal particles. Finally, the diester of FDCA can be synthesized via the oxidative esterification of HMF in one-pot reaction over supported Au catalysts.

### 6.7.2 *Future Outlook*

Although significant achievements have been obtained in the catalytic synthesis of FDCA and its derivatives, further improvements are still needed in chemical aerobic oxidation of HMF for realizing the industrially large-scale and economical production of FDCA and its derivatives as follows:

(1) One challenge for the aerobic oxidation of HMF over supported noble metal catalysts is the instability of the catalysts. The stability of the catalyst can be improved from several aspects such as enhancing the interaction of metal nanoparticles and the supports, confining the metal nanoparticles in pores with a limited size to avoid their aggregation, improving the catalytic activity with full conversion and 100% selectivity to avoid catalyst deactivation by adsorption of byproducts, or by using bimetallic or trimetallic nanoparticles to improve the ability to prevent their poisoning;
(2) Most of the current methods are performed in water in the presence of excess base, resulting in the process being less green and more expensive. More importantly, the neutralization of the salt of FDCA further increases the operating costs and produces additional salt byproducts, making the process less-desirable and less cost-effective. Therefore, much more effort is needed to develop more environmentally-friendly catalytic systems that can effectively promote base-free oxidation of HMF into FDCA.
(3) Compared to noble metal catalysts, non-noble-metal catalysts are much less studied. From the viewpoint of practical application, the development of efficient, low-cost, and stable transition metal catalysts such as Co, Fe and Mn towards aerobic oxidation of HMF into FDCA is therefore very important.

(4) The use of carbohydrates or lignocelluloses as the feedstocks for the one-pot production of FDCA is strongly desirable, as these feedstocks are cheap and abundant, which will decrease the production cost. To realize this goal, multifunctional catalysts, combining the metal sites and solid acid/base catalysts should be carefully designed. It is suggested that the acid/base sites used for the dehydration of carbohydrates into HMF should be located in hydrophilic environments and the metal sites for oxidation reaction should be located in hydrophobic environments, thus the carbohydrates will easily be absorbed on the acid/base sites and the intermediate HMF will easily desorb from the hydrophobic metal size to be oxidized.

(5) Very few studies have been performed to examine the kinetics and reaction mechanisms of catalytic synthesis of FDCA either from HMF or direct from carbohydrates. Deep understanding of the intrinsic kinetics and mechanisms will provide insights into rational design of more efficient and stable catalysts on atomic levels for the oxidation of HMF into FDCA or one-pot oxidative conversion of carbohydrates into FDCA.

(6) Last, but not the least, catalytic synthesis of FDCA on a large scale is extremely important. Many current routes are technically feasible, but economically prohibitive. Development of energetically and economically viable processes is a long-standing task, which involves interdisciplinary problems of chemistry, material science, and process engineering.

**Acknowledgements** This work was supported by the National Natural Science Foundation of China (No. 21203252) and the funding offered by the China scholarship council (201408420018).

## References

1. Corma A, Iborra S, Velty A. Chemical routes for the transformation of biomass into chemicals. Chem Rev. 2007;107(6):2411–502.
2. Bozell JJ. Connecting biomass and petroleum processing with a chemical bridge. Science. 2010;329(5991):522–3.
3. Gallezot P. Conversion of biomass to selected chemical products. Chem Soc Rev. 2012;41(4):1538–58.
4. Morais ARC, da Costa Lopes AM, Bogel-Łukasik R. Carbon dioxide in biomass processing: contributions to the green biorefinery concept. Chem Rev. 2014;115(1):3–27.
5. Liguori F, Moreno-Marrodan C, Barbaro P. Environmentally friendly synthesis of γ-valerolactone by direct catalytic conversion of renewable sources. ACS Catal. 2015;5(3):1882–94.
6. Katryniok B, Paul S, Dumeignil F. Recent developments in the field of catalytic dehydration of glycerol to acrolein. ACS Catal. 2013;3(8):1819–34.
7. Beerthuis R, Rothenberg G, Shiju NR. Catalytic routes towards acrylic acid, adipic acid and ε-caprolactam starting from biorenewables. Green Chem. 2015;17(3):1341–61.
8. Amaniampong PN, Jia X, Wang B, Mushrif SH, Borgna A, Yang Y. Catalytic oxidation of cellobiose over $TiO_2$ supported gold-based bimetallic nanoparticles. Catal Sci Tech. 2015;5(4):2393–405.

9. Zhang J, Li J, Tang Y, Lin L, Long M. Advances in catalytic production of bio-based polyester monomer 2, 5-furandicarboxylic acid derived from lignocellulosic biomass. Carbohyd Polym. 2015;130:420–8.
10. Hanson SK, Baker RT. Knocking on wood: base metal complexes as catalysts for selective oxidation of lignin models and extracts. Acc Chem Res. 2015;48(7):2037–48.
11. Ma R, Xu Y, Zhang X. Catalytic oxidation of biorefinery lignin to value-added chemicals to support sustainable biofuel production. ChemSusChem. 2015;8(1):24–51.
12. Zakrzewska ME, Bogel-Łukasik E, Bogel-Łukasik R. Ionic liquid-mediated formation of 5-hydroxymethylfurfural: a promising biomass-derived building block. Chem Rev. 2010;111(2):397–417.
13. Zhang ZH, Liu B, Lv KL, Sun J, Deng KJ. Green Chem. 2015;16:2762–70.
14. Lan J, Lin J, Chen Z, Yin G. Transformation of 5-hydroxymethylfurfural (HMF) to maleic anhydride by aerobic oxidation with heteropolyacid catalysts. ACS Catal. 2015;5(4):2035–41.
15. Werpy T, Petersen G, Aden A, Bozell J, Holladay J, White J, Manheim A, Eliot D, Lasure L, Jones S. Top value added chemicals from biomass. Volume 1: Results of screening for potential candidates from sugars and synthesis gas. U. S. Department of Energy report. Pacific Northwest National Laboratory/U.S. Department of Energy: Oak Ridge, TN. 2004.
16. Haworth WN, Jones WG, Wiggins LF. The conversion of sucrose into furan compounds. Part II. Some 2: 5-disubstituted tetrahydrofurans and their products of ring scission. J Chem Soc. 1945;1–4.
17. Jacquel N, Saint-Loup R, Pascault JP, Rousseau A. Bio-based alternatives in the synthesis of aliphatic–aromatic polyesters dedicated to biodegradable film applications. Polymer. 2015;59:234–42.
18. Rajendran S, Raghunathan R, Hevus I, Krishnan R, Ugrinov A, Sibi MP, Sivaguru J. Programmed photodegradation of polymeric/oligomeric materials derived from renewable bioresources. Angew Chem Int Ed. 2015;54(4):1159–63.
19. Sousa AF, Vilela C, Fonseca AC, Matos M, Freire CS, Gruter GJM, Silvestre AJ. Biobased polyesters and other polymers from 2, 5-furandicarboxylic acid: a tribute to furan excellency. Polym Chem. 2015;6(33):5961–83.
20. Eerhart AJJE, Faaij APC, Patel MK. Replacing fossil based PET with biobased PEF; process analysis, energy and GHG balance. Energy Environ Sci. 2012;5(4):6407–22.
21. Gandini A, Silvestre AJ, Neto CP, Sousa AF, Gomes M. The furan counterpart of poly (ethylene terephthalate): an alternative material based on renewable resources. J Polym Sci Part A: Polym Chem. 2009;47(1):295–8.
22. Gandini A. Furans as offspring of sugars and polysaccharides and progenitors of a family of remarkable polymers: a review of recent progress. Polym Chem. 2010;1(3):245–51.
23. Davis ME. Heterogeneous catalysis for the conversion of sugars into polymers. Top Catal. 2015;58(7–9):405–9.
24. Lewkowski J. Synthesis, chemistry and applications of 5-hydroxymethyl-furfural and its derivatives. Arkivoc. 2001;1:17–54.
25. Nagarkar SS, Chaudhari AK, Ghosh SK. Selective $CO_2$ adsorption in a robust and water-stable porous coordination polymer with new network topology. Inorg Chem. 2012;12:12572–6.
26. Rose M, Weber D, Lotsch BV, Kremer RK, Goddard R, Palkovits R. Biogenic metal–organic frameworks: 2, 5-furandicarboxylic acid as versatile building block. Microporous Mesoporous Mater. 2013;181:217–21.
27. Fittig R, Heinzelmann H. Ueber neue Derivate der Schleimsaure. Chem Ber. 1876;9:1198.
28. Gonis G, Amstutz ED. The preparation of furan-2, 5-dicarboxylic acid. J Org Chem. 1962;27(8):2946–7.
29. Miura T, Kakinuma H, Kawano T, Matsuhisa H. Method for producing furan-2, 5-dicarboxylic acid U.S. Patent. 2008; US-7411078.
30. Holade Y, Morais C, Servat K, Napporn TW, Kokoh KB. Toward the electrochemical valorization of glycerol: Fourier transform infrared spectroscopic and chromatographic studies. ACS Catal. 2013;3(10):2403–11.

31. Ciriminna R, Palmisano G, Pagliaro M. Electrodes functionalized with the 2, 2, 6, 6-tetramethylpiperidinyloxy radical for the waste-free oxidation of alcohols. ChemCatChem. 2015;7(4):552–8.
32. Grabowski G, Lewkowski J, Skowroński R. The electrochemical oxidation of 5-hydroxymethylfurfural with the nickel oxide/hydroxide electrode. Electrochim Acta. 1991;36:1995–1995.
33. Vuyyuru KR, Strasser P. Oxidation of biomass derived 5-hydroxymethylfurfural using heterogeneous and electrochemical catalysis. Catal Today. 2012;195(1):144–54.
34. Chadderdon DJ, Xin L, Qi J, Qiu Y, Krishna P, More KL, Li W. Electrocatalytic oxidation of 5-hydroxymethylfurfural to 2, 5-furandicarboxylic acid on supported Au and Pd bimetallic nanoparticles. Green Chem. 2014;16(8):3778–86.
35. Cha HG, Choi KS. Combined biomass valorization and hydrogen production in a photoelectrochemical cell. Nature chem. 2015;7(4):328–33.
36. Thomas SM, DiCosimo R, Nagarajan V. Biocatalysis: applications and potentials for the chemical industry. Trends Biotechnol. 2002;20(6):238–42.
37. Krystof M, Perez-Sanchez M, de Maria PD. Lipase-mediated selective oxidation of furfural and 5-hydroxymethylfurfural. ChemSusChem. 2013;6(5):826–30.
38. Hanke PD. Enzymatic oxidation of HMF. USA Patent. 2009;2009/023174 A2
39. Koopman F, Wierckx N, de Winde JH, Ruijssenaars HJ. Efficient whole-cell biotransformation of 5-(hydroxymethyl) furfural into FDCA, 2, 5-furandicarboxylic acid. Bioresource Tech. 2010;101(16):6291–6.
40. Dijkman WP, Groothuis DE, Fraaije MW. Enzyme-catalyzed oxidation of 5-hydroxymethylfurfural to furan-2, 5-dicarboxylic acid. Angew Chem Int Ed. 2014;53(25):6515–8.
41. Dijkman WP, Binda C, Fraaije MW, Mattevi A. Structure-based enzyme tailoring of 5-hydroxymethylfurfural oxidase. ACS Catal. 2015;5(3):1833–9.
42. Partenheimer W, Grushin VV. Synthesis of 2, 5-diformylfuran and furan-2, 5-dicarboxylic acid by catalytic air-oxidation of 5-hydroxymethylfurfural. unexpectedly selective aerobic oxidation of benzyl alcohol to benzaldehyde with metal-bromide catalysts. Adv Synth Catal. 2001;343(1):102–11.
43. Saha B, Dutta S, Abu-Omar MM. Aerobic oxidation of 5-hydroxylmethylfurfural with homogeneous and nanoparticulate catalysts. Cat Sci Technol. 2012;2(1):1.
44. Hansen TS, Sadaba I, Garcia-Suarez EJ, Riisager A. Cu catalyzed oxidation of 5-hydroxymethylfurfural to 2,5-diformylfuran and 2,5-furandicarboxylic acid under benign reaction conditions. Appl Catal A-Gen. 2014;456:44–50.
45. Verdeguer P, Merat N, Gaset A. Oxydation catalytique du HMF en acide 2, 5-furane dicarboxylique. J Mol Catal. 1993;85(3):327–44.
46. Rass HA, Essayem N, Besson M. Selective aqueous phase oxidation of 5-hydroxymethylfurfural to 2, 5-furandicarboxylic acid over Pt/C catalysts: influence of the base and effect of bismuth promotion. Green Chem. 2013;15(8):2240–51.
47. Ait Rass H, Essayem N, Besson M. Selective aerobic oxidation of 5-HMF into 2,5-furandicarboxylic acid with Pt catalysts supported on $TiO_2$-and $ZrO_2$-based supports. ChemSusChem. 2015;8(7):1206–17.
48. Niu W, Wang D, Yang G, Sun J, Wu M, Yoneyama Y, Tsubaki N. Pt nanoparticles loaded on reduced graphene oxide as an effective catalyst for the direct oxidation of 5-hydroxymethylfurfural (HMF) to produce 2, 5-furandicarboxylic acid (FDCA) under mild conditions. Bull Chem Soc Jpn. 2014;87(10):1124–9.
49. Davis SE, Houk LR, Tamargo EC, Datye AK, Davis RJ. Oxidation of 5-hydroxymethylfurfural over supported Pt, Pd and Au catalysts. Cataly Today. 2011;160(1):55–60.
50. Vinke PV, Van Dam HE, Van Bekkum H. Platinum catalyzed oxidation of 5-hydroxymethylfurfural. Stud Surf Sci Catal. 1990;55:147–58.
51. Vinke PV, Van der Poel W, Van Bekkum H. On the oxygen tolerance of noble metal catalysts in liquid phase alcohol oxidations the influence of the support on catalyst deactivation. Stud Surf Sci Catal. 1991;59:385–94.

52. Raouf F, Taghizadeh M, Yousefi M. Influence of CaO-ZnO supplementation as a secondary catalytic bed on the oxidative coupling of methane. React Kinet Mech Catal. 2014;112(1):227–40.
53. Miao Z, Wu T, Li J, Yi T, Zhang Y, Yang X. Aerobic oxidation of 5-hydroxymethylfurfural (HMF) effectively catalyzed by a $Ce_{0.8}Bi_{0.2}O_{2-\delta}$ supported Pt catalyst at room temperature. RSC Adv. 2015;5(26):19823–9.
54. Lilga MA, Hallen RT, Gray M. Production of oxidized derivatives of 5-hydroxymethylfurfural (HMF). Top Catal. 2010;53(15–18):1264–9.
55. Siankevich S, Savoglidis G, Fei Z, Laurenczy G, Alexander DT, Yan N, Dyson PJ. A novel platinum nanocatalyst for the oxidation of 5-Hydroxymethylfurfural into 2, 5-Furandicarboxylic acid under mild conditions. J Catal. 2014;315:67–74.
56. Siyo B, Schneider M, Pohl MM, Langer P, Steinfeldt N. Synthesis, characterization, and application of PVP-Pd NP in the aerobic oxidation of 5-hydroxymethylfurfural (HMF). Catal Lett. 2014;144(3):498–506.
57. Gawade AB, Tiwari MS, Yadav GD. Biobased green process: selective hydrogenation of 5-hydroxymethyl furfural (HMF) to 2, 5-dimethyl furan (DMF) under mild conditions using Pd-$Cs_{2.5}H_{0.5}PW_{12}O_{40/K-10}$ clay. ACS Sustain Chem Eng. 2016;4(8):4113–23.
58. Zhang ZH, Zhen JD, Liu B, Lv KL, Deng KJ. Selective aerobic oxidation of the biomass-derived precursor 5-hydroxymethylfurfural to 2, 5-furandicarboxylic acid under mild conditions over a magnetic palladium nanocatalyst. Green Chem. 2015;17(2):1308–17.
59. Liu B, Ren YS, Zhang ZH. Aerobic oxidation of 5-hydroxymethylfurfural into 2, 5-furandicarboxylic acid in water under mild conditions. Green Chem. 2015;17(3):1610–7.
60. Mei N, Liu B, Zheng JD, Lv KL, Tang DG, Zhang ZH. A novel magnetic palladium catalyst for the mild aerobic oxidation of 5-hydroxymethylfurfural into 2, 5-furandicarboxylic acid in water. Cataly Sci Tech. 2015;5(6):3194–202.
61. Hutchings GJ. Vapor phase hydrochlorination of acetylene: correlation of catalytic activity of supported metal chloride catalysts. J Catal. 1985;96(1):292–5.
62. Haruta M, Kobayashi T, Sano H, Yamada N. Novel gold catalysts for the oxidation of carbon monoxide at a temperature far below 0°C Chem Lett. 1987;(2):405–8.
63. Pan M, Brush AJ, Pozun ZD, Ham HC, Yu WY, Henkelman G, Mullins CB. Model studies of heterogeneous catalytic hydrogenation reactions with gold. Chem Soc Rev. 2013;42(12):5002–13.
64. Wittstock A, Wichmann A, Bäumer M. Nanoporous gold as a platform for a building block catalyst. ACS Catal. 2012;2(10):2199–215.
65. Casanova O, Iborra S, Corma A. Biomass into chemicals: aerobic oxidation of 5-hydroxymethyl-2-furfural into 2, 5-furandicarboxylic acid with gold nanoparticle catalysts. ChemSusChem. 2009;2(12):1138–44.
66. Albonetti S, Lolli A, Morandi V, Migliori A, Lucarelli C, Cavani F. Conversion of 5-hydroxymethylfurfural to 2, 5-furandicarboxylic acid over Au-based catalysts: Optimization of active phase and metal–support interaction. Appl Catal B: Environ. 2015;163:520–30.
67. Zhang MZ, Zhang YB, Pan XQ, Wu TX, Zhang B, Li JW, Yang X, Ting Y, Zhang ZD, Yang XG. Superior catalytic performance of $Ce_{1-x}Bi_xO_{2-\delta}$ solid solution and Au/$Ce_{1-x}Bi_xO_{2-\delta}$ for 5-hydroxymethylfurfural conversion in alkaline aqueous solution. Catal Sci Tech. 2015;5(2):1314–22.
68. Liu JX, Ma JP, Cai JY, Ma H, Du ZT, Xu J. Advances in catalytic synthesis of bio-based dicarboxylic acid. Science China Chem. 2015;45(5):526–32.
69. Gorbanev YY, Klitgaard SK, Woodley JM, Christensen CH, Riisager A. Gold-catalyzed aerobic oxidation of 5-hydroxymethylfurfural in water at ambient temperature. ChemSusChem. 2009;2(7):672–5.
70. Wu Y, Wang DS, Zhou G, Yu R, Chen C, Li YD. Sophisticated construction of Au islands on Pt-Ni: an ideal trimetallic nanoframe catalyst. J Am Chem Soc. 2014;136(33):11594–7.
71. Rasul S, Anjum DH, Jedidi A, Minenkov Y, Cavallo L, Takanabe K. A highly selective copper-indium bimetallic electrocatalyst for the electrochemical reduction of aqueous $CO_2$ to CO. Angew Chem Int Ed. 2015;127(7):2174–217.

72. Pasini T, Piccinini M, Blosi M, Bonelli R, Albonetti S, Dimitratos N, Hutchings GJ. Selective oxidation of 5-hydroxymethyl-2-furfural using supported gold–copper nanoparticles. Green Chem. 2011;13(8):2091–9.
73. Albonetti S, Pasini T, Lolli A, Blosi M, Piccinini M, Dimitratos N, Cavani F. Selective oxidation of 5-hydroxymethyl-2-furfural over $TiO_2$-supported gold-copper catalysts prepared from preformed nanoparticles: effect of Au/Cu ratio. Catal Today. 2012;195(1):120–6.
74. Villa A, Schiavoni M, Campisi S, Veith GM, Prati L. Pd-modified Au on carbon as an effective and durable catalyst for the direct oxidation of HMF to 2, 5-furandicarboxylic acid. ChemSusChem. 2013;6(4):609–12.
75. Gupta NK, Nishimura S, Takagaki A, Ebitani K. Hydrotalcite-supported gold-nanoparticle-catalyzed highly efficient base-free aqueous oxidation of 5-hydroxymethylfurfural into 2, 5-furandicarboxylic acid under atmospheric oxygen pressure. Green Chem. 2011;13(4):824–7.
76. Zope BN, Davis SE, Davis RJ. Influence of reaction conditions on diacid formation during Au-catalyzed oxidation of glycerol and hydroxymethylfurfural. Top Catal. 2012;55(1–2):24–32.
77. Wan XY, Zhou CM, Chen JS, Deng WP, Zhang QH, Yang YH, Wang Y. Base-free aerobic oxidation of 5-hydroxymethyl-furfural to 2, 5-furandicarboxylic acid in water catalyzed by functionalized carbon nanotube-supported Au–Pd alloy nanoparticles. ACS Catal. 2014;4(7):2175–85.
78. Artz J, Mallmann S, Palkovits R. Selective Aerobic Oxidation of HMF to 2, 5-Diformylfuran on Covalent Triazine Frameworks-Supported Ru Catalysts. ChemSusChem. 2015;8(4):672–9.
79. Nie JF, Xie JH, Liu HC. Efficient aerobic oxidation of 5-hydroxymethylfurfural to 2, 5-diformylfuran on supported Ru catalysts. J Catal. 2013;301:83–91.
80. Wang SG, Zhang ZH, Liu B, Li JL. Environmentally friendly oxidation of biomass derived 5-hydroxymethylfurfural into 2, 5-diformylfuran catalyzed by magnetic separation of ruthenium catalyst. Ind Eng Chem Res. 2014;53(14):5820–7.
81. Gorbanev YY, Kegnæs S, Riisager A. Selective aerobic oxidation of 5-hydroxymethylfurfural in water over solid ruthenium hydroxide catalysts with magnesium-based supports. Catal Lett. 2011;141(12):1752–60.
82. Gorbanev YY, Kegnæs S, Riisager A. Effect of support in heterogeneous ruthenium catalysts used for the selective aerobic oxidation of HMF in water. Top Catal. 2011;54(16–18):1318–24.
83. Ståhlberg T, Eyjólfsdóttir E, Gorbanev YY, Sádaba I, Riisager A. Aerobic oxidation of 5-(hydroxymethyl) furfural in ionic liquids with solid ruthenium hydroxide catalysts. Catal Lett. 2012;142(9):1089–97.
84. Davis SE, Zope BN, Davis RJ. On the mechanism of selective oxidation of 5-hydroxymethylfurfural to 2, 5-furandicarboxylic acid over supported Pt and Au catalysts. Green Chem. 2012;14(1):143–7.
85. Davis SE, Benavidez AD, Gosselink RW, Bitter JH, De Jong KP, Datye AK, Davis RJ. Kinetics and mechanism of 5-hydroxymethylfurfural oxidation and their implications for catalyst development. J Mol Catal A: Chem. 2014;388:123–32.
86. Zope BN, Hibbitts DD, Neurock M, Davis RJ. Reactivity of the gold/water interface during selective oxidation catalysis. Science. 2010;330:74–87.
87. Saha B, Gupta D, Abu-Omar MM, Modak A, Bhaumik A. Porphyrin-based porous organic polymer-supported iron (III) catalyst for efficient aerobic oxidation of 5-hydroxymethyl-furfural into 2, 5-furandicarboxylic acid. J Catal. 2013;299:316–20.
88. Gao LC, Deng KJ, Zheng JD, Liu B, Zhang ZH. Efficient oxidation of biomass derived 5-hydroxymethylfurfural into 2, 5-furandicarboxylic acid catalyzed by Merrifield resin supported cobalt porphyrin. Chem Eng J. 2015;270:444–9.
89. Jain A, Jonnalagadda SC, Ramanujachary KV, Mugweru A. Selective oxidation of 5-hydroxymethyl-2-furfural to furan-2, 5-dicarboxylic acid over spinel mixed metal oxide catalyst. Catal Comm. 2015;58:179–82.

90. Wang SG, Zhang ZH, Liu B. Catalytic Conversion of Fructose and 5-Hydroxymethylfurfural into 2, 5-Furandicarboxylic Acid over a Recyclable $Fe_3O_4$–$CoO_x$ Magnetite Nanocatalyst. ACS Sustain Chem Eng. 2015;3(3):406–12.
91. Kröger M, Prüße U, Vorlop KD. A new approach for the production of 2, 5-furandicarboxylic acid by *in-situ* oxidation of 5-hydroxymethylfurfural starting from fructose. Top Catal. 2000;13(3):237–42.
92. Ribeiro ML, Schuchardt U. Cooperative effect of cobalt acetylacetonate and silica in the catalytic cyclization and oxidation of fructose to 2, 5-furandicarboxylic acid. Catal Comm. 2003;4(2):83–6.
93. Yi G, Teong SP, Li X, Zhang Y. Purification of biomass-derived 5-hydroxymethylfurfural and its catalytic conversion to 2, 5-furandicarboxylic acid. ChemSusChem. 2014;7(8):2131–5.
94. Teong SP, Yi G, Cao X, Zhang Y. Poly-benzylic ammonium chloride resins as solid catalysts for fructose dehydration. ChemSusChem. 2014;7(8):2120–4.
95. Yi G, Teong SP, Zhang Y. The direct conversion of sugars into 2, 5-furandicarboxylic acid in a triphasic system. ChemSusChem. 2015;8(7):1151–5.
96. Taarning E, Nielsen IS, Egeblad K, Madsen R, Christensen CH. Chemicals from renewables: aerobic oxidation of furfural and hydroxymethylfurfural over gold catalysts. ChemSusChem. 2008;1(1–2):75–8.
97. Casanova O, Iborra S, Corma A. Biomass into chemicals: one pot-base free oxidative esterification of 5-hydroxymethyl-2-furfural into 2, 5-dimethylfuroate with gold on nanoparticulated ceria. J Catal. 2009;265(1):109–16.
98. Deng J, Song HJ, Cui MS, Du YP, Fu Y. Aerobic oxidation of hydroxymethylfurfural and furfural by using heterogeneous $Co_xO_y$-N@C Catalysts. ChemSusChem. 2014;7(12):3334–40.

# Chapter 7
# Production of Glucaric/Gluconic Acid from Biomass by Chemical Processes Using Heterogeneous Catalysts

**Ayumu Onda**

**Abstract** Gluonic acid is a platform chemical that can be derived from biomass for making plastics and food additives, while glucaric acid is expected to be a sustainable precursor for producing adipic acid. This chapter provides an overview of the chemical processes for the production of gluconic acid and glucaric acid from mono-saccharides and poly-saccharides with a focus on heterogeneous catalysts. Because catalytic conversions of biomass are usually in aqueous solutions, the solid catalysts must be water tolerant. The oxidation of glucose into gluconic acid has been well-investigated using supported noble metals, such as Pt, Pd, and Au catalysts and those bimetallic catalysts, such as PdAu and PtBi. For the heterogeneous catalytic oxidation of glucose in aqueous solutions, three hypotheses can be used to explain catalyst deactivation: (i) poisoning of the catalyst by oxygen or hydrogen, (ii) poisoning of the catalyst by products, and (iii) metal sintering and metal leaching. Bimetallic catalysts with appropriate compositions have higher activity than those of monometallic catalysts to inhibit poisoning of oxygen, hydrogen, and acidic products. Supported Au catalyst also shows high yields of gluconic acid in aqueous solutions even without pH control. On the other hand, compared with oxidation of glucose to gluconic acid, oxidation into glucaric acid is a challenging reaction because an oxidation step from gluconic acid to glucaric acid is slow. Among tested catalysts, supported Pt and PtCu catalysts show relatively high glucaric acid yields. Direct production of gluconic acid from polysaccharides using bifunctional catalysts is also discussed. Bifunctional catalysts, such as sulfonated activated-carbon supported platinum, have acidic active sites and redox actives sites that can provide efficient conversion of gluconic acid and glucaric acid from polysaccharides.

**Keywords** Gluconic acid • Glucaric acid • Catalytic hydrothermal process • Heterogeneous catalyst • Bifunctional catalyst • Sugar • Polysaccharide

---

A. Onda (✉)
Research Laboratory of Hydrothermal Chemistry, Faculty of Science, Kochi University, Kochi, Japan
e-mail: aonda@kochi-u.ac.jp



Z. Fang et al. (eds.), *Production of Platform Chemicals from Sustainable Resources*, Biofuels and Biorefineries 7, DOI 10.1007/978-981-10-4172-3_7

## 7.1 Production of Gluconic Acid from Glucose Over Heterogeneous Catalysts

Biomass is a promising sustainable chemical resource [1, 2]. Gluconic acid and glucaric acid, as high value-added chemical compounds, are mainly obtained from selective oxidation of glucose (Scheme 7.1). Gluconic acid is a carboxylic aldose and an important industrial product in the food and pharmaceutical industries or as water-soluble cleansing agent [3]. Many methods, including biochemical, heterogeneously catalytic, chemical and electrochemical procedures have been proposed to effectively oxidize glucose into gluconic acid. At present, gluconic acid is mainly produced by fermentation processes and enzymatic processes using glucose oxidase and glucose dehydrogenase [3, 4]. However, biochemical methods have some disadvantages, such as slow reaction rate, low space time efficiency, and difficulty in the separation of the enzyme from the product. To overcome these issues, inorganic heterogeneous catalysts, such as supported noble metal catalysts for oxidizing monosaccharides with oxygen or air have been proposed [5–12]. Here, some reported chemical processes using heterogeneous catalysts are introduced. The oxidation of glucose to gluconic acid has been investigated using supported Pt, Pd, and Au catalysts [13–16] or their bimetallic catalysts, such as PtPd, PdAu, PtPdBi, PtAu, PtBi, and PdTe [17–24].

### *7.1.1 Pd and Pt Monometallic Catalysts*

In aqueous solutions, oxidation of D-glucose to D-gluconic acid was carried out by oxygen or air at atmospheric pressure of monometallic, such as Pd and Pt, catalysts at 30–80 °C [5]. Abbadi et al. reported results using 5% Pt/C catalyst at 50 °C as shown in Fig. 7.1 and Table 7.1 [9]. Glucose was converted under 0.2 atm of oxygen partial pressure and the main product was gluconic acid with initial rates being almost independent of pH control. However, in reaction without pH control, conversion of glucose and yield of gluconic acid plateaued to be under 25% and 15%, respectively. The pH in reaction media decreased to about 2.6 because of production of gluconic acid. On the other hand, when the pH of reaction media was controlled by adding KOH, conversion of glucose and yield of gluconic acid increased with increasing of the pH. Below pH = 7, poisoning of the catalyst was markedly observed. The catalytic performance of Pt/C catalyst in the oxidation of glucose into

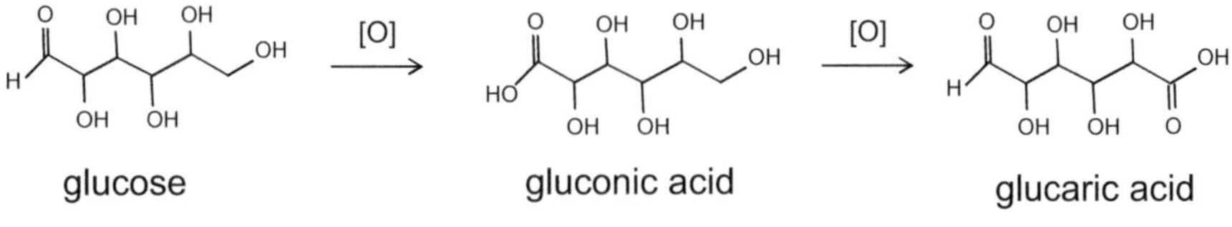


**Scheme 7.1** Oxidation of glucose into gluconic acid and into glucaric acid

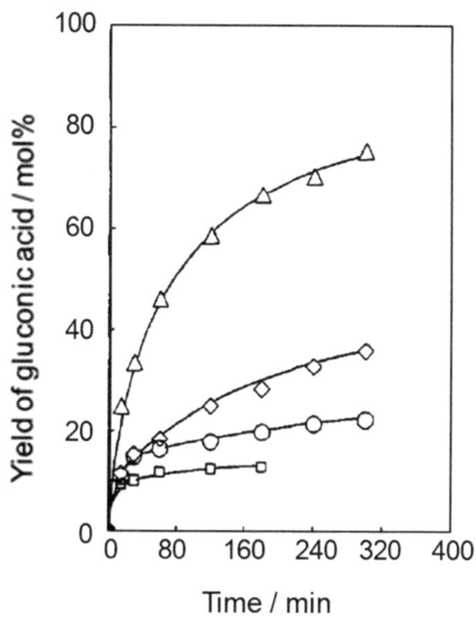


**Fig. 7.1** Effects of pH on the oxidation of D-glucose to D-gluconic acid. □ No pH control, ○ pH = 5, ◇ pH = 7, △ pH = 9. Reaction conditions: 0.05 mol·L$^{-1}$ glucose aqueous solution 80 mL, 5%Pt/C catalyst 40 mg, reaction temperature 50 °C, total pressure 1.0 atm ($P(O_2)$ = 0.2 atm) (Reprinted from Ref. [9], Copyright © 1995, with permission from Elsevier)

**Table 7.1** Product distributions in the oxidation of D-glucose over Pt/C catalyst

| pH | Conv. (%) | Yield (%) gluconic acid | Sel. (%) gluconic acid | 2-keto-D-gluconic acid | Guluronic acid | Glucaric acid | Arabinonic acid | Oxalic acid |
|---|---|---|---|---|---|---|---|---|
| 9 | 85 | 75 | 89 | 0.3 | 4.2 | 1.2 | 0.6 | 4.5 |
| 7 | 45 | 36 | 79 | 2 | 2.2 | 0.4 | 0.7 | 3.9 |
| 5 | 28 | 22 | 81 | 1 | 0.7 | – | 0.4 | 3.7 |
| No control | 23 | 14 | – | 0.4 | 0.6 | – | 0.3 | 3.8 |

Reprinted from Ref. [9], Copyright © 1995, with permission from Elsevier
Reaction conditions: 0.05 mol L$^{-1}$ glucose aqueous solution 80 mL, 5%Pt/C catalyst 40 mg, reaction temperature 50 °C, total pressure 1.0 atm (P ($O_2$) = 0.2 atm), reaction time = 6 h

gluconic acid showed a strong dependence on the pH of the reaction media. Product distribution was hardly affected by the pH control. The selectivity of the catalyst for the oxidation of D-glucose toward D-gluconic acid was reasonably good and ranged from 80% in acidic conditions to almost 90% in alkaline conditions. The by-products, 2-keto-D-gluconic acid and oxalic acid were formed and those authors speculated that the termination of the reaction without pH control was caused by strong adsorption of the reaction products on catalysts. Besson et al. [25, 26] examined the effect of palladium particle size on the gluconic acid yield in the glucose oxidation over Pd/C catalyst. The catalyst with Pd particles larger than 3 nm converted almost 100% of glucose, whereas Pd/C catalyst with Pd particles smaller than 2 nm gave about 60% of maximum conversion. This could indicate particle size dependence on the catalytic stability for oxygen poisoning in the oxidation as the smaller particles probably have stronger affinity for oxygen. For the heterogeneously catalytic oxidation of glucose in aqueous solution, three hypotheses to explain the deactivation of catalysts are suggested.

(i) Deactivation of the catalyst by poisoning due to oxygen or hydrogen adsorption. Noble metals have high stability for the oxidation, but even Pd and Pt are chemisorbed by oxygen, which can cause deactivation. Chemisorption of hydrogen is also often a cause of the deactivation of Pd and Pt catalysts. In the case of Pd and Pt, the phenomenon is reversible in the sense that the catalyst can be reactivated by hydrogen (or oxygen) and another reducing (or oxidizing) agent, such as glucose. When the reduced Pt/C catalyst is contacted with the substrate before the introduction of $O_2$ gas, Pt/C catalyst show much higher catalytic activity than for catalysts contacted with substrate after introduction of $O_2$ gas. Therefore, catalytic activity seems to depend on $O_2$ adsorption on the catalysts [27].
(ii) Deactivation of the catalyst by poisoning due to adsorption of products formed such as gluconate species. The active sites are at the surface of the Pd and Pt particles and the chemisorption of products formed can cause deactivation. In the case of chemisorption of gluconate species, the phenomenon can be reversible in the sense that the catalyst can be reactivated by alkaline treatment. Therefore, in Fig. 7.1, higher pH results in higher glucose conversion.
(iii) Deactivation of catalyst due to metal sintering and metal leaching. In aqueous solvent systems, metal leaching is a serious problem. Both Pt and Pd have relatively high stability for sintering and leaching under aqueous conditions, even when the solution is hot and acidic. However, it has been reported [28] that in solutions of D-ribose and D-xylose, both platinum and aluminum from Pt/$Al_2O_3$ catalysts undergo leaching with Pt leaching being much larger than Al for the reaction of D-ribose, since D-ribose is favorable for metal complexing. Sintering of metal particles also causes a decrease in catalytic activity . The support materials and preparation methods are important for inhibiting the growth of metal particle sizes during reaction and pretreatment [26, 29].

### *7.1.2 Pd-M and Pt-M Bimetallic Catalysts*

Alloy and bimetallic catalysts have been examined since the 1950s, and both alloys and bimetallic catalysts have been investigated for scientific interest in terms that the electronic and ensemble effects have on the catalytic activity and selectivity.

Many important catalytic processes that use bimetallic catalysts have been developed. The Pt–Re, Pt–Ir, and Pt–Sn catalysts are very important in petroleum reforming [30, 31], however, at present, their improved catalytic activities or product selectivity compared with monometallic catalysts is still not well understood.

For the heterogeneous catalytic oxidation of glucose in aqueous solution, the catalytic activities and selectivities of supported monometallic palladium catalysts in the oxidation processes increase by addition of metals such as Bi, Tl, Co, Sn, and Pb [21, 24, 26]. Karski et al. studied the influence of added heavy metals on palladium supported catalysts regarding their selectivity for glucose oxidation [23]. The highest selectivity for the reaction of liquid phase oxidation of glucose was achieved with bismuth.

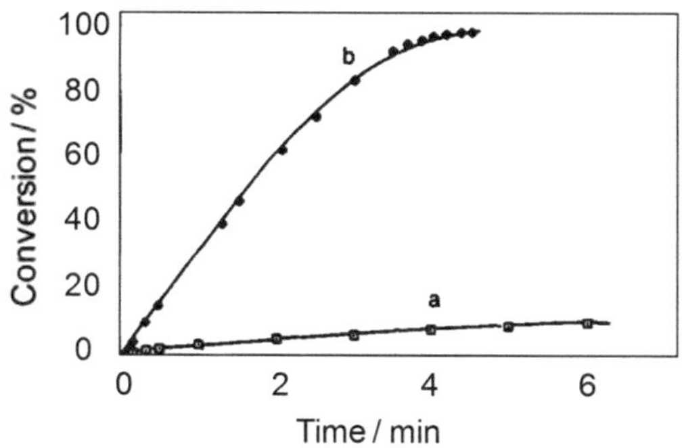


**Fig. 7.2** Glucose conversion as a function of time for a catalytic ratio [glucose]/[Pd] = 3150. Reaction conditions: 0.05 mol $L^{-1}$ glucose aqueous solution 80 mL, 5%Pd/C (**a**) and Pd-Bi/C (**b**) catalyst 40 mg, reaction temperature 50 °C, total pressure 1.0 atm (P ($O_2$) = 0.2 atm) (Reprinted from Ref. [26], Copyright © 1995, with permission from Elsevier)

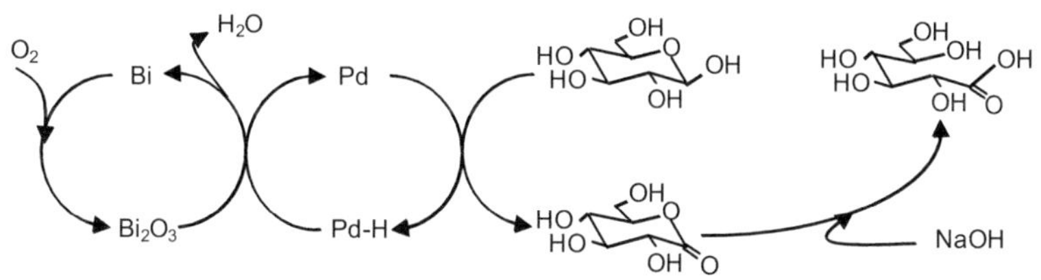


**Scheme 7.2** The mechanism of glucose oxidation into gluconate on Pd-Bi catalyst (Reprinted from Ref. [26], Copyright © 1995, with permission from Elsevier)

Fig. 7.2 shows the oxidation of glucose over Pd/C and Pd-Bi/C catalyst [26]. The rate of glucose oxidation was about 20 times larger on Pd-Bi/C than on Pd/C. The selectivity to D-gluconate was also improved to be 99.8% at a conversion of 99.6%. In recycling of the catalyst, there was no bismuth leaching and the activity and selectivity were almost constant. Besson et al. suggested the mechanism of glucose oxidation over Pd-Bi/C catalyst and interpreted the results in terms of bismuth playing the role of protecting palladium from over-oxidation because Bi has a stronger affinity for oxygen than Pd, which was evident from calorimetric measurements [33]. Furthermore, oxygen bonded to Bi easily reacts with hydrogen bonded to Pd, which would maintain the catalytic activity of $Pd^0$, as shown in Scheme 7.2.

## *7.1.3 Supported Au Catalysts*

For the oxidation of glucose, supported gold catalysts are interesting because of their high activity and stability under low pH conditions. Biella et al. reported that the oxidation of D-glucose into D-gluconic acid occurred using 1%Au/C catalyst [6]. As shown in Fig. 7.3, the catalytic activity of gold catalyst depends on pH for

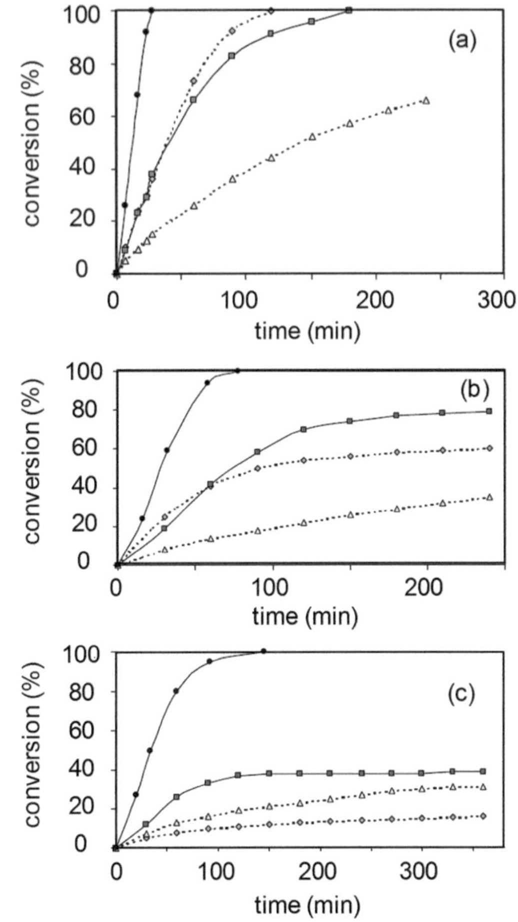


**Fig. 7.3** Glucose oxidation over 1%Au/C (•), 5%Pt/C (△), 5% Pd-5%Bi/C (■), and 1%Pt-4%Pd-5%Bi/C (◆) catalysts under pH-controlled conditions by NaOH. Reaction conditions: glucose 4 wt%; glucose/M = 1000; $O_2$ flow = 20 ml/min; T = 50 °C: (**a**) the pH was maintained at 9.5; (**b**) the pH was maintained at 8; (**c**) the pH was maintained at 7 (Reprinted from Ref. [6], Copyright © 2002, with permission from Elsevier)

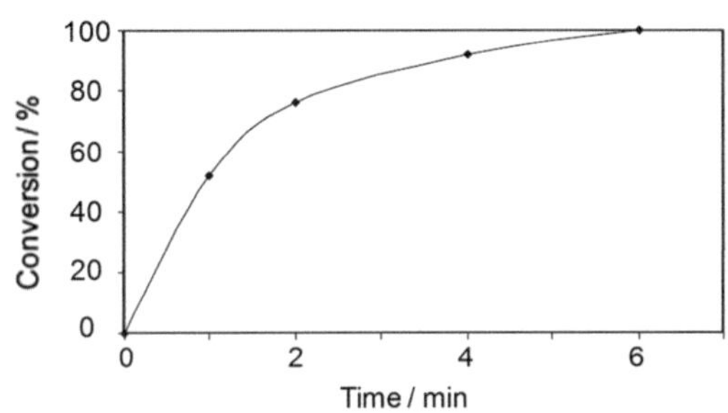


**Fig. 7.4** Glucose oxidation at uncontrolled pH conditions using 1%Au/C catalyst. Reaction conditions: glucose 4 wt%; glucose/M = 1000; $pO_2$ = 300 kPa; T = 100 °C (Reprinted from Ref. [6], Copyright © 2002, with permission from Elsevier)

both Pt/C and Pd-Bi/C catalysts. The Au/C catalyst showed much higher activity than Pt/C and Pd-Bi catalysts, and its selectivity was comparable with Pd-Bi/C bimetallic catalysts. Moreover, Au/C catalyst gave 100% conversion of glucose under uncontrolled pH conditions and supported gold catalyst are active at pH as low as 2.5 as shown in Fig. 7.4 [6]. In recycling, gold is more stable towards

**Table 7.2** Catalytic activities of supported monometallic and bimetallic catalysts on carbon in uncontrolled pH experiments

| Metals | Conversion (%) | TOF ($h^{-1}$) |
|---|---|---|
| Au | 11 | 51 |
| Pt | 13 | 60 |
| Pd | <2 | <2 |
| Rh | <2 | <2 |
| Au-Pt (1:1) | 64 | 295 |
| Au-Pd (1:1) | 20 | 92 |
| Au-Rh (1:1) | <2 | <2 |


Reaction conditions: 1 wt% noble metal on carbon; $T = 70$ °C; $P(O_2) = 300$ kPa; glucose/Au = 3000; reaction time = 6.5 h

**Table 7.3** Oxidation with monometallic and bimetallic catalysts in the controlled pH experiments

| Metals | Type of catalyst | Conversion (%) | TOF ($h^{-1}$) |
|---|---|---|---|
| Au | Supported | 43 | 17,200 |
| Au-Pt (2:1) | Supported | 44 | 17,600 |
| Au | Unsupported | 12 | 4600 |
| Pt | Unsupported | <2 | <500 |
| Pd | Unsupported | 5 | 2000 |
| Au-Pt (2:1) | Unsupported | 26 | 10,500 |
| Au-Pd (1:1) | Unsupported | 28 | 11,600 |


Reaction conditions: pH 9.5. Glucose/Au = 20,000; $T = 50$ °C; reaction time = 2.5 h

deactivation than platinum and palladium catalysts. However, even for Au/C catalysts, in the case of uncontrolled pH, ICP analysis shows that 10% of the metal is lost after two runs and about the 70% of metal is lost after six runs.

Comotti et al. compared the results of monometallic catalysts Au, Pt, Pd and Rh, and bimetallic catalysts in the aerobic oxidation of D-glucose into D-gluconic acid [32, 34]. Experiments of the catalytic reactions were performed under 300 kPa of $O_2$ partial pressure and at 50–90 °C using supported catalysts having average particle sizes of metals as 3–5 nm. Table 7.2 summarizes the catalytic activities of supported monometallic and bimetallic catalysts under uncontrolled pH experiments. Comparing activities of mono-metal catalysts, Au and Pt catalysts show much higher activity than Rh and Pd catalysts. On the other hand, the activity of bimetallic Au particles combined with Pd and Pt show higher activity than monometallic Au, Pd, and Pt catalysts. However, Au-Rh catalysts have lower activity than monometallic Au catalyst. As shown in Table 7.3, in the presence of alkali at a controlled pH of 9.5, the catalytic activity of Au for oxidation of glucose to gluconate is higher than activities of Pd and Pt, whereas Au-Pt and Au-Pd bimetallic catalystic activities improved over that of monometallic gold.

As mentioned above, Au has catalytic activity so that there are many reports for the oxidation of glucose over supported gold catalysts [33]. However, one problem for Au-containing catalysts is the sintering of Au particles, because many metals become soluble in hot water under specific conditions. For example, unsupported gold particles without a protecting agent were initially as active as Au/C catalysts at 30 °C in basic solution, but the as-is particles became sintered and their particle sizes changed from 3.6 to 10 nm. To circumvent this problem, mesoporous materials with about 5 nm channels are effective as supports. Gold particles supported onto mesoporous materials remain unchanged [19, 32]. Metal–oxides ($Al_2O_3$, $ZrO_2$, $TiO_2$, $CeO_2$) are also suitable supports for Au particles in which catalytic activity increases as particle size decreases [8].

Another effective method to prepare stable Au particles is to encapsulate them with polymers. Au particles protected by polymers show good product selectivity, but they are often less active than an Au/$TiO_2$ catalyst prepared by a deposition-precipitation method [34]. To increase their catalytic activity, a series of not supported but PVP polymer-protected Au alloy nanoparticles have been synthesized: Ag–Au, Au–Pd, Au–Pt, and Au–Pt–Ag [12]. Trimetallic Au–Pt–Ag (70:20:10) particles with an average diameter of about 2 nm were several times more active than monometallic and bimetallic particles having nearly the same size.

Hydrogen peroxide ($H_2O_2$) has higher activity than molecular oxygen or air for the oxidation of glucose over Au/$Al_2O_3$ catalyst [35]. Microwave-assisted oxidation of sugars, such as glucose, galactose, mannose, maltose, and cellobiose is effective with hydrogen peroxide over Au/$Al_2O_3$ and Au/$CeO_2$ catalysts [36]. Although oxidation with $H_2O_2$ can lead over-oxidation, the conversion and selectivity of glucose to gluconate exceeds 99% using 1 equivalent of $H_2O_2$ per unit of glucose even in aqueous solutions with high glucose concentrations (ca. 50 wt%) at 40 °C. Even at room temperature around 25 °C, glucose oxidation with $H_2O_2$ at pH 9 using a suspended Au/$SiO_2$ catalyst showed an 85% selectivity to gluconate at 100% conversion with ultrasound [37].

## 7.2 Production of Glucaric Acid Over Heterogeneous Catalysts

Glucaric acid is one of the value-added chemicals that can be obtained from biomass, and it is a promising raw material for the production of biodegradable polymers [38]. It can be an important precursor of adipic acid, which is conventionally produced from fossil resources (Scheme 7.3). However, production of glucaric acid from glucose has many challenges.

Current synthesis methods of glucaric acid are far from satisfactory. For example, chemical oxidation using strong oxidants such as nitric acid to oxidize glucose are non-selective and expensive [39]. Moreover, processes that use nitric acid as oxidant result in $NO_x$ emissions [39]. In other methods, production of glucaric acid

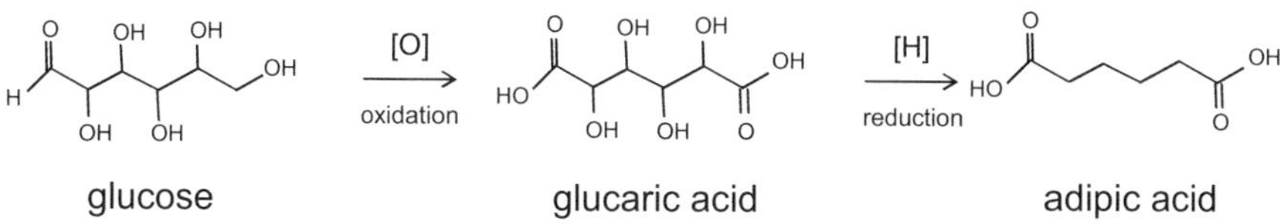


**Scheme 7.3** Possible route of adipic acid production from glucose via glucaric acid

is through electrochemical means or by stoichiometric oxidation using additives. Significant amounts of additives and formation of toxic by-products and inorganic salts are the major issues in terms of the sustainability and environmental compatibility of these methods. The use of environmentally-friendly oxidants such as air, molecular $O_2$ or dilute $H_2O_2$ and solid catalysts are advantageous. Recently, 2, 2, 6, 6-tetramethyl-1-piperidinyloxy (TEMPO) as catalyst for selective oxidation of primary alcohols of glucose to glucaric acid has been studied attracted for which the yield of glucaric acid was over 85% for the oxidation of glucose [40]. However, high cost and difficulty in recycling TEMPO are issues preventing this method from reaching industrial scale. To overcome these issues, inorganic heterogeneous catalysts, such as supported noble metal catalysts and bimetallic catalysts, are expected to be effective for oxidizing monosaccharides into glucaric acid with oxygen or air [41–44].

### 7.2.1 *Productions of Glucaric Acid Using Solid Catalysts*

In 1981, Dirkx et al. reported that glucaric acid could be formed from glucose with carbon-supported platinum catalysts [27]. Experiments were performed using 5%Pt/C catalyst in aqueous solutions of gluconic acid and glucaric acid at 55 °C and at a pH of 10 controlled by adding NaOH. In the oxidation, glucose rapidly disappeared within 30 min and the yield of gluconic acid was about 60% in 15 min. In contrast, the formation rate of glucaric acid from gluconic acid was slower than that of gluconic acid. The yield of glucaric acid was about 50% in 120 min, and it almost remained unchanged up to 300 min reaction time. Their results indicated that the oxidation of gluconic acid to glucaric acid is apparently a slow step and it is not easy to obtain high yields of glucaric acid. In addition, they suggested that most of the by-products formed were from gluconic acid in the production of glucaric acid from glucose, and that L-guluronic acid was one of the by-products [27]. In consequence, the oxidation step of the intermediate gluconic acid into glucaric acid is slow and low selective is obtained in the glucaric acid production from glucose.

Lee et al. reported the oxidation of glucose into glucaric acid using supported 5 wt% platinum catalysts [45]. Under 6.2 bar $O_2$ partial pressure, glucose was immediately converted into gluconic acid and then glucaric acid was slowly produced as shown in Fig. 7.5. The highest yield of glucaric acid was 74% under 14 bar $O_2$ gas partial pressure. The Pt/C catalyst was stable after five cycles without leaching of Pt.

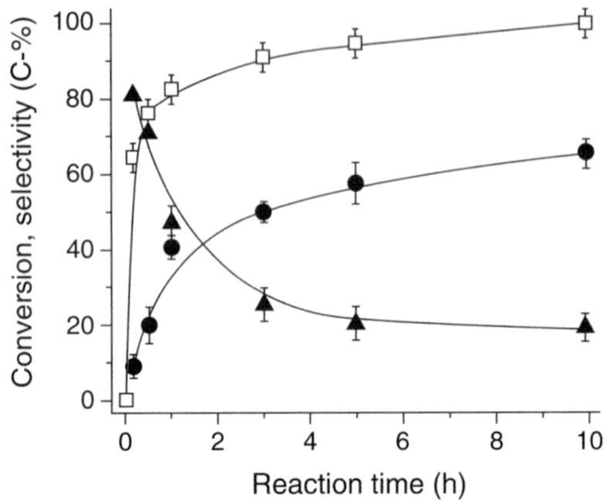


**Fig. 7.5** Glucose conversion and product selectivity for glucose oxidation in water on a Pt/C catalyst as a function of reaction time (□ glucose conversion; ▲ gluconic acid selectivity; ● glucaric acid selectivity). Reaction conditions: 10 wt% glucose solution 20 mL, 80 °C, 6.2 bar $O_2$, glucose/Pt (mol/mol) = 54 (Reprinted from Ref. [45], Copyright © 2016, with permission from Royal Society of Chemistry)

The product selectivity depended markedly on not only partial pressure of $O_2$ gas, but also on the support materials and pH. Pressurized oxygen can enhance the catalytic activity for the oxidation of intermediate gluconic acid, and support materials and pH can increase selectivity into glucaric acid by improving desorption of glucaric acid and inhibit formation of by-products. As shown in Table 7.4, glucaric acid yield under 14 bar $O_2$ gas partial pressure was clearly higher than yields obtained under 6 bar and 3 bar $O_2$ partial pressure which indicates that high pressure oxygen promotes oxidation of gluconic acid into glucaric acid. Pt/C catalyst in acidic aqueous medium using $H_2SO_4$ gave gluconic acid as a major product, while in basic solutions using NaOH, it gave significant amounts of low carbon chain carboxylic acids as by-products, such as dicarboxylic acids, tartronic acid and oxalic acid. When pH was not controlled, glucaric acid yield was higher than other acids in the solution controlled by NaOH and $H_2SO_4$. In addition, as shown in Fig. 7.6, carbon was an effective support in the oxidation of glucose, and Pt/C catalyst showed higher catalytic activity and glucaric acid yield than Pt/$Al_2O_3$ and Pt/$SiO_2$ catalysts, which might be due to the adsorption ability for substrates.

Bimetallic catalysts, such as MnPt, CoPt, FePt, AuPt, and PtCu, are reported as effective catalysts for oxidation of glucose into glucaric acid as shown in Table 7.4 [43–47]. Jin et al. reported highly catalytic performance of bimetallic PtCu supported on titania for oxidation of glucose into glucaric acid with 1 bar of oxygen gas in alkali media [43]. Table 7.5 shows the conversion and glucaric acid selectivity for oxidation of gluconate over Pt, Pd, Cu, Co, and bimetallic catalysts. The major products formed from gluconate oxidation were glucaric acid, tartronic acid, and oxalic acid, and monocarboxylic acids such as glyceric acid, lactic acid, glycolic acid, and formic acid. The product selectivity was almost the same for all catalysts. Pt and Pd catalysts showed markedly higher catalytic activity than Cu and Co cata-

**Table 7.4** Comparison of glucaric acid yields in the glucose oxidation using various monometallic and bimetallic catalysts

| Catalysts | T (°C) | $O_2$ (bar) | Reaction Time (h) | (Initial pH) Base or Acid | Glucaric acid (Gluconic acid) Yield (%) | Ref. Year |
|---|---|---|---|---|---|---|
| Pt/C | 55 | 1 | 2 | (10) NaOH | 50 (–) | [44] 1981 |
| $Pt/SiO_2$ | 80 | 5 | 5 | no | 34 (–) | [47] 2010 |
| $MnPt_{19}/SiO_2$ | 80 | 5 | 5 | no | 38 (–) | [47] 2010 |
| $CoPt_{20}/SiO_2$ | 80 | 5 | 5 | no | 34 (–) | [47] 2010 |
| $FePt/SiO_2$ | 80 | 5 | 5 | no | 28 (–) | [47] 2010 |
| Pt/C | 80 | 5 | 5 | no | 57 (–) | [47] 2010 |
| $AuPt/TiO_2$ | 112 | 27 | 5 | no | 70 (–) | [46] 2011 |
| $Pt/TiO_2$ | 60 | 1 | 6 | NaOH | 27 (–) | [43] 2015 |
| $PtCu/TiO_2$ | 60 | 1 | 4 | NaOH | 32 (–) | [43] 2015 |
| $Cu/TiO_2$ | 60 | 1 | 6 | NaOH | 0.9 (–) | [43] 2015 |
| Pt/C | 80 | 3 | 3 | no | 23 (54) | [45] 2016 |
| Pt/C | 80 | 6 | 3 | no | 47 (27) | [45] 2016 |
| Pt/C | 80 | 14 | 3 | no | 58 (21) | [45] 2016 |
| Pt/C | 80 | 6 | 1 | (0.6) $H_2SO_4$ | 21 (58) | [45] 2016 |
| Pt/C | 80 | 6 | 1 | (7.2) No | 41 (48) | [45] 2016 |
| Pt/C | 80 | 6 | 1 | (13) NaOH | 9 (59) | [45] 2016 |
| Pt/C | 70 | 14 | 10 | no | 65 (9) | [45] 2016 |
| Pt/C | 80 | 14 | 10 | no | 74 (6) | [45] 2016 |
| Pt/C | 90 | 14 | 10 | no | 62 (2) | [45] 2016 |

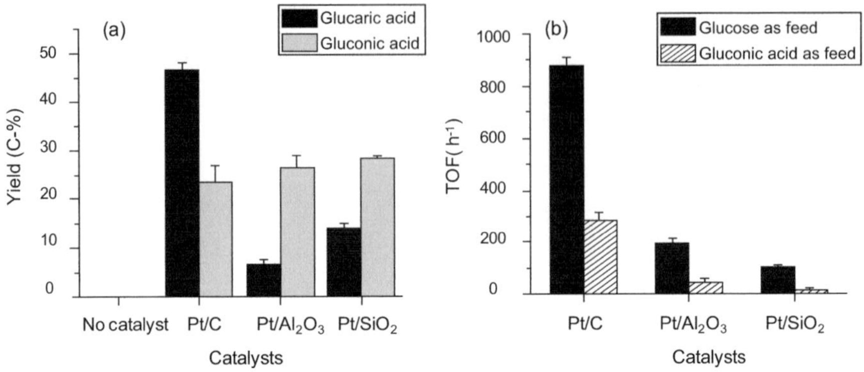


**Fig. 7.6** Gluconic and glucaric acid yields over three catalysts. Reaction conditions: 10 wt% glucose solution 20 mL, 80 °C, 6.2 bar $O_2$, glucose/Pt (mol/mol) = 54. TOFs were calculated at 20% conversion (Reprinted from Ref. [45], Copyright © 2016, with permission from Royal Society of Chemistry)

**Table 7.5** Comparison of bimetallic catalysts for the oxidation of sodium gluconate

| | | Selectivity (%) | | | |
|---|---|---|---|---|---|
| Catalyst | Conversion (%) | Glucaric acid | Tartronic acid | Oxalic acid | Others |
| $Pt/CeO_2$ | 28 | 47 | 31 | 6.4 | 16 |
| $Pt/TiO_2$ | 71 | 38 | 28 | 5.8 | 12 |
| $Pd/CeO_2$ | 51 | 37 | 27 | 10.9 | 16 |
| $Pd/TiO_2$ | 41 | 44 | 28 | 8.3 | 12 |
| $Cu/CeO_2$ | 1.6 | 29 | 31 | 11 | 19 |
| $Cu/TiO_2$ | 2.8 | 31 | 30 | 9.9 | 21 |
| $Co/CeO_2$ | 5.5 | 22 | 39 | 16 | 15 |
| $PtCu/CeO_2$ | 72 | 32 | 24 | 6.6 | 16 |
| $PtCo/CeO_2$ | 51 | 37 | 27 | 11 | 18 |
| $PtPd/CeO_2$ | 41 | 37 | 28 | 10 | 15 |
| $PtCu/TiO_2$ [a] | 100 | 32 | 27 | 20 | 11 |

Reprinted from Ref. [43], Copyright © 2016, with permission from Elsevier
Reaction conditions: 3 g sodium gluconate, 1.0 g NaOH, 0.1 g solid catalyst, metal loading 2 wt%, T: 60 °C. Others of selectivity: glyceric, glycolic, and formic acids. Reaction time: 6 h
[a]Reaction time: 4 h

lysts. For bimetallic catalysts, PtCu showed higher catalytic activity than PtPd and PtCo catalysts. The catalyst support also affected the catalytic properties as well as the glucose oxidation. $TiO_2$ showed better promotional effect for the oxidation into glucaric acid than $CeO_2$.

Colmenares et al. reported on the photo-oxidation of glucose to glucaric acid in the presence of $TiO_2$ as catalyst [48]. Experiments were performed typically as follows: 150 mL of a mixed solution of $H_2O$:acetonitrile = 10:90, 150 mg of $TiO_2$ catalyst, and 0.42 mmol of glucose were added into a batch reactor at 30 °C for 10 min UV irradiation time. The conversion and selectivity depended significantly on the

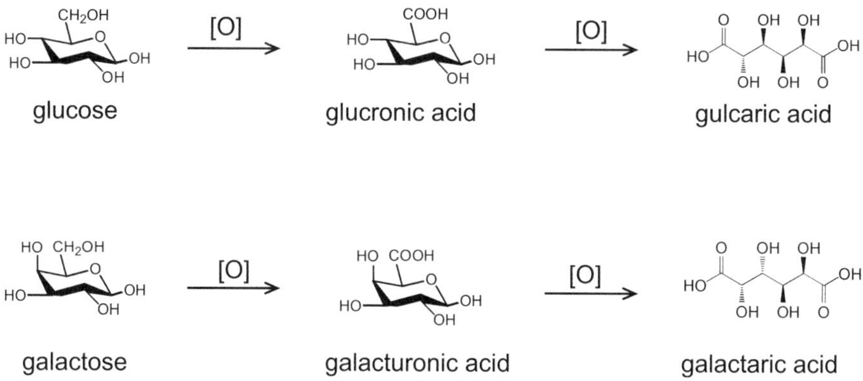


**Scheme 7.4** Formation of glucaric acid and galactaric acid from sugars via glucuronic acid and galacturonic acid

nature of $TiO_2$ catalysts and the ratio of $H_2O$ to acetonitrile. In 100% $H_2O$ solvent, conversions of glucose were relatively high, but the selectivities of glucaric acid and gluconic acid were almost zero. In contrast, in a mixed-solvent system of $H_2O$:acetonitrile = 10:90, $TiO_2$ catalyst prepared by the sol-gel method gave 13% selectivity of glucaric acid and 14% selectivity of gluconic acid at 32% conversion of glucose. Although decarboxylation proceeded readily during photocatalytic oxidation, significant amounts of gluconic acid and glucaric acid were observed by the photocatalytic oxidation of glucose. The glucaric acid yield was low, but improvements of the photocatalytic method can be expected, as well as those of the conventional oxidation methods.

## 7.2.2 *Oxidation of Uronic Acid Using Solid Catalysts*

Uronic acids, such as glucuronic and galacturonic shown in Scheme 7.4, can be obtained from hemicelluloses and pectin. Uronic acids have a carboxyl group at the $C_6$ carbon site similar to glucose and galactose. So, it is expected to be easier to prepare glucaric acid from uronic acid than from aldose sugars. Rautiainen et al. reported that uronic acid could be used as raw material for the synthesis of aldaric acids over $Au/Al_2O_3$ catalysts [49]. Commercially available D-gluconic acid and D-galacturonic acid were used as reactants. Typically, they added 50 mL of $H_2O$, 25 mg of $Au/Al_2O_3$ catalyst, 2.58 mmol of uronic acid (glucuronic acid and galacturonic acid), and 100 mL $min^{-1}$ of flowing $O_2$ into a batch reactor. The reaction temperature and pH were 25–60 °C and 8–10, respectively. As shown in Table 7.6, the pH had a significant effect on the initial reaction rate. At pH 10 at 60 °C, a TOF value close to 8000 $h^{-1}$ was measured for glucuronic acid oxidation. The reaction rate was comparable with those of the oxidation of glucose and galactose into gluconic acid and galactonic acid. When the pH was 8 or 9 at 60 °C, the TOF was about

**Table 7.6** Oxidation of uronic acids and monosaccharides into corresponding aldaric and aldonic acids with $Au/Al_2O_3$[a]

| Entry | Substrate | pH | T (°C) | Specific activity (mmol per g Au per min) | TOF ($h^{-1}$) | Selectivity (%) |
|---|---|---|---|---|---|---|
| 1[b] | Glucuronic acid | 10 | 25 | 103 | 2270 | >99 |
| 2 | Glucuronic acid | 10 | 60 | 297 | 7920 | >99 |
| 3 | Glucuronic acid | 9 | 60 | 238 | 5970 | >99 |
| 4 | Glucuronic acid | 8 | 60 | 130 | 3250 | >99 |
| 5 | Galacturonic acid | 10 | 60 | 286 | 6840 | >99 |
| 6 | Glucose | 10 | 60 | 391 | 10,300 | >99 |
| 7 | Galactose | 10 | 60 | 276 | 6900 | >99 |
| 8 | Glucuronic acid | 10 | 60 | 191 | n/a | >99 |



[a]Reaction conditions: 2.6 mmol substrate, 0.09 mol% Au, 50 ml water, 100 ml min_1 $O_2$. Conversion >99% unless otherwise stated. [b]Conversion 95%. Catalyst recycled by washing with water and drying. TOF not calculated due to changed dispersion

3000 $h^{-1}$ or 6000 $h^{-1}$, respectively. The TOF decreased with decreasing the pH. The apparent activation energy for glucuronic acid oxidation was dependent on pH which can be attributed to the higher energy barrier for desorption of acids at lower pH.

## 7.3 Bifunctional Catalysts for Direct Production of Gluconic Acid

Heterogeneous catalytic processes for oxidation of monosaccharides with oxygen or air into sugar acids have been developed using Pd-Bi/C and Au/C catalysts as described previously. However, polysaccharides represent the largest sources of organic compounds in nature. Therefore, direct production of sugar acids from polysaccharides is of great significance as necessary for sustainable society. Development of bifunctional catalysts for the hydrolysis of polysaccharides into monosaccharides and the oxidation of monosaccharides into sugar acids can be anticipated. As known examples of bifunctional catalysts, noble metal supported solid acid catalysts, such as Pt/zeolites and Pt/sulfated-zirconia, have been reported to be effective for the dehydroisomerization of butane into isobutene and the isomerization of alkanes in gas-solid catalytic reactions [57–60]. On the other hand, for polysaccharide conversions in hot water, these bifunctional catalysts are easily destroyed or dissolved into water at low pH that is attributed to the production of organic acids. The selective hydrolysis of cellulose with a sulfonated carbon catalyst and a sulfonated activated-carbon ($AC\text{-}SO_3H$) catalyst and polyoxometalate catalysts have been reported [50–56, 61]. Carbon materials and POM complexes are highly stable in hot and acidic water. In this section, some bifunctional catalysts,

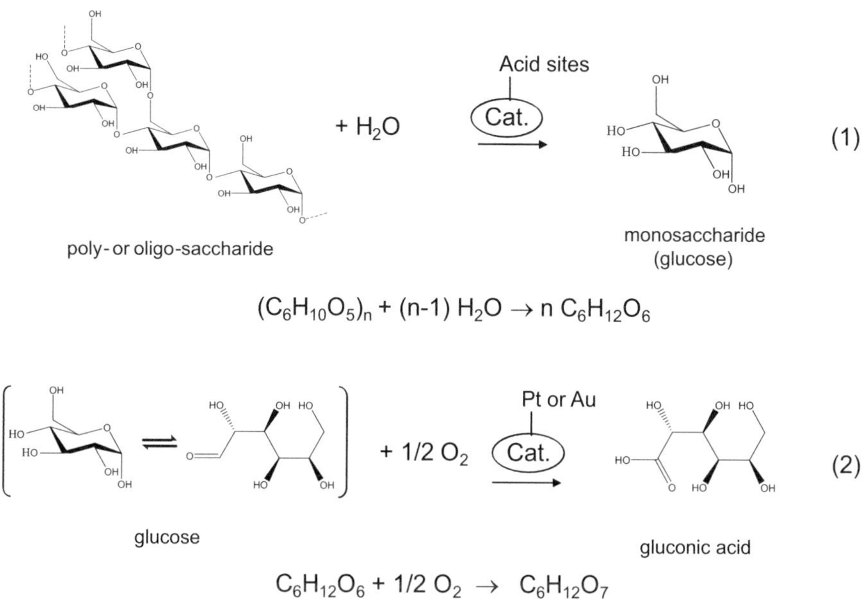


**Scheme 7.5** Conversion of polysaccharides, such as starch, cellobiose, cellulose and sugars into gluconic acid over bifunctional catalysts such as Pt/AC-$SO_3H$

such as acidic carbon materials and polyoxometalate catalysts supported noble metal catalysts are introduced as being effective for direct conversion of polysaccharides into gluconic acid (Scheme 7.5) [62].

### *7.3.1 Bifunctional Sulfonated Activated-Carbon Supported Platinum Catalyst*

Gluconic acid can be produced from polysaccharides, such as starch and cellobiose, at relatively mild reaction conditions at 120 °C under 1 bar of air by a one-pot process using a bifunctional sulfonated activated-carbon supported platinum (Pt/AC-$SO_3H$) catalyst. Pt/AC-$SO_3H$ was made by a two-step method that firstly prepared Pt/AC by impregnation, and sulfonation of the Pt/AC that was carried out under concentrated sulfuric acid. Then, the material was hydrothermally treated at 200 °C. The obtained Pt/AC-$SO_3H$ catalyst had strong acidic sulfo groups, platinum fine particles, and high surface area of 226 $m^2\ g^{-1}$ and it was highly water-tolerant even under hot acidic water. The loading amounts of platinum was 4 wt%. As shown in Fig. 7.7, XRD patterns indicated the metallic state of the bulk of platinum particles and the platinum particle sizes were 4.7 nm for Pt/AC-$SO_3H$ catalyst as estimated by the diffraction peaks of XRD.

**Fig. 7.7** XRD patterns of (**a**) activated-carbon, (**b**) AC-$SO_3H$, (**c**) Pt/AC, (**d**) Pt/AC-$SO_3H$

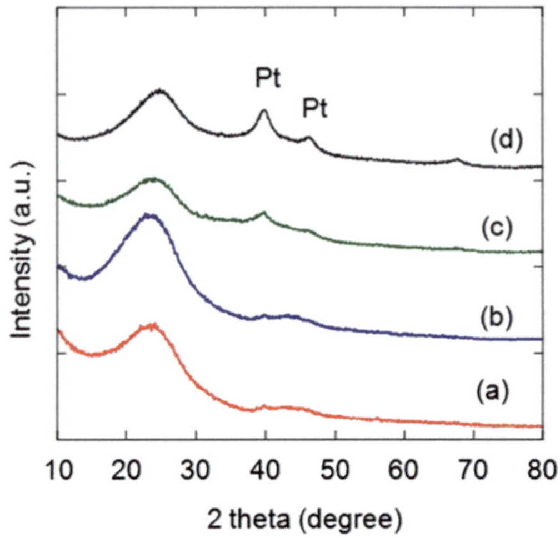


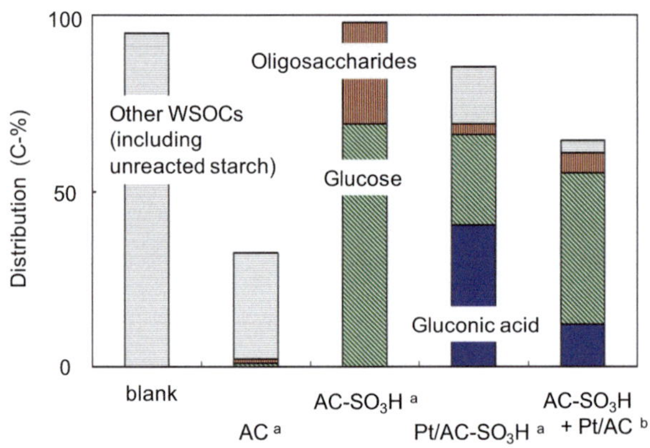


**Fig. 7.8** Reaction of starch at 120 °C for 24 h. Starch 45 mg, water 5.0 mL, [a]catalyst 50 mg. [b]mixed catalyst 100 mg (50 mg of Pt/AC with 50 mg of AC-$SO_3H$)

## *7.3.2 Conversion of Starch*

As shown in Fig. 7.8, activated carbon (AC) catalyst with carboxylic acid groups gave very low amounts of glucose and 68 C-% of the remaining fractions were adsorbed onto AC. In contrast, sulfonated activated-carbon (AC-$SO_3H$) catalyst gave glucose in 69 C-% yields and oligo-saccharides in 29 C-% yields. Glucose yields increased with increasing reaction time over the AC-$SO_3H$ catalyst, and the glucose yield reached values over 90% (Fig. 7.9a).

On the other hand, as shown in Fig. 7.8, the Pt/AC-$SO_3H$ catalyst gave gluconic acid in 40 C-% yields. The active sites for the glucose oxidation into gluconic acid are the platinum particles, because the catalysts without platinum gave no amounts of gluconic acid. After reaction, the loading amounts of platinum was 4.0 wt% determined by ICP and the average platinum particle size was estimated to be

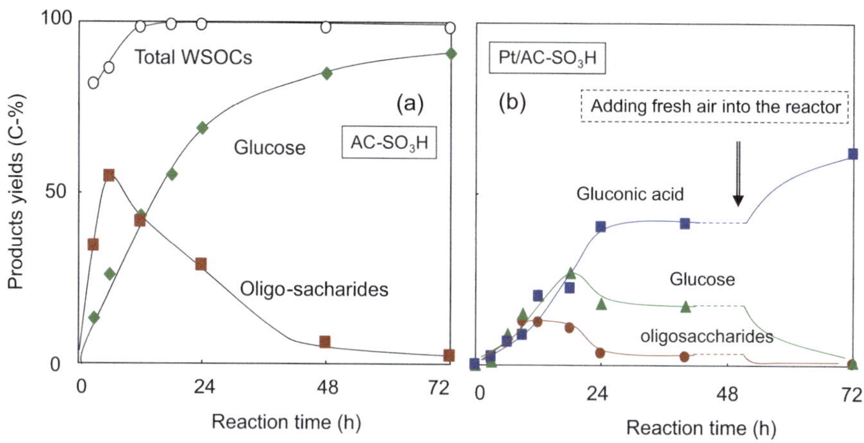


**Fig. 7.9** The changes in product yields during the reaction of starch at 120 °C over the AC-$SO_3H$ (**a**) and the Pt/AC-$SO_3H$ (**b**) catalysts. Starch 45 mg, water 5.0 mL, and catalyst 50 mg

4.5 nm by XRD, which were almost the same values as the fresh catalyst. This means that the platinum particles on sulfonated activated-carbon were mostly neither leached nor sintered during the reaction. Fig. 7.9b shows changes in product yields in starch hydrolysis over Pt/AC-$SO_3H$ catalyst. The yields of oligosaccharides, glucose, and gluconic acid increased linearly with an increase in reaction time up to 9 h. Then, gluconic acid was produced and became the main product at 24 h. Molar ratios of gluconic acid to total S atoms and total Pt atoms in the catalyst were about 8 and 9, respectively, which indicates that the sulfo groups and platinum particles catalytically accelerated the hydrolysis of starch into glucose and the oxidation of glucose into gluconic acid. The pH values before and after the starch reaction over the Pt/AC-$SO_3H$ catalyst were, nearly neutral (pH = 6) and acidic (pH about 3), respectively. The decrease in pH value was mainly due to the production of gluconic acid, because there was almost no sulfuric ions detected an ion chromatography. The catalyst was separated by filtration, and then it was used repeatedly. Furthermore, as shown in Fig. 7.9b, the yield of gluconic acid increased to about 60 C-% by adding fresh air into the reactor. These results show that the Pt/AC-$SO_3H$ catalyst remained mostly active under the reaction conditions. As a consequence, the catalytic active sites, such as Pt nano particles and sulfo acidic groups, were stable and effective for direct conversion of starch into gluconic acid in aqueous solutions at 120 °C.

### *7.3.3 Conversion of Various Polysaccharides*

As shown in Fig. 7.10, conversion of pullulan, which is a polysaccharide of glucose with α-glycoside bonds and is easily soluble in water at room temperature, was accelerated by Pt/AC-$SO_3H$ catalyst, as well as starch. Product yields from

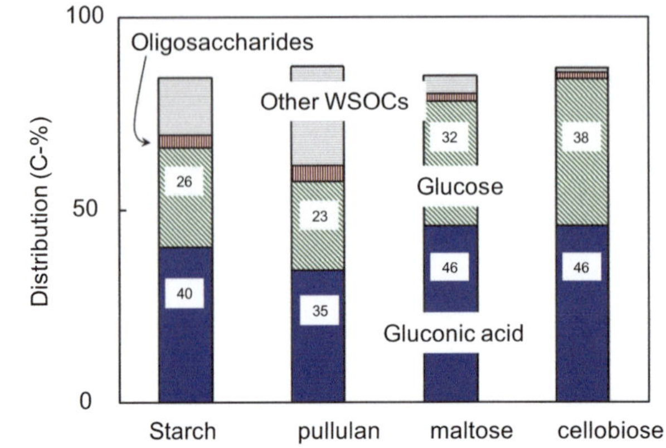


**Fig. 7.10** Catalytic conversion of starch, pullulan, maltose and cellobiose at 120 °C for 24 h with Pt/AC-$SO_3H$ catalyst. Saccharides 45 mg, water 5.0 mL, and catalyst 50 mg

cellobiose were almost same as those from maltose. These results show that Pt/AC-$SO_3H$ catalyst has bifunctional catalytic properties for direct production of gluconic acid from hydrothermally soluble poly- and di-saccharides even when they have a β-1,4 glycoside bond.

### *7.3.4 Comparison of Pt/AC-$SO_3H$ Catalyst to Mixed Catalyst of AC-$SO_3H$ with Pt/AC*

To consider the catalytic properties of Pt/AC-$SO_3H$, AC-$SO_3H$ catalyst and Pt/AC catalyst were mixed and used as catalyst for starch conversion under air. As shown in Fig. 7.8, hydrolysis products with the mixed catalyst were in significantly smaller amounts than those with the Pt/AC-$SO_3H$ catalyst. The mixed catalyst gave glucose in 43 C-% yield and gluconic acid in 11 C-% yield and large amounts of adsorption of about 30 C-%, whereas, the Pt/AC-$SO_3H$ catalyst gave gluconic acid in 40 C-% and relatively small amounts of adsorption of about 10 C-% under the same reaction conditions. It can be suggested that the location of platinum nano-particles on acidic support is effective for direct production of gluconic acid from polysaccharides and to inhibit strong adsorption of polysaccharides onto neutral carbon supports as well as to promote desorption of gluconate from the platinum particles.

As a consequence of Sect. 7.3.1, a new bifunctional Pt/AC-$SO_3H$ catalyst was prepared by the impregnation and sulfonation method . The catalyst was highly water-tolerant and showed catalytic properties for the hydrolysis of polysaccharides and sequentially the air-oxidation into gluconic acid in a one pot process with hot water. Au/AC-$SO_3H$ catalyst showed an increased yield of gluconic acid from starch and cellobiose, and the results are being prepared as an original paper.

**Table 7.7** Catalytic performances of Au catalysts loaded on various supports for cellobiose conversion at 145 °C

| Catalyst | Surface area/ $m^2 g^{-1}$ | Conversion (%) | Selectivity (%) | | Ref. |
|---|---|---|---|---|---|
| | | | Glucose | Gluconic acid | |
| $Au/Al_2O_3$[a] | 277 | 93 | 6.5 | 19 | [18] |
| Au/MCM-41[a] | 612 | 97 | 17 | 20 | [18] |
| Au/H-ZSM-5[a] | 285 | 77 | 39 | 41 | [18] |
| Au/MgO[a] | 106 | 100 | 0 | 10 | [18] |
| Au/AC[a] | 1200 | 38 | 82 | 14 | [18] |
| Au/graphite[a] | 4 | 48 | 40 | 31 | [18] |
| Au/XC-72[a] | 163 | 58 | 64 | 21 | [18] |
| Au/CNT[a] | 122 | 91 | 0 | 60 | [18] |
| Au/CNT[b] | – | 84 | 0 | 86 | [63] |
| $Au/Cs_{1.2}H_{1.8}PW_{12}O_{40}$[b] | 35 | 97 | 0 | 99 | [63] |
| $Au/Cs_{2.2}H_{0.8}PW_{12}O_{40}$[b] | 56 | 96 | 1 | 95 | [63] |
| $Au/Cs_{3.0}PW_{12}O_{40}$[b] | 83 | 95 | 0 | 85 | [63] |
| Au/H-ZSM-5[b] | – | 45 | 0 | 76 | [63] |
| $Au/Al_2O_3$[b] | – | 95 | 0 | 31 | [63] |
| Au/CNT[c] | – | 91 | – | 72 | [64] |
| Au/carbon xerogel_air[c] | 624 | 73 | – | 78 | [64] |
| Au/carbon xerogel[c] | 522 | 77 | – | 53 | [64] |
| $Au/TiO_2$[d] | – | 85 | 10 | 70 | [65] |
| $Au\text{-}Cu/TiO_2$[d] | – | 85 | 5 | 80 | [65] |

Reaction conditions: Au loading of catalysts =0.5 wt% [a] or 1.0 wt% [b, c, d]; T = 145 °C; cellobiose concentration = 15 mmol $L^{-1}$ [a] or 12 mmol $L^{-1}$ [b, c] or 28 mmol $L^{-1}$ [d], $P(O_2)$ = 5 bar [a, b, c] or 10 bar [d]

## 7.3.5 *Cellobiose Conversion into Gluconic Acid Over Various Gold Catalysts*

Gold nanoparticles have highly catalytic performance for the oxidation of glucose into gluconic acid as mentioned in Sect. 7.1. Some researchers have reported on the catalytic properties of supported gold catalysts for production of gluconic acid from cellobiose at 145 °C under pressurized oxygen. Because cellobiose is easily hydrolyzed under hydrothermal conditions at 145 °C, various kinds of support materials can be used. Specific surface area and concentration of gold nanoparticles and catalytic properties of support materials are important for selective production of gluconic acid from cellobiose.

Tan et al. reported on the selective oxidation of cellobiose to gluconic acid with carbon supported Au nanoparticles in the presence of oxygen in water [18]. Table 7.7 shows a summary of the conversion of cellobiose over Au fine particles on carbon materials and zeolites, at 145 °C under 5 bar of oxygen. The catalytic activity and selectivity markedly depended on the type of the support. Au/CNT catalysts showed the highest yields of gluconic acid. The CNT was a carbon nanotube treated with concentrated $HNO_3$. Au catalysts using the other carbon supports, such as AC, graphite, and XC-72, showed lower cellobiose conversions and gluconic acid selectivities, although they gave high selectivity for the hydrolysis products of glucose and gluconic acid. The CNT support might play a role not only to hydrolyze cellobiose, but also to improve the oxidation catalytic activity of Au particles catalysts.

In contrast, Au/$Al_2O_3$, Au/MgO and Au/MCM-41, MCM-41 are mesoporous materials of silica that had high catalytic activity, but gave high yields of by-products, which would be formed due to catalytic properties of the supports. Au/H--ZSM-5 catalyst showed relatively high catalytic activity and selectivity as well as carbon supported Au catalysts, which would be due to the acidic property of H-ZSM-5.

The same group reported on the catalytic behavior of new bifunctional catalysts that were protonated polyoxometalate supported gold [63]. The reaction conditions were almost the same as ref. [18] in water at 145 °C under 5 bar of $O_2$, except for loading amounts of gold. TEM observation clarified that the mean-sizes of the Au nanoparticles were in the range 2.5–3 nm for all samples. Au/$Cs_{1.2}H_{0.8}PW_{12}O_{40}$ catalyst exhibited an excellent gluconic acid yield of 97%. A further increase in Cs content to x = 2.2 and 3.0 decreased the selectivity from 99% of x = 1.2 to 95% of x = 2.2 and 85% of x = 3.0 although conversions of cellobiose hardly depended on the Cs content (x). Au/$Cs_{3.0}PW_{12}O_{40}$ catalyst is not protonated and has no strong acidity, but it gave high yields (81%) of hydrolysis products, which suggests that the strong acidity of the support materials is not effective for hydrolysis of cellobiose. The results are unexpected and may be due to the high reaction temperature (145 °C). In addition, in Table 7.7 the supports with basic sites, such as MgO and $Al_2O_3$, gave large amounts of by-products, which would indicate that protonating treatment of the support enhances the selectivity of gluconic acid to inhibit production of by-products. Compared with 1% Au/CNT catalyst reported in ref. [63] and 0.5% Au/CNT catalyst reported in ref. [18] (Table 7.7), 1% Au/CNT catalyst gave significantly higher gluconic acid selectivity (86%) and yield (72%) than 0.5% Au/CNT catalyst. These results show that by-product would be produced over the support, but not over the gold particles. As shown in Table 7.7, Au/$Cs_{3-x}H_xPW_{12}O_{40}$ catalysts with the larger surface area had the higher selectivity of gluconic acid and Au/CNT catalysts with the larger loading amounts of gold exhibited the higher selectivity of gluconic acid, which might indicate that the concentration of gold particles is important for gluconic acid selectivity from cellobiose. Eblagon et al. [64] reported on the cellobiose conversion into gluconic acid using carbon supported Au catalyst under almost the same reaction conditions as those of Wang's group [63]. They showed that oxidized carbon supported Au catalysts with the higher concentrations of functional groups, such as phenolic groups, had higher selectivity of gluconic acid. It can be considered that the effect of the functional group on the selectivity is due to not only adsorption ability mentioned by Eblagon et al. [64], but also due to surface acidity that inhibits production of by-products as suggested by the overview given in this chapter.

Amaniampong et al. reported the catalytic behavior of new bimetallic Au-Cu catalysts [65]. As shown in Table 7.7, Au/$TiO_2$ catalysts allow high gluconic acid yields comparable with those of the Au/CNT catalyst and bimetallic AuCu/$TiO_2$ catalyst increased the selectivity and yield of gluconic acid. Bimetallic AuCu particles might have higher catalytic properties than monometallic` Au particle in the glucose oxidation. Although there needs to be attention given to the leaching of Cu species, the results are interesting for future development of this reaction system.

## 7.4 Conclusions and Future Outlook

Gluonic acid and glucaric acid can be expected to become platform chemicals for making useful products, such as plastics and food additives from biomass. Especially, glucaric acid can be a promising sustainably-produced chemical as a route to production of adipic acid. The highly selective oxidation of glucose using monometallic and bimetallic catalysts, such as Au, PtPd, PdAu, and, PtBi supported on carbon, has been studied by many researchers. Supported Au catalyst is one of the appropriate catalysts for glucose oxidation into gluconic acid, because Au is highly water tolerant and has suitable catalytic activity for the oxidation. For supports of Au catalysts, carbon materials with acidic functional groups, such as phenol groups, are generally appropriate for the oxidation of sugars. Titania is also an appropriate material as support. However, the development of the production of glucaric acid from glucose is still not sufficient. A practical process will require progress in new oxidation catalysts that have high activity for the selective oxidation of both a primary alcohol group and an aldehyde group into carboxyl groups in the chemical structures of sugars. On the other hand, the direct conversions of natural polysaccharides are still challenging processes. Practical processes will require progress in new bifunctional catalysts with oxidation active sites, acidic active sites, and they must have water tolerant properties.

## References

1. Huber GW, Iborra S, Corma A. Synthesis of transportation fuels from biomass: chemistry, catalysts, and engineering. Chem Rev. 2006;106:4044–98.
2. Chheda JN, Huber GW, Dumesic JA. Liquid-phase catalytic processing of biomass-derived oxygenated hydrocarbons to fuels and chemicals. Angew Chem Int Ed. 2007;46:7164–83.
3. Ramachandran S, Fontanille P, Pandey A, Larroche C. Gluconic acid: properties, applications and microbial production. Food Technol Biotechnol. 2006;44:185–95.
4. Godjevargova T, Dayal R, Turmanova S. Gluconic acid production in bioreactor with immobilized glucose oxidase plus catalase on polymer membrane adjacent to anion-exchange membrane. Macromol Biosci. 2004;4:950–6.
5. Besson M, Gallezot P. Selective oxidation of alcohols and aldehydes on metal catalysts. Catal Today. 2000;7:127–41.
6. Biella S, Prati L, Rossi M. Selective oxidation of D-glucose on gold catalyst. J Catal. 2002;206:242–7.

7. Mirescu A, Prüße U. Selective glucose oxidation on gold colloids. Catal Commun. 2006;7:11–7.
8. Ishida T, Kinoshita N, Okatsu H, Akita T, Takei T, Haruta M. Influence of the support and the size of gold clusters on catalytic activity for glucose oxidation. Angew Chem Int Ed. 2008;47:9405–8.
9. Abbadi A, van Bekkum H. Effect of pH in the Pt-catalyzed oxidation of D-glucose to D-gluconic acid. J Mol Catal A. 1995;97:111–8.
10. Onda A, Ochi T, Kajiyoshi K, Yanagisawa K. A new chemical process for catalytic conversion of D-glucose into lactic acid and gluconic acid. Appl Catal A. 2008a;343:49–54.
11. Abbadi A, Gotlieb KF, van Bekkum H. Study on solid acid catalyzed hydrolysis of maltose and related polysaccharides. Starch. 1998;50:23–8.
12. Zhang H, Toshima N. Glucose oxidation using Au-containing bimetallic and trimetallic nanoparticles. Cat Sci Technol. 2013;3:268–78.
13. Dirkx JMH, van der Baan HS, van den Broen JMAJJ. Preparation of D-glucaric acid by oxidation of D-gluconic acid-catalyzed by platinum on carbon. Carbohydr Res. 1977;59:63–72.
14. Banu M, Venuvanalingam P, Shanmugam R, Viswanathan B, Sivasanker S. Sorbitol hydrogenolysis over Ni, Pt and Ru supported on NaY. Top Catal. 2012;55:897–907.
15. Abbadi A, Makkee M, Visscher W, van Veen JAR, van Bekkum H. Effect of pH in the Pd-catalyzed oxidation of D-glucose to D-gluconic acid. J Carbohydr Chem. 1993;12:573–87.
16. Onal Y, Schimpf S, Claus P. Structure sensitivity and kinetics of D-glucose oxidation to D-gluconic acid over carbon-supported gold catalysts. J Catal. 2004;223:122–33.
17. Liang X, Liu CJ, Kuai P. Selective oxidation of glucose to gluconic acid over argon plasma reduced $Pd/Al_2O_3$. Green Chem. 2008;10:1318–22.
18. Tan XS, Deng WP, Liu M, Zhang QH, Wang Y. Carbon nanotube-supported gold nanoparticles as efficient catalysts for selective oxidation of cellobiose into gluconic acid in aqueous medium. Chem Commun. 2009;46:7179–81.
19. Ma CY, Xue WJ, Li JJ, Xing W, Hao ZP. Mesoporous carbon-confined Au catalysts with superior activity for selective oxidation of glucose to gluconic acid. Green Chem. 2013;15:1035–41.
20. Worz N, Brandner A, Claus P. Platinum–Bismuth-catalyzed oxidation of glycerol: kinetics and the origin of selective deactivation. J Phys Chem C. 2010;114:1164–72.
21. Wenkin M, Touillaux R, Ruiz P, Delmon B, Devillers M. Influence of metallic precursors on the properties of carbon-supported bismuth-promoted palladium catalysts for the selective oxidation of glucose to gluconic acid. Appl Catal A. 1996;148:181–99.
22. Wenkin M, Renard C, Ruiz P, Delmon B, Devillers M. Promoting effects of bismuth in carbon-supported bimetallic Pd–Bi catalysts for the selective oxidation of glucose to gluconic acid. Stud Surf Sci Catal. 1997;110:517–26.
23. Karski S, Paryjczak T, Witonska I. Selective oxidation of glucose to gluconic acid over bimetallic Pd-Me catalysts (Me = Bi, Tl, Sn, Co). Kinet Catal. 2003;44:618–22.
24. Witonska I, Karski S, Frajtak M. Supported Pd–Te catalysts for selective oxidation of glucose to gluconic acid in liquid phase. Przem Chem. 2011;90:475–80.
25. Besson M, Gallezot P, Lahmer E, Fleche G, Fuertes P, Kosak JR. In: Johnson TA, editor. Catalysis of organic reactions, vol. 53. New York: Marcel Dekker; 1993. p. 169–80.
26. Besson M, Gallezot P, Lahmer E, Flbche G, Fuertes P. Catalytic oxidation of glucose on bismuth-promoted palladium catalysts. J Catal. 1995;152:116–21.
27. Dirkx JMH, van der Baan HS. The oxidation of gluconic acid with platinum on carbon as catalyst. J Catal. 1981;67:14–20.
28. Venema ER, Peters JA, van Bekkum H. Platinum-catalyzed oxidation of aldopentoses to aldaric acids. J Mol Catal. 1992;77:75–85.
29. Gallezot P, de Mésanstourne R, Christidist Y, Mattioda G, Schouteeten A. Catalytic oxidation of glyoxal to glyoxylic acid on platinum metals. J Catal. 1992;133:479–85.

30. Botman MJP, de Vreugd K, Zandbergen HW, de Block R, Ponec V. The effect of alloying Pt with Re on the intermediates in hydrocarbon reactions: reactions of 2,2-dimethylbutane. J Catal. 1989;116:469–79.
31. Betizeau C, Leclercq G, Maurel R, Bolivar C, Charcosset H, Frety R, Tournayan L. Platinum-rhenium-alumina catalysts: III catalytic properties. J Catal. 1976;45:179–88.
32. Comotti M, Pina CD, Rossi M. Mono- and bimetallic catalysts for glucose oxidation. J Mol Catal A. 2006;251:89–92.
33. Besson M, Gallezot P, Pinel C. Conversion of biomass into chemicals over metal catalysts. Chem Rev. 2014;114:1827–70.
34. Mirescu A, Berndt H, Martin A, Prüsse U. Long-term stability of a 0.45% $Au/TiO_2$ catalyst in the selective oxidation of glucose at optimized reaction conditions. Appl Catal A. 2007;317:204–9.
35. Saliger R, Decker N, Prüsse U. D-Glucose oxidation with $H_2O_2$ on an $Au/Al_2O_3$ catalyst. Appl Catal B. 2011;102:584–9.
36. Omri M, Pourceau G, Becuwe M, Wadouachi A. Improvement of gold -catalyzed oxidation of free carbohydrates to corresponding aldonates using microwaves. ACS Sustain Chem Eng. 2016;4:2432–8.
37. Bujak P, Bartczak P, Polanski JJ. Highly efficient room-temperature oxidation of cyclohexene and d-glucose over nano gold $Au/SiO_2$ in water. J Catal. 2012;295:15–21.
38. Kiely DE, Chen L, Lin T. Hydroxylated nylons based on unprotected esterified D-glucaric acid by simple condensation reactions. J Am Chem Soc. 1994;116:571–8.
39. Smith TN, Hash K, Davey CL, Mills H, Williams H, Kiely DE. Modifications in the nitric acid oxidation of D-glucose. Carbohydr Res. 2012;350:6–13.
40. Merbouh N, Bobbit JM, Bruckner C. 4-AcNH-tempo-catalyzed oxidation of aldoses to aldaric acids using chlorine or bromine as terminal oxidants. J Carbohydr Chem. 2001;21:65–77.
41. Ibert M, Fuertès P, Merbouh N, Fiol-Petit C, Feasson C, Marsai F. Improved preparative electrochemical oxidation of D-glucose to D-glucaric acid. Electrochim Acta. 2010;55:3589–94.
42. Bin D, Wang H, Lia J, Wang H, Yin Z, Kang J, He B, Li Z. Controllable oxidation of glucose to gluconic acid and glucaric acid using an electrocatalytic reactor. Electrochim Acta. 2014;130:170–8.
43. Jin X, Zhao M, Shen J, Yan W, He L, Thapa PS, Ren S, Subramaniam B, Chaudhari RV. Exceptional performance of bimetallic $Pt_1Cu_3/TiO_2$ nanocatalysts for oxidation of gluconic acid and glucose with $O_2$ to glucaric acid. J Catal. 2015;330:323–9.
44. Ibert M, Fuertès P, Merbouh N, Feasson C, Marsai F. Evidence of benzilic rearrangement during the electrochemical oxidation of D-glucose to D-glucaric acid. Carbohydr Res. 2011;346:512–8.
45. Lee J, Saha B, Vlachos DG. Pt catalysts for efficient aerobic oxidation of glucose to glucaric acid in water. Green Chem. 2016;18:3815–22.
46. Murphy VJ, Shoemaker J, Zhu G, Archer R, Salem F, Dias EL. US Patent. 2011. 2011/0306790A1.
47. Boussie TR, Dias EL, Fresco ZM, Murphy VJ, Shoemaker J, Archer R, Jiang H. US Patent. 2010. 2010/0317823A1.
48. Colmenares JC, Magdziarz A, Bielejewska A. High-value chemicals obtained from selective photo-oxidation of glucose in the presence of nanostructured titanium photocatalysts. Bioresour Technol. 2011;102:11254–7.
49. Rautiainen S, Lehtinen P, Chen J, Vehkamaki M, Niemela K, Leskela M, Repo T. Selective oxidation of uronic acids into aldaric acids over gold catalyst. RSC Adv. 2015;5:19502–7.
50. Dhepe PL, Ohashi M, Inagaki S, Ichikawa M, Fukuoka A. Hydrolysis of sugars catalyzed by water-tolerant sulfonated mesoporous silicas. Catal Lett. 2005;102:163–9.
51. Takagaki A, Tagusagawa C, Domen K. Glucose production from saccharides using layered transition metal oxide and exfoliated nanosheets as a water-tolerant solid acid catalyst. Chem Commun. 2008;14:5363–5.

52. Onda A, Ochi T, Yanagaisawa K. Selective hydrolysis of cellulose into glucose over solid acid catalysts. Green Chem. 2008b;10:1033–7.
53. Onda A, Ochi T, Yanagaisawa K. Hydrolysis of cellulose selectively into glucose over sulfonated activated-carbon catalyst under hydrothermal conditions. Top Catal. 2009;52:801–7.
54. Suganuma S, Nakajima K, Kitano M, Yamaguchi D, Kato H, Hayashi S, Hara M. Hydrolysis of cellulose by amorphous carbon bearing $SO_3H$, COOH, and OH groups. J Am Chem Soc. 2008;135:12757–93.
55. Kobayashi H, Komanoya T, Hara K, Fukuoka A. Water-tolerant mesoporous-carbon-supported ruthenium catalysts for the hydrolysis of cellulose to glucose. ChemSusChem. 2010;3:440–3.
56. Shimizu K, Furukawa H, Kobayashi N, Itaya Y, Satsuma A. Effects of Brønsted and Lewis acidities on activity and selectivity of heteropoly acid-based catalysts for hydrolysis of cellobiose and cellulose. Green Chem. 2009;11:1627–32.
57. Pirngruber GD, Seshan K, Lercher JA. Dehydroisomerization of n-butane over Pt–ZSM5 (I): effect of the metal loading and acid site concentration. J Catal. 1999;186:188–200.
58. Komatsu T, Ikenaga H. Dehydroisomerization of butane into isobutene on Pt–Sn intermetallic compounds supported on H-SAPO-11. J Catal. 2006;241:426–34.
59. Ebitani K, Konishi J, Hattori H. Skeletal isomerization of hydrocarbons over zirconium oxide promoted by platinum and sulfate ion. J Catal. 1991;130:257–67.
60. Iglesia E, Soled ST, Kramer GM. Isomerization of alkanes on sulfated zirconia: promotion by Pt and by adamantyl hydride transfer species. J Catal. 1993;144:238–53.
61. Deng W, Zhang Q, Wang Y. Polyoxometalates as efficient catalysts for transformations of cellulose into platform chemicals. Dalton Trans. 2012;41:9817–31.
62. Onda A, Ochi T, Yanagaisawa K. New direct production of gluconic acid from polysaccharides using a bifunctional catalyst in hot water. Catal Comm. 2011;12:421–5.
63. An D, Ye A, Deng W, Zhang Q, Wang Y. Selective conversion of cellobiose and cellulose into gluconic acid in water in the presence of oxygen, catalyzed by polyoxometalate-supported gold nanoparticles. Chem Eur J. 2012;J18:2938–47.
64. Eblagon K, Pereira M, Figueiredo J. One-pot oxidation of cellobiose to gluconic acid. Unprecedented high selectivity on bifunctional gold catalysts over mesoporous carbon by integrated texture and surface chemistry optimization. Appl Catal B. 2016;184:381–96.
65. Amaniampong P, Booshehri A, Jia X, Dai Y, Wang B, Mushrif S, Borgna A, Yang Y. High-temperature reduction improves the activity of rutile $TiO_2$ nanowires-supported gold-copper bimetallic nanoparticles for cellobiose to gluconic acid conversion. Appl Catal A. 2015;505:16–27.

# Chapter 8
# Production of 1,4-Diacids (Succinic, Fumaric, and Malic) from Biomass

**Qiang Li and Jianmin Xing**

**Abstract** The 1,4-diacids including succinic, malic, and fumaric acids are naturally occurring C4 building blocks that are regarded as major metabolic intermediates in most eukaryotic or prokaryotic microorganisms. During the past decade, 1,4-diacids have established themselves as forerunners of the biorefinery platform chemicals. There is significant global interest in producing bulk chemicals from renewable resources using microorganisms. Compared with conventional production methods derived from fossil fuel feedstocks, the synthesis of 1,4-diacids from sugars or natural biomass using microorganism producers is highly efficient and has the potential to reduce world dependence on fossil fuels. This chapter reviews microorganism producers, cultivation, separation technologies, alternative substrates of lignocellulosic biomass, and integration strategies to provide analysis of the strategies and economics of 1,4-diacid commercial-scale production.

**Keywords** C4 dicarboxylic acids • Microbial production • Biomass • Cultivation strategies • Downstream processing • *In situ* product recovery • Techno-economics challenges

## 8.1 Introduction

### *8.1.1 Platform Chemical Production Using a Biorefinery Concept*

Platform chemicals act as the elementary materials for synthesizing many chemical intermediates and useful polymers. Presently, most marketable platform chemicals are synthesized by petrochemical processes that use crude oil, natural gas, or coal as

---

Q. Li (✉) • J. Xing
National Key Laboratory of Biochemical Engineering, Institute of Process Engineering, Chinese Academy of Sciences, Beijing, China

University of Chinese Academy of Sciences, Beijing, China
e-mail: qiangli@ipe.ac.cn; jmxing@ipe.ac.cn



Z. Fang et al. (eds.), *Production of Platform Chemicals from Sustainable Resources*, Biofuels and Biorefineries 7, DOI 10.1007/978-981-10-4172-3_8

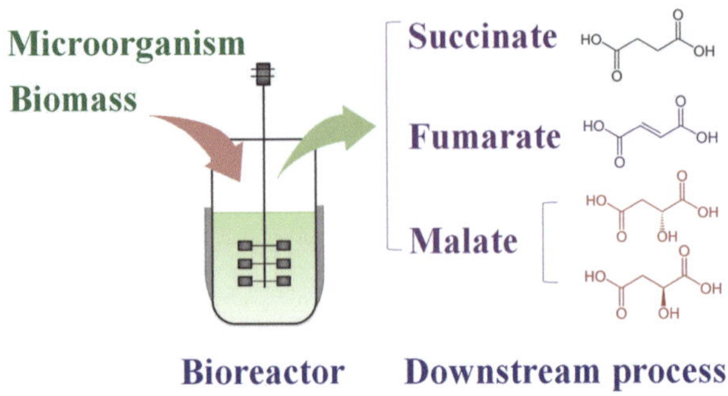


**Fig. 8.1** Production of succinic, malic, and fumaric acids using a biorefinery concept

the starting materials [1]. C4 platform chemicals mainly consist of carboxylic acids, dicarboxylic acids (such as succinic, malic, and fumaric acids), alcohols, amino (di) carboxylic acids, diols, diamines, and dienes [2]. In the past 10 years, biorefinery of C4 platform chemical from biomass has been considered as a substitute for petrochemical-based approaches, which would lower the risk of energy security, limit use of fossil oils, and reduce environment problems associated with greenhouse gas emissions.

C4 platform chemical production from bio-resources includes efficient improvement of natural strains or engineered microorganisms for fermentation, optimization of separation and purification, integration of suitable unit operations. Price-competitive and sustainable resource-derived feedstocks are important for the entire cost evaluation. Succinic, together with fumaric and malic acids, have been well-known as the important building block chemicals that can be made from sugars or biomass substrates [3]. Figure 8.1 shows a bio-based strategy for the refinery of these three platform organic acids. Improvement of industrial strains capable of making the target products at high concentration, yields, and productivity is the first task. The optimization of downstream processes plays an important role in reducing the total production cost which is the major competing factor compared with existing petrochemical processes.

### 8.1.2 *Current State and Perspectives of C4 Dicarboxylic Acids—Succinic, Malic, and Fumaric Acids*

Existing commercial production of C4 dicarboxylic acids remains dependent on petrochemical production methods. The application extends to agricultural, pharmaceutical, food, adsorbents, and polymer industries [4]. C4 dicarboxylic acids, together with their derivatives, have a $15 \times 10^9$ US dollar market as bulk chemicals [5].

*The U.S. Department of Energy* has listed succinic acid as one of the top twelve building block chemicals for over 10 years [3]. *U.S. Food and Drug Administration* has approved succinic acid as a safe chemical. Succinic acid acts as an anticarcinogenic agent and as an insulinotropic agent in the pharmaceutical area.

In the chemical industry, succinic acid is a precursor for the production of numerous high value compounds such as 1,4-butanediol, N-methyl-2-pyrrolidone, γ-butyrolactone, tetrahydrofuran, and 2-pyrrolidone. Succinic acid is a precursor to many specialized polymers. For example, polybutylene succinic acid (PBS) is a promising biopolymer [3–5]. The production of succinic acid by microorganisms has been widely studied not only in traditional rumen bacteria, but also in many other host organisms. The major metabolic pathway uses the reductive branch of the tricarboxylic acid (TCA) cycle.

The growing succinic acid market is stimulating the commercialization of bio-based succinic acid. Many famous companies, such as Reverdia, Myriant, Succinity and BioAmber, have already touched the stage of industrialization.

Fumaric acid is generally used as beverage ingredient and food acidulant [6], or antibacterial agent in the feed industry. Fumaric acid can be polymerized to biodegradable polymers. Fumaric acid is regarded as a key intermediate in the production of L-malic acid and L-aspartic acid, which are used in sweeteners, beverages and other areas [7]. Bio-based fumaric acid was studied previously, but biorefinery of fumaric acid is being replaced by petrochemical synthesis methods in the market. *Rhizopus* species are well-known as native producers of fumaric acid via TCA pathway. Until now, few efforts have been made to improve the production processes of *Rhizopus* sp. through metabolic engineering, which suggests that this field is rich in potential opportunities. Many other microorganisms, including *S. cerevisiae* and genetic modified *E. coli*, have been engineered for fumaric acid production.

Malic acid has various applications in the food, beverage, metal cleaning, pharmaceuticals and plastics industries [3, 8]. The naturally occurring Malic acid is L-form, whereas L- and D-malic acid are prepared synthetically as a racemic mixture. Malic acid can be formed in metabolic cycles in both the TCA and glyoxalate cycles; it offers cells with carbon skeletons or energy. Because the current chemical synthetic method makes a racemic mixture of L- and D-malic acid, biorefinery of malic acid remains superior in terms of stereo-selectivity. Malic acid can be mainly produced by fungal species. However, when fungal fermentation reaches the pilot or industrial scale, some difficulties will need to be overcome, such as oxygen transfer and especially morphological problems. Potential toxin formation is another question. An alternative way to produce L-malic acid is biorefinery by well-characterized microorganisms, such as yeast and *E. coli*.

Downstream processes require separation and purification of the 1,4-diacid products, which remains the main economic hurdles in the commercialization of bio-based production. Downstream unit operations of the C4 dicarboxylic acids from the fermentation broth have much more similarities. Adsorption, precipitation methods and *in situ* product recovery strategies are widely applied to downstream processes. Separation and purification techniques, including precipitation, extraction, adsorption, electrodialysis and membrane separation, have been effectively used for the recovery of bio-based C4 dicarboxylic acids. One challenge to successful separation of 1,4-diacids from crude broths is how to apply purification technologies on the industrial-scale that have good productivity and yields as those in the laboratory.

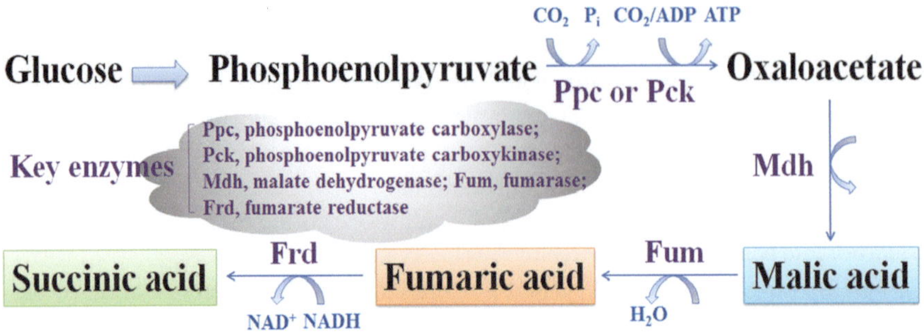


**Fig. 8.2** Anaplerotic pathways that produce succinic, fumaric, and malic acids from glucose

For years, these technological challenges are considered as frontline issues of an economically-sustainable biorefinery.

## 8.2 Upstream Processing

In spite of many benefits and the growing attention to biotechnology, common commercial routines in biorefinery processes are constrained by time and cost. Various methods and tools are essential to the natural or engineered strains required in which scientists and engineers to develop effective production processes.

### *8.2.1 Microbial Producers*

Most prokaryotic and eukaryotic microorganisms produce succinic, malic, and fumaric acids as naturally major metabolic intermediates as shown in Fig. 8.2. Different kinds of genetic changes and modification have been made to wild type microorganisms, and many engineered strains have been retained in the current 1,4-diacid production in bioreactors. Host strain engineering research is carried out with the following objectives: (1) to decrease by-products, (2) to improve both redox availability and energy efficiency, (3) to recognize the blocks to the target formation through diagnostic research.

Succinic acid can be made as a fermentation end-product or as an intermediate of the central metabolism via the reductive TCA branch [9]. Traditional microorganisms for succinic acid production are well known as *Anaerobiospirillum succiniciproducens*, *Actinobacillus succinogenes*, *Mannheimia succiniciproducens* and recombinant *Escherichia coli* strains (Table 8.1) [10–12]. Theoretically, two molecules of succinic acid are synthesized from one molecule of monosaccharide anaerobically. Carbon dioxide consumption (one mole $CO_2$ per mole succinic acid) makes the synthesis environmentally-friendly [13, 14]. *Corynebacterium*

**Table 8.1** Typical microorganisms used for succinic acid fermentation

| General producer | Type | Species |
|---|---|---|
| Natural | Bacteria | *A. succinogenes* |
| | | *A. succiniciproducens* |
| | | *M. succiniciproducens* |
| | | *Bacteroides fragilis* |
| | | *Enterococcus flavescens* |
| | | *Klebsiella pneumoniae* |
| | | *Succinivibrio dextrinosolvens* |
| | Fungi | *Aspergillus niger* |
| | | *Paecilomyces varioti* |
| | | *Penicillium simplicissimum* |
| Engineered | Bacteria | *E. coli* |
| | | *C. glutamicum* |
| | Yeast | *S. cerevisiae* |

**Table 8.2** Typical microorganisms used for fumaric acid fermentation

| Producer | Type | Species |
|---|---|---|
| Natural/ engineered | Fungi | *R. arrhizus* |
| | | *R. formosa* |
| | | *R. nigricans* |
| | | *R. oryzae* |
| | Bacteria | *E. coli* |

*glutamicum* is a quick-growing microorganism used for a long period in industrial production of organic acids, amino acids, and nucleic acids. *C. glutamicum*, can produce organic acids such as lactic, succinic, and acetic acid using glucose as a carbon source with high-density cells and high productivity [15]. Besides *C. glutamicum*, genome sequence of *Saccharomyces cerevisiae* is clearly and physiologically well characterized, and it can make 1,4-diacids at very low pH cultivation conditions. Efforts to genetically modify *S. cerevisiae* for succinic acid production have been reported [16]. Genetic improvement and industrial utilization of *S. cerevisiae* have been established, which makes *S. cerevisiae* suitable for the succinic acid refinery.

An inclusive range of microorganisms produces fumaric acid as a naturally occurring intermediate in the TCA branch, or as a product in the urea cycle. Succinate dehydrogenase can convert succinate to fumaric acid by oxidation. Fumarase can convert fumarate to malate. Some microorganisms make fumaric acid as the fermentative product. Particularly, fungi are recognized as fumaric acid producers [17]. Fumaric acid producers are normally identified as *Rhizopus*, *Cunninghamella*, *Mucor*, and *Circinella* species, among which *Rhizopus* species (Table 8.2) is one of the best-known producers. *Rhizopus* species can yield fumaric acid in both aerobic and anaerobic cultivation conditions. However, it should be

**Table 8.3** Typical microorganisms used for malic acid fermentation

| Producer | Type | Species |
|---|---|---|
| Natural/ engineered | Fungi | *Aspergillus flavus* |
| | | *A. niger* |
| | | *A. oryzae* |
| | | *Aureobasidium pullulans* |
| | Bacteria | *Escherichia coli* |
| | | *Thermobifida fusca* |
| | Yeast | *Saccharomyces cerevisiae* |
| | | *Zygosaccharomyces rouxii* |

noted that not all *R. oryzae* species are able to make fumaric acid. *R. oryzae* can be categorized into two categories: type I merely forms fumaric acid, with little or no lactic acid production; type II forms lactic acid with little or no fumaric acid production [18].

Natural malic acid producers at high production levels such as *Aspergillus flavus*, are presently disqualified as a result of special growth requests and some mycotoxins. Because *A. oryzae* is a close relative or even one ecotype of *A. flavus*, the high-level malic acid production with a GRAS status (generally regarded as safe) can be integrated with current large-scale production capability [19]. Genetic engineering has been used to improve malic acid production capacity of fungi, bacteria and yeasts. Table 8.3 summarizes several representative microorganisms used for malic acid production.

### *8.2.2 Metabolic Engineering Toward Higher Yield*

As intermediates of the TCA cycle, succinic, fumaric and malic acids are end products if sugar or glycerol are used as the carbon source [20]. Many pathways are used for these three 1,4-diacids fermentation such as the reductive branch of the TCA cycle, the oxidative TCA cycle, or the glyoxylate pathway. Figure 8.3 shows the TCA pathway and reductive carboxylation pathway to succinic, fumaric and malic acids accumulation of *E. coli*, and succinic acid production pathway of *A. succinogenes*. The formation of one mole of succinic acid through the reductive pathway is considered to fix one mole of $CO_2$ and consume two moles of nicotinamide adenine dinucleotide hydride (NADH). One mole of glucose can furnish two moles of NADH by the glycolytic pathway. Consequently, the redox balance as well as high yield is thoroughly linked to the level of NADH. Thus, the current metabolic fluxes and regulation tools require enhanced schemes of optimized metabolic engineering of *E. coli* for succinic acid production. Systematic engineering has been applied to develop succinic acid biorefineries [20–22].

*A. succinogenes*, *A. succiniciproducens*, *M. succiniciproducens* are recognized as natural producers. Many efforts have been put afford to improve the yield by

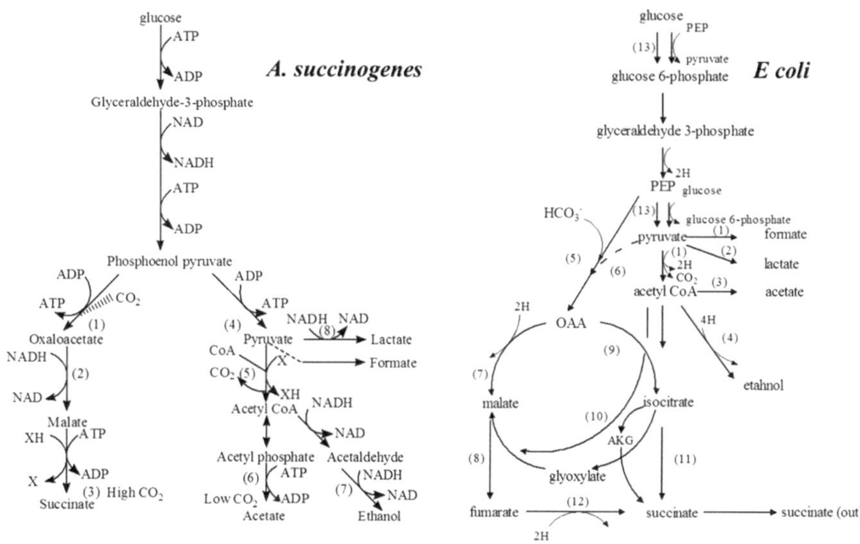


**Fig. 8.3** TCA cycle pathway and reductive carboxylation pathway to succinic, fumaric, and malic acids accumulation of *E. coli*: *1* pyruvate formate lyase (PFL); *2* lactate dehydrogenase (LDH); *3* ack-pta; *4* alcohol dehydrogenase (ADH); *5* phosphoenolpyruvate carboxylase (PEPC); *6* pyruvate carboxylase (PC); *7* malate dehydrogenase (MDH); *8* fumarase; *9* citric synthesis; *10* malic enzyme; *11* isocitric lyase (LCL); *12* fumarate reductase (FRD); *13* glucose phosphotransferase. Succinic acid production pathway by *A. succinogenes*: *1* phosphoenolpyruvate carboxykinase (PEPCK); *2* malate dehydrogenase (MDH); *3* FRD; *4* pyruvate kinase (PK); *5* pyruvate oxidase/reductase; *6* acetate kinase (AK); *7* alcohol dehydrogenase (ADH); *8* LDH

fermentation engineering [23]. Because *E. coli*, *C. glutamicum*, and *S. cerevisiae* are not natural producers, fully genetic engineered pathways are required for improving facilities to produce succinic acid. In the improvement of metabolic engineering, many strategies have been taken for host strains' application. Several attempts have been made to overcome the existing limits of host strains, that include introduction of an ATP-dependent glucose transport system, knockout of ADHE or LDH encoding genes that leading to by-product formation pathway, positive change of the ATP formation and control of pyruvate accumulation [24]. Better understanding of metabolic circumstance of native succinic acid strains and heterologous host strains is critical to pursue economic and biotechnological production of succinic acid. Much work needs to be done to realize the preferred target.

Internal redox balance for example, enhances overexpression of key enzymes, and amends pathways of by-product formation as one approach that leads to high levels of production. Taking overexpression of genes that code for anaplerotic enzymes in the resultant strain *C. glutamicum* ΔldhA-pCRA717 for example, when pyruvate carboxylase activity is enhanced, the production rate of succinic acid by the mutant strain is 1.5-fold higher than that of the control parental strain. When the dry cell weight of the mutant strain reached 50 g/L, succinic acid could be produced as well with the recurrent addition of $Na_2CO_3$ and sugar under oxygen deprivation.

Succinic acid titers reached almost 150 g/L, and the yield was over 0.9 g/g with a productivity of over 3 g/L/h [25].

The usage of *S. cerevisiae* has been intensively investigated for the production of ethanol. Because succinic acid is not normally produced at high levels in *S. cerevisiae*, metabolic engineering provides an important strategy to increase succinic acid formation. Many studies so far focus on approaches using the TCA cycle in the route of oxidative or glyoxylate shunt leading to synthetic succinic acid. At present, the highest level succinic acid fermentation is 43 g/L, with a productivity of 0.45 g/L/h in *S. cerevisiae* (Reverdia) [26]. Yan et al. [27] modified the *S. cerevisiae* pathway for succinic acid manufacture. The engineered *S. cerevisiae* could produce 6.17 ± 0.34 g/L of succinic acid via the fabricated path. The succinic acid production increased to over 8 g/L when GPD1 was deleted. After regulation by biotin and urea, succinic acid titer was nearly 10 g/L with a yield of over 0.3 mol/mol glucose. By suitable $CO_2$ supplement in the bioreactor, the genetic modified strain could produce almost 13 g/L succinic acid with a yield of over 0.2 mol/mol glucose. The whole fermentation process was conducted at pH 3.8. Thus, the planned engineering scheme is effective for the production of free acid at low pH values. The ability to grow and produce succinic acid at low pH is a good economic advantage of yeast strains compared with other succinic acid producing microorganisms.

C3 together with C1 mechanism relating to the $CO_2$ fixation is catalyzed by pyruvate carboxylase under aerobic cultivation conditions, which explains the high yields and productivity in fumaric acid production by microorganisms [6]. This type of $CO_2$ fixation process can lead to the formation of oxaloacetic acid; thus, intermediates in C4 citrate cycle can be reserved for biorefinery during aerobic cultivation conditions. When the nitrogen resource becomes restrictive, the growth phase ends, while utilization of the glucose and $CO_2$ fixation processes continue, which makes C4 1,4-diacids continue to accumulate. Theoretically, the maximal yield is two moles of fumaric acid per mole of glucose consumed. Reductive pyruvate carboxylation results in two moles of $CO_2$ fixation. Strains cannot produce ATP for the purposes of maintenance and transport if the reductive pyruvate carboxylation is the only pathway. Consequently, the TCA cycle is also active during fumaric acid production [17]. Metabolic engineering has been conducted to fumaric acid production by non-fumaric acid producers. The chief difficulty of fumaric acid fermentation by *Rhizopus* species is their morphology control in bioreactor. The *Rhizopus* species are less genetically accessible than *S. cerevisiae* and *E. coli*.

For the route from glucose to malic acid, adenosine triphosphate (ATP) is neutral. This process can lead to the fixation of one mole of $CO_2$ per mole of malic acid. For this case, the maximum theoretical yield is two moles malic acid per mole of glucose [8]. This kind of cytosolic pathway is usually denoted as the reductive tricarboxylic acid (rTCA) pathway, which is utilized by many *Aspergillus* species that natively produce malic acid. Yeast synthesizes malic acid by either cytosolic or mitochondrial pathway. The enzymes that are intricate in the malic acid metabolic pathway have an impact on the production amount of malic acid such as fumarase overexpression in the TCA cycle, malate dehydrogenase overexpression in the reductive TCA cycle, the malate synthase in glyoxylate pathway, the pyruvate carboxylase in the reductive pathway, and even the cytosolic malate dehydrogenase. Nakayama et al. characterized malic acid production mechanism in sake yeast, and found high-level production of malic acid by yeast could be realized by the suppression of mitochondrial activity [28]. In bacteria *T. fusca* muC, the synthesis pathway of malic acid was shown to be derive of from phosphoenolpyruvate to oxaloacetate, followed by the reduction process from oxaloacetate to malate. The yield of malic acid by the engineered *T. fusca* muC-16 was increased by 47.9% compared with the parent strain muC. The muC-16 strain can grow on 100 g/L cellulose with 62.76 g/L malic acid in batch fermentation [29]. In another attempt, the phosphoenolpyruvate (PEP) carboxykinase of *M. succiniciproducens* MBEL55E was overexpressed in recombinant *E. coli*. The final malic acid production was about 10 g/L after only 12 h aerobic fermentation [23].

### *8.2.3 Cofactor Engineering of Strains*

Metabolic engineering of the 1,4-diacid producer can be used to debottleneck the 1,4-diacid pathway, improve product transportation, and enhance paths straightly involved in the 1,4-diacid manufacture [30]. However, the carbon flux shift to the 1,4-diacid route is insufficient to govern the formation of some metabolite by-products, although it is frequently used to disallow pathways involved in by-product formation. Moreover, optimization of the yield and productivity can be realized by arrangement of the genetic operations.

Evolution coupled with genetic modification is performed to grow overproducing strains for succinic acid biorefineries. To keep the redox or ATP stability, Singh et al. cloned a gene encoding the ATP-generating PEPCK enzyme, to eliminate the pathway for forming ethanol and acetate to improve succinic acid production in the ldhA, pflB, ptsG mutant strain. The recombinant strains can make succinic acid with a 60% increased production compared with the control *E. coli* [31]. In engineered *E. coli* KLPPP, phosphoenolpyruvate carboxykinase was overexpressed with the deletion of lactate dehydrogenase, phosphotransacetylase–acetate kinase, pyruvate formate lyase, and pyruvate oxidase genes. The recombinant strain showed a high conversion yield of succinic acid on galactose (1.2 mol/mol) compared to glucose (0.48 mol/mol). The concentration of succinic acid was over 22 g/L, and the molar yield was about 1.1 mol/mol total sugar [32]. One type of *E. coli* strain with ptsG mutation and homologous or heterologous (cyanobacterial) ppc gene expression was engineered for production of succinic acid. Oxaloacetate was the precursor of succinic acid. As a result of ppc gene overexpression, the PEP carboxylase activity was enhanced, and then the PEP carboxylation ability of producing oxaloacetate was improved. The recombinant strain had good performance in succinic acid fermentation, especially using glucose and xylose mixtures as the carbon source [33].

It is not easy to produce fumaric acid via genetic modified microorganisms. From another perspective, it provides possible chance or solution for improving fumaric acid fermentation by recombinant strains. In the study of Xu et al., both malate dehydrogenase (RoMDH) and fumarase (RoFUM1) originating from *R. oryzae* was introduced into *S. cerevisiae* [34]. In their study, when pyruvate carboxylase was overexpressed, a novel cytosolic fumaric acid biosynthetic pathway was reconstructed.

*E. coli* can produce fumaric acid. However, carbon flux to acetyl-CoA, acetate, formate, and succinate must be restricted or blocked. Besides *E. coli* and *Pichia pastoris*, *Rhizopus spp.* has also been genetically modified. However, most *Rhizopus* species are unaffected by antibiotics, such as ampicillin, chloramphenicol and streptomycin. Uracil auxotrophic isolates of *Rhizopus spp.* can usually be utilized by genetic changes; however, it is not easy to isolate the target. Compared with *E. coli* or *S. cerevisiae*, there are seldom effective genetic methods available for *Rhizopus spp.* Thus, it has been largely unexplored for fumaric acid production via genetic alteration of *Rhizopus* spp. There is one disadvantage of fumaric acid production by *E. coli*: the cultivation has to be conducted at near-neutral pH because the acidic circumstances prohibit the growth of organisms. A low-pH fermentative process would be advantageous especially for separation processing. Thus, many *Rhizopus* and *S. cerevisiae* species would be useful host organisms, which are tolerant of acidic conditions.

As production hosts, *Aspergillus* offers several prospective advantages when scaling up, one of which is that the strains can be genetically modified to produce high levels of malic acid. A general genetic tool always comprises available markers, selective promoters, and gene knockout/transformation methods. Most importantly, a number of species' genome sequences should be annotated. High-throughput screening methods can be used to evaluate established mutagenesis systems [35, 36]. Besides *S. cerevisiae* and *E. coli,* there is no native replicating extrachromosomal vectors; thus, the unstable artificial ones are usually selected and developed. However, engineered *A. oryzae* has been engineered by overexpression of native genes for improvement of malic acid production. In this way, rTCA pathway was conjunct with a C4 dicarboxylic acid transporter. Malic acid productivity increased 27%. By *A. oryzae* NRRL 3488 overexpressing three genes, the final malic acid titer was 154 g/L in 164 h. The productivity was 0.94 g/L/h, and the conversion yield of glucose was almost 1.4 mol/mol, which was a very high malic acid production in the literature [37].

## 8.3 Fermentation Process Engineering

Optimization and enhancement of fermentation process is determined by technical maturity and economics. Fermentation performance dictates the whole strategy and effectiveness of the biorefinery process [38]. Main performance factors including concentration, yield, productivity, and process efficiency, whereas techno-economic analysis can provide a direct data-driven understanding of the fermentation cost.

Titer is defined as the fermentative 1,4-diacid concentration in the fermentation broth. Titer is normally stated in the unit of g/L (grams of 1,4-diacid/liter of fermentation broth). High fermentative titers result in importantly cheap costs from the upstream to the downstream stage with less energy and substance consumption. Productivity is normally stated in the unit of g/L/h (grams product/liter of fermentation broth/time), which is defined as the production rate of 1,4-diacid per bioreactor volume. The productivity factor controls the overall fermentative volume for the achievable capacity goal. An increase in productivity decreases the number or volume of requisite fermenters and otherwise reduces the related operation capital. Productivity improvement of a designed plant can help increase the plant capacity utilization. Yield is defined as the conversion efficiency from one or more feedstock fraction to the target 1,4-diacid product. The unit is g/g sugar or mol/mol sugar. The yield answers the question of the number of feedstocks required for a particular amount of the 1,4-diacid production. The theoretical maximum to the product yield is determined by the stoichiometry of product pathway and the connected metabolic pathways. Because a fraction of the feedstock is necessary for metabolic growth and energetic systems maintenance, the maximum theoretical yield is usually unachievable. Fermentations with low yields not only increase the cost of net feedstock, but also raise the downstream costs because vacant feedstocks or surplus by-products must be separated from the crude broth.

Fermentation process engineering involves assessing many process parameters, such as media proposal, seed cultivation and inoculum, pH and temperature control, gas transfer, time control and feedstock feeding approach. Some scale-dependent factors, such as partial pressure of $CO_2$ for anaerobic cultivation, and agitation speed, are studied to assess the producers' sensitivity for commercial-scale conditions, which is necessary for guaranteeing probable performance across scales. Sometimes, different engineered strains are used for different scale-up conditions in developing a process on the industrial scale.

Table 8.4 summarizes common feed strategies for succinic, fumaric and malic acids production. Process engineering approaches can be used to change the fermentative efficiency, but usually this is not enough. Media components and fermentation control are the straightforward methods to obtain a high-level manufacture of 1,4-diacids. As the seed and preferred fermentation ensilage constantly have a range of nutrient components, the effect of one single component and the relationship between different constituents need careful study and experimental design. Factor screening can be performed by some methods such as response surface methodology (RSM), Plackett-Burman design (PBD), or central composite design (CCD), to find the most significant aspects.

**Table 8.4** Common feed strategies for succinic, fumaric and malic acids production

| Feed strategy | Main principles |
|---|---|
| Constant | Constant set rate, cell specific growth rate gradually reduces, cell mass rises linearly |
| Speed variation | Accelerated speed in cultivation process (gradient or linearity), specific growth rate is altering |
| Index feeding | Feed speed rises exponentially, specific growth rate remains relentless, the cell mass rises exponentially |
| pH-governed | Cell growth state valuation based on the pH, and control the glucose feed speed to regulate the pH to a constant rate |
| Dissolved oxygen -governed | Dissolved oxygen (DO) controlling, adjusted the feed velocity of the feedstocks |
| Dissolved oxygen-stat | To preserve a constant DO by adjusting the dissolved oxygen, the mixing speed and feeding speed |
| $CO_2$ release rate (CER)-control | The nutrition feed speed is adjusted by identifying the CER and the feedstocks utilization estimation |
| Biomass response | Controlling the quantity of feedstocks by measuring the cell mass and adjusting the nutrient consumption |

### *8.3.1 Production of Succinic, Fumaric and Malic Acids Acid from Sugar*

Most research on 1,4-diacid fermentation uses sugars as the feedstock. Glucose is always used as one type of carbon source in most organisms. Use of glucose as a carbon source in cells is by either aerobic or anaerobic fermentation. All of these processes follow from an earlier metabolic pathway named glycolysis. The first step of glycolysis is the glucose phosphorylation by hexokinase, which creates glucose 6-phosphate.

Except glucose, many other monosaccharides, disaccharides, even polysaccharides are generally used in succinic acid production. Effect of various sugars on the fermentation of succinic acid has been broadly tested [13–15]. For the production of fumaric acid, glucose is not the only appropriate monosaccharide for *Rhizopus* strains [39]. Xylose, fructose, can be used and converted as well, but the yields are relatively low. Among the disaccharides, maltose can be transformed into fumaric acid while lactose is slow, and sucrose cannot yield fumaric acid. Fumaric acid can be produced by *R. arrhizus* NRRL 2582 using starch because the strain has amylase. The performance of the fermentation is not efficient as that with glucose as the carbon source, but it is still cost effective [40]. Carbon sources for malic acid fermentation include starch, molasses, and the cellulosic hydrolytes such as crop stock hydrolyte. If the fermentation can be scaled-up, a green process would be realized for crop stock conversion, which can be used in food or chemical fields. Joint production of malic acid and succinic acid is possible with a sugar tolerant yeast named *Zygosaccharomyces rouxii* V19. The maximum produced amount of malic acid is about 75 g/L, and the glucose consumption yield is about 33 8% yield with 0.5% supplement of glutamic acid, when a little succinic acid and malic acid are added to the media [36].

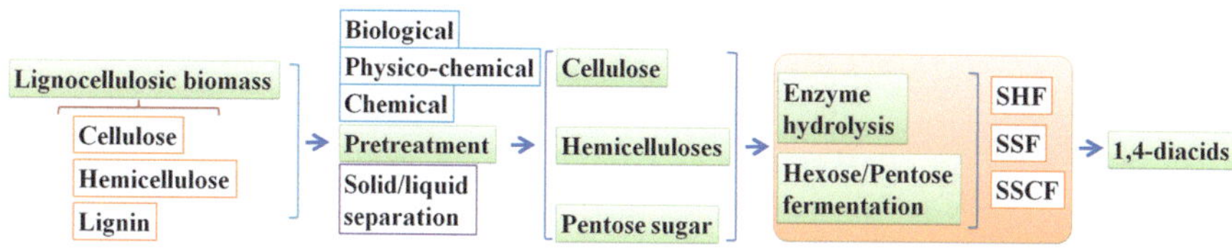


**Fig. 8.4** Multi-stage cellulosic C4 diacids production

## *8.3.2 Alternative Substrates from Lignocellulosic Biomass*

There is a major focus worldwide on conversion of renewable lignocellulosic biomass to platform chemicals [41]. It is forecast in 2025 that nearly 30% of the raw materials used for chemical production will be from biomass resources [42]. The effective exploitation of renewable sources can play a key role in the economic viability of cellulosic 1,4-diacids. Figure 8.4 shows cellulosic C4 diacids production processes. Lignocellulose is the main structure of plants that consists of cellulose, hemicellulose and lignin (Fig. 8.4) whereas cellulosic feedstocks should be non-food based crop residues, wood residues, dedicated energy crops, and industrial wastes. It is therefore, a big challenge to transform the sugars derived from these feedstocks into 1,4-diacids since there is much variability in their cellulosic content. The conversion processes usually contain five major steps: biomass type, biomass pretreatment, choice or production of saccharolytic enzymes-cellulases and hemicellulases, fermentation of hexoses or pentoses, and downstream recovery. New technologies can provide a broad range of residual feedstocks that are suitable for sustainable 1,4-diacids fermentation with less costs or by-products [43].

An eco-friendly process for conversion of polysaccharides to monomeric sugars is via hydrolysis enzymes, i.e., hemicellulases, which promote cellulose hydrolysis by exposing the cellulose fibers. Nearly 50 years have passed since the hydrolysis process of cellulose was conducted by cellulolytic enzymes [44]. After pretreatment by different techniques, biomass can be used as the feedstocks using various saccharification and fermentation strategies. Separate hydrolysis and fermentation (SHF) is used widely as a separate hydrolysis and fermentation process marked as a traditional two-step procedure. Firstly, the lignocellulose is hydrolysed by the special enzymes. Secondly, the reducing sugars produced are fermented to 1,4-diacids.

Simultaneous saccharification and fermentation (SSF) trials have been conducted on the saccharification of the lignocellulosic biomass and the subsequent conversion of the sugars to 1,4-diacids. Biomass, enzymes and fermentation microbe were placed in the same reactor. A successful SSF process relies on many factors such as pH, substrate, temperature, agitation, and feed strategy. Simultaneous saccharification and co-fermentation (SSCF) has been known as one of the possible choices for xylose-rich lignocellulosic materials conversion. Numerous genetically engineered microorganisms have been used to test SSCF for end products glucose and xylose. Efficient process design plays an important role. When a single unit is used and several different operations are performed for these integration processes, it is possible for the engineer to increase the performance efficiency of the coupled process [45, 46].

The problem with hydrolysates fermentation by *E. coli* is that the pentose in hydrolysis mixtures is usually delayed that tends to incomplete consumption [47, 48]. If the preferred feedstock, such as glucose, has a certain concentration in the media, the passage and metabolism of other sugars can be inhibited until the glucose is exhausted in many cases. However, fermentation of xylose to succinate can be enhanced via ATP supply in metabolically engineered *E. coli* K12 [49]. A mutant strain *E. coli* AFP184 can simultaneously consume xylose and glucose for succinic acid fermentation [50]. Moreover, detoxification is usually necessary to decrease many toxic components that are formed in the pretreatment process, although the extra detoxification increases operation cost. Efficient fermentation process for making succinic acid from corn stalk hydrolysates can be realized by a ptsG mutant *E. coli* [51–56]. Fungi can grow in hydrolysates, which originate from the acid hydrolysis of corn straw. Through a two-stage corn straw utilization strategy for making fumaric acid by *R. oryzae*, the final fumaric acid production can reach about 28 g/L, with 0.35 g/g yield and 0.33 g/L/h productivity, respectively [40].

Solid-state fermentation by filamentous fungi, particularly direct bioconversion or coupled hydrolysis and fermentation from lignocellulosic feedstock, has potential to be an important biotechnology for fumaric, and malic acid production [57]. *Aspergillus* species can make malic acid from thin stillage [58]. Li et al. reported that *Rhizopus Delemar* can produce malic acid at high levels from corn straw hydrolytes [59]. Consequently, malic acid yield can be improved by the regulation of the metabolic network of the host strain. *R. Delemar* can make over 120 g/L malic acid after 60 h fermentation. Zou et al. reported that a novel *A. pullulans* can make polymalic acid (PMA) and malic acid from hydrolysates of xylose and corncob. The results showed that fed-batch cultivation with xylose in a five-Liter fermenter achieved a high PMA titer (about 80 g/L) with about 91 g/L of malic acid after hydrolysis. The maximum PMA productivity was 0.52 g/L/h [60].

Biomass derived bio-oil have been investigated intensively as an alternative of diesel and gasoline fuels. Production of bio-oil via pyrolysis/thermochemical conversion of biomass has attracted much attention. Bio-oil has a high energy density and has potentials for partial replacement of fuels. However, the oxygen content of bio-oil is as high as 50 wt/wt%, the water content of bio-oil is 15–30 wt/wt%, and the C/H ratio is high; thus, it cannot be used directly as transportation fuel. Deep deoxygenation is regarded as one of key upgrading technologies, which is critical to the treatment of bio-oil. Adding water can help separate bio-oil into two fractions, an aqueous fraction (AP-bio-oil) together with a heavy organic fraction. The heating value of the organic fraction is always bigger than the value of crude bio-oil as a result of its low oxygen content. So, it is a nice method to recovery bio-oil from the aqueous phase and the organic phase before it is upgraded. However, there are many different components in AP-bio-oil. The AP-bio-oil is able to offer many carbon sources and some nitrogen sources to support microbe growth. The transgenic *E. coli* named MG-PYC can live in modified M9 media, which contains 20 v/v%

AP-bio-oil, and produces 0.38 g/L succinic acid. When glucose at a concentration of 4 g/L in the media is used by addition of 20 v/v% AP-bio-oil, succinic acid production increased from 1 to 2.4 g/L. When the carbon source was replaced by enzymatic hydrolysates of corn stock, the final succinic acid production was 10 g/L. This media by addition of 212.5 v/v% AP-bio-oil can result in about 12 g/L succinic acid concentration [61]. The paper industry produces paper solid waste (PSW) as part of its processing pulp. PSW can be used for fumaric acid production. The filamentous fungus *R. oryzae* 1526 can make fumaric acid through submerged and solid state fermentation. The results showed that the PPSW can provide carbon source and trace elements for the growth of the fungus *R. oryzae* and fumaric acid can be formed during fermentation [62, 63].

There are carbohydrates in lignocellulose, but it is difficult to use, because the pretreatment process is expensive and the conversion rate to bioproducts is low. However, with the development and improvement of pretreatment technology and microbial metabolic engineering, lignocellulose is regarded as an important biomaterial resource for value-added chemical production. Lignocellulose is more competitive as substrate than sugar or starch for 1,4-diacid production in the future. Cellulose can be effectively dissolved by ionic liquids (ILs) in mild conditions in recent studies. In this process, the linkage between cellulose, hemicelluloses and other components are broken. In addition, pretreatment by ILs can destroy the crystallinity of cellulose, which renders the biomaterials more susceptible to the hydrolysis process by enzymes. Among them, most types of wood such as hardwood and softwood can completely dissolve in 1-allyl-3-methylimidazolium chloride (AmimCl). Various biomass pretreated by ILs has been studied for production of biochemicals or biofuels. In the study of Wang et al. [64], both pinewood and corn stover were pretreated by AmimCl and then used as a media for succinic acid production. Results confirmed that cellulose can be effectively extracted from pinewood, and pinewood was in degraded into a uniform pulp with the help of AmimCl. The enzymatic hydrolysis rate of pinewood extraction reached over 72%. Moreover, corn stover can be effectively treated by AmimCl pretreatment integrated with either team explosion or hot compressed water. Extract of pinewood can make over 20 g/L succinic acid. The average yield was about 0.4 g/g biomass. Workflow calculations revealed that pinewood pretreated with ILs can reach a 57% theoretical yield of succinic acid.

Macroalgae are regarded as "Third generation biomass", and are viable renewable feedstocks for biochemicals or biofuels. Compared to many land-based crop materials, macroalgae show many attractive features: (i) land is not required for the cultivation of macroalgae, (ii) fertilizers are not required and (iii) freshwater is not needed so that it does not compete with food-crops. Moreover, due to lack of hemicellulose and lignin, few pretreatment steps are needed to release fermentative sugars, which can indeed reduce the operation costs. In the study of Bai et al., the engineered *E. coli* BS002 was constructed to make succinic acid from *Laminaria japonica* (a typical macroalgae, Fig. 8.5a) [65]. After several pretreatment steps coupled with enzymatic hydrolysis, *L. japonica* hydrolysates can produce about 10.3 g/L glucose and 10.1 g/L mannitol. *L. japonica* hydrolysates were used for

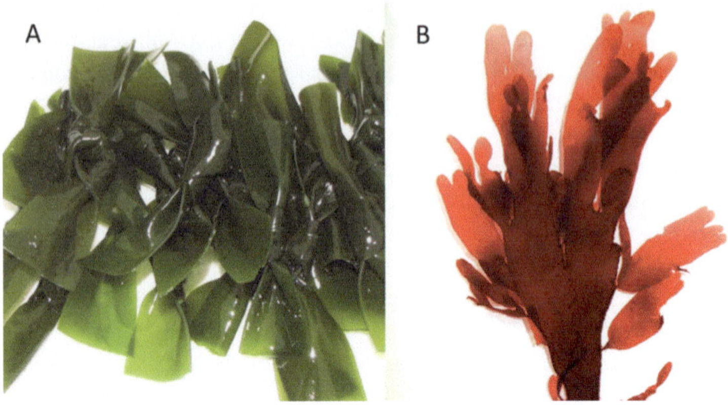


**Fig. 8.5** Succinic acid production from macroalgae, which is termed 3rd generation biomass, (**a**) *Laminaria japonica*, (**b**) *Palmaria palmate*

succinic acid fermentation by the engineered *E. coli* BS002 as the substrate. Finally, about 18 g/L succinic acid was made from the hydrolysates by a 72 h dual-phase-fermentation strategy to obtain a yield of 1.24 mol/mol total sugar, which was 73% of the maximum theoretical yield. Succinic acid was similarly produced by engineered *E. coli* KLPPP using red macroalgae (*Palmaria palmate*, Fig. 8.5b). The *P. palmata* hydrolysates were full of glucose and galactose, which were used effectively for succinic acid fermentation. The production of succinic acid was about 22 g/L, and the yield was 1.13 mol/mol total sugars, which was 66% of the maximum theoretical yield. These results demonstrate macroalgae has novel and economical features for the production of biochemical [32, 65].

### *8.3.3 Cultivation Strategies with High Production Levels*

Simple anaerobic cultivation does not facilitate cell growth and metabolism of many microbes such as *E. coli*. Dual-phase fermentation strategy, which is defined as aerobic cell growth followed by anaerobic fermentation, has been developed to enhance the cell mass accumulation and subsequent 1,4-diacid production [9]. Engineered *E. coli* strain was conducted in the dual-phase fermentation process. Cell mass was a significant signal for the time of transition from aerobic to anaerobic phase. Great cell density in the bioreactor by the DO-control strategy is usually used to improve the cell mass of the cell during the aerobic stage. In a dual-phase fermentation process in bioreactor, initial aerobic growth of *E. coli* SD121 facilitates the subsequent anaerobic production of succinic acid by ptsG mutation. The final concentration of succinic acid was about 60 g/L with a yield of 0.87 g/g sugar [33]. Jiang et al. studied the effect of growth phase feeding methods on succinic acid production by *E. coli* AFP111 [66]. The physiological state of the aerobically grown cells was critical for the subsequent anaerobic fermentation of succinic acid. For example, some environmental and physiological factors of the aerobic phase played a key role in the improvement of succinic acid production. A novel membrane-bioreactor-electrodialysis system was built for succinic acid production with high concentration, productivity and yield. In this system, ultrafiltration membrane was applied to attach fermentation and separation process, and to clarify the fermentation broth. Under the optimized conditions, biomass concentration and succinic acid concentration were over 40 g/L and 14 g/L, respectively, which were 20 times higher compared with the batch cultivation process in the study of Meynial-Salles et al. [67]. More *in situ* product recovery strategies are discussed in Sect. 8.4.3.

Up until now, for *Rhizopus* strains, fermentation processes described have made 126 g/L fumaric acid with 1.38 g/L/h productivity and the conversion yield of glucose was 0.97 g/g. This required optimization factors such as pH, aeration, and carbonate/$CO_2$ supply [7, 17]. Limitations of the *Rhizopus* strains are pH tolerance and morphology. Especially, the accessibility for genetic engineering and versatility to the alternative carbon sources are other limitations. The energetic mechanism of fumaric acid exportation by *Rhizopus* should be carefully investigated for host strains. In a SSF process with starchy materials (the total sugar concentration of cornstarch was 100 g/L.), 2-deoxyglucose-resistant mutant strains of *R. Oryzae* can make as high as 44 g/L fumaric acid. During the SSF process with degermed corn powder (100 g/L initial total sugar) by DG-3, the maximum fumaric acid concentration was about 33 g/L with 0.44 g/L/h productivity [39].

Malic acid is usually made by a "two-step fermentation" process. Fumaric acid is made and then converted to malic acid by fumarate hydratase. In the synthesis pathway, malic acid can be catalyzed from fumaric acid synthesized absolutely via a chemical synthetic method. A one-step-fermentation is always an optimal strategy since it makes malic acid using normal sugars as carbon resource. Furthermore, $CO_2$ fixation also remains in the fermentation process of malic acid. During fermentation process by *Saccharomyces* spp., L-malic acid can be made by the fumarate pathway that was catalyzed by cytosolic fumarase or mitochondrial fumarase, or by oxaloacetic acid that was catalyzed by malate dehydrogenase (MDH), but the titer was as low as 0.5–1 g/L [35]. A continuous system was set up for L-malic acid conversion from fumaric acid in a novel micro-reactor with surface-immobilized fumarase, which indicated the possibilities for micro-reactor-based biotransformation [68].

## 8.4 Downstream Processing

It is critical to decrease the production cost of the biorefinery process so as to compete with traditional petrochemical production. In general, over 60% of the total production cost (TPC) is made by the downstream processing whose target is absolute separation of the products from the crude broth [69]. Solid/liquid separation is usually the first downstream processing unit operation carried out by centrifugation or microfiltration, commonly followed by ultrafiltration to separate cells or cell fragments, proteins and other solid compounds from the supernatant broth. Different unit operations have been used, for example, precipitation processes with calcium hydroxide or ammonia, traditional membrane separation processes such as electrodialysis, and predispersed solvent extraction or reactive extraction. Adsorption and crystallization unit operations are usually used for the final purification of the isolated 1,4-diacids. These unit operations are described and their features are discussed with their possible application for the industrial separation of C4 1,4-diacids crystals.

### *8.4.1 Main Separation Unit Operations*

The classical manufacturing approach for the separation of carboxylic acids from crude fermentation broth is a precipitation process using calcium oxide or calcium hydroxide. With the help calcium oxide or calcium hydroxide, the calcium salts of 1,4-diacid can be filtered from the aqueous broth and then reacted with sulfuric acid. By this means, the free 1,4-diacid is purified. By-product is a large amount of gypsum. It is a remarkable fact that many other organic acids in the broth will be precipitated together with succinic acid, fumaric acid and malic acid at the same time [69, 70]. Precipitation with ammonia has been widely studied for succinic acid separation.

Ion-exchange remain the main operation for separation and purification of 1,4-diacids from aqueous broths. Ion-exchange is mainly used for the removal of residual anions or cations. Generally, ion-exchange steps are usually applied in extraordinary purification stages as selectivity is not high enough, thus, the product yields remain low.

Liquid–liquid extraction is a widely-used technique in the separation of fermentation-based 1,4-diacids on the lab scale. However, liquid–liquid extraction is not widely applied on the large-scale as most traditional extraction reagents show unfavorable distribution coefficients for different 1,4-diacids. To improve the extraction yield and the selectivity, many researchers use different reactive components for the liquid–liquid extraction of 1,4-diacids from an aqueous phase, such as amines, which can dissolve in miscible organic solvents. Amines have a great affinity for negatively charged components because of their high basicity, which makes them appropriate for the extraction. The pre-treatment of fermentation broth as well as the preparation of succinic acid ammonium using ammonia in the re-extraction can be used to improve the extraction yield of succinic acid produced by microbial fermentation. The amine reacts with the 1,4-diacids at the interface between the organic phase and aqueous phase, which leads to the reaction of amine–1,4-diacid–complexes. These complexes are then solubilized into the organic phase. The mechanism consists of proton transfer and ion pair construction relying on the types of amines and the organic solvents. Predispersed solvent extraction (PDSE) is a novel method for separating solutes from extremely dilute solution by solvent extraction quickly. The use of colloidal liquid aphrons (CLAs, micron-sized solvent droplets surrounded by a thin aqueous film) in predispersed solvent extraction may ameliorate the problems such as emulsion formation, reduction of interfacial mass transfer and low interfacial mass transfer areas in solvent extraction process.

Electrodialysis is a favored method in the food field for removal of citric acid from juices. The application of electrodialysis for downstream separation of 1,4-diacids has been investigated on the lab-scale [69]. Shortcomings of electrodialysis are its high energy consumption, the cost of the membrane material, and the low selectivity and capacity. Another question is the problem of binary ions, which are hard to treat via the electrodialysis membrane. Membrane fouling is another problem presently in the continuous electrodialysis process.

Membrane-based technology, together with centrifugation, has been generally used for the removal of solids from liquid solutions. This process can result in low operation cost and energetic consumption. Investigators have reported the application of filtration in the separation of proteins, acids, and other natural organic compounds which is a feasible strategy for clarifying complex solid–liquid mixture systems. Moreover, several membrane systems can be conducted on fermentative systems to recycle strain cells and obtain product. In integrated systems, permeant can be drawn off at times, which can reduce the concentrations of the target products around cells. In integrated systems, separation is coupled with fermentation without coagulant, so it is widely used in the centrifugation process.

As one of the oldest but effective processes for the preparation of 1,4-diacid crystals, crystallization process can be used as a final purification step. As succinic, fumaric and malic acids are solid forms in room temperature, direct crystallization might provide the desired target in solid form without extra unit operations. However, the product yield is low because many salts of 1,4-diacids remain as residuals in the crystal broth. The low-purity product is hard to use as monomer for many polymerization reactions. To obtain a high purity (>99%) of 1,4-diacids, further purification process is needed to remove residual ionic impurities.

Examination of small differences on the molecular-scale structure of organic acids has had an effect on understanding the mechanisms of adsorption. Hwang et al. studied the adsorption behavior of C4-dicarboxylic acids at a hematite/water interface [72], and indicated that orientation of the two carboxylic groups and $pK_a$ values of the organic acids substantially affected the adsorption density as well as the position and characteristics of the pH adsorption edges. Succinic acid has higher molecular flexibility. Fumaric acid with transconfiguration appears to bind to hematite mainly as a deprotonated outer-sphere complex using only one carboxylic group.

### 8.4.2 *Separation and Purification from the Crude Broth*

Most of the proposed process schemes described above for the separation of 1,4-diacids from crude broth do not have sufficiently high (ca. 95%) selectivity. Since the final purification results determine one important cost factor, the

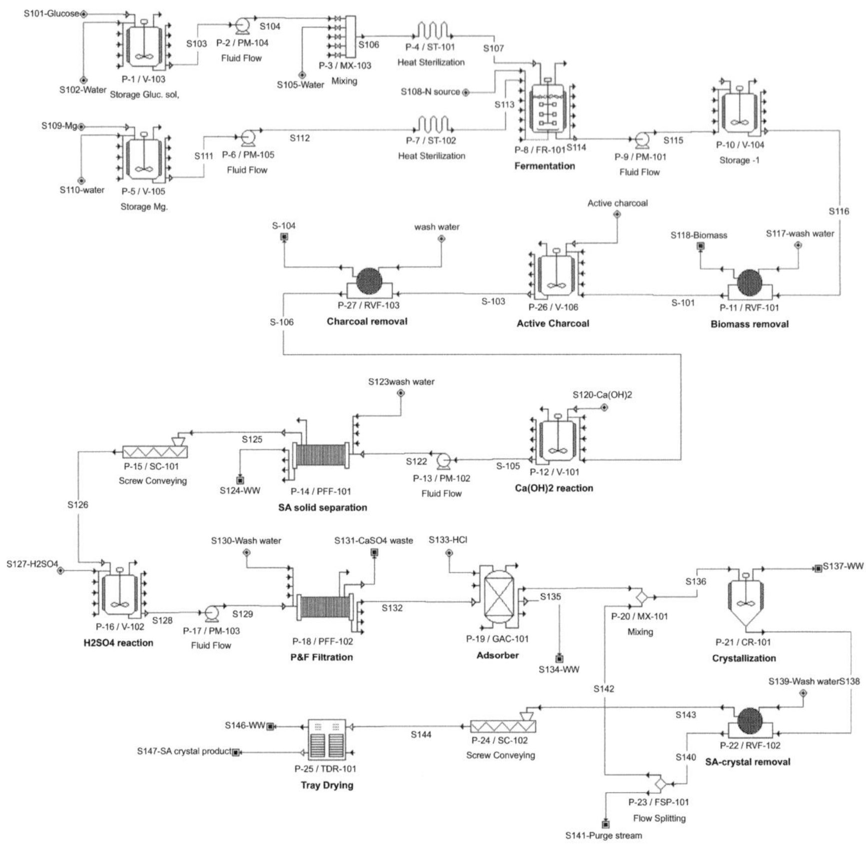


**Fig. 8.6** Batch and fed-batch succinic acid fermentation processes by software SuperPro Designer

production of by-products such as acetic acid, formic acid, pyruvic acid in the TCA branch should be reduced to a minimum level by metabolic engineering of the microbe producers and optimization of the cultivation processes. The contaminants of the post-fermentation broth usually include left-over glycerol, enzymes, other proteins, salts and other acids formed in the process of microbial cultivation and metabolism. Thus, the final standard of bioconversion success is determined by the purification process. Figure 8.6 shows batch and fed-batch succinic acid fermentation processes simulated by software SuperPro Designer. It can be seen that unit operations of the downstream process are numerous and important.

One effective unit operation can influence the entire recovery yield of the downstream process. In crystallization, carboxylic acids show varied distribution between dissociated and free-acid forms at various pH values. Undissociated 1,4-diacids have various solubilities. Li et a.l [73] found that the solubility of free succinic acid was less than 30 g/L at 4 °C and pH 2.0. In this condition, the other by-product acids, such as lactic, acetic and formic acids, were totally water miscible. So succinic acid crystallization from fermentation broth was realized at 4 °C and at low pH, and by-product acids remained in the solution. By this one-step recovery strategy, succinic acid could be selectively crystallized with 70% yield and 90% purity, respectively.

Ultrafiltration can be another important unit to clarify the succinic acid fermentation broths [74]. Various ultrafiltration membranes such as PES (100 kDa, PES 30 kDa, PES 10 kDa) and RC (10 kDa) have been tested and the results show ultrafiltration is achievable for succinic acid from fermentation broth [75], for which 99.6% microorganism cells could be removed from the broth and proteins could be separated well. The highest removal rate of proteins was 87% for membrane PES10 kDa. A clear permeate was obtained after ultrafiltration compared to centrifugation. Study of membrane fouling mechanisms in ultrafiltration revealed that many are fouled by concentration polarization or the cake layer by resistance-in-series model. Hermia's model as well as its four individual sub-models was used to analyze the predominant fouling mechanism. Results indicated that the fouling of RC and PES (30 kDa) was controlled by the blocking mechanism, while PES (100 kDa) was controlled by the intermediate blocking and PES (10 kDa) was controlled by the cake layer.

The most common method used for fumaric acid capture is precipitation. $CaCO_3$ suspension has always been used as a neutralizing reagent as its $Ca^{2+}$ salt slowly dissolves during the period of fumaric acid production. $CaCO_3$ also results in viscosity and foaming problems as a result of the less solubility of calcium fumarate. As a result of the higher solubility of sodium fumarate and magnesium fumarate, fumarate production using neutralizing reagents, such as $Na_2CO_3$ or $MgCO_3$ can lead to alternative downstream operations. If $CaCO_3$ acts as a neutralizer, the broth is firstly filtered to remove the cells, and then acidified to obtain the acid crystals. When $Na_2CO_3$ or $MgCO_3$ is used, heating is avoided because the cell might be possibly reused.

Membrane separation can be regarded as one of the recovery steps in the process of removal of organic acid from simulated or actual fermentation broths [70]. Staszak et al. evaluated concentration and separation of fumaric acid from simulated and actual fermentation broth by ceramic membranes [71]. They showed that the retention of fumaric salt rose with the increasing pH values in the feed solution. The retention of uncharged chemicals such as fumaric acid or glycerol, were less than 2%.

Nanofiltration and bipolar electrodialysis in a hybrid system has been proposed to obtain fumaric acid from fermentation broths [76], which showed that fumaric acid can be isolated and concentrated efficiently, with no additional costs related to waste generation and regeneration. Adsorption can be used as an alternative method for recovering the low-concentration fumaric acid in the filtrate after crystallization. Acetone was applied to effectively desorb fumaric acid from the column packed with activated carbon. The desorption process was followed with acetone removal by evaporation and washing with water to produce highly purified fumaric acid crystals. Both activated carbon and acetone can be recovered and reused in the repeated adsorption – desorption process [77].

**Table 8.5** Platforms of process integration for succinic acid production by *in situ* product recovery (ISPR) methods

| Microbe | ISPR routine | Production (g/L) | Productivity (g/L'h) | Yield (g/g) | Time (h) | Ref. |
|---|---|---|---|---|---|---|
| *C. glutamicum* | Membrane for cell recycling | 146 | 3.17 | 0.92 | 46 | [25] |
| *A. succiniciproducens* | Electrodialysis | 84 | 10.4 | 0.89 | 350 | [66] |
| *A. succinogens* | *In situ* adsorption | 3.96 | 0.11 | 0.44 | 36 | [80] |
| *A. succinogens* | Expanded bed adsorption | 145.2 | 1.3 | 0.52 | 126 | [81] |
| *E. coli MG-PYC (pTrchisA-pyc, 4ldhA)* | Membrane separation | 73 | 0.56 | 0.68 | 130 | [82] |

Different configurations of microbial electrolysis desalination cell have been examined for the production of malic acid. The microbial electrolysis desalination and chemical-production cell (MEDCC) has a less internal resistance. And among the reactors, the anode-biofilm population had a high *Geobacter* percentage [78]. MEDCC as a new system for malic acid production was successfully coupled to biological electrodialysis with bipolar membranes. The specific electricity usage of the MEDCC was around 10% to 30% of the electricity consumption in this electrodialysis process [79].

### 8.4.3 In Situ *Product Recovery (ISPR)*

Many integrated biotechnology platforms have been developed for sustainable processes [69, 70]. Inhibition caused by accumulative acidic products is a chief cause for small production of a target product in fermentation broths. The concentration of the product can be increased by a method called *in situ* product removal strategy and coupled systems. Table 8.5 summarizes integrated bioprocesses for the production of succinic acid. The significance of the integrated process is understood as "good integration = high efficiency". Initial assistance from process engineering concerning commercially viable proposal principles is a vibrant element of many technological platforms. Successful platforms of process integration can provide streamlined workflows for the "design-build-check-learn" cycles, which not only encourage multidisciplinary collaboration, but also fundamental understanding of biological ISPR processes [1].

An appropriate ISPR technique should be selected according to the physical and chemical properties of the target products and the process fermentation process. *A. succinogenes* 130Z can make succinic acid in a batch fermentation process. The batch cultivation is effective carried out with alkaline anion-exchange adsorbents as the solid pH neutralizing reagent. Alkaline anion exchange adsorbent is used for *in situ* removal of succinic acid from the broth and the hydroxyl groups of the acids react with the alkaline anion exchange adsorbent at the same time [80]. Meynial-Salles et al. established a novel fermentation-membrane-separation system that produced a concentrated succinic acid broth with over 80 g/L titers by the strain *A. succiniciproducens* [67]. For this system, fouling of membranes as well as flux throughput in the modules should be carefully monitored. Li et al. [81] applied an ISPR technique to *A. succinogenes* fed-batch fermentation (Fig. 8.7, ISPR unit coupled in yellow frame). Expanded bed adsorption using anion exchange adsorbents was used to directly remove succinic acid from the bioreactor efficiently. The united fermentation-separation system could eliminate inhibitory effects of succinic acid for continuous high level production. The coupled process could prolong cell growth cycle and reduce the number of downstream steps. The final succinic acid production was over 140 g/L, and the conversion yield was about 0.5 g/g glucose. However, the adsorption coupled fermentation generally yielded a huge amount of wastewater and eluting/regeneration reagents. A unique membrane separation based fermentation system was set up for succinic acid production with less water and energy consumption [82]. In this combined system, product inhibition was improved by *in situ* recovering acids and reloading the fresh broth. Cell density remained high for a longer time up to 130 h, and succinic acid production rose to over 70 g/L. Succinic acid could be crystallized with 90% recovery. HPLC and $^1$H NMR analyses showed that the purity of the separated solid succinic acid surpassed 99%.

Fumarate salts can cause inhibitory effects on fermentation to some extent [83]. Production of soluble sodium or potassium fumarate can inhibit cell growth and metabolism when the titers of fumarate reach about 40 g/L by *R. arrhizu*. It is clear that free fumaric acid can inhibit strain production more seriously than dissociated sodium fumarate. When the pH is low, produced fumaric acid by the fungi inertly diffuses back over the plasma membrane, thus reducing the intracellular pH until the fermentation becomes terminated. In an integrated process of fermentation and crystallization, upon cooling and seeding, fumaric acid can directly crystallize from fermentation broths which can actually lead to an extreme decrease in the consumption of neutralizing base required during fermentation and of acidification in the product recovery process [84, 85]. However, one difficulty of *in situ* separation via *Rhizopus* species is the morphological problem of the fungi clumps, which are inclined to grow on the bioreactor walls and even on the stirrers. Morphology problems lead to separation unit constraints. Thus, much effort is put into improving morphology of the fungi producers.

ISPR processes have shown greatly benefit in improving production, recovery yield and productivity of the fermentation processes. Compared to the *in situ* precipitation strategy, the efficiencies of many other separation units are not high, and some are much more complicated. The industrial achievability of ISPR processes remains to be recognized systematically, and thorough knowledge of the integration process needs further investigation for application.

In summary, Table 8.6 compares common separation methods and unit operations used for 1,4-diacid purification in downstream processes.

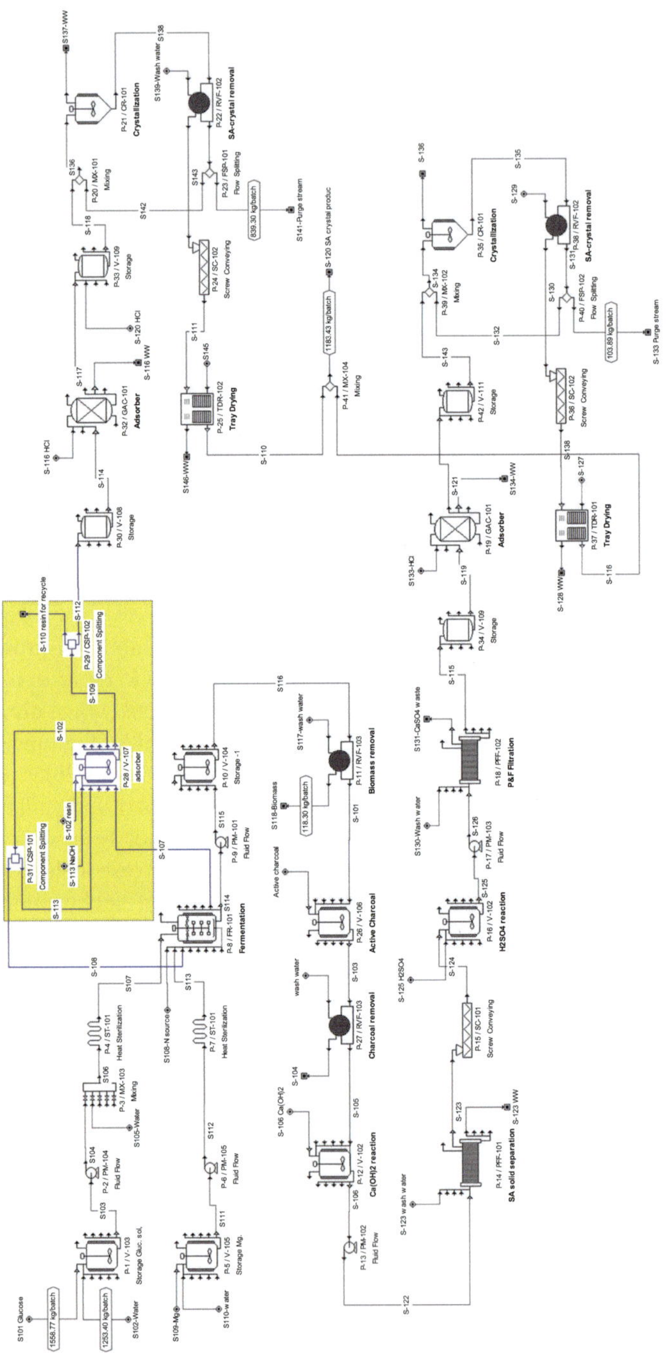

**Fig. 8.7** Fermentation and ISPR process for the succinic acid production designed by software SuperPro Designer

**Table 8.6** Downstream processing strategies

| Unit operation | Benefits | Drawbacks |
|---|---|---|
| Precipitation | Low-cost precipitants and few technological obstacles | Large amounts of precipitants; accumulation of by-products |
| Adsorption | Operation and scale-up simple and routine | Large amounts of acids, alkalis and water consumption; regeneration of resin required; low selectivity; |
| Extraction | High capacity and low- energy consumption | Broth pretreatment required; many expensive extraction agents |
| Membrane separation/ Electrodialysis | Relatively mild conditions; suitable for continuous or *in situ* separation | High energy consumption; expensive membranes; low selectivity and contamination |
| Crystallization | Operation and scale-up simple and routine without additional reagents | Low yield and purity; recrystallization or refining unit required |
| *In situ* separation | Integration with the fermentation process; eliminates product inhibition | Relatively complicated processes; regeneration of separation sorbent is required |

## 8.5 Final Remarks

### *8.5.1 Techno-Economics Challenges*

Design and realization of an adequate process for the production of 1,4-diacids from fermentation broth depends not only on the scalability and robustness, but also on the overall recovery yield and total costs. A suitable factory scheme must embody a plan that can make a product with profit. Cost sensitivity as well as many other factors can allow for optimization of plant factors to minimize the production cost. This sensitivity analysis can provide information about the range of possible outcomes and about the extent to which the outcome responds to thinkable variations in uncertain parameters. Although many biomass-derived C4 1,4-diacid production processes are successful on the lab-scale, it is still challenging for large-scale operations. Much effort is necessary for developing a thorough understanding of biomass, stable engineered strain and unit operations enhancement in upstream and downstream processes. Researchers will need to focus on biomass conversion with competitive and sustainable approaches.

Process simulation is challenging and critical for the understanding of the bio-based process designs of C4 1,4-diacid production from biomass. Taking techno-economics of biomass pretreatments as an example, one economical pretreatment approach for a feedstock may not be suitable for another type of biomass, even though the bioresources belong to the same species. Hindrances of the current pretreatment methods contain inadequate recovery of cellulose or lignin, accumulation of by-products which result in inhibitory effect on the end-product fermentation, high chemical/energy consumption, and unwanted by-product production. Most of the techno-economic studies are based on process simulations for C4 1,4-diacid production from biomass to make acid-soluble lignin content and produce fermentative sugars (glucose, xylose or arabinose).

### 8.5.2 Conclusions and Future Outlook

The scientific literature provides many cases of microorganisms modified to make C4 1,4-diacids. Traditionally, the gentle way to the commercialization stage is realized by typical mutagenesis, fermentation process, separation and purification. The fully integrated technology platform leverages a realistic methodology to strain improvement coupled to the next generation tools in synthetic biology, systematic engineering, high-throughput screening (HTS), fermentation enhancement, and efficient separation tools. Developing methods, skills, and data bases that can be quickly integrated and used to improve a potential production platform are necessary. The traditional petrochemical synthetic methods mainly depend on crude oil as the feedstock. Delivery periods and costs, available storage factory, weather and international markets, availability of cash and economy can all affect the price of oil. Nevertheless, presently, both WTI crude oil price and Brent crude oil price are about US $45, which are much less than the crude oil prices in 2008. Concerning global energy security and the environmental sustainability, a safe and reliable biorefinery plant for platform chemical production will attract growing attention as an alternative to the traditional petrochemical factory.

The market availability for succinic acid and its derivatives is predicted to be 245,000 tons per year [20]. Actually, petrochemical industry dominates most commercial succinic acid production. Extraordinarily, the noteworthy development in the economically competitive route of succinic acid production from renewable biomass has being carried out years ago. Presently, industrial plant of succinic acid from natural feedstocks has been realized in the world [3]. Right now, succinic acid is being produced commercially through the fermentation of glucose from renewable feedstock by several companies. Commercial biotechnological succinic acid production is reported by Requette and Bioamber. In Spain, Succinity is a joint venture between BASF SE and Corbion Purac., which produces 10,000 tons succinic acid per year. In USA, Myriant, together with ThyssenKrupp Uhde, has a 13,600 t/y capacity. A joint venture between DNP Green Technology, ARD, and Mitsui & Co. in France, named BioAmber, built a demonstration plant with 3000 t/y capacity in 2015, whose biorefinery processes for producing succinic acid will lead to a total reduction in $CO_2$ emissions and a 60% reduction of energy consumption (www.bio-amber.com). "Bio-succinic acid makes the big time." On March, 2016, BioAmber announced that she has successfully started up its first commercial-scale platform for bio-based succinic acid production, bringing a reliable source of the versatile bio-based material to the market (www.icis.com). In Italy, there is also a 10,000 t/y succinic acid capacity in Reverdia [26], which is a joint venture between Royal DSM N.V. and Roquette Frères. Organisms used by these manufacturers include *Basfia succiniciproducens*, *E. coli*, *S. cerevisiae*, et al. According to a survey report from MarketsandMarkets (M&M, an international market research and management advisory company established in 2001) [86], the market of succinic acid is anticipated to grow at a speed of 18.7% from 2011 to 2016. The worldwide market of succinic acid in terms of profits is evaluated to be worth over $182 million in 2010, and is likely to be worth $496 million by 2016. Asia-Pacific is the third succinic acid consuming region and will probable to be the largest rising market in the future due to the strong demand from developing countries such as India and China. The increasing demand of succinic acid is promoting scientists and engineers to develop cost effective synthesis routes to support its overgrowing markets.

Commercially fumaric acid is presently only produced by catalytic isomerization from petroleum-derived maleic acid, resulting in a conversion yield of over 90%. Maleic anhydride can be transformed into maleic acid through the catalytic oxidation of hydrocarbons including benzene, butene and butane. According to Roa Engel et al. [6], annual production of maleic anhydride is crudely 1.8 million tons, 3% of which is used for the refining of fumaric acid. The amount of fumaric acid produced is roughly 90,000 tons per year. Prices of the petroleum-based fumaric acid are estimated to be about $1–2 per kg or $0.7–0.8 per lb. The rough price is nearly 10% higher than the price of maleic anhydride, which is used as the raw material.

Malic acid has a prospective market. It is a crude material for the manufacture of polymalic acid (PMA). PMA is a new biodegradable polymer similar to polylactic acid, however, it relies on the economic and process feasibility of the manufacture. Many bulk retailers report the prices of malic acid at about $0.9 per kg [57]. Racemic malic acid can be made by the double hydration of maleic anhydride, and the U.S. production capacity was about 5000 tons in 2000 [87]. Chiral resolution of the racemic mixture gives the (S)-enantiomer that can be selectively obtained from the fumaric acid fermentation process. The production of L-malic acid has already been industrialized. At present, many companies pay attention to the L-malic acid production as it is an intermediate-volume compound. Its annual world production is roughly 40,000 tons [7]. It is assessed that the demand request of L-malic acid could be up to 200,000 tons per year. The main global companies producing malic acid include Bartek (Canada), CIA Quimica (Mexico), Croda colloids/Tate & Lyle PLC (UK), Fuso/Kawasaki Kasei/Kyowo Hakko Kogyo (Japan), Lonza (Switzerland), Merck KGaA (Germany), Rifa Industrial/Yongsan Chemicals (Korea), Thirumalai Chemicals (India), and Nanjing Kokhai Biotechnical/Changzhou Changmao (China). Presently, the raw materials for the synthetic processes are derived from crude oil or natural gas. Nevertheless, the increased demand for food and drug safety will compete with industrial routes in the not too distant future. Microbial production is a potential option for economic manufacture of C4 1,4-diacids.

## References


1. Barton NR, Burgard AP, Burk MJ, Crater JS, Osterhout RE, Pharkya P, Steer BA, Sun J, Trawick JD, Van Dien SJ, Yang TH, Yim H. An integrated biotechnology platform for developing sustainable chemical processes. J Ind Microbiol Biotechnol. 2015;42:349–60.
2. Dumeignil F, Capron M, Katryiok B, Wojcieszak R, Löfberg A, Girardon JS, Desset S, MArin MA, Duhamel LJ, Paul S. Biomass-derived platform molecules upgrading through catalytic processes: yielding chemicals and fuels. J Jpn Petrol Inst. 2015;58:257–73.
3. Becker J, Lange A, Fabarius J, Wittmann C. Top value platform chemicals: bio-based production of organic acids. Curr Opin Biotechnol. 2015;36:168–75.
4. Jang YS, Kim B, Shin JH, Choi YJ, Choi S, Song CW, Lee J, Park HG, Lee SY. Bio-based production of C2–C6 platform chemicals. Biotechnol Bioeng. 2012;109:2437–59.
5. McKinlay JB, Vieille C, Zeikus JG. Prospects for a bio-based succinate industry. Appl Microbiol Biotechnol. 2007;76:727–40.
6. Roa Engel CA, Straathof AJJ, Zijlmans TW, van Gulik WM, van der Wielen LAM. Fumaric acid production by fermentation. Appl Microbiol Biotechnol. 2008;78:379–89.
7. Xu Q, Li S, Huang H, Wen JP. Key technologies for the industrial production of fumaric acid by fermentation. Biotechnol Adv. 2012;30:1685–96.
8. Liu YH, Song JN, Tan TW, Liu L. Production of fumaric acid from L-malic acid by solvent engineering using a recombinant thermostable fumarase from *Thermus thermophilus* HB8. Appl Biochem Biotechnol. 2015;175:2823–31.
9. Thakker C, Martínez I, San KY, Bennett GN. Succinate production in *Escherichia coli*. Biotechnol. 2012;7:213–24.
10. Cheng KK, Zhao XB, Zeng J, Zhang JA. Biotechnological production of succinic acid: current state and perspectives. Biofuels Bioprod Biorefin. 2012;6:302–18.
11. Cao YJ, Zhang RB, Sun C, Cheng T, Liu YH, Xian M. Fermentative succinate production: an emerging technology to replace the traditional petrochemical processes. Biomed Res Int. 2013;11:1–12.
12. Song H, Lee SY. Production of succinic acid by bacterial fermentation. Enzym Microb Technol. 2006;39:352–61.
13. Beauprez JJ, Mey MD, Soetaert WK. Microbial succinic acid production: natural versus metabolic engineered producers. Process Biochem. 2010;45:1103–14.
14. Li J, Zheng XY, Fang XJ, Liu SW, Chen KQ, Jiang M, Wei P, Ouyang PK. A complete industrial system for economical succinic acid production by *Actinobacillus succinogenes*. Bioresour Technol. 2011;102:6147–52.
15. Bechthold I, Bretz K, Kabasci S, Kopitzky R, Springer A. Succinic acid: a new platform chemical for biobased polymers from renewable resources. Chem Eng Technol. 2008;31:647–54.
16. Pinazo JM, Domine ME, Parvulescu V, Petru F. Sustainability metrics for succinic acid production: a comparison between biomass-based and petrochemical routes. Catal Today. 2015;239:17–24.
17. Roa Engel CA, van Gulik WM, Marang L, van der Wielen LAM, Straathof AJJ. Development of a low pH fermentation strategy for fumaric acid production by *Rhizopus oryzae*. Enzym Microb Technol. 2011;48:30–47.
18. Wang XY, Chen J, Quinn PJ. Reprogramming microbial metabolic pathways. Dordrecht: Springer; 2012. p. 225–40.
19. Knuf C, Nookaew I, Brown SH, McCulloch M, Berry A, Nielsen J. Investigation of malic acid production in *Aspergillus oryzae* under nitrogen starvation conditions. Appl Environ Microbiol. 2013;79:6050–8.
20. Chen XZ, Zhou L, Tian KM, Kumar A, Singh S, Prior BA, Wang ZX. Metabolic engineering of *Escherichia coli*: a sustainable industrial platform for bio-based chemical production. Biotechnol Adv. 2013;31:1200–23.
21. Hong KK, Kim JH, Yoon JH, Park HM, Choi SJ, Song GH, Lee JC, Yang YL, Shin HK, Kim JN, Cho KH, Lee JH. O-Succiny-L-homoserine-based C4-chemical production: succinic acid,

homoserine lactone, γ-butyrolactone, γ-butyrolactone derivatives, and 1,4-butanediol. J Ind Microbiol Biotechnol. 2014;41:1517–24.
22. Yang L, Lübeck M, Ahring BK, Lübeck PS. Enhanced succinic acid production in *Aspergillus saccharolyticus* by heterologous expression of fumarate reductase from Trypanosoma brucei. Appl Microbiol Biotechnol. 2016;100:1799–809.
23. Moon SY, Hong SH, Kim TY, Lee SY. Metabolic engineering of *Escherichia coli* for the production of malic acid. Biochem Eng J. 2008;40:312–20.
24. Cheng KK, Wang GY, Zeng J, Zhang JA. Improved succinate production by metabolic engineering. Biomed Res Int J. 2013;3:1–12.
25. Okino S, Noburyu R, Suda M, Jojima T, Inui M, Yukawa H. An efficient succinic acid production process in a metabolically engineered *Corynebacterium glutamicum* strain. Appl Microbiol Biotechnol. 2008;81(3):459–64.
26. Ahn JH, Jang YS, Lee SY. Production of succinic acid by metabolically engineered microorganisms. Curr Opin Biotechnol. 2016;42:54–66.
27. Yan DJ, Wang CX, Zhou JM, Liu YL, Yang MH, Xing JM. Construction of reductive pathway in *Saccharomyces cerevisiae* for effective succinic acid fermentation at low pH value. Bioresour Technol. 2014;156:232–9.
28. Nakayama S, Tabata K, Oba T, Kusumoto K, Mitsuiki S, Kadokura T, Nakazato A. Characteristics of the high malic acid production mechanism in *Saccharomyces cerevisiae* sake yeast strain No. 28. J Biosci Bieng. 2012;114:281–5.
29. Deng Y, Mao Y, Zhang XJ. Metabolic engineering of a laboratory-evolved *Thermobifida Fusca* muC strain for malic acid production on cellulose and minimal treated lignocellulosic biomass. Biotechnol Prog. 2016;32:14–20.
30. Kamm B. Production of platform chemicals and synthesis gas from biomass. Angew Chem Int Ed. 2007;46:5056–8.
31. Singh A, Cher Soh K, Hatzimanikatis V, Gill RT. Manipulating redox and ATP balancing for improved production of succinate in *E. coli*. Metab Eng. 2011;13:76–81.
32. Olajuyin AM, Yang MH, Liu YL, Mu TZ, Tian JN, Adaramoye OA, Xing JM. Efficient production of succinic acid from *Palmaria palmata* hydrolysate by metabolically engineered *Escherichia coli*. Bioresour Technol. 2016;214:653–9.
33. Wang D, Li Q, Yang MH, Zhang YJ, Su ZG, Xing JM. Efficient production of succinic acid from corn stalk hydrolysates by a recombinant *Escherichia coli* with ptsG mutation. Process Biochem. 2011;46:365–71.
34. Xu G, Liu L, Chen J. Reconstruction of cytosolic fumaric acid biosynthetic pathways in *Saccharomyces cerevisiae*. Microb Cell Factories. 2012;24:11–24.
35. Yéramian N, Chaya C, Suárez Lepe JA. L-(−)-malic acid production by *Saccharomyces* spp. during the alcoholic fermentation of wine (1). J Agric Food Chem. 2007;55:912–9.
36. Kazuya Taing OT. Production of malic and succinic acids by sugar-tolerant yeast *Zygosaccharomyces rouxii*. Eur Food Res Technol. 2007;224:343–7.
37. Brown SH, Bashkirova L, Berka R, Chandler T, Doty T, McCall K, McCulloch M, McFarland S, Thompson S, Yaver D, Berry A. Metabolic engineering of *Aspergillus oryzae* NRRL 3488 for increased production of L-malic acid. Appl Microbiol Biotechnol. 2013;97:8903–12.
38. Morais AR, Lukasik RB. Green chemistry and the biorefinery concept. Sustain Chem Process. 2013;1:18.
39. Deng YF, Li S, Xu Q, Gao M, Huang H. Production of fumaric acid by simultaneous saccharification and fermentation of starchy materials with 2-deoxyglucose-resistant mutant strains of *Rhizopus oryzae*. Bioresour Technol. 2012;107:363–7.
40. Xu Q, Li S, Fu YQ, Tai C, Huang H. Two-stage utilization of corn straw by *Rhizopus oryzae* for fumaric acid production. Bioresour Technol. 2010;101:6262–4.
41. Girisuta B, Kalogiannis KG, Dussan K, Leahy JJ, Hayes MHB, Stefanidis SD, Michailof CM, Lappas AA. An integrated process for the production of platform chemicals and diesel miscible fuels by acid-catalyzed hydrolysis and downstream upgrading of the acid hydrolysis residues with thermal and catalytic pyrolysis. Bioresour Technol. 2012;126:92–100.

42. Yabushita M, Kobayashi H, Fukuoka A. Catalytic transformation of cellulose into platform chemicals. Appl Catal B-Environ. 2014;145:1–9.
43. Kiran EU, Trzcinski AP, Ng WJ, Liu Y. Enzyme production from food wastes using a biorefinery concept. Waste Biomass Valor. 2014;5:903–17.
44. Isikgora FH, Becer CR. Lignocellulosic biomass: a sustainable platform for the production of bio-based chemicals and polymers. Polym Chem-UK. 2015;6:4497–559.
45. Kıran EU, Trzcinskia AP, Liu Y. Platform chemical production from food wastes using a biorefinery concept. J Chem Technol Biotechnol. 2014;90:1364–79.
46. Menon V, Rao M. Trends in bioconversion of lignocellulose: biofuels, platform chemicals & biorefinery concept. Prog Energ Combust. 2012;38:522–50.
47. Bradfield MFA, Mohagheghi A, Salvachúa D, Smith H, Black BA, Dowe N, Beckham GT, Nicol W. Continuous succinic acid production by *Actinobacillus succinogenes* on xylose-enriched hydrolysate. Biotechnol Biofuels. 2015;8:181–97.
48. Bradfield MFA, Nicol W. Continuous succinic acid production from xylose by *Actinobacillus succinogenes*. Bioprocess Biosyst Eng. 2016;39:233–44.
49. Liu R, Liang LY, Chen KQ, Ma JF, Jiang M, Wei P, Ouyang PK. Fermentation of xylose to succinate by enhancement of ATP supply in metabolically engineered *Escherichia coli*. Appl Microbiol Biotechnol. 2012;94(4):959–68.
50. Andersson C, Hodge D, Berglund KA, Rova U. Effect of different carbon sources on the production of succinic acid using metabolically engineered *Escherichia coli*. Biotechnol Prog. 2007;23:381–8.
51. Salvachúa D, Mohagheghi A, Smith H, MFA B, Nicol W, Black BA, Biddy MJ, Dowe N, Beckham GT. Succinic acid production on xylose-enriched biorefinery streams by *Actinobacillus succinogenes* in batch fermentation. Biotechnol Biofuels. 2016;9:28–42.
52. Corona-González RI, Varela-Almanza KM, Arriola-Guevara E, Martínez-Gómez AJ, Pelayo-Ortiz C, Toriz G. Bagasse hydrolyzates from *Agave tequilana* as substrates for succinic acid production by *Actinobacillus succinogenes* in batch and repeated batch reactor. Bioresour Technol. 2016;205:15–23.
53. Zhao Y, Cao WJ, Wang Z, Zhang BW, Chen KQ, Ouyang PK. Enhanced succinic acid production from corncob hydrolysate by microbial electrolysis cells. Bioresour Technol. 2016;202:152–7.
54. Tan JP, Jahim JM, Wu TY, Harun S, Kim BH, Mohammad AW. Insight into biomass as a renewable carbon source for the production of succinic acid and the factors affecting the metabolic flux toward higher succinate yield. Ind Eng Chem Res. 2014;53:16123–34.
55. Akhtar J, Ani Idris A, Aziz RA. Recent advances in production of succinic acid from lignocellulosic biomass. Appl Microbiol Biotechnol. 2014;98:987–1000.
56. Cok B, Tsiropoulos I, Roes AL, Patel MK. Succinic acid production derived from carbohydrates: an energy and greenhouse gas assessment of a platform chemical toward a bio-based economy. Biofuel Bifprod Bior. 2014;8:16–29.
57. Mondala AH. Direct fungal fermentation of lignocellulosic biomass into itaconic, fumaric, and malic acids: current and future prospects. J Ind Microbiol Biotechnol. 2015;42:487–506.
58. West TP. Malic acid production from thin stillage by *Aspergillus* species. Biotechnol Lett. 2011;33:2463–7.
59. Li XJ, Liu Y, Yang Y, Zhang H, Wang HL, Wu Y, Zhang M, Sun T, Cheng JS, Wu XF, Pan LJ, Jiang ST, Wu HW. High levels of malic acid production by the bioconversion of corn straw hydrolyte using an isolated *Rhizopus Delemar* strain. Biotechnol Bioproc E. 2014;19:478–92.
60. Zou X, Yang J, Tian X, Guo MJ, Li ZH, Li YZ. Production of polymalic acid and malic acid from xylose and corncob hydrolysate by a novel *Aureobasidium pullulans* YJ 6–11 strain. Process Biochem. 2016;51:16–23.
61. Wang CX, Thygesen A, Liu YL, Li Q, Yang MH, Dang D, Wang Z, Wan YH, Lin WG, Xing JM. Bio-oil based biorefinery strategy for the production of succinic acid. Biotechnol Biofuels. 2013;6:74.

62. Das RK, Brar SK, Verma M. Potential use of pulp and paper solid waste for the bio-production of fumaric acid through submerged and solid state fermentation. J Clean Prod. 2016;112:4435–44.
63. Zhang X, Wang X, Shanmugam KT, Ingram LO. L-malate production by metabolically engineered *Escherichia coli*. Appl Environ Microbiol. 2011;77:427–34.
64. Wang CX, Yan DJ, Li Q, Sun W, Xing JM. Ionic liquid pretreatment to increase succinic acid production from lignocellulosic biomass. Bioresour Technol. 2014;172:283–9.
65. Bai B, Zhou JM, Yang MH, Liu YL, Xu XH, Xing JM. Efficient production of succinic acid from macroalgae hydrolysate by metabolically engineered *Escherichia coli*. Bioresour Technol. 2015;185:56–61.
66. Jiang M, Liu SW, Ma JF, Chen KQ, Yu L, Yue FF, Xu B, Wei P. Efficient production of succinic acid from macroalgae hydrolysate by metabolically engineered *Escherichia coli*. Appl Environ Microbiol. 2015;76(4):1298–300.
67. Meynial-Salles I, Dorotyn S, Soucaille P. A new process for the continuous production of succinic acid from glucose at high yield, titer, and productivity. Biotechnol Bioeng. 2008;99:129–35.
68. Gorazd Stojkovič G, Plazl I, Žnidaršič-Plazl P. L-Malic acid production within a microreactor with surface immobilised fumarase. Microfluid Nanofluid. 2011;10:627–35.
69. Cheng KK, Zhao XB, Zeng J, Wu RC, Xu YZ, Liu DH, Zhang JA. Downstream processing of biotechnological produced succinic acid. Appl Microbiol Biotechnol. 2012;95:841–50.
70. Curcio E, Profio GD, Drioli E. Recovery of fumaric acid by membrane crystallization in the production of L-malic acid. Sep Purif Technol. 2003;33:63–73.
71. Staszak K, Woźniak M, Sottek M, Karaś Z, Prochaska K. Removal of fumaric acid from simulated and real fermentation broth. J Chem Technol Biotechnol. 2015;90:432–40.
72. Hwang YS, Lenhart JJ. Adsorption of C4-dicarboxylic acids at the hematite/water interface. Langmuir. 2008;24:13934–43.
73. Li Q, Wang D, Wu Y, Li WL, Zhang YJ, Xing JM, Su ZG. One step recovery of succinic acid from fermentation broths by crystallization. Sep Purif Technol. 2010;72:294–300.
74. Wang CX, Li Q, Tang H, Zhou W, Yan DJ, Xing JM, Wan YH. Clarification of succinic acid fermentation broth by ultrafiltration in succinic acid bio-refinery. J Chem Technol Biotechnol. 2013;88:444–8.
75. Wang CX, Li Q, Tang H, Yan DJ, Zhou W, Xing JM, Wan YH. Membrane fouling mechanism in ultrafiltration of succinic acid fermentation broth. Bioresour Technol. 2012;116:366–71.
76. Woźniak MJ, Prochaska K. Fumaric acid separation from fermentation broth using nanofiltration (NF) and bipolar electrodialysis (EDBM). Sep Purif Technol. 2014;125:179–86.
77. Zhang K, Zhang LJ, Yang ST. Fumaric acid recovery and purification from fermentation broth by activated carbon adsorption followed with desorption by acetone. Ind Eng Chem Res. 2014;53:12802–8.
78. Liu GL, Zhou Y, Luo HP, Cheng X, Zhang RD, Teng WK. A comparative evaluation of different types of microbial electrolysis desalination cells for malic acid production. Bioresour Technol. 2015;198:87–93.
79. Liu GL, Luo HP, Wang HH, Wang BW, Zhang RD, Chen SS. Malic acid production using a biological electrodialysis with bipolar membrane. J Membr Sci. 2014;471:179–84.
80. Li Q, Li WL, Wang D, Liu BB, Tang H, Yang MH, Liu Q, Xing JM, Su ZG. pH neutralization while succinic acid adsorption onto anion-exchange resins. Appl Biochem Biotechnol. 2010;160(2):438–45.
81. Li Q, Wang D, Hu GY, Xing JM, Su ZG. Integrated bioprocess for high-efficiency production of succinic acid in an expanded-bed adsorption system. Biochem Eng J. 2011;56:150–7.
82. Wang CX, Ming W, Yan DJ, Zhang CC, Yang MH, Liu YL, Zhang Y, Guo BH, Wan YH, Xing JM. Novel membrane-based biotechnological alternative process for succinic acid production and chemical synthesis of bio-based poly (butylene succinate). Bioresour Technol. 2014;156:6–13.
83. Engel CR, Straathof A, Van Gulik W, Van Der Wielen L. Integration of fermentation and crystallisation to produce fumaric acid. New Biotechnol. 2009;25S:S173.

84. Chen ST, Duh SC, Wang KT. Facile synthesis of L-malic acid by a consecutive enzymatic reaction. Biotechnol Lett. 1994;16:355–8.
85. Prochaska K, Woźniak-Budych MJ. Recovery of fumaric acid from fermentation broth using bipolar electrodialysis. J Membr Sci. 2014;469:428–35.
86. MarketsandMarkets. Succinic acid market by applications & geography—global trends & forecasts (2011–2016); 2012.
87. Miltenberge K. Hydroxycarboxylic acids, aliphatic. Ullmann's Encyclopedia of Industrial Chemistry. Wiley-VCH, Weinheim; 2000.

# Part IV
# Production of Alcohols

# Chapter 9
# Production of Sorbitol from Biomass

**José R. Ochoa-Gómez and Tomás Roncal**

**Abstract** Sorbitol is a natural occurring sugar alcohol with a current industrial demand of about 2,000,000 $t.year^{-1}$ showing its huge worldwide commercial interest, encompassing uses in chemical, food, textiles, pharmaceutical, and health care and cosmetic industries. The current interest for substituting oil derived chemicals by biomass derived ones has boosted the interest in sorbitol production because in 2004 it was identified by US Department of Energy as one of the 12 top chemicals derived from carbohydrates which could be potentially used as platform chemicals for producing valuable chemical intermediates and materials for industry, and inclusion of sorbitol in the listing currently remains. This review analyzes both sorbitol's current market and its potentiality as a platform chemical. Subsequently, current state of sorbitol production by chemical, electrochemical and biotechnological methods is revised, and includes a key issue for industrial success: its recovery and purification. Finally, some prospects about the direction of future research for overcoming current bottlenecks for further development are discussed.

**Keywords** Sorbitol chemical production • Sorbitol as a platform chemical • Sorbitol electrosynthesis • Sorbitol biotechnological production • D-glucose reduction

## Abbreviations

| | |
|---|---|
| BET | Brunauer–Emmett–Teller, authors of the BET theory that is the basis for an analysis technique for the measurement of the specific surface area of a material |
| ATP | adenosine 5′-triphosphate |
| ED | electrodialysis |
| GFOR | D-glucose-fructose oxidoreductase |

J.R. Ochoa-Gómez (✉) • T. Roncal
TECNALIA, Division of Energy and Environment, Biorefinery Department, Parque Tecnológico de Álava, Leonardo Da Vinci, 11, 01510 Miñano, Spain
e-mail: jramon.ochoa@tecnalia.com



Z. Fang et al. (eds.), *Production of Platform Chemicals from Sustainable Resources*, Biofuels and Biorefineries 7, DOI 10.1007/978-981-10-4172-3_9

| | |
|---|---|
| GHSV | gas hourly space velocity, $h^{-1}$ |
| HRTEM | high-resolution transmission electron microscopy |
| $NAD^+$ | oxidized form of nicotinamide adenine dinucleotide |
| NADH | reduced form of nicotinamide adenine dinucleotide |
| NADP | Nicotinamide adenine dinucleotide phosphate |
| NCNT | nitrogen doped carbon nanotubes |
| NP | nanoparticles |
| RPBR | continuous recycle packed-bed reactor |
| TOF | turnover frequency, $s^{-1}$ or $h^{-1}$ |
| TOS | time on stream |
| WHSV | weight hourly space velocity, $h^{-1}$ |

## 9.1 Introduction

Sorbitol (D-glucitol, D-sorbitol, D-glucohexane-1,2,3,4,5,6-hexol, CAS Number 50-7-4) was discovered in 1872 by the French chemist Boussingault in the berries of the mountain ash (*Sorbus aucuparia L.*) and is now known to occur naturally in a wide range of fruits and berries. During several decades it was only obtained from natural sources and scarcely consumed at prices corresponding to a fine chemical with almost no practical interest outside of the medicinal practice. However, during the Great Depression, the Atlas Powder Co. (since 1971 ICI Americas Inc.) undertook a program for investigating production methods and industrial uses of the higher polyhydric alcohols. This led to the operation of a pilot plant during 1935 and 1936 and finally, as a consequence of the rapidly growing demand, to the construction of a facility for manufacturing 1400 t.year$^{-1}$ of sorbitol and mannitol by electroreduction of D-glucose resulting from corn starch [1]. A few years later the electrochemical process was discontinued and displaced by the more economical high-pressure catalytic hydrogenation, which is currently the single worldwide operated chemical process for sorbitol manufacturing.

Since that time, the industrial importance of sorbitol has been increasingly higher becoming a commodity as shown by a current worldwide demand of about 2000 kt.year$^{-1}$ and a price around 0.55–0.65 \$.kg$^{-1}$ as a 70% syrup, the more usual commercial form. This importance is likely to grow in the coming future due to the current worldwide trend for moving from the petro-economy to the bioeconomy, requiring the use of biomass-derived chemicals as building blocks for manufacturing the intermediates and materials needed for maintaining the standard of living. In fact, sorbitol was identified by US Department of Energy as a top 12 chemical derived from carbohydrates that could be used as a platform chemical [2], and inclusion of sorbitol in the list currently remains [3].

Undeniably, the growing industrial importance of sorbitol is attracting strong interest in improving methods in its production as well as in looking into new processes. The objective of this chapter is to give an overview of sorbitol production processes. Section 9.2 is devoted to justify the interest in continuous development of sorbitol production methods as well as in improvement of the current one by highlighting the industrial importance of sorbitol from two perspectives: the current one and the future one. The former is set forth by describing its market and applications (Sect. 9.2.1) while the latter (Sect. 9.2.2) by means of its growing use as a platform building block for manufacturing valuable chemicals such as, e.g., isosorbide, lactic acid, sorbitan esters, lower glycols such as ethylene glycol, propylene glycols and glycerol, hydrocarbons and aromatics.

Section 9.3 is devoted to describe the different sorbitol production methods: hydrogenation of D-glucose, electrochemical reduction of D-glucose and biotechnological conversion of fructose. Chemical production of sorbitol by hydrogenation of D-glucose (Sect. 9.3.1) is by far the most studied process, because it is the primary industrial process. After a short process description, the three main pillars supporting further developments of this production technology are addressed: catalysts, operation mode and alternative biomass raw materials. Catalyst performance and stability play a key role in process feasibility. Hydrogenation conditions are reviewed using catalyst nature as leitmotif, from the oldest nickel Raney catalysts to the more recent supported Ruthenium-based catalysts. While the more usual operation mode is still the batch one (actually semi-batch because $H_2$ pressure is kept constant by continuous feeding of hydrogen) due to both good use of the catalyst and good temperature control, two driving forces are pushing for use of continuous processes: (i) avoid the catalyst removal for recycling needed in batch methods leading to increased recovery costs and progressive catalyst deactivation, and (ii) the growing sorbitol market requiring high productivities. Finally, the use of alternative biomass raw materials, mainly the direct conversion of cellulose into sorbitol, is discussed due to its very likely future industrial importance because: (a) the growing demand for sorbitol requires increasing amounts of D-glucose which currently is obtained from starch, a key polysaccharide for food production, and consequently alternative non-food D-glucose containing raw materials are needed, with cellulose being an obvious alternative because it can be obtained from the broadly available lignocellulosic materials; and (b) boosting the use of sorbitol as a platform chemical will require a further decrease in its manufacturing cost (for instance, for manufacturing fuels) and one way to lower the cost is by manufacturing sorbitol directly from cellulose in a one-pot process, avoiding the step of hydrolysis to D-glucose and purification of the resulting syrup.

Section 9.3.2 discusses the electrochemical production of sorbitol from D-glucose. Bibliography (both papers and patents) devoted to this issue is negligible in comparison with that related to chemical synthesis. However, while currently not used industrially, it could have some industrial importance in the future taking advantage of development of new cell designs, ion exchange membranes, anodes and decorated porous cathodes in other fields such as in chlor-alkali industry and fuel cells. These advancements could allow working at high apparent current densities while keeping both a low cell voltage and a very high sorbitol selectivity resulting in both lower CAPEX and OPEX, thus making the electrochemical production of sorbitol not only competitive at some production capacities, but also attractive due to safe issues as absence of hydrogen handling and operation at ambient pressure.

Section 9.3.3 deals with the biotechnological production of sorbitol, which could be used at industrial scale in the future, provided that high space-time-yields, high sorbitol concentration in culture broth and a straightforward downstream procedure for sorbitol recovery are achieved. Production methods using the bacterium most extensively studied, *Zymomonas mobilis*, are described but also those using Lactic Acid Bacteria such as *Lactobacillus plantarum* and *Lactobacillus casei*.

Recovery and purification are essential points in industrial production. Consequently, this key issue for the economic feasibility of a chemical process is discussed in Sect. 9.3.4. Finally, the chapter ends with Sect. 9.4 devoted to conclusions and prospects highlighting the key issues for further process development, such as catalyst improvement and implementation of continuous processes.

## 9.2 Sorbitol Industrial Importance

### *9.2.1 Sorbitol Market*

The sorbitol market was about 1830 kt in 2013 and is expected to grow at 3.6% CAGR from 2014 to 2020, until a total of about 2337 kt affording a global market of around USD 3.9 billion by 2020 [4]. It is sold in both liquid and solid forms, with liquid sorbitol marketed as a 70 wt% aqueous syrup accounting for about 83% of demand in 2013. Shares by end-use segments in 2013 were: cosmetics and personal care, 32.8%; food, 29.4%; chemical end-use segment, 24%; and pharmaceuticals, 6%, with chemical end-use segment expected to have above average growth rates up to 2020 as a consequence of its increasingly use as a platform chemical [5].

By application, "diabetic and dietetic food and beverages" is the largest use and was valued over USD 400 million in 2013 (24% market share), due to the increasing importance of functional foods with low calorie and sugar free ingredients. Toothpaste was the second largest application segment accounting for more than 20% of the total volume, and is expected to grow significantly because the high refractive index of sorbitol allows its use as crystals in transparent gels. Vitamin C accounted for about 15% of the total market. Other applications include surfactants (in the form of wetting and foaming agents, dispersants, detergents, and emulsifiers), tobacco for providing a mild effect in sniff, softener and color stabilizer in textiles, softener in leather industries and rigid polyurethane foams manufacturing. Key players are Roquette, Cargill and Archer Daniels Midland which together accounted for a market share of over 70% in 2013.

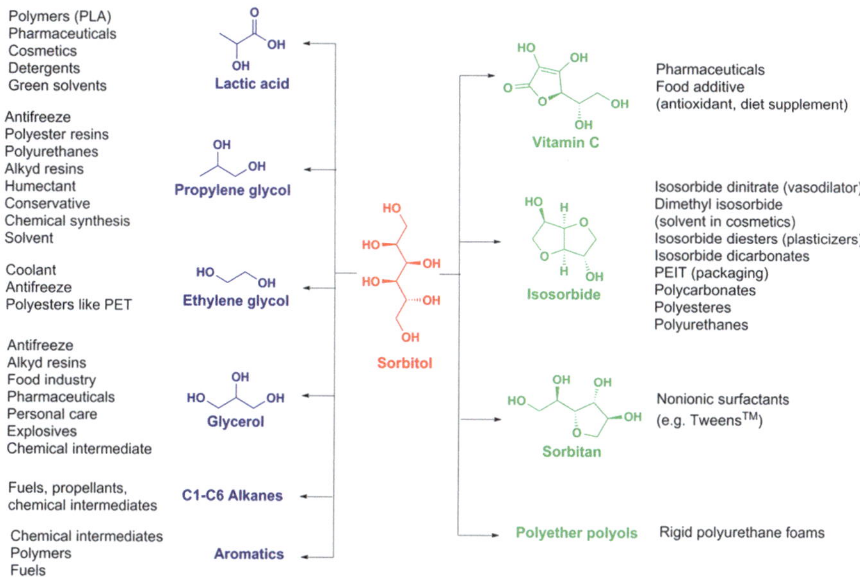


**Fig. 9.1** Sorbitol platform. *Green*: currently manufactured. *Blue*: potentially obtainable in the future on an industrial scale

## 9.2.2 *Sorbitol as a Platform Chemical*

Chemicals that are currently manufactured or could be manufactured from sorbitol, and their applications, are depicted in Fig. 9.1.

Interest in using sorbitol as a platform chemical was shown as early as 1928 by IG Farbenindustrie AG [6] which developed a process for dehydrating sorbitol. Although the chemicals obtained were not identified it is obvious in the light of current knowledge that isosorbide was one of them. A few of the important chemicals currently being produced from sorbitol are:

- Vitamin C has been produced until recently through the Reichstein-Grüssner process, wherein D-sorbitol is converted to L-ascorbic using a first fermentation step resulting in sorbose followed by several chemical steps. Industrial production involves the use of two first fermentation steps leading to the key intermediate 2-keto-L-gulonic acid that is finally chemically converted into vitamin C [7].
- Isosorbide is obtained by a double dehydration of sorbitol via sorbitan using acidic catalysts. Isosorbide is also a chemical platform with increasing industrial interest as shown by the new manufacturing facility put into full-scale operation by Roquette Group in Lestrem (France) with a 20,000 t.year$^{-1}$ production capacity [8]. Isosorbide applications encompassing the synthesis of isosorbide dinitrate, used in pharmaceuticals as a vasodilator, dimethyl isosorbide used today as a solvent in cosmetics, isosorbide diesters (plasticizers) and dicarbonates

(polymer synthesis), and polymers such as PEIT (poly(ethylene terephthalate-co-isosorbide terephthalate)) for packaging, other polyesters for inks, toners, powder coatings, packaging and durable goods, polycarbonates, in substitution of bisphenol A, for durable goods and optical media, epoxy resins for paints and polyurethanes for foams and coatings [9–12].

- Sorbitan, as a mixture of 1,4-anhydrosorbitol, 1,5-anhydrosorbitol and 1,4,3,6-dianhydrosorbitol with the first being the predominant product, is manufactured by single dehydration of sorbitol but at a lower temperature than that leading to isosorbide. It is used as raw material for manufacturing non-ionic surfactants: sorbitan esters (marketed with the trade mark of Spans) and ethoxylated sorbitan esters (marketed with the trade name of Tweens), both useful as solubilizers and emulsifiers in food, pharmaceuticals and cosmetics.
- Sorbitol-based polyether polyols, which are considered to be the universal polyols for rigid polyurethane foams, useful in applications such as thermoinsulation, wood imitations, packaging, flotation materials and so on [13].

Other chemicals could be produced in the future according to increasing literature related to the production of industrially valuable sorbitol derivatives:

- Glycols, such as ethylene glycol, glycerol and propylene glycol. Hydrogenolysis of sorbitol resulting in ethylene glycol and propylene glycols was described as early as 1940 [14]. This reaction proceeds better in alkaline medium [15] probably because C-C cleavage by retro-aldols reactions is a key step to lower glycols [16]. Catalysts based in many metals, such as Ni, Cu, Ru, Pt and Pd, have been used, with Ru showing the highest catalytic activity [17]. Reported reactions are carried out at 423–523 K and 1.4–8 MPa [18] leading to sorbitol conversions of 70–100% with total $C_2$-$C_3$ glycols selectivities of 60–80%. In fact, Global BioChem Technology Group produces industrially 1,2-propylene glycol, ethylene glycol, 1,2-propanediol and 2,3-butanediol from sorbitol since 2008 in a 200 kt.year$^{-1}$ facility at Changchun (China) [19].
- Lactic acid is used for manufacturing green solvents, such as ethyl lactate, and in the synthesis of poly(lactic acid) (PLA), which is increasingly used for producing biodegradable packaging. Likewise, it is also a chemical platform from which other valuable chemicals such us acrylic acid, acetaldehyde, 2,3-pentanedione, pyruvic acid and 1,2-propanediol can be obtained [20]. Racemic lactic acid (non-useful for PLA synthesis but useful for the other above mentioned uses) can be obtained by alkaline hydrothermal conversion of sorbitol in 39.5% yield at 553 K using a NaOH/sorbitol molar ratio of 2 [21].
- Alkanes, ranging from $C_1$ to $C_6$, can be produced by aqueous-phase dehydration/hydrogenation involving a bi-functional pathway in which sorbitol is repeatedly dehydrated by a solid acid ($SiO_2$–$Al_2O_3$) or a mineral acid (HCl) catalyst and then hydrogenated on a metal catalyst (Pt or Pd) [22]. A yield of hexane as high as 56% at 50 g.L$^{-1}$ of aqueous sorbitol, 523 K and 4 MPa of $H_2$ using a Pt/$NbOPO_4$ catalyst has been reported [23].
- Aromatics, together with cyclic-hydrocarbons are the key components of jet fuel. Conventional technologies need further aromatization and isomerization for

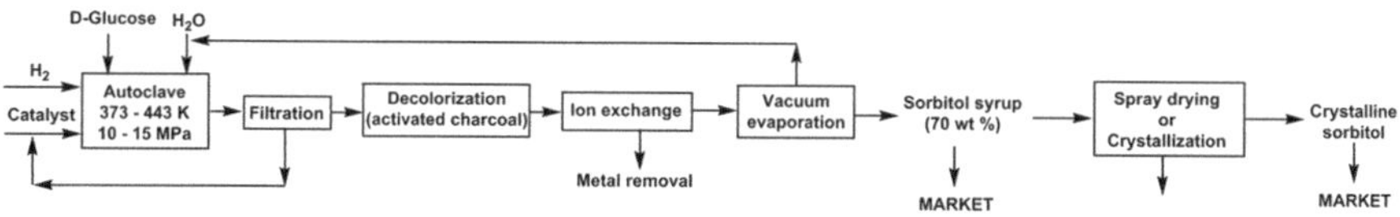


**Fig. 9.2** Schematic flow diagram for sorbitol production by D-glucose hydrogenation

producing these chemicals. However, a 40.4 wt% yield oil containing 80.0% of aromatics (69.7%: Toluene, xylenes, alkylbenzenes, indenes and naphtalenes) and cyclic-hydrocarbons (10.30%: Methyl-cyclopentane, cyclohexane, methyl-cyclohexane, ethyl-cyclohexane) is produced via aqueous catalytic hydrodeoxygenation of sorbitol over a Ni-HZSM-5/SBA-15 catalyst in a fixed-bed reactor at 593 K, WHSV of 0.75 $h^{-1}$, GHSV of 2500 $h^{-1}$ and 4.0 MPa of hydrogen pressure [24].

## 9.3 Sorbitol Production from Biomass

### *9.3.1 Chemical Production of Sorbitol*

Sorbitol production by hydrogenation of aqueous solutions of D-glucose using metal-based reducing catalysts such as Ni is a well-established process [25, 26]. A schematic diagram of a typical process is depicted in Fig. 9.2. The D-glucose, resulting e.g. from starch hydrolysis, for hydrogenation must be of the highest purity to prevent the catalyst becoming poisoned. Ion exchange, carbon treatment and/or crystallization are techniques normally used to achieve the necessary purity.

In a typical process [27] the sugar syrup containing 30–60 wt% D-glucose is reacted in a batch slurry reactor at pH 7–9 with hydrogen at high temperature (typically, 373–443 K) and high pressure (typically, 3–15 MPa) in the presence of a suitable hydrogenation catalyst, with Raney-type nickel catalysts being the most used until recently, in an amount of 2.5 to 12 wt% relative to the D-glucose solution. For batch production, reaction times are in the order of 1–3 h depending on reaction conditions. When using a Raney-nickel catalyst suitable reaction pH range is 8.0–9.0, but a pH of about 7.5 is used in industrial production for preventing D-glucose isomerization into mannose leading to mannitol by hydrogenation. The resulting crude sorbitol aqueous solution is treated with activated charcoal for decolorization, subjected to an ion exchange treatment to remove dissolved metal catalyst, and then vacuum evaporated to obtain a 70 wt% sorbitol syrup, which may be spray dried or further concentrated to be crystallized to obtain crystalline sorbitol. Handling and storage after hydrogenation must be carried out according to good manufacturing practice to ensure microbiological problems are avoided. Continuous processes have been tested since the 1980s [28]. A D-glucose conversion of 100% and a selectivity to sorbitol higher than 99% are typically obtained.

**Catalysts** Nickel-based catalysts are still those mostly industrially used due to their low costs in comparison with other suitable metal-based catalysts later herein discussed, such as ruthenium-based catalysts. However, soon after its industrial use its drawbacks became apparent: deactivation by sintering, leaching into the reaction mixture leading to enhanced purification costs, and poisoning. The industry required more stable and active catalysts. Strategies for achieving such a goal have been: (a) Addition of promoters to Raney nickel catalysts; (b) Use of supports for nickel catalysts; and (c) New catalysts, with Ru-based ones being largely preferred.

**(a) Addition of Promoters to Raney Nickel Catalysts** It has been found that the addition of some metal promoters enhances notably the activity and stability of Raney Nickel catalysts. Thus, Court et al. [29] prepared catalysts from $Ni_{2-x}M_xAl_3$, (where M = Cr, Fe, Co, Cu, Mo and $x \leq 0.4$) and found that the promoter metal, except cobalt, favors aluminum retention in the catalyst and ascribed at least partially the increase of the nickel activity observed to this high residual aluminum content. The activity of Cr or Mo promoted catalysts was about twice that of the unpromoted Raney nickel. Cobalt and copper present in small amounts did not change the properties of the Ni-Al system.

Gallezot et al. [30] used Mo, Cr, Fe and Sn as promoters for Raney nickel catalysts. No data about conversions and selectivities were reported because the study was aimed at studying the activity of catalysts by measuring initial conversion rates. The activities of Mo- and Cr-promoted catalysts decrease only very slightly with recycling due to surface poisoning by cracking products formed in side reactions. Fe- and Sn-promoted catalysts deactivated very rapidly due to leaching away from the surface, but while iron was leached to the liquid phase tin remained in the catalyst micropores. They suggested an acid-base catalyst concluding that promoters in a low-valent state on the nickel surface act as Lewis adsorption sites for the oxygen atom of the carbonyl group, which is then polarized favoring its hydrogenation via a nucleophilic attack on the carbon atom by hydride ions. The same conclusion was reached by Li et al. [31] studying the activity of a Ni–B/$SiO_2$ amorphous catalyst, without and with metal promoters, in the liquid phase hydrogenation of a 50 wt% aqueous solution of D-glucose at 373 K, 4.0 MPa and 6 h in a stainless steel autoclave containing 1 wt% of catalyst vs. D-glucose. The addition of W, Mo and Cr increased conversion from 30% to 35%, 42% and 49%, respectively, with the optimum contents of W, Mo and Cr being 10, 5 and 10 wt%, respectively. Above said concentrations a decrease in the activity was observed since too many Ni active sites were covered by these oxides. Selectivities were closed to 100%.

Hoffer et al. [32] studied the influence on Raney-type Ni catalysts of Mo and Cr/Fe as promoters in the hydrogenation of an aqueous solution of D-glucose (10 wt%) using a three-phase slurry reactor at 4.0 MPa of $H_2$ and 393 K. Selectivity to sorbitol was >99%. The promoter concentrations (wt%) were different at bulk (1.4% Mo, 2.4%/3.6% Cr/Fe) and surface (8.7% Mo, 3.8%/5.9% Cr/Fe) catalyst, being essentially as oxides. It was shown that promoters enhanced and stabilized the BET surface area leading to an increase of the catalysts activity. The BET surface areas increased from 56 $m^2\ g_{cat}^{-1}$ in the Raney nickel catalyst to 77 and 112 $m^2\ g_{cat}^{-1}$ in the

Mo and Cr/Fe promoted catalysts, respectively. The activity of catalysts increased accordingly, from 0.35 $kg^{-1}$ $s^{-1}$ to 0.50 and 0.90 $kg^{-1}.s^{-1}$, respectively. However, even when the performance of catalyst is normalized to its surface area, the promoted systems still exhibit an enhanced activity [33]. Consequently, the enhancement of the reaction rate by the promoters was ascribed to the promoters being more electropositive than Ni and acting as adsorption sites for D-glucose, which generates an ionized species that is susceptible to attack by hydrogen. Catalyst stability was also enhanced by promoters as shown by a lower loss in activity (30% and 16% for Mo- and Cr/Fe-promoted catalysts, respectively) than that in unpromoted catalyst (48%) after 3 recycles. However, promoters were unable to keep long term stability as shown by the continuous loss in activity, which after 5 recycles were 45.6% and 40.1% for Mo- and Cr/Fe-promoted catalysts, respectively. Mo was not leached while Fe was severely leached (27%) while Cr was slightly leached (1.7%), although according to authors loss in activity was not mainly due to Fe loss but to poisoning of the active sites by D-gluconic acid, which was formed via a Cannizzaro-type reaction of D-glucose induced by an alkaline environment in the presence of nickel.

**(b) Use of Supports for Nickel Catalysts** Use of supports both to increase metal dispersion and to have large exposed surface area has revealed to be a good strategy for improving Ni activity and stability. Thus, a number of carriers such as C, $SiO_2$, $TiO_2$, $Al_2O_3$, $ZrO_2$, and mixtures thereof, have been tested. Kusserow et al. [34] studied both catalyst preparation method (precipitation, impregnation, sol-gel and template syntheses) and support nature ($SiO_2$, $TiO_2$, $Al_2O_3$ and carbon) in the batch hydrogenation of a 40–50 wt% aqueous D-glucose solution to sorbitol at 393 K and 12 MPa $H_2$, for 5 h using a catalyst load of 1 wt% vs. D-glucose, comparing the results with a Ni commercial catalyst (Ni(66.8 wt%)/$SiO_2$, KataLeuna GmbH Catalysts, Leuna, Germany). Regarding the specific activities (μg of sorbitol.$g^{-1}$ nickel.$s^{-1}$), they found that the activity follows the sequence $Al_2O_3$ (100–210) > $TiO_2$ (80–170) > $SiO_2$ (5–50) ≈ commercial catalyst > C (0–15). Generally, catalysts prepared by impregnation had a higher activity than those prepared by incipient wetness. On the other hand, the selectivity (expressed as the amount of by-products) followed the sequence $SiO_2$ (2.1–4.1 wt%) > C (3.4–6.3 wt%) > $Al_2O_3$ (5.5–7.5 wt%) > $TiO_2$ (~8 wt%). The same research group [35] demonstrated the importance of particle size for preventing Ni leaching. They prepared Ni catalysts with different metal loadings (5, 10, and 20 wt%) by impregnation with nickel ethylenediamine complexes, which had small nickel particles (mean diameter: 2–3 nm) even for high metal loadings (20 wt%). These catalysts showed almost no nickel leaching compared with the commercial Ni/$SiO_2$ above described in D-glucose hydrogenation. Unfortunately, they were slightly less active (TOF (number of D-glucose molecules converted per second and per surface Ni site): 2 to 10 × $10^{-3}$ $s^{-1}$) than the commercial nickel/silica catalyst (TOF: 14 × $10^{-3}$ $s^{-1}$) and gave lower yields to sorbitol (3–42% compared to 60%). Catalyst activity was strongly dependent on pretreatment conditions performed. Calcination before reduction led to higher conversion (19–45%) and selectivity to sorbitol (81–92%) than direct reduction without calcination pretreatment (conversion: 10–16% and selectivity: 21–59%) after 5 h of reaction time, respectively. The difference was ascribed to complete decomposition of the nickel ethylenediamine precursor being achieved only when calcination pretreatment was performed.

Geyer et al. [36] studied the performance of Ni catalysts supported on $ZrO_2$, $TiO_2$, $ZrO_2/TiO_2$, $ZrO_2/SiO_2$ and $MgO/Al_2O_3/SiO_2$, in comparison with a $Ni/SiO_2$ catalyst, all of them prepared by precipitation. Batch hydrogenation tests were carried out with a 50 wt% D-glucose aqueous solution for 4 h at 393 K and 12 MPa $H_2$ using a catalyst concentration of 1.5 wt%. Ni contents in catalysts ranged between 39.9 and 51.2 wt%. No conversion data were reported. However, sorbitol yields were as follows: $Ni/ZrO_2/SiO_2$ (97.8%) > $Ni/TiO_2$ (97.5%) > $Ni/ZrO_2/TiO_2$ (96.7%) > $Ni/ZrO_2$ (93.2%) > $Ni/SiO_2$ (92.4%) > $Ni/MgO/Al_2O_3/SiO_2$ (83.9%), while TOFs ($10^{-3}$ $s^{-1}$): $Ni/TiO_2$ (44.8) > $Ni/ZrO_2/TiO_2$ (35.3%) > $Ni/ZrO_2$ (28.2) > $Ni/ZrO_2/SiO_2$ (26.5) > $Ni/SiO_2$ (18.2) > $Ni/MgO/Al_2O_3/SiO_2$ (14.3). Assuming that yields between 96.7% and 97.8% were equal due to the analytical uncertainty it can be concluded that only the $TiO_2$ containing catalysts had a better catalytic performance than the commercial $Ni/SiO_2$ catalyst. Surprisingly, no correlation between the metal dispersion and specific hydrogenation activity was found which was ascribed to the probable hydrogenation activity of the catalysts being determined not only by dispersion but also by a metal support interaction. The stability of the metal dispersion under reaction conditions was investigated with selected catalysts by a treatment in the reaction mixture over a period of 100 h. $Ni/ZrO_2$ catalyst had a higher stability of the Ni dispersion than $Ni/SiO_2$ due to a reduced leaching of the $ZrO_2$ support.

High activity and stability for the continuous hydrogenation of 10 wt% D-glucose to sorbitol in aqueous solution in a stainless steel fixed-bed reactor were reported by Li et al. [37] using 60%Ni/AlSiO catalysts prepared by the co-precipitation method, in which AlSiO were the composite supports with different $Al_2O_3/SiO_2$ mass ratios. They found that the catalyst with an $Al_2O_3/SiO_2$ mass ratio of 4 (Ni/AlSiO-4) in the support hydrothermically treated at 423 K exhibited the high hydrothermal stability when the supported nickel particles were small (about 5.2 nm) and highly dispersed, confirming the importance of particle size in catalyst stability. D-glucose conversions and sorbitol selectivities were 100% between 373 and 413 K at 4 MPa of $H_2$ pressure and a WHSV of 1 $h^{-1}$. In comparison, a $Ni/SiO_2$ catalyst (particle size 15–30 nm) deactivated quickly due to the fast aggregation of supported Ni particles. Increase in activity was attributed to the higher $H_2$ uptake of smaller Ni particles.

**(c) New Catalysts** Due to their higher activity and better selectivity in various hydrogenation processes, the Ni-metalloid amorphous alloy catalysts have been studied as potential substitutes for Raney Ni catalysts. Li et al. [38] prepared a novel skeletal Ni-P amorphous alloy catalyst (Raney Ni-P, containing 68 wt% Ni, 25 wt% Al and 7 wt% P) by alkali leaching a Ni-P-Al amorphous alloy obtained by the rapid quenching technique of a melting solution (1673 K) containing 48.2 wt% Ni metal, 48.7 wt% Al metal, and 3.1 wt% red P. The performance of this catalyst compared with a Raney Ni one in the liquid phase D-glucose hydrogenation was tested under the following experimental conditions: catalyst loading, 3 wt% vs. D-glucose;

D-glucose concentration, 50 wt%; 393 K; $P_{H2}$ 4.0 MPa; stirring rate, 1200 rpm; reaction time, 6.0 h. While selectivity to sorbitol was 99.5% for both catalysts, conversions and TOFs were 55.8% and 0.40 $s^{-1}$ for Raney Ni-P catalyst, and 17.2% and 0.11 $s^{-1}$ for the Raney Ni one. Authors concluded that this huge increase in performance was apparently the result of promotion of Ni-active sites by phosphorus leading to a significantly higher density of Ni surface atoms. The amount of Ni dissolved in the reaction mixture was less than 1.0 ppm, $< 7.3 \times 10^{-3}$ wt% of the initial Ni content in the catalyst. The catalyst was used repetitively until an abrupt decrease in the activity was observed. No significant decrease in activity was observed in the first 5 cycles of the hydrogenation. It was also found that the deactivated catalyst could be easily regenerated by treating it again in a 6.0 M NaOH solution, suggesting the idea that the Ni leaching was not the main cause of deactivation. Unfortunately, no data about Ni leaching with the number of reaction cycles were given.

As mentioned above in **(a)**, Li et al. [31] reported the activity of a Ni–B/$SiO_2$ amorphous catalyst (6.4 wt% Ni, no data about B content) in the liquid phase hydrogenation of a 50 wt% aqueous solution of D-glucose at 373 K, 4.0 MPa of $H_2$ and 6 h in a stainless steel autoclave containing 1 wt% of catalyst vs. D-glucose. The amorphous catalyst exhibited much higher activity than other Ni-based catalysts, such as the corresponding crystallized Ni–B/$SiO_2$ catalyst, the Ni/$SiO_2$ catalyst and the commercial Raney Ni one. Thus, TOF was two times higher for amorphous Ni–B/$SiO_2$ (0.024 $s^{-1}$) than for Raney Ni (0.0135 $s^{-1}$). While selectivities were near to 100%, conversions were strongly dependent on catalyst characteristics and reaction conditions, ranging from 30% to 82%. Catalyst activity was kept almost constant for 5 cycles and only less than 1.0 ppm Ni in the product mixture was detected by ICP, 0.027% of the initial Ni content in the catalyst, showing the increase in stability of the Ni–B/$SiO_2$ amorphous catalyst. Besides the high dispersion of the Ni–B/$SiO_2$ amorphous catalyst, its high activity was mainly attributed to its favorable structural characteristics, such as the high concentration of coordinately unsaturated sites and the strong union between Ni active sites, and the electronic interaction between Ni and B making Ni electron-rich. The increased electronic density on Ni weakens the adsorption of D-glucose on catalyst surface, via the donation to Ni of one electron pair from the oxygen of the carbonyl group, thereby favoring the competitive adsorption of hydrogen against the C = O group. Therefore, more hydrogen could be adsorbed on the Ni–B/$SiO_2$ amorphous catalyst which in turn enhanced its hydrogenation activity.

However, substitution of Ni-based catalysts by other typical hydrogenation supported noble metal-based catalysts seems to be the more promising way for further process development from an industrial standpoint. Catalysts based on platinum, palladium, rhodium, and ruthenium have been tested [34, 39]. Taking into account that, on one hand, the order of activity for D-glucose hydrogenation found was Ru > Ni > Rd > Pd, with specific activities of Ru catalysts being about 50 times higher than those of Ni catalysts; and, on the other hand, that ruthenium is not leached under the usual reaction conditions for D-glucose hydrogenation [39], Ru based catalysts have been by far the more studied with the aim of designing more active Ru-based catalysts with the lowest possible cost to compensate the high cost of Ru.

The superior performance of supported Ru-based catalysts in comparison with other Pt-group metals for hydrogenating carbonyl groups into the corresponding alcohols in water solutions has been discussed by Michel and Gallezot [40]. They identified two mechanisms, both involving water. The first mechanism is related to the interactions, via hydrogen bonds, between the C=O group adsorbed on the Ru surface and adjacent adsorbed water molecules, lowering the energy barriers leading to an easier hydrogenation of carbonyl groups by dissociated hydrogen. The second mechanism is that the dissociation of adsorbed water on the ruthenium surface increases the surface concentration in hydrogen atoms, thus favoring the hydrogenation reaction. They concluded that more studies are needed for determining the respective importance of these two mechanisms and, mainly, to confirm the role of water as a source of hydrogen.

Hoffer et al. [33] studied carbon supported Ru catalysts as better alternatives for Raney-type Ni catalysts. They found that the Ru/C catalysts had higher activities, Ru does not leach, and the activity is proportional to the Ru surface area and independent of the preparation method. A selectivity higher than 98% (no conversions given) was obtained in a three-phase slurry reactor at 393 K and 4.0 MPa hydrogen pressure using a 10 wt% D-glucose solution. All Ru catalysts were at least two times more active than the Ni catalysts per kg of catalyst. The strong stability of Ru-based catalysts in comparison with Ni-based ones was unequivocally demonstrated by Kusserow et al. [34] by operating a continuous process for 1150 h time on stream using a Ru(0.47 wt%)/$Al_2O_3$ catalyst, as later discussed in the next section. No Ru losses were detected.

Guo et al. [41] prepared a Ru-B amorphous alloy catalyst in the form of ultrafine particles. The Ru-B catalysts were more active than that of Co-B, Ni-B amorphous catalysts as well as Raney Ni catalysts for the D-glucose hydrogenation. The Ru-B amorphous catalyst exhibited higher activity than its corresponding crystallized Ru-B and pure Ru powder catalysts, showing the promoting effects of both the amorphous structure and the electronic interaction between the metallic Ru and the alloying B. A 50 wt% aqueous solution of D-glucose was converted in 2 h into sorbitol in 95.1% conversion and 100% selectivity to sorbitol using an amorphous $Ru_{88.9}B_{11.1}$ (amounts by weight) catalyst (1 wt% vs. D-glucose) at 353 K and 4.0 MPa of $H_2$ pressure. Under same conditions, Ni-B, Co-B, crystallized Ru-B, pure Ru, and a commercial Raney Ni (3 wt% vs. D-glucose) catalysts gave conversions of 58.8%, 47.8%, 23.1%, 22.2% and 16.7%, respectively.

Catalyst development until 2013 has been reviewed by Zhang et al. [42], including Ru catalysts. Studies have been devoted to determine the influence on Ru-based catalyst activity of preparation methods, supports ($SiO_2$, γ-$Al_2O_3$, MCM-41) and Ru precursors (Ru acetate and Ru trichloride). After analyzing the bibliography Zhang et al. concluded there are three key points that can serve for further development: (a) pore and BET surface area (related to catalyst dispersion – size): catalyst performance increases as both increase; (b) preparation methods, having an important influence of catalyst performance; and (c) catalyst life: a critical issue due to the high cost of Ru, so that studies should be focused on improving catalyst life through developing effective Ru carriers and cheap catalyst regeneration methods.

**Table 9.1** Hydrogenation of D-glucose to sorbitol over ZSM-5, Ru/C and Ru/ZSM-5 catalysts [43]

| Catalysts[a] | SBET ($m^2.g^{-1}$) | D[b] (nm) | Acid ($mmol.g^{-1}$) | Dispersion | Conversion (%) | Sorbitol yield (%) | TOF ($h^{-1}$) |
|---|---|---|---|---|---|---|---|
| Ru(5 wt%)/C | – | – | – | 0.23 | 81.3 | 61.7 | 13 |
| ZSM-5 | 358 | 0.55 | 3.6 | – | 30.2 | 1.2 | 2 |
| Ru/ ZSM-5-TF | 383 | 0.52 | 3.1 | 0.54 | 99.6 | 99.2 | 32 |
| Ru/ ZSM-5-MS | 339 | 0.26 | 2.1 | 0.46 | 98.4 | 95.8 | 15 |
| Ru/ ZSM-5-AT | 345 | 0.66 | 3.4 | 0.48 | 98.8 | 97.4 | 18 |

Hydrogenation conditions: 25 wt% D-glucose solution, 2 h, 4.0 MPa $H_2$, 393 K, catalyst concentration: 4 wt% vs. D-glucose

[a]$SiO_2/Al_2O_3$ ratio: 38, except for Ru/ZSM-5-AT: 30

[b]Average pore diameter

Thus, Guo et al. [43] studied the influence of preparation method and the support nature in the hydrogenation of D-glucose to D-sorbitol over Ru/ZSM-5 catalysts. An incipient wetness impregnation method and a one-step template-free process (Ru/ZSM-5-AT catalyst) were used. For the conventional impregnation method both mesoporous ZSM-5 supports created by desilication in alkaline medium (Ru/ ZSM-5-AT catalyst) and commercial microporous (Ru/ZSM-5-MS catalyst) were tested. The hydrogenation of a 25 wt% D-glucose aqueous solution was performed for 2 h at 4.0 MPa of $H_2$ and 393 K using a catalyst concentration of 4 wt% relative to D-glucose, and compared with those obtained using the HZSM-5 zeolite and a commercial Ru(5 wt%)/C as catalysts. Catalyst characteristics and hydrogenation results are given in Table 9.1.

As shown in Table 9.1, Ru catalyst prepared by the one-step template-free process was highly dispersed in the ZSM-5 framework and exhibited higher catalytic activity than others, with conversions and yields exceeding those obtained with catalysts prepared by the conventional impregnation method with microporous or desilicated ZSM-5 supports, and also with the support alone and the commercial Ru/C catalyst. The catalytic activity increased with increasing Ru loading up to 4.1 wt% but, more importantly, the formation of undesirable by-products, i.e. D-fructose and D-mannitol, was significantly minimized with increased metal loading, as shown by the increase in sorbitol selectivity from 15.9% when the content of Ru was 1.2 wt%, to 99.6 when it was 4.1 wt%. From an industrial standpoint two features shown by Ru/ZSM-5-TF are very significant. On one hand, it leads to a selectivity of 99.6%, meaning a simpler purification procedure, while that obtained with Ru/ZSM-5-MS and Ru/ZSM-5-AT catalysts were 97.4% and 98.6% respectively. On the other hand, it showed high stability against leaching and poisoning and could be reused several times, meaning a longer life span than that of the other catalysts. The catalyst regeneration proved to be key for maintenance of catalyst performance. When Ru/ZSM-5-TF was only dried at 393 K in air, and reused without further treatment, the sorbitol yield decreased from 99.2 to 89.2% after five runs. However, when the spent catalyst was washed with water and subsequently washed with ethanol or acetone three times each, there was no obvious loss of the catalytic reactivity. Consequently, the deactivation was probably due to the accumulation of organic and inorganic species adsorbed on the catalyst surface. These results clearly showed that Ru/ZSM-5-TF was effective for the hydrogenation of D-glucose and had good stability during the reaction process. Their excellent catalytic behavior and stability were ascribed to the extensive dispersion of the Ru particles, the strong interaction between the Ru species and the ZSM-5 support, as well as its suitable surface acidity-basicity balance. That is to say, the catalytic performance is strongly dependent on the catalyst preparation method and its method of maintenance.

Mishra et al. [44] studied the performance of HY zeolite supported ruthenium nanoparticles catalysts using a HY zeolite with Si/Al ratio = 80 (referred to as HYZ). Catalysts were prepared by using conventional impregnation–reduction method using $NaBH_4$ in ethanol as reducing agent. The activity tests were carried out with a 20 wt% aqueous solution of D-glucose for 20 min at 393 K and 5.5 MPa $H_2$ using a Ru(1 wt%)/HYZ catalyst in a concentration of 2.5 wt% relative to D-glucose and comparing results with Ru(1 wt%)/NiO-$TiO_2$ and Ru(1 wt%)/$TiO_2$ catalysts previously studied by the same research group [45] and a Ru(5 wt%)/C commercial catalyst. Metal dispersions were 23.6, 8.6, 4.4 and 6.7 for Ru(1 wt%)/HYZ, Ru(1 wt%)/NiO-$TiO_2$, Ru(1 wt%)/$TiO_2$ and Ru(5 wt%)/C, respectively. Hydrogenation tests were carried out with a 20 wt% aqueous solution of D-glucose for 20 min at 393 K and 5.5 MPa $H_2$ with a catalyst level of 2.5 wt% relative to D-glucose. Ru(1 wt%)/HYZ showed a higher conversion (19.4%), selectivity (97.6%) and TOF (1275 $h^{-1}$) than the other catalysts. TOF was 1.08-, 1.19- and 5.4-fold higher that with Ru(1 wt%)/NiO-$TiO_2$, Ru(1 wt%)/$TiO_2$ and Ru(5 wt%)/C catalysts, respectively. After further optimization a 100% conversion and a 98.7% selectivity were achieved under the same experimental conditions by increasing reaction time up to 2 h. It was concluded that the acidity (mild acidity) of zeolite support plays an important role in increasing both the selectivity and the activity to sorbitol.

Aho et al. [46] have studied the influence of metal dispersion in the semi-batch hydrogenation of a 18 g.$L^{-1}$ D-glucose aqueous solution at 393 K and 1.9 MPa hydrogen pressure over several Ru/C catalysts. Those with ruthenium particle sizes between 1.2 and 10 nm were investigated. All were active with selectivity to sorbitol being 87–96%, except for the 10 nm catalyst (28.8%). TOF was maximum for ruthenium nanoparticles of 2.9 nm.

Lazaridis et al. [47] reported the hydrogenation/hydrogenolysis of D-glucose over Pt and Ru catalysts supported on activated micro/mesoporous carbon (AC), at 1.6 MPa and 453 K and low D-glucose concentration (2.7 wt%). The effects of metal content (1–5 wt%), method of metal pre-treatment/reduction ($H_2$ at 623 K or $NaBH_4$) and reaction time (1–12 h) were studied. All the Ru and Pt/AC catalysts were very active (conversion ≥97%) with the Pt/AC catalysts being also very selective toward sorbitol (selectivity ≥90%) irrespective of metal content, method of reduction and reaction time. However, the maximum selectivity was 95%, a low value in comparison with that industrially desired due to the high temperature used causing hydrogenolysis resulting in various lower sugar alcohols, such as 1,2,5,6-hexanetetrol, arabinitol, threitol, glycerol and 1,2-propanediol. Conversely, the Ru/AC catalysts exhibited a wide range of sorbitol selectivity values (55–93%) depending on metal loading, method of reduction and reaction time. Thus, the 93% selectivity was only achieved with a metal content of 3 wt% and a reaction time of 1 h using $NaBH_4$ as a metal reducing agent. Under the same experimental conditions, the sorbitol selectivity increased when the reaction time decreased from 12 h to 1 h and when Ru loading increased from 1 wt% to 3–5 wt%. Moreover, the sorbitol selectivity was higher for the hydrogen reduced catalysts than for those reduced with $NaBH_4$. The higher selectivity of the Pt/AC catalysts toward sorbitol could be related to the abundance of well formed, single crystal Pt nanoparticles (2–6 nm) compared to the less crystalline Ru/AC catalysts. Thus, using HRTEM measurements the co-existence of small crystalline metallic Ru nanoparticles together with amorphous $Ru(O)_x^{\delta+}$ species within relatively larger aggregates (20–30 nm) was identified. Based on this Ru morphology, the authors proposed a model of formation of protonic acid sites $Ru(O)_xH^+$, as well as of protons released in the aqueous solution from the $Ru(O)_xH^+$ species, in the presence of hydrogen. Thus, $Ru(O)_xH^+$ species are formed by spillover of the formed hydrogen atoms by the dissociative adsorption of molecular $H_2$ on the metallic Ru(0) nanoparticles to adjacent $Ru(O)_x^{\delta+}$ species, with simultaneous electron transfer from the H atoms to the positively charged Ru-O species. The authors proposed that hydrogenation of glucose to sorbitol occurred by the dissociative adsorption of molecular $H_2$ on the metallic Ru(0) nanoparticles and the subsequent addition of the pair of hydrogen atoms to the hemiacetal group of glucose, leading to the cleavage of the C-O bond and the formation of the hydroxyl group, as it generally accepted [18, 42]. However, the protonic $Ru(O)_xH^+$ acidic sites as well as the released $H^+$ in solution induced the conversion of the sorbitol to 1,2,5,6-hexanetetrol via dehydration and hydrogenation reactions. Likewise, this acidic environment induced hydrogenolysis of the C3-C4 bond of sorbitol resulting in glycerol, which in turn can be then converted to 1,2-propanediol via dehydration and hydrogenation, while this molecule can be also formed via hydrogenolysis of the C3-C4 bond of 1,2,5,6-hexanetetrol. According to this mechanism, the selectivity to sorbitol of the Ru/AC catalysts could be strongly improved by means of a catalyst preparation method preventing the formation of $Ru(O)_x^{\delta+}$ species.

The good performance and industrial possibilities of Ru-B amorphous alloys as catalysts for sorbitol synthesis from D-glucose previously shown by Guo et al. [41] and discussed in this section have been recently confirmed by Wang et al. [48] as well as the importance of a suitable carrier. They synthesized a highly dispersed Ru–B amorphous alloy catalyst by loading Ru–B amorphous alloy nanoparticles (NPs) on a matrix consisting of a highly ordered mesoporous silica nanospheres (MSNS) externally covered by methyl groups but internally grafted by aminopropyl groups. The amino and methyl groups acted synergistically as effective functionalities for highly dispersing Ru–B NPs within the pore channels of the mesoporous host. Such catalyst was used in the liquid-phase D-glucose hydrogenation to

D-sorbitol under the following conditions: 50 wt% D-glucose aqueous solution, 3.0 MPa $H_2$, 373 K, 3 wt% catalyst relative to D-glucose, reaction time of 80 min. The amorphous $Ru_{89}$–$B_{11}$/$NH_2$&$CH_3$-MSNS catalyst with a Ru load of 4.4 wt% provided a 100% D-glucose conversion with 100% selectivity to sorbitol while conversions were 4.5%, 23% and 22% for a commercial Raney Ni catalyst, a commercial Ru(5 wt%)/C catalyst and the crystallized $Ru_{89}$–$B_{11}$/$NH_2$&$CH_3$-MSNS catalyst, respectively. The amorphous catalyst kept its activity for 9 cycles, showing good potential for industrial use.

These researchers also compared the activity of this catalyst with those of amorphous Ru-B alloys deposited on pure MSNSs and mono-functionalized MSNSs ($NH_2$-MSNS and $CH_3$-MSNS). Activities decreased in the order Ru–B/$NH_2$&$CH_3$-MSNS > Ru–B/$NH_2$-MSNS > Ru–B/$CH_3$-MSNS > Ru–B/MSNS, i.e. in the order in the same order than the particles dispersion as shown by the surface areas. Moreover, TEM images showed as Ru-B NPs were located differently depending on the matrix type. Thus, in the Ru–B/MSNS catalyst Ru–B nanowires within the channels of MSNSs were observed probably due to the agglomeration of the Ru–B NPs located inside the pore channels together to form nanowires during the preparation process. Furthermore, a portion of Ru–B NPs was situated on the external surface of MSNS because some particles were larger than the pore size of MSNS. For Ru–B/$NH_2$-MSNS, nanowires were absent, which was attributed to the coordination effect of amino groups in the interior of pores on $Ru^{3+}$ ions. Reduction of these coordinated $Ru^{3+}$ ions with $BH4^-$ led to the location of Ru–B NPs inside the mesoporous channels. However, some large NPs were still observed on the external surface. Large NPs located on the external surface were absent for Ru–B/$CH_3$-MSNS due to the high hydrophobic external surface, but Ru–B nanowires appeared between the walls of MSNSs, which was ascribed to the absence of stabilizing groups inside the pore channels. Moreover, Ru–B NPs-aggregates were observed in Ru–B/$CH_3$-MSNS due to the fall of Ru–B NPs from the pore channels of support. For Ru–B/$NH_2$&$CH_3$-MSNS no aggregated Ru–B NPs were observed, which was explained as due to both the stabilizing effect of amino groups on $Ru^{3+}$ ions and the high hydrophobic external surface deriving from methyl groups. Consequently, the superior activity of the amorphous $Ru_{89}$–$B_{11}$/$NH_2$&$CH_3$-MSNS catalyst was attributed to this difference in Ru-B location. The location of non-aggregated Ru–B NPs inside the pore channels together with the increased D-glucose concentration in the channels owing to the microreactor effect [49, 50] would lead to an increase in the collision frequency between reactants and Ru active sites and, therefore, to an increase in catalyst activity. As pointed out by authors, the results of their study demonstrate that region-selectively functionalizing the surface of ordered mesoporous materials with diverse groups allows for molecular-level fine-tuning of catalytic performance of the introduced guest materials.

Dabbawala et al. [51] studied the aqueous phase hydrogenation of D-glucose to sorbitol using Ru supported on amine functionalized nanoporous hypercrosslinked polystyrene polymer based catalysts (Ru/AFPS) prepared by simple impregnation-chemical reduction method. The interest of this study relies also in that it represents one of the first studies about utilization of porous polymeric supports to stabilize Ru nanoparticles in D-glucose hydrogenation. Catalyst performance tests were carried out using a 4 wt% D-glucose aqueous solution at 353–383 K and 3.5–6.0 MPa $H_2$ for 60 min, with a catalyst concentration of 5 wt% relative to D-glucose. At 373 K and 5.5 MPa $H_2$, the catalytic activity of Ru(5 wt%)/AFPS catalyst was much higher than those of Ru(5 wt%)/Polystyrene, Ru(5 wt%)/C, Ru(5 wt%)/$TiO_2$, Ru(5 wt%)/$Al_2O_3$ and Ru(5 wt%)/$SiO_2$ as shown by a TOF of 230 $h^{-1}$, about 2.5-fold higher than that of the unfunctionalized support (Ru(5 wt%)/PS) and about 1.3 times higher than the second more active catalyst (Ru(5 wt%)/C). Likewise, the Ru(5 wt%)/AFPS catalyst led to a 67.6% sorbitol yield, the higher of all catalysts studied, with a selectivity to sorbitol of 98%, equal to that obtained with Ru(5 wt%)/PS and higher than those obtained with the remaining catalysts. This higher catalytic performace was attributed to the functional amine groups present in AFPS polymer surface which stabilize Ru nanoparticles and enhance the dispersion of Ru nanoparticles, provide a better wettability thereby improving the mixing with water compared to Ru/PS catalyst. On the other hand, the AFPS catalyst nanoporous structure with its wide range of pores was considered to be of key importance. Thus, the presence of micropores in AFPS catalyst leads to a better control of Ru nanoparticles size, while the presence of mesopores allows substrate molecules to easily approach the Ru active metal center without diffusion limitations. Ru(5 wt%)/AFPS was reused up to five times without significant loss in activity and selectivity. This stability was attributed to the strong interaction between support and active Ru species.

**Operation Mode** It is worth mentioning that although most production methods described in the literature are referred to as batch processes, most of them operate at constant hydrogen pressure such that hydrogen is continuously fed as it is consumed which makes them semi-batch processes. Batch (semi-batch) processes have a good use of the catalyst as well as good temperature control as main advantages which explain their wide industrial use for sorbitol production. However, the catalyst removal for recycling is a great disadvantage leading to increased costs and progressive catalyst deactivation. The importance of the reactor type on catalysts stability in D-glucose hydrogenation has been highlighted by Doluda et al. [52] who have shown that the use of common batch and shaker type reactor systems results in high losses of the initial catalysts, due to catalysts grinding on the reactor impeller and reactor walls and during the catalysts separation and washing. Besides, selectivity is also negatively affected by formation of gluconic acid due to the presence of trace oxygen. On the other hand, the continuous increase in sorbitol demand requires more productive processes. Consequently, the development of continuous methods is imperative due to the higher space-time yields and the absence of an expensive catalyst separation step.

Boyers and Flushing [53] disclosed a method for the catalytic hydrogenation of carbohydrates in the presence of suspended finely divided ruthenium carrier catalysts, method which could be continuously carried out although no example with this operation was given. However, the use of a finely divided catalyst suspension catalyst involved a complicated filtration separation and made the quantitative recovery of the precious catalyst difficult.

Lepper and Schütt [54] described a continuous method for the hydrogenation of carbohydrates in the presence of a catalyst solid bed of Ru(2 wt%)/C in lumps (cylindrical form, diameter 2 mm, length 2 to 5 mm). The process involved the use of two series-connected reactors with a capacity of 4.2 liters and each having an inside diameter of 70 mm. Working a 423 K and a hydrogen pressure of 25.0 MPa, 10,400 kg of a 50 wt% aqueous solution of D-glucose were converted into sorbitol in 1550 h (1.7 kg-D-glucose.$h^{-1}.m^{-2}$) with a conversion of 99% and a selectivity of 99.7%. The catalyst activity was 0.30 mol-D-glucose.$h^{-1}.g_{Ru}{}^{-1}$, corresponding to a turnover frequency (TOF) of 30.3, 3.4 times larger than that obtained using a ruthenium catalyst in suspension.

Selectivity of Lepper and Schütt process is within the desired one but conversion while being as high as 99% was below the target of 100% for decreasing purification costs. Looking for this goal Déchamp et al. [55] studied the continuous D-glucose hydrogenation to sorbitol in a trickle-bed reactor in the presence of kieselguhr-supported nickel catalysts with cocurrent downflow mode in the temperature range 343–403 K and in the pressure range 4–12 MPa, but catalyst performance decreased with time due to continuous leaching of both nickel and support. Similar catalyst deactivation was observed by Tukac [56] with same reactor type in the temperature range 388–438 K at 0.5–10 MPa starting from a 40 wt% aqueous solution of D-glucose and using commercial supported nickel catalysts.

Ruthenium-based catalysts are a clear alternative to nickel for achieving 100% conversions but process conditions must be carefully chosen to avoid deactivation as shown by Arena [57] who used Ru/$Al_2O_3$ catalysts that became deactivated because of the presence of iron and sulfur impurities and because the physical properties of the alumina support were modified. To overcome this problem, Gallezot et al. [58] studied the same process at 373 K under 8 MPa of hydrogen but using ruthenium catalysts (1.6–1.8 wt%) supported on active charcoal pellets (cylinders of 0.8 mm-diameter for minimizing internal diffusion limitation) instead of nickel catalysts, with the metal under the form of 1-nm particles homogeneously distributed throughout the support. The reactor was a trickle-bed one of stainless steel (length, 330 mm; internal diameter, 15.8 mm; internal volume, 60 $cm^3$) and the feed was a 40 wt% aqueous solution of D-glucose (0.17 kg-D-glucose.$h^{-1}.m^{-2}$). D-glucose conversion was 100% and sorbitol selectivity 99.2% while catalyst activity was 1.1 mol-D-glucose $h^{-1}$ $g_{Ru}{}^{-1}$, corresponding to a TOF of 111.2, 3.67 times larger than that reported by Lepper and Schütt with the Ru(2 wt%)/C in lumps [54] and under milder conditions, very likely due to the lower catalyst particle size and its better dispersion on the support. It was found that selectivity strongly depended on residence time, with the highest one obtained just when 100% conversion was achieved. Above this point, mannitol was formed by epimerization of sorbitol, as well as iditol and galactitol (both by C-2 and C-4 epimerization of sorbitol, respectively) and arabitol (by hydrogenolysis of sorbitol). The catalyst activity was stable over several weeks and no leaching of ruthenium was detected.

Longer tests on continuous hydrogenation of D-glucose to sorbitol have been reported by Kusserow et al. [34] who in a study took more than 1100 h time on stream (TOS) and investigated the deactivation of an industrial Ni(66.8 wt%)/$SiO_2$ catalyst (1100 TOS) and a Ru(0.47 wt%)/$Al_2O_3$ catalyst (1150 TOS) with a particle size of 1.9 nm. Experiments were carried out in a trickle bed reactor (Catatest, Vinci Technologies, France, no dimensions data provided), under the following conditions: 40 wt% D-glucose solution was delivered at a flow rate of 40 $mL.h^{-1}$ to the reactor packed with 40 mL of catalyst, $H_2$ flow was 23 $L.h^{-1}$ at a pressure of 8.0 MPa, reaction temperature was 353 K. At the beginning of the experiment, the conversion of D-glucose was 99.3% for the nickel catalyst and 99.9% for the ruthenium catalyst. Selectivity to sorbitol was 98.8% and 99.0%, respectively and remained unchanged during the experiments for both catalysts. The conversion dropped below 98% after 770 h TOS for the Ni catalyst, and after 1080 h TOS for the Ru catalyst. After 1100 h TOS for Ni catalyst and 1150 h for Ru catalyst, conversions were 95% and 97%, respectively. No leaching of Ru could be detected by ICP-OES (detection limit 80 $\mu g.L^{-1}$). Conversely, the Ni level in the product production ranged between 0.005 and 0.05 wt% in the first 250 h TOS, while it was constant (~0.008 wt%) over the remaining reaction time. The nickel content of the catalyst decreased by ~25% (from 66.8% to 50.5 wt%) during the experiment with a visible disintegration of the support, not surprising for silica under hydrothermal conditions. The BET surface decreased by 60% (from 180 to 85 $m^2.g^{-1}$), the fraction of mesopores increased while that of micropores decreased. It was concluded that the permanent drag-out of nickel and silica with the product solution was caused by the hot and acidic reaction media (pH 4 in the product solution) and the chelating properties of D-glucose and sorbitol. The small deactivation observed in Ru catalyst was attributed to poisoning by metal impurities leaching out of the working material (stainless steel) of the reactor.

Aho et al. [59] studied the continuous hydrogenation of D-glucose over carbon supported ruthenium catalysts in trickle flow reactors (12.5 mm inner diameter and 120 mm length) operated in co-current mode at 402 K and 2 MPa of hydrogen pressure. Long-term (100 h) stability testing of a commercial Ru/C catalyst and Ru/nitrogen doped carbon nanotubes (NCNT) catalysts was performed. It was found that Ru/NCNT worked without significant deactivation for more than 100 h, while the commercial catalyst underwent substantial deactivation within this period. The increased stability of the NCNT supported Ru catalyst was ascribed to stronger support–Ru interactions as compared to the Ru/C catalyst. No ruthenium leaching was observed for the Ru/NCNT catalyst, while leaching from the commercial activated carbon catalyst was found to be significant. Catalyst activities were similar to those reported by Gallezot et al. [58] but this process needs further optimization because maximum D-glucose conversion and selectivity were low relative to the industrial requirements: 91.4% and 98.2%, respectively, with mannitol and fructose as the main impurities. Likewise, D-glucose concentration in the aqueous feed was very low, 3.5 wt%, 11–14.5-fold less than in current industrial batch processes, leading to an economic penalty in the evaporation step needed for achieving the sorbitol concentration (70 wt%) in the commercial product. Conversion could be increased by increasing both the hydrogen pressure and the residence time, but care must be taken with the increase of the latter because could lead to a higher amount of impurities.

**Alternative Biomass Raw Materials** Although D-glucose is the only raw material industrially used for production of sorbitol, it is obvious that a large cost reduction could be achieved by starting directly from a D-glucose source that is directly converted into sorbitol in a one-pot reaction. The idea is to carry out simultaneously the hydrolysis of the D-glucose-containing polysaccharide and the hydrogenation of the free D-glucose, a process requiring the combined use of both acid and metal catalysts, for hydrolysis and hydrogenation, respectively.

Starting directly from lignocellulosics would be an ideal scenario for economic production of sorbitol. However, due to the compositional complexity of this biomass, selectivity is not high and a mixture of sugar alcohols is obtained leading to enhanced separation costs. Thus, milled Japanese cedar wood chips containing 40.9 wt%, 24.8 wt%, and 33.4 wt% of cellulose, hemicellulose and lignin, respectively, were directly converted into sugar alcohols using 4%Pt/C (93 wt% relative to chips mass) as a catalyst in water without any acid catalysts at 463 K under a $H_2$ pressure of 5 MPa [60]. Total sugar conversion was 94.1% in 16 h while sugar-alcohol yields were 36%, 12.6%, 2.3%, 6.9% and 4.3% for sorbitol, mannitol, galactitol, xylitol and arabitol, respectively. Total sugars conversion decreased to 44% when unmilled chips were used, showing the huge importance of both surface area, as the hydrolysis step to sugars is a heterogeneous process, and the decrease of the cellulose crystallinity due to the milling.

Therefore, attempts have been focused on polymeric D-glucose precursors, i.e., starch and cellulose. Cellobiose (4-*O*-β-D-glucopyranosyl-D-glucose) has been also converted directly to sorbitol, but it is more a simple model compound resembling cellulose because it is the repeat unit of cellulose ((1,4-*O*-β-glucopyranosyl)$_{n-1}$-D-glucose). Consequently, its conversion will not be discussed herein. Readers can, e.g., consult references [61, 62].

**Starch** The simultaneous hydrolysis and hydrogenation of starch has been known for a long time, Thus, in 1950 Hartstra et al. [63] reported a process in which a sorbitol yield of 60% was obtained in 75 min by hydrogenating a 37.5 wt% starch water suspension in the presence of 0.01–1 wt% relative to the amount of starch of a Lewis acid such as magnesium chloride, nickel sulfate or stannous chloride, and a nickel-kieselguhr hydrogenation catalyst (10 wt% based on starch amount) under a hydrogen pressure of 5–20 MPa at a temperature of 433–473 K. The sorbitol was purified by crystallization to 97% purity.

A much more effective process than that of reference [63] for sorbitol production from starch was developed by Jacobs and Hinnekens [64]. Aqueous 10–30 wt% corn starch was 100% batch-wise converted within 1 h into sorbitol with a selectivity higher than 95% at 403–453 K and 5.5 MPa of $H_2$ using Ru (3 wt%) supported on H-USY zeolite as a catalyst. The amount of Ru metal was 0.12–0.36 wt% relative to starch. The acidic zeolite support catalyzes the starch hydrolysis to D-glucose which is then hydrogenated to sorbitol on the ruthenium surface. Catalyst was 17-fold recycled with no loss in activity.

Some processes have also been developed starting from starch hydrolysates mainly composed of D-glucose but also containing D-glucose oligomers. Hydrogenation is carried out with both nickel [65] and ruthenium catalysts [66] under the same conditions used starting from D-glucose. However, a mixture of sorbitol, maltitol and hydrogenated oligomers, such as maltotriitol, is produced boosting the sorbitol separation and purification costs. In fact, hydrogenated starch hydrolysates are marketed as such and used as sweeteners and humectants.

**Cellulose** Cellulose is an especially attractive alternative raw material due to its huge abundance and because it does not interfere with the food chain, unlike starch. The hydrolytic hydrogenation of cellulose to sorbitol has been reviewed by Van de Vyver et al. [67], Zhang et al. [42], Yabushita et al. [68] and Li et al. [69]. The reaction consists of two steps: (i) hydrolysis of cellulose to D-glucose, and (ii) hydrogenation of glucose to sorbitol, with the step (i) being the limiting step of the overall reaction. Consequently, it is not strange that, on one hand, efforts have been focused on developing catalytic systems with an acidic component able to speed up the hydrolysis of cellulose, and, on the other hand, the performance of heterogeneous catalysts is very dependent on the presence of acidic functional groups on the supports. An obvious strategy is that used by Balandin et al. in its pioneering work [70, 71], consisting of using mineral soluble acids together with a hydrogenation catalyst such as Ru/C. They reported a total 82% yield of mannitol and sorbitol (no specific sorbitol yield was reported) under 7 MPa of hydrogen and 433 K in 2 h in the presence of sulfuric acid. Although this strategy has been followed by several research groups [72, 73], using both mineral acids and heteropolyacids together with hydrogenation catalysts containing Pd, Pt or Ru, soluble acids result in a more difficult separation procedure, generate a large amount of waste sludge in the acid neutralization step and also can cause corrosion of the reactor.

Therefore, the use of heterogeneous catalysts combining a metal hydrogenation catalyst and an acidic support is being investigated as a more industrially efficient and sustainable alternative. Thus, microcrystalline cellulose (0.8 wt%) was hydrogenated into sorbitol in a 25% yield in 24 h using Pt(2.5 wt%)/$\gamma$-$Al_2O_3$ as a catalyst at 5 MPa and 463 K [74, 75]. The ability of the catalyst for hydrolyzing cellulose to D-glucose was thought to the due to the acidic surface of the support. Also, the metal seems to participate directly in the cellulose hydrolysis by increasing the $H^+$ concentration by heterolytic $H_2$ dissociation on its surface [76]. Han and Lee [77] reported one-pot conversion of cellulose to sorbitol using catalyst combining metal nanoparticles (Ni, Pd, Pt or Ru) and sulphonic groups on the surface of an activated charcoal (M/AC-$SO_3H$) support in a neutral aqueous solution. Using a 4.2 g.$L^{-1}$ aqueous suspension of ball-milled cellulose and a 40 wt% catalyst load relative to cellulose, a 95% cellulose conversion and a 71.1% sorbitol yield were obtained at 438 K in 36 h with Ru(10 wt%)/AC-$SO_3H$. The Ru-to-S ratio and the metal load were shown to be important parameters for effective conversion. No deactivation was observed even after 5 repeated reactions. Pt/AC-$SO_3H$ showed a comparable catalytic activity to Ru/AC-$SO_3H$ while Pd/AC-$SO_3H$ and Ni/AC-$SO_3H$ gave poorer sorbitol yields.

Water has been reported as a key solvent because experiments with Ru/C, i.e. with a non-acidic support, as a catalyst as well as substitution of water by other solvents have shown the need of hot water for hydrolyzing cellulose which is thought to be due to the in situ production of $H^+$ [78]. Nevertheless, the key role of acidic supports has been highlighted by Deng et al. [79] who found that Ru supported on carbon nanotubes (CNT) was the best catalyst for obtaining sorbitol in a 36% yield from cellulose. By using $NH_3$-TPD and $H_2$-TPD characterizations the authors concluded that plenty of acid sites and unique hydrogen species over the Ru/CNT were important for sorbitol formation.

The insolubility of cellulose in the aqueous reaction media means that reaction proceeds in heterogeneous phase and consequently cellulose particle size and crystallinity are important factors as shown by Ribeiro et al. [80] who using Ru/AC catalysts obtained a cellulose conversion of 36% in 5 h with a sorbitol of 40% when microcrystalline cellulose was the raw material, while a conversion of 90% and a sorbitol selectivity close to 80% were obtained if the catalyst was ball-milled together with cellulose. The catalyst showed excellent stability after repeated use. Disruption of cellulose crystallinity can be achieved by several methods such as ball-milling, rod-milling and planetary ball-milling in less than 1 h while jet-milling is not effective [81].

All reported processes starting from cellulose lead to sorbitol yields below 70%, with selectivities generally well below 80%. Typically, lower alcohols, such as xylitol, glycerol, propylene glycol, ethylene glycol and methanol, resulting from hydrogenolysis of sorbitol and soluble oligomers are also formed, together with some amounts of mannitol, coming from hydrogenation of mannose which in turn in formed by epimerization of D-glucose, and sorbitan and isosorbide, both resulting from sorbitol dehydration. Consequently, cellulose is still far away of being a good raw material for industrial production of sorbitol due, on one hand, to its extremely low water solubility and low reactivity leading to a low productivity, and, on the other hand, to the low sorbitol selectivity obtained so far relative to the industrial processes currently used.

To overcome these drawbacks attempts have been focused on using ionic liquids (IL) as reaction media because a number of them solubilize cellulose [82] by breaking the hydrogen bonds preventing cellulose solubilization, thereby increasing the cellulose hydrolysis rate and consequently the productivity of the process. For example, 1-butyl-3-methylimidazolium chloride (BMImCl) dissolves up to 25 wt% of cellulose, while 1-allyl-3-methylimidazolium chloride and 1-ethyl-3-methylimidazolium acetate dissolve 14.5 wt% and 16 wt% at 353 K and 363 K, respectively. Thus, Ignatyev et al. [83] demonstrated that a combination of a heterogeneous Pt or Rh catalyst and a homogeneous Ru catalyst can completely convert cellulose to a sorbitol/D-glucose mixture in 1-butyl-3-methyl imidazolium chloride ([BMIM]Cl). A full cellulose conversion with a 74% sorbitol yield were achieved at 423 K, 3.5 MPa of $H_2$ in 48 h with a cellulose concentration of 5 wt% using water (60 $\mu L.g^{-1}$ cellulose) and KOH (0.34 wt% relative to cellulose) as co-catalysts and a mixture of a heterogeneous (Rh(5%)/C catalyst, 7.6 wt% relative to cellulose, and a homogeneous one ($HRuCl(CO)(PPh_3)_3$, 6.6 wt% relative to cellulose. The homogeneous catalyst was needed due to the low solubility of hydrogen in ionic liquids. No reaction was observed in its absence, even when pure D-glucose was used as raw material. Important drawbacks of this process are the use of a homogeneous precious metal based complex catalyst which difficult product separation and the low hydrogen solubility, inherent to IL, leading to long reaction times. These problems together with the added difficult for separation the polar sorbitol from the polar IL represent strong pitfalls for industrial use in the short and medium term.

### 9.3.2 Electrochemical Production of Sorbitol

The electrochemical production of sorbitol was one of the first industrial processes for producing an organic chemical by electrosynthesis. Anodic reaction was the water oxidation resulting in oxygen and protons while cathodic reaction was the reduction of D-glucose to sorbitol as main product and mannitol as a byproduct. Anodic and cathodic compartments were separated each other by a diaphragm (Fig. 9.3). In 1937 the Atlas Powder Company (since 1971 ICI Americas Inc.) built a facility for manufacturing 1400 $t.year^{-1}$ of sorbitol and mannitol by electroreduction of D-glucose (Fig. 9.4) resulting from corn starch [1].

The electrochemical reactor (cell) was an open-top rectangular tank 3.96 m long, 1.83 m wide and 0.91 m deep, fitted with amalgamated-lead cathodes and lead anodes. 35 anodes and 36 cathodes, each one of 1.66 $m^2$, were connected in monopolar mode in each cell. Anodic and cathodic compartments were separated by a diaphragm of unglazed porcelain. The anolyte was a diluted aqueous solution of sulfuric acid while the catholyte was an aqueous solution of sodium hydroxide, sodium sulfate which was needed for increasing electrical conductivity for lowering power consumption without increasing pH excessively, and D-glucose. The plant contained 12 cells working batchwise at 85 $A.m^{-2}$ and room temperature at a voltage of 20 $V.cell^{-1}$. In the alkaline medium D-glucose isomerized partially to mannose and both D-glucose and mannose were reduced to sorbitol and mannitol, respectively. For both improving sorbitol mass transfer from the bulk solution to the cathode surface and controlling reaction temperature, the D-glucose solution was continuously circulated through the cell and a double pipe counter-current cooling coil. The productivity of the cell was later improved by working at 400 $A.m^{-2}$ at 358–373 K [84].

After batch completion, water was partially removed from the catholyte by vacuum evaporation leaving a residue of sorbitol, mannitol and sodium sulfate. The residue was treated with hot ethyl alcohol for dissolving sorbitol and mannitol. The sulfate was filtered out the hot solution and the filtrate was sent to a crystallizer to remove the mannitol by crystallization and subsequent centrifugal separation. Mannitol was purified by repeated recrystallization from distilled water. The alcoholic sorbitol solution was evaporated to remove the ethyl alcohol resulting in

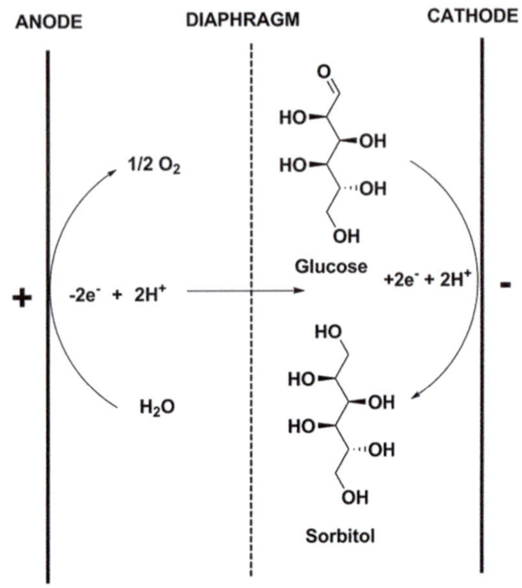


**Fig. 9.3** Anodic and cathodic reactions in the electrosynthesis of sorbitol from D-glucose using divided cells

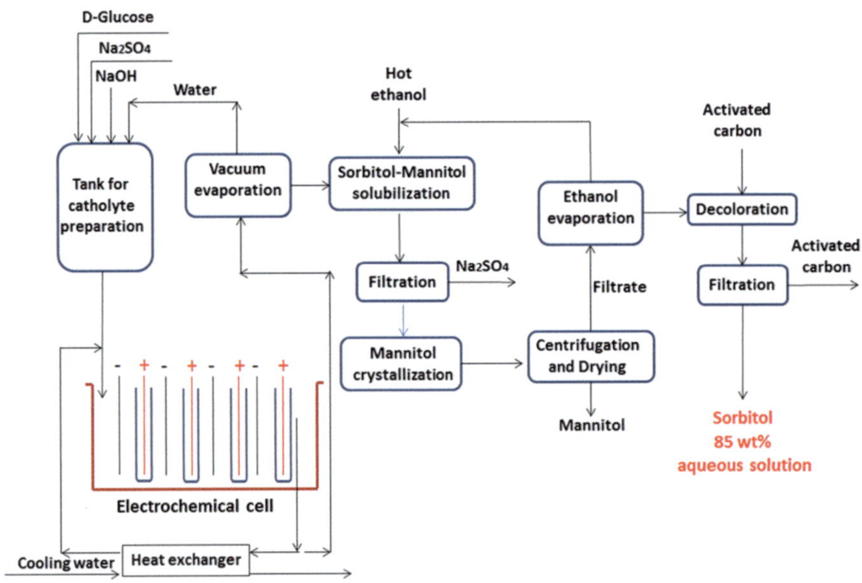


**Fig. 9.4** Flow diagram for Atlas Powder Company electrochemical production of sorbitol from cornstarch-derived D-glucose

85 wt% water solution of sorbitol which was decolorized with activated carbon, filtered and drummed for shipment. D-glucose and sodium sulfate were present in small amounts as impurities.

As known, the electrochemical process was displaced a few years later by the more cost-effective high-pressure catalytic hydrogenation. Since then, the bibliography (both papers and patents) devoted to electrochemical synthesis is extremely low in comparison with that related to chemical synthesis and even microbiological synthesis, with efforts being focused on reducing the manufacturing costs. Cost reduction requires decreasing power consumption and/or increasing cell productivity by obtaining simultaneously a valuable chemical at the anode (paired synthesis).

Reduction in power consumption could be dramatic if an undivided cell could be used due to the decrease in inter-electrode gap as well as the removal of the voltage drops due to the anolyte and the diaphragm. However, the aldehyde group in D-glucose is more easily oxidizable than water which requires the use of an anodic depolarizer of lower oxidation potential than that of D-glucose. Thus, Hefti and Kolb [85] developed a method in which sorbitol was synthesized by electroreducing an aqueous solution of D-glucose and sodium sulphite in an undivided cell with an amalgamated lead cathode and a graphite anode. The anodic oxidation of sulphite to sulfate prevents the D-glucose oxidation. Chemical and current yields were 95% and 90%, respectively. However, 0.8 kg of sodium sulphite.$kg^{-1}$ of D-glucose was needed increasing significantly the raw materials costs. Other strong drawback of this procedure was the low current density used (50 A.$m^{-2}$) requiring an electrode area 8-fold higher than that in the Atlas Powder process.

The paired electro-oxidation and electroreduction of D-glucose for manufacturing gluconic acid and sorbitol, respectively, has been reported [86, 87]. As depicted in Fig. 9.5 for a divided cell, the cathodic reaction is the direct electroreduction of D-glucose to sorbitol while the anionic one is an indirect electro-oxidation in which the bromine, electrogenerated by electro-oxidation of bromide, oxidizes chemically D-glucose to gluconic acid regenerating at the same time the bromide ion, which in this way is used cyclically. Both undivided and divided cells can be used. Thus, an undivided packed-bed electrode flow reactor fitted with a Raney Ni powder cathode and a graphite chip anode was reported by Park el al [86]. Electrolyte was an aqueous solution of 1.6 M D-glucose and 0.4 M $CaBr_2$. Working at 333 K, pH 5–7, a current of 250 A.$kg^{-1}$ of nickel powder and a flow rate of 6 L.$h^{-1}$ current yields were 100% for both gluconic acid and sorbitol. However, these favorable results were obtained by circulating a low electrical charge corresponding to 20% of the theoretical one leading to a low D-glucose conversion of 20%. On the other hand, this type of cell is difficult to scale-up for industrial purposes.

Li et al. [87] used a divided filter press cell in which the Pb sheet cathode was separated from a dimensionally stable anode (DSA) by a cation exchange membrane. Gluconic acid and sorbitol were obtained in 90% chemical yields and over 80% current efficiencies by paired electrolysis of D-glucose at 333 K, 500 A.$m^{-2}$ and an electrical charge equal to 110% of the theoretical one. The anolyte was 66.7 wt% D-glucose and 2 wt% NaBr in water, and the catholyte 66.7 wt% D-glucose,

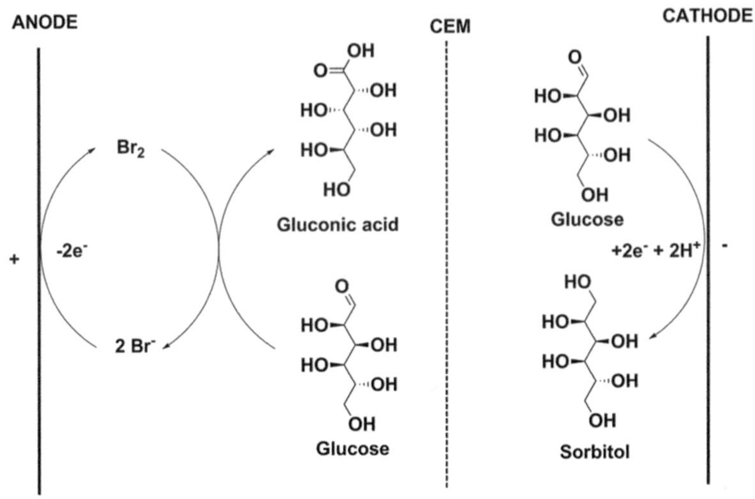


**Fig. 9.5** Paired electrosyntheis of gluconic acid and sorbitol from D-glucose. *CEM* Cation exchange membrane

2.5 wt% NaOH and 2.5 wt% $Na_2SO_4$ in water, respectively. The anodic process is an three-step indirect one in which bromide anion is electro-oxidized to bromine in a first step, which oxidizes chemically D-glucose to gluconolactone in a second step, step in which the bromide anion is regenerated. Finally, gluconolactone is hydrolyzed to gluconic acid in a third step. The cathodic reaction is the direct electroreduction of D-glucose to sorbitol. A low 4 V cell voltage leads to a power consumption of 1.6 $kWh.kg^{-1}$-sorbitol.

The process of Li et al. [87] is easy to scale-up but about 1.1 kg of gluconic acid is produced per kg of sorbitol, which is a major drawback because the small gluconic acid market (about 5%) compared to that of sorbitol. Therefore, this process is only suitable for low volume sorbitol facilities able to produce an amount of gluconic acid easily absorbed by the market. An anodic reaction yielding a chemical of similar market volume than sorbitol is needed for large scale sorbitol production through this procedure.

### *9.3.3 Biotechnological Production of Sorbitol*

An interesting alternative to the chemical production methods is based on biotechnology, and involves the use of enzymes and microorganisms as (bio)catalysts to convert biomass-derived feedstocks into sorbitol. It is generally accepted that biotechnological methods show a number of advantageous properties compared with chemical methods, such as ambient temperature and pressure operation, which reduces energy costs, and a high selectivity and specificity, which minimizes by-product generation.

A complete review about biotechnological production of sorbitol was previously published by Silveira and Jonas [88], containing the knowledge up to 2002 on this topic.

***Zymomonas mobilis*, a Bacterium Able to Produce Sorbitol** The possibility of producing ethanol using other microorganisms different to *Saccharomyces cerevisiae*, which is the yeast used on the industrial scale, has been considered. Among potential candidates, special attention has been paid to *Z. mobilis*, a gram-negative bacterium that can be found in sugar rich plant materials and fermented plant juices. *Z. mobilis* shows several advantages over yeasts as a bioethanol producing microorganism: (1) a higher sugar uptake rate; (2) a higher growth rate; (3) a higher ethanol yield; (4) a lower biomass production; (5) a higher ethanol tolerance; (6) it does not require controlled addition of oxygen during the fermentation; and (7) it is amendable to genetic manipulation. These advantageous features have allowed this bacterium to be considered as a potential platform for future biorefineries [88]. The only limitation of *Z. mobilis* compared to the yeast is that its range of usable carbon sources is restricted to D-glucose, fructose and sucrose.

*Z. mobilis* shows an unusual property regarding carbohydrate metabolism: it uses the Entner-Doudoroff pathway anaerobically to degrade D-glucose [89]. In this process, 1 mol D-glucose is converted to almost 2 mol, the one of ethanol and the other of $CO_2$, that is, it is able to produce ethanol, at near theoretical levels. Fructose and sucrose, the latter following hydrolysis to D-glucose + fructose, are metabolized through the same pathway. Carbohydrate metabolism also generates 1 mol ATP and results in a limited biomass formation, accounting for not more than 2–5% of the carbohydrate consumed.

It is known that *Z. mobilis* converts up to 95% of added D-glucose or fructose to ethanol, but when grown on sucrose the alcohol yield is greatly reduced. As some strains produced levan when cultured on sucrose, it was first considered that part of the sugar was directed to the synthesis of this fructose polysaccharide. However, the amount of levan produced accounted for only a small proportion of the missing carbon. Further studies showed that the fraction of sucrose not dedicated to ethanol or levan synthesis resulted in the production of an additional product: sorbitol [90, 91].

Sorbitol was shown to be solely derived from the fructose, but it was only produced when both fructose and D-glucose were present. Very interestingly, sorbitol formation was accompanied by an equimolar production of gluconic acid, which suggested that two enzymes coupled together by an unknown cofactor to catalyze these conversions [92].

The actual mechanism was soon elucidated, showing that a previously unknown enzyme was responsible for both steps, the oxidation of D-glucose to glucono-δ-lactone and the reduction of fructose to sorbitol [93]. The enzyme, named as D-glucose-fructose oxidoreductase (GFOR; EC1.1.199), contains a tightly bound NADP as a cofactor and acts through a classical ping-pong mechanism, where

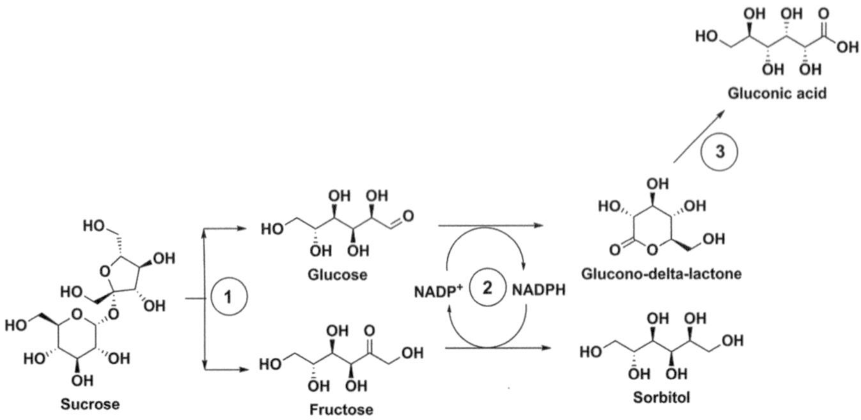


**Fig. 9.6** Mechanism of sucrose conversion into sorbitol and gluconic acid by *Z. mobilis*. *1*: Invertase; *2*: GFOR; *3*: gluconolactonase

D-glucose is first converted to gluconolactone, which leaves the enzyme, and then fructose is reduced to sorbitol. GFOR is a homotetramer located at the periplasm, whose physiological function is proposed to be the regulation of osmotic stress of the cell when grown on high sugar concentrations. Additionally, it was found that gluconolactone leaving GFOR was converted to gluconic acid by the enzyme gluconolactonase, although the hydrolysis reaction may also occur spontaneously, which makes GFOR-catalyzed reaction being almost irreversible *in vivo* and *in vitro*.

A schematic representation of the steps and enzymes involved in sugar conversion into sorbitol and gluconic acid is shown in Fig. 9.6.

The principal advantage of GFOR catalyzing D-glucose/fructose conversion into gluconic acid/sorbitol is twofold. On the one hand, it is a self-regenerating redox enzyme system and, on the other hand, it does not require the exogenous addition of expensive and unstable adenine nucleotide cofactors due to the NADP being tightly bound to the enzyme.

**Biotechnological Production of Sorbitol Using** *Z. mobilis* In usual co-fermentation of D-glucose and fructose by *Z. mobilis*, sorbitol accumulates in the culture medium (the bacterium cannot metabolize it) while gluconic acid is metabolized through the Entner-Doudoroff pathway, resulting finally in the production of ethanol and $CO_2$. So, it was proposed that in a cell-free system or with the purified GFOR it would be possible to avoid gluconic acid metabolism and ethanol formation, and only obtain the primary products, that is, sorbitol and gluconic acid. According to this strategy, a toluene-permeabilized suspension of *Z. mobilis* cells, previously grown and concentrated, was used to treat mixtures of D-glucose and fructose [94, 95]. Cell permeabilization was intended to increase their permeability and remove soluble co-factors and high energy compounds needed to carry out gluconic acid metabolism, so avoiding its conversion into ethanol and other products. With this system, concentrations as high as 290 g.L$^{-1}$ sorbitol were achieved in batch, reaching yields close to 95%. The use of Ca-alginate immobilized cells instead of free cells rendered similar results in batch and, for the first time, a continuous process was developed showing only a small loss of enzyme activity (less than 5%) after 120 h operation.

Alternative methods to toluene treatment of cells were evaluated to block ethanol production, finding that a drying treatment of cells appeared to selectively inactivate enzymes responsible for the conversion of D-glucose or fructose to ethanol, while GFOR and gluconolactonase retained their activities [96].

Other different permeabilization methods for *Z. mobilis* cells were developed. One of them was described by Bringer-Meyer and Sahm [97], involving freezing at 253 K and thawing at room temperature, and resulting in a production of sorbitol of 233 g.L$^{-1}$ from a mixture of 234 g.L$^{-1}$ D-glucose and 234 g.L$^{-1}$ fructose, that is, an almost quantitative conversion was achieved (98.5% yield), with an excellent productivity of 46.6 g.L$^{-1}$.h$^{-1}$.

Permeabilization of *Z. mobilis* cells using cationic detergents such as cetyltrimethylammonium bromide (CTAB), increased sorbitol production from 240 to 295 g.L$^{-1}$ from equimolar fructose and D-glucose mixtures up to 60% (w/v), and avoided ethanol production. With these permeabilized cells, in a two-stage continuous process with κ-carrageenan-immobilized and polyethylenimine-hardened cells, no significant decrease in the conversion yield (>98%) was observed after 75 days [98]. Both the use of *Z. mobilis* CTAB-permeabilized cells in the production of sorbitol and the method for cell immobilization were later patented by Rehr and Sahm [99, 100]. Operational stability of the system could be enhanced by either drying or adding polyols to κ-carrageenan beads, which increased their rigidity, achieving the best results with a two-stage continuous packed bed reactor, where a sorbitol concentration of 178.6 g.L$^{-1}$ with a productivity of 25 g.L$^{-1}$.h$^{-1}$ was obtained [101].

In the course of reaction, gluconic acid formation leads to a pH decrease causing enzyme inhibition, an event that must be avoided through alkali addition to the medium. As a feasible alternative, the coupling of an electrodialysis unit to a system containing permeabilized and immobilized *Z. mobilis* cells, was evaluated in order to efficiently remove gluconic acid and prevent acidification and enzyme inhibition, allowing an evident improvement in the stability of the enzyme, which maintained its reaction rate unchanged for 60 h of operation [102].

Most of the previous work on biological sorbitol production involves the use of permeabilized *Z. mobilis* cells as biocatalyst. However, with the aim of reducing production costs, the use of intact, non-permeabilized cells, was also proposed [103, 104]. Applying this method, Silveira et al. [104] found that the use of up to 650 g.L$^{-1}$ of an equimolar mixture of D-glucose and fructose resulted in an almost complete bioconversion to sorbitol and gluconic acid, without ethanol formation and reaching yields over 91% for both products. This result was explained by the effects that the high substrate concentration could cause on the cells, mainly the loss of cell viability due to the high osmotic pressure and the inhibition of the normal ethanologenic metabolism, resulting in a preferential utilization of substrates via GFOR.

Therefore, the use of free-untreated cells of *Z. mobilis*, despite their slightly lower yields compared to those obtained with permeabilized cells, appeared as an attractive option.

The biotechnological production of sorbitol using *Z. mobilis* suffers a problem related to the relatively high cost of fructose compared to product value. As a result, the use of alternative, low cost, substrates has been considered and evaluated.

A feasible alternative is sucrose, a disaccharide composed of two monosaccharides that are co-substrates of GFOR. Toluene-permeabilized *Z. mobilis* cells were co-immobilized in calcium-alginate beads with invertase, and applied to the production of gluconic acid and sorbitol from sucrose in a RPBR [105]. Maximum productivities for sorbitol of 5.20 $g.L^{-1}.h^{-1}$ were achieved at a dilution rate of 0.053 $h^{-1}$ and a sucrose concentration of 20% when recirculated at the rate of 1200 $mL.h^{-1}$. The co-immobilized enzymes were reported to remain stable for 250 h in the RPBR without any loss of activity.

Another substrate checked was Jerusalem artichoke, an inulin-rich plant source. Inulin is a fructan, that is, a fructose-rich polysaccharide, so it could substitute pure fructose as GFOR co-substrate. Therefore, toluene permeabilized *Z. mobilis* cells, co-immobilized with inulinase (the inulin depolymerizing enzyme) in calcium-alginate beads were used to convert D-glucose and Jerusalem artichoke into gluconic acid and sorbitol [106]. In a RPBR, the maximum productivity for sorbitol was found to be 26 $g.L^{-1}.h^{-1}$, showing the co-immobilized enzymes full stability for 250 h without any loss of activity.

Sugar cane molasses, which is a by-product of the sugar industry, has been considered as an attractive substrate for sorbitol production [107]. Molasses contains not only a high sucrose concentration, but in addition other important substances beneficial for the fermentation process and it has low cost. Cazetta et al. [107] carried out an optimization study for sorbitol production by *Z. mobilis* in sugar cane molasses, finding that the best conditions for sorbitol production were 300 $g.L^{-1}$ total reducing sugars, where around 14 $g.L^{-1}$ sorbitol were produced, only representing a 23% of the fructose utilized. This reduced sorbitol production was attributed to the high concentration of salts present in sugar cane molasses that may have raised the osmotic pressure above acceptable levels, reducing cell viability and suppressing sorbitol production.

Two low-cost feedstocks, inulin and cassava starch, have been tested to produce sorbitol [108]. The process involved the use of a commercial glucoamylase enzyme for the simultaneous saccharification of inulin and starch into high titer D-glucose and fructose hydrolysate, replacing the expensive and not commercially available inulinase enzyme. Conversion was carried out using immobilized whole cells of a recombinant GFOR over-expressing *Z. mobilis* strain, achieving a titer of 180 $g.L^{-1}$ sorbitol in batch [108].

Following a strategy of introducing metabolic engineering technologies into biotechnological production of sorbitol, another recombinant *Z. mobilis* strain over-expressing GFOR was constructed, showing a specific activity at least twofold greater than that in the wild type strain [109]. Addition of divalent metal ions, especially $Zn^{2+}$, to freezing and thawing permeabilized recombinant cells was found to improve the bioconversion process, by inhibiting the Entner-Doudoroff pathway enzymes, resulting in a drastic reduction in ethanol and a significant increase in sorbitol yields. Using an equimolar D-glucose/fructose mixture as substrate, ethanol production was reduced from 16.7 to 1.8 $g.L^{-1}$ and sorbitol yield was increased to virtually 100%, up to 161.1 $g.L^{-1}$.

Besides D-glucose, GFOR can also accept other alternative donor substrates, including monosaccharides as xylose and galactose, or disaccharides as maltose and lactose, which are oxidized to their corresponding aldonic acids. Conversion of a mixture of lactose and fructose into lactobionic acid and sorbitol by permeabilized *Z. mobilis* cells carrying GFOR and gluconolactonase enzymes was studied, using either mobilized or Ca-alginate immobilized cells. Under optimal operating conditions, an almost quantitative conversion of 350 mM fructose into sorbitol can be achieved [110].

As an alternative to the use of *Z. mobilis* cells, either permeabilized or not, as biocatalyst, some attempts were made to develop a process employing cell-free GFOR [111]. This strategy was explained by the higher enzyme concentrations that could be applied and by the lack of mass transfer limitations that could be expected using free GFOR, which might increase productivity. However, the free enzyme was found to rapidly inactivate during the time course of its own catalytic action, requiring supplementation with thiol-protecting agents to increase GFOR stability.

A continuous process in an ultrafiltration membrane reactor was developed using a crude cell extract of *Z. mobilis*, where enzyme inactivation was almost completely avoided by adding to reaction medium weak bases, such as Tris or imidazol, to neutralize gluconic acid produced, and dithiothreitol to protect thiol groups [112]. This system was reported to operate over a time period of more than 250 h without significant decrease in substrate conversion or enzyme activity, resulting in a sorbitol productivity of 4.37 $g.L^{-1}.h^{-1}$ and a yield close to 40%. The use of a tangential ultrafiltration loop reactor allowed increasing substrate conversion and productivity up to more than 85% and 5 $g.L^{-1}.h^{-1}$, respectively, from 3 M sugar [113].

A compilation of the main achievements in biotechnological production of sorbitol using *Z. mobilis* is shown in Table 9.2. Sorbitol synthesis with *Z. mobilis* is a biocatalytic process, and not a fermentation, contrary to some reports. Biocatalyst is GFOR enzyme, which is bound to bacterial cells, so it can be considered as a whole-cell biocatalyst. The process comprises two separate steps: (1) biocatalyst generation and (2) biocatalytic reaction. The step of biocatalyst generation is carried out by culture of the microorganism to obtain as much biomass as possible to be used thereupon in the biocatalysis. Therefore, the first step is certainly a true fermentation.

Space-time yields shown in Table 9.2 actually correspond to productivities of the second step, the biocatalytic step, and not to that of the whole process. As substrates and time consumed in biocatalyst generation are not taken into account, the actual productivities of the whole process could be considered to be lower than those values shown in Table 9.2.

**Table 9.2** Examples of biotechnological production of sorbitol using *Zymomonas mobilis*[a]

| Biocatalyst | Permeabilization | Bioreactor | Substrate | Sorbitol concentration ($g.L^{-1}$) | Sorbitol yield (%) | Space-time yield ($g.L^{-1}.h^{-1}$) | Specific productivity ($g.g^{-1}.h^{-1}$) | Ref. |
|---|---|---|---|---|---|---|---|---|
| Free cells | Toluene | Batch, stirred tank | 300 $g.L^{-1}$ D-glucose, 300 $g.L^{-1}$ fructose | 290 | 94.3 | 18.9 | – | [95] |
| Free cells | Freezing and thawing | Batch, stirred tank | 234 $g.L^{-1}$ D-glucose, 234 $g.L^{-1}$ fructose | 233 | 98.5 | 46.6 | 1.08 | [97] |
| Free cells | CTAB[b] | Batch, stirred tank | 300 $g.L^{-1}$ D-glucose, 300 $g.L^{-1}$ fructose | 295 | 98 | – | – | [98] |
| Ca-alginate co-immobilized cells and invertase | Toluene | Continuous recycle packed-bed | 200 $g.L^{-1}$ sucrose | 98 | 92 | 5.20 | – | [105] |
| Ca-alginate co-immobilized cells and inulinase | Toluene | Continuous recycle packed-bed | 200 $g.L^{-1}$ D-glucose and Jerusalem artichoke | 74.3 | – | 26 | – | [106] |
| Free cells | CTAB[b] | Batch, stirred tank | 281 $g.L^{-1}$ D-glucose, 292 $g.L^{-1}$ fructose | 285 | 97.0 | 47.5 | 0.6 | [99] |
| κ-Carrageenan immobilized cells | CTAB[b] | Two-stage continuous packed bed | 300 $g.L^{-1}$ D-glucose, 300 $g.L^{-1}$ fructose | 178.6 | – | 25 | 1.6 | [101] |
| Crude cell extract | – | Continuous tangential ultrafiltration loop | 3 M sugar | – | >85 | 5.25 | – | [113] |

| | | | | | | | | |
|---|---|---|---|---|---|---|---|---|
| Free cells | – | Batch, stirred tank | 300 g.L$^{-1}$ D-glucose, 300 g.L$^{-1}$ fructose | 279 | 92.1 | – | 1.0 | [103] |
| Free cells | – | Batch, stirred tank | 325 g.L$^{-1}$ D-glucose, 325 g.L$^{-1}$ fructose | 300 | 91 | 45 | 1.5 | [104] |
| Free cells | – | Batch, stirred tank | Molasses, 300 g.L$^{-1}$ reducing sugars | 13.87 | 23 | 0.38 | 1.9 | [107] |
| Free recombinant GFOR over-expressing cells; Zn2+ suppl. | Freezing and thawing | Batch, stirred tank | 157.1 g.L$^{-1}$ D-glucose, 158.9 g.L$^{-1}$ fructose | 161.1 | 100.3 | 61.2 | 3.06 | [109] |
| Immobilized recombinant GFOR over-expressing cells | – | Batch, stirred tank | Inulin and cassava starch, glucoamylase | 180 | – | – | – | [108] |

[a]Temperatures for bioconversions between 33 and 312 K, with most at 312 K. pH: 6.2–6.5
[b]*CTAB* Cetyltrimethylammonium bromide

Largest space-time yields are associated, not surprisingly, to high cell (biocatalyst) loads, greater than 15 g cells (dry weight).$L^{-1}$ [97, 99, 101, 104, 109] as occurs in any catalytic reaction. In turn, specific productivities of *Z. mobilis* catalyst, which give an idea of its specific activity, depend primarily on the strain used, but also on reaction conditions, including whether cells are permeabilized or not, and the permeabilization method. Specific productivities, when available, are reported to be between 0.6 and 1.9 g.$g^{-1}$.$h^{-1}$ (Table 9.2). As expected, overexpression of the GFOR enzyme results in an increased specific productivity, amounting to more than 3 g.$g^{-1}$.$h^{-1}$ [108].

**Other Microorganisms Producing Sorbitol** In addition to *Z. mobilis*, only a few microorganisms have been described as natural sorbitol producers. An example is the methanol-utilizing yeast *Candida boidinii*, which was shown to produce sorbitol from D-glucose using an intact cell system, with methanol as the energy source for generating NADH for the reduction of D-glucose to sorbitol. With this system, maximum amounts of sorbitol of 8.8 and 19.1 g.$L^{-1}$ were obtained from 20 g.$L^{-1}$ of D-glucose and fructose, respectively [114]. Sorbitol was directly produced from fructose in a reduction catalyzed by a sorbitol dehydrogenase enzyme [115].

Another yeast, *Saccharomyces cerevisiae* strain ATCC 36859, has also been shown to be able to produce sorbitol when cultured in Jerusalem artichoke juice [116]. The yeast produced ethanol from the beginning of the process, and sorbitol production occurred only after D-glucose depletion. When the juice was supplemented with 3% yeast extract, sorbitol concentration reached a 4.6%, with a yield of 0.259 g.$g^{-1}$ sugar consumed.

Engineering strategies have been applied to some lactic acid bacteria in order to obtain new sorbitol producers. All of these strategies involve a few manipulations of carbon and energy metabolism, including overexpression of the key enzymes for converting a substrate to sorbitol, blocking re-utilization of the produced sorbitol, decreasing other by-product synthesis, and improving redox balance [117–119].

Lactic acid bacteria do not produce sorbitol at detectable levels, but some of them have the ability to use sorbitol as a carbon source through the genes contained in the sorbitol operon. In *Lactobacillus case*i and *L. plantarum* two D-sorbitol-6-phosphate dehydrogenase-encoding genes have been found, which catalyze oxidation of sorbitol-6-phosphate to fructose-6-phosphate. As this reaction is reversible, the possibility of sorbitol production from fructose exists, although it is very unlikely because both genes are tightly controlled by catabolite repression and substrate induction.

So, in order to avoid this and to revert the sorbitol catabolic pathway toward sorbitol synthesis, a strain of *L. casei* was constructed by integration of a D-sorbitol-6-phosphate dehydrogenase-encoding gene (*gutF*) in the chromosomal lactose operon, and the resulting recombinant strain was shown to produce small amounts of sorbitol from D-glucose with a yield of 2.4% [117]. Subsequent inactivation of the L-lactate dehydrogenase gene led to an increase in sorbitol production (yield, 4.3%), suggesting that the engineered route provided an alternative pathway for $NAD^+$ regeneration. However, the recombinant *L. casei* strain suffered from two drawbacks that diminished sorbitol production: after D-glucose depletion sorbitol was reutilized, and mannitol was also produced. The first problem was avoided by deleting the *gutB* gene, responsible for uptake and reutilization of the synthesized sorbitol, and the second one by inactivating the *mtlD* gene, encoding the enzyme catalyzing the conversion of fructose-6-phosphate to mannitol-1-phosphate. The resulting strain was able to convert lactose into sorbitol with a yield of 9.4% using an optimized fed-batch system and whey permeate as a substrate [118].

A similar strategy was used with *L. plantarum*, which was metabolically engineered to produce sorbitol by constitutive overexpression of the two sorbitol-6-phosphate dehydrogenase genes (*srlD1* and *srlD2*) in a mutant strain deficient for both L- and D-lactate dehydrogenase activities [118]. Using resting cells under pH control with D-glucose as substrate, sorbitol yield approached to 65%, which is close to the maximal theoretical value of 67%.

Nevertheless, reported sorbitol production by all these microorganisms, though possible, falls far below that obtained with *Z. mobilis*, so they are not currently alternatives.

### 9.3.4 Recovery and Purification of Sorbitol

Main by-products in sorbitol production [33, 120] are depicted in Fig. 9.7. D-glucose can isomerize to fructose and mannose, especially in alkaline medium. Hydrogenation of mannose yields mannitol while hydrogenation of fructose yields mannitol as well as sorbitol. L-iditol can be formed by sorbitol isomerization, mainly at the end of the reaction due to the high concentration of sorbitol. Gluconic acid can be formed by D-glucose oxidation but its formation can be prevented by deoxigenating the reaction mixture before reaction. Additionally, xylitol and lower sugar alcohols could be formed by hydrogenolysis of sorbitol mainly at high temperatures, but they are not present under usual industrial conditions.

Formation of such a by-products can be largely prevented by carefully selecting the reaction conditions as shown by the fact herein reported that in the catalytic hydrogenation of D-glucose a 100% D-glucose conversion and a higher than 99% selectivity to sorbitol are obtained. The typical procedures for recovery and purification of sorbitol manufactured by catalytic hydrogenation and electroreduction of D-glucose have been depicted in Figs. 9.2 and 9.3, respectively. Also, chromatographic methods are currently industrially used for separating mixtures of sorbitol and mannitol, such as the PuriTech's ION-IX continuous countercurrent technology [121] by means of which a typical blend of mannitol/sorbitol with a content of mannitol between 20% and 50% is separated to produce both chemicals in a purity ≥98%.

Relative to the biotechnological production, the sorbitol recovery and purification is only addressed in a few papers. Ferraz et al. [122] reported the use of an electrodialysis (ED) system coupled to the bioreactor to simultaneously remove

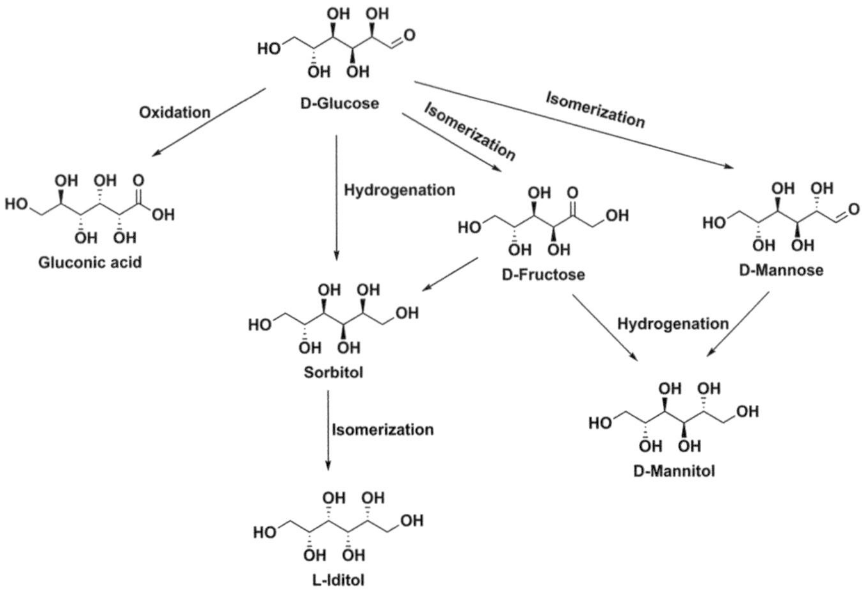


**Fig. 9.7** Main by-products in sorbitol production

gluconic acid from the medium as it is produced. Gluconic acid is a weak acid (pKa 3.86) meaning that at the pH (6.2–6.5) of the medium it is fully dissociated and, therefore, it is obvious that it can be separated and concentrated in the concentration compartment of a 2-compartments ED cell, while sorbitol remains in the diluted compartment. However, we consider that full removal of gluconic acid by ED is not economically feasible. Below a concentration of 1–5 $g.L^{-1}$ the sorbitol solution must be treated by ion exchange for separating the residual gluconic acid. Chun and Rogers [94] reported the separation of the products using a basic anion exchange resin and a solution of $Na_2B_4O_7/H_3BO_3$ as eluent, and Silveira et al. [123] proposed a method for recovering sorbitol and sodium gluconate by selective precipitation of sodium gluconate with organic solvents like methanol and ethanol. Anyway, it is obvious that a full recovery procedure of sorbitol from the complex biotech reaction media has not developed yet, and further and intensive research must be carried out to achieve this goal. One possibility to investigate is the use of the aqueous two-phase extraction technique, taking advantage of the salting-out effect created by saturating the aqueous solution with a highly soluble salt. This method has given very good results for recovering and concentrating 2,3-butanediol from a fermentation broth using an ethanol/phosphate system [124]. A 98.1% 2,3-butanediol recovery was achieved.

## 9.4 Conclusions and Future Outlook

Three reported processes for sorbitol production have been reviewed: (i) catalytic hydrogenation of D-glucose, (ii) electroreduction of D-glucose and, (iii) biotechnological production from fructose using the enzyme D-glucose-fructose oxidoreductase (GFOR) from *Z. mobilis*.

While the electroreduction of D-glucose was the first process industrially operated, it was soon replaced by the catalytic hydrogenation of the same raw material using Raney Ni catalysts due to both the high power consumption and low productivity of the former. In fact, the latter is the process exclusively used currently at industrial level due of course to its higher productivity, but especially because a 100% D-glucose conversion is achieved with a selectivity of sorbitol higher than 99%, which simplified strongly the purification process. The early problems related to nickel deactivation by sintering, leaching into the reaction mixture and poisoning have been largely solved by using some promoters, such as Cr and Mo, developing Ni supported catalysts to increase metal dispersion and large exposed surface area resulting in improved Ni activity and stability, and using new catalysts, with ruthenium-based ones being the most used due to the much higher activity of Ru (up to 50-fold higher than Ni) and because Ru is not leached. These features make the catalytic hydrogenation of D-glucose an extremely solid production process which will be difficult to be displaced. However, in the next 5 years sorbitol worldwide demand will exceed 2300 kt and it is expected to grow continuously which represents a challenge for increasing the productivity at lower costs. This challenge will be the driving force for further improving the catalytic chemical production of sorbitol through three ways in which research efforts should be focused on:

1. Developing more efficient continuous processes, which potentially are also more environmentally friendly as shown by LCA analysis [27];
2. Designing new more efficient and stable catalysts for working both at lower temperatures and pressures than present. It is worth to mention that hydrogenation pressures used in sorbitol production are much higher (10–15 MPa) than those usually employed in most industrial hydrogenation processes (<1.5 MPa). Ru-based catalysts with enhanced activity and stability by introducing metal promoters and modified supports appear to be the more promising way; and
3. Developing industrially cost-effective methods for using directly the polymeric sources of D-glucose as raw materials, mainly cellulose because it does not interfere with the food chain. Taking into account both the low cellulose concentrations in the current reported processes and that cellulose hydrolysis is the limiting step the one pot option does not appear to be the appropriate choice. Probably, a two-step process involving a first step of cellulose fast hydrolysis and a second one of hydrogenation of the hydrolysate. The fast hydrolysis step could be carried out by the method developed by Fang [125] consisting of putting lignocellulose biomass in pure water in concentrations up to 51.5% and heating rapidly the mixture to 603–676 K at 19–42 MPa. As a result, 89 ~ 99% of the lignocellulose biomass is dissolved and rapidly hydrolyzed to saccharides in 3.38 ~

21.79 s, or by the Ultra-Fast Hydrolysis (UFH) method reported by Cantero et al. [126] using a continuous facility with instantaneous heating and cooling and with reaction residence times as low as 20 milliseconds. At 673 K and 23 MPa cellulose was hydrolyzed into soluble sugars with a selectivity of 96% on a carbon basis at a residence time of 30 milliseconds with an extremely low selectivity to 5-hydroxymethylfurfural, a known inhibitor of fermentations, of 0.01%. Reducing the residence time of a continuous process to milliseconds is a key breakthrough in process intensification opening the possibility of moving from the conventional reactors ($m^3$ volume) to microreactors ($cm^3$ volume).

Since its replacement by the catalytic hydrogenation of D-glucose, the research efforts devoted to the electrochemical production of sorbitol from D-glucose have been scarce and basically devoted to look for an anodic reaction producing a valuable chemical, ascorbic acid, for reducing production costs. However, the sorbitol and ascorbic acid market volumes are so unbalanced that this is not a suitable approach for revival of sorbitol electrochemical production. However, the electrochemical method has several important advantages. Namely, it operates at ambient pressure, it does not use hydrogen and it works at a lower temperature than the chemical hydrogenation. In other words, it is attractive due to safety issues associated with hydrogen handling and the electrochemical process operates at ambient pressure. Moreover, the strong drawbacks that led to its displacement by the catalytic hydrogenation, i.e. high power consumption and low productivity, could be overcome by the great advances in electrochemical technology since that time. Thus, the use of decorated (with Ni and Ru) porous cathodes has shown to be very effective in the reduction of carbonyl moieties to alcohol ones [127] and could allow to work at higher apparent current densities, i.e. higher productivities, while keeping both a low cathode potential and a very high sorbitol selectivity resulting in both lower cost and OPEX. On the other hand, much lower power consumption could be achieved by using the zero gap cells [128] in which both anode and cathode are in contact with the membrane, i.e. inter-electrode gap is equal to membrane thickness (between ~150 and 250 μm, depending on membrane type). Consequently, the electrochemical production of sorbitol could have a niche in low volume production plants (below 10,000 $t.year^{-1}$) where the facility needed in a catalytic hydrogenation process for in situ hydrogen production is not economically feasible.

Regarding the sorbitol biotechnological production, the use of GFOR from *Z. mobilis* constitute currently the best method. However, it still bears some weaknesses that make it uncompetitive with respect to chemical methods. One of them is that the enzyme requires a co-substrate in addition to the substrate (fructose) to be converted into sorbitol. Therefore, when D-glucose is the co-substrate, as usually occurs, one mole of gluconic acid is obtained per mole of sorbitol produced. This means that, in order to favor the cost-effectiveness of the biotechnological process, it would be necessary to find a suitable outlet for such gluconic acid, which allows increasing its market well above its current volume. Perhaps, other donor substrates alternative to D-glucose could be favorably introduced, providing that their resulting aldonic acids have a larger market than that of gluconic acid. Another factor that hinders progress in the application of this biotechnological process is that the substrate is fructose, a monosaccharide having a relatively high cost compared to product value. So, the use of low cost substrates, such as molasses or inulin/fructan-rich feedstocks, alternative to pure or semipure fructose preparations, must be a priority. Likewise, the application of recombinant DNA technologies could drive progress in this field, through the improvement of the catalytic properties of GFOR and its large-scale production to allow its use as an efficient biocatalyst. Finally a critical issue so that the biotechnological production of sorbitol has an industrial possibility is the development of a cost-effective separation and purification procedure. This is a current major bottleneck for further development. Few papers are devoted to aspects of separation and purification of sorbitol. Researchers in the field need to be aware of its key importance and increase dramatically their efforts to develop an industrially acceptable solution.

## References


1. Taylor RL. Sorbitol from D-glucose by electrolytic reduction. Chem Met Eng. 1937;44:588–91.
2. Werpy T, Petersen G, editors. Top value added chemicals from biomass; Vol. 1: results of screening for potential candidates from sugars and synthesis gas. Oak Ridge: U.S. Department of Energy; 2004.
3. Bozell JJ, Petersen GR. Technology development for the production of biobased products from Biorefinery carbohydrates – the US department of Energy's "Top 10" revisited. Green Chem. 2010;12:539–54.
4. Grand View Research. Global sorbitol market. https://www.grandviewresearch.com/press-release/global-sorbitol-market (2015). Last accessed 15 May 2016.
5. Radiant Insights Inc.. Sorbitol market size, price analysis, research report, 2020. http://www.radiantinsights.com/research/sorbitol-market (2015). Last accessed 15 May 2016.
6. The manufacture and production of valuable products from sorbitol. GB Patent 301655 (1928).
7. Pappenberger G, Hohmann HP. Industrial production of L-ascorbic acid (vitamin C) and D-isoascorbic acid. Adv Biochem Eng Biotechnol. 2014;143:143–88.
8. Laird K. Roquette brings world's largest isosorbide production unit on stream. Plastics today. http://www.plasticstoday.com/roquette-brings-worlds-largest-isosorbide-production-unit-stream/85076483422068 (2015). Last accessed 15 May 2016.
9. Rose M, Palkovits R. Isosorbide as a renewable platform chemical for versatile applications—Quo Vadis? ChemSusChem. 2012;5:167–76.
10. Bersot JC, Jacquel N, Saint-Loup R, Fuertes P, Rousseau A, Pascault JP, Spitz R, Fenouillot F, Monteil V. Efficiency increase of poly (ethylene terephthalate-co-isosorbide terephthalate) synthesis using bimetallic catalytic systems. Macromol Chem Phys. 2011;212:114–2120.
11. Ochoa-Gómez JR, Gil-Río S, Maestro-Madurga B, Lorenzo-Ibarreta L, Gómez de Miranda O. Improved method for manufacturing 1,4:3,6-dianhydrohexitol di(alkyl carbonate)s. US Patent Application 20150336978A1 (2015).
12. Fenouillot F, Rousseau A, Colomines G, Saint-Loup R, Pascault JP. Polymers from renewable 1,4:3,6-dianhydrohexitols (isosorbide, isomannide and isoidide): A review. Prog Polym Sci. 2010;35:578–622.
13. Ionescu M. Polyether polyols for rigid polyurethane foams. In:Chemistry and technology of polyols for polyurethanes. Shawbury: Rapra Technology Limited; 2005. p. 343–4.

14. DuPont de Nemours. Improvements in or relating to the manufacture of polyhydric alcohols. GB Patent 528064 (1940).
15. Zhao L, Zhou JH, Sui ZJ, Zhou XG. Hydrogenolysis of sorbitol to glycols over carbon nanofiber supported ruthenium catalyst. Chem Eng Sci. 2010;65:30–5.
16. Du W-C, Zheng L-P, Shi J-J, Xia S-X, Hou Z-Y. Production of $C_2$ and $C_3$ polyols from D-sorbitol over hydrotalcite-like compounds mediated bi-functional Ni–Mg–Al–Ox catalysts. Fuel Process Technol. 2015;139:86–90.
17. Guo X, Guan J, Li B, Wang X, Mu X, Liu H. Conversion of biomass-derived sorbitol to glycols over carbon materials supported Ru-based catalysts. Scientific Reports 5: Article number 16451 (2015).
18. Ruppert AM, Weinberg K, Palkovits R. Hydrogenolysis goes bio: from carbohydrates and sugar alcohols to platform chemicals. Angew Chem Int Ed. 2012;51:2564–601.
19. Pang J, Zheng M, Sun R, Wang A, Wang X, Zhang T. Synthesis of ethylene glycol and terephthalic acid from biomass for producing PET. Green Chem. 2016;18:342–59.
20. Dusselier M, Wouwe PV, Dewaele A, Makshina E, Sels BF. Lactic acid as a platform chemical in the biobased economy: the role of chemocatalysis. Energy Environ Sci. 2013;6:1415–42.
21. Ramírez-López C, Ochoa-Gómez JR, Gil-Río S, Gómez-Jiménez-Aberasturi O, Torrecilla-Soria J. Chemicals from biomass: synthesis of lactic acid by alkaline hydrothermal conversion of sorbitol. J Chem Technol Biotechnol. 2011;86:867–74.
22. Huber GW, Dumesic JA. An overview of aqueous-phase catalytic processes for production of hydrogen and alkanes in a biorefinery. Catal Today. 2006;111:119–32.
23. Xi J, Xia Q, Shao Y, Ding D, Yang P, Liu X, Lu G, Wang Y. Production of hexane from sorbitol in aqueous medium over $Pt/NbOPO_4$ catalyst. Appl Catal B-Environ. 2016;181:699–706.
24. Weng Y, Qiu S, Ma L, Liu Q, Ding M, Zhang Q, Zhang Q, Wang T. Jet-fuel range hydrocarbons from biomass-derived sorbitol over Ni-HZSM-5/SBA-15 catalyst. Catalysts. 2015;5:2147–60.
25. Müller J, Hoffmann U. Verfahren zur Darstellung von hochwertigen Alkoholen durch katalytische Reduktion von Zuckerarten mit Wasserstof. DE Patent 544 666 (1925).
26. Müller J, Hoffmann U. Verfahren zur Darstellung von hochwertigen Alkoholen durch katalytische Reduktion von Zuckerarten mit Wasserstof. DE Patent 554 074 (1926).
27. Gericke D, Ott D, Matveeva VG, Sulman E, Aho A, Murzin DY, Roggan S, Danilova L, Hessel V, Loebg P, Kralisch D. Green catalysis by nanoparticulate catalysts developed for flow processing? Case study of D-glucose hydrogenation. RSC Adv. 2015;5:15898–908.
28. Schiweck H, Bär A, Vogel R, Schwarz E, Kunz M, Dusautois C, Clement A, Lefranc C, Lüssem B, Moser M, Peters S. Sugar alcohols. In: Ullmann's Encyclopedia of Industrial Chemistry. Wiley-VCH Verlag GmbH & Co. KGaA; 2012.
29. Court J, Damon JP, Masson J, Wierzchowski P. Hydrogenation of D-glucose with bimetallic catalysts (NiM) of Raney type. Stud Surf Sci Catal. 1988;41:189–96.
30. Gallezot P, Cerino PJ, Blanc B, Flèche G, Fuertes P. D-glucose hydrogenation on promoted Raney-nickel catalysts. J Catal. 1994;146:93–102.
31. Li H, Li H, Deng J-F. D-glucose hydrogenation over Ni–B/$SiO_2$ amorphous alloy catalyst and the promoting effect of metal dopants. Catal Today. 2002;74:53–63.
32. Hoffer BW, Crezee E, Devred F, Mooijman PRM, Sloof WG, Kooyman P, van Langeveld AD, Kapteijn F, Moulijn JA. The role of the active phase of Raney-type Ni catalysts in the selective hydrogenation of D-glucose to d-sorbitol. Appl Catal A-General. 2003;253:437–52.
33. Hoffer BW, Crezee E, Mooijman PRM, van Langeveld AD, Kapteijn F, Moulijn JA. Carbon supported Ru catalysts as promising alternative for Raney-type Ni in the selective hydrogenation D D-glucose. Catal Today. 2003;79–80:35–41.
34. Kusserow B, Schimpf S, Claus P. Hydrogenation of D-glucose to Sorbitol over Nickel and Ruthenium Catalysts. Adv Synth Catal. 2003;345:289–99.

35. Schimpf S, Louis C, Claus P. Ni/$SiO_2$ catalysts prepared with ethylenediamine nickel precursors: Influence of the pretreatment on the catalytic properties in D-glucose hydrogenation. Appl Catal A. 2007;318:45–53.
36. Geyer R, Kraak P, Pachulski A, Schödel R. New catalysts for the hydrogenation of D-glucose to sorbitol. Chem Ing Tech. 2012;84:513–6.
37. Li S, Chen H, Shen J. Preparation of highly active and hydrothermally stable nickel catalysts. J Colloid Interface Sci. 2016;447:68–76.
38. Li HX, Wang WJ, Deng JF. D-glucose hydrogenation to sorbitol over a skeletal Ni-P amorphous alloy catalyst (Raney Ni-P). J Catal. 2000;191:257–60.
39. Wisniak J, Simon R. Hydrogenation of D-glucose, fructose, and their mixtures. Ind Eng Chem Prod Res Dev. 1979;18:50–7.
40. Michel C, Gallezot P. Why is ruthenium an efficient catalyst for the aqueous-phase hydrogenation of biosourced carbonyl Compounds? ACS Catal. 2015;5:4130–2.
41. Guo H, Li H, Zhua J, Ye W, Qiao M, Dai W. Liquid phase D-glucose hydrogenation to d-glucitol over an ultrafine Ru-B amorphous alloy catalyst. J Mol Catal A Chem. 2003;200:213–21.
42. Zhang J, Li J-B, Wu S-B, Liu Y. Advances in the catalytic production and utilization of sorbitol. Ind Eng Chem Res. 2013;52:11799–815.
43. Guo X, Wang X, Guan J, Chen X, Qin Z, Mu X, Xian M. Selective hydrogenation of D-D-glucose to D-sorbitol over Ru/ZSM-5 catalysts. Chin J Catal. 2014;35:733–40.
44. Mishra DK, Dabbawala AA, Parka JJ, Jhung SH, Hwang J-S. Selective hydrogenation of D-glucose to D-sorbitol over HY zeolite supported ruthenium nanoparticles catalysts. Catal Today. 2014;232:99–107.
45. Mishra DK, Lee J-S, Chang JS, Hwang J-S. Liquid phase hydrogenation of D-glucose to D-sorbitol over the catalyst (Ru/NiO–$TiO_2$) of ruthenium on a NiO-modified $TiO_2$ support. Catal Today. 2012;185:104–8.
46. Aho A, Roggan S, Simakova OA, Salmi T, Murzin DY. Structure sensitivity in catalytic hydrogenation of D-glucose over ruthenium. Catal Today. 2015;241:195–9.
47. Lazaridis PA, Karakoulia S, Delimitis A, Comanc SM, Parvulescu VI, Triantafyllidis KS. D-glucose hydrogenation/hydrogenolysis reactions on noble metal(Ru, Pt)/activated carbon supported catalysts. Catal Today. 2015;257:281–90.
48. Wang S, Wei W, Zhao Y, Li H, Li H. Ru–B amorphous alloy deposited on mesoporous silica nanospheres: an efficient catalyst for D-glucose hydrogenation to D-sorbitol. Catal Today. 2015;258:327–36.
49. Santiso EE, George AM, Turner CH, Kostov MK, Gubbins KE, Buongiorno-Nardelli M, Sliwinska-Bartkowiak M. Adsorption and catalysis: The effect of confinement on chemical reactions. Appl Surf Sci. 2005;252:766–77.
50. Pan X, Fan Z, Chen W, Ding Y, Luo H, Bao X. Enhanced ethanol production inside carbon-nanotube reactors containing catalytic particles. Nat Mater. 2007;6:507–11.
51. Dabbawala AA, Mishra DK, Hwang J-S. Selective hydrogenation of D-glucose using amine functionalized nanoporous polymer supported Ru nanoparticles based catalyst. Catal Today. 2016;265:163–73.
52. Doluda V, Grigorev M, Matveeva V, Sulman E, Sulman M, Lakina N, Molchanov V, Rebrov EV. Evaluation of D-glucose hydrogenation catalysts stability in different reactor systems. WSEAS Transact Biol Biomed. 2016;13:44–51.
53. Boyers GG, Flushing NY. Hydrogenation of mono- and disaccharides to polyols. US Patent 2,868,847 (1959).
54. Lepper H, Schütt H. Process for the continuous preparation of polyhydric alcohols. US Patent 4,520,211 (1985).
55. Déchamp N, Gamez A, Perrard A, Gallezot P. Kinetics of D-glucose hydrogenation in a trickle-bed reactor. Catal Today. 1995;24:29–34.
56. Tukac V. D-glucose hydrogenation in a trickle-bed reactor. Collect Czechoslov Chem Commun. 1997;62:1243–8.

57. Arena BJ. Deactivation of ruthenium catalysts in continuous D-glucose hydrogenation. Appl Catal A General. 1992;87:219–29.
58. Gallezot P, Nicolaus N, Flèche G, Fuertes P, Perrad A. D-glucose hydrogenation on ruthenium catalysts in a trickle-bed reactor. J Catal. 1998;180:51–5.
59. Aho A, Roggan S, Eränen K, Salmia T, Murzin DU. Continuous hydrogenation of D-glucose with ruthenium on carbon nanotube catalysts. Cat Sci Technol. 2015;5:953–9.
60. Yamaguchi A, Sato O, Mimura N, Hirosaki Y, Kobayashi H, Fukuoka A, Shirai M. Direct production of sugar alcohols from wood chips using supported platinum catalysts in water. Catal Commun. 2014;54:22–6.
61. Li J, Soares HSMP, Moulijn JA, Makkee M. Simultaneous hydrolysis and hydrogenation of cellobiose to sorbitol in molten salt hydrate media. Cat Sci Technol. 2013;3:1565–72.
62. Almeida JMAR, Da Vià L, Carà D, Carvalho Y, Romano PN, Peña JAO, Smith L, Sousa-Aguiar EF, Lopez-Sanchez JA. Screening of mono- and bi-functional catalysts for the one-pot conversion of cellobiose into sorbitol. Catal Today. 2016; In press, corrected proof available.
63. Hartstra L, Bakker L, van Westen HA. Hydrogenation of carbohydrates. US Patent 2518235 (1950).
64. Jacobs P, Hinnekens H. Single-step catalytic process for the direct conversion of polysaccharides to polyhydric alcohols. EP Patent 0329923 (1989).
65. Kasehagen L. Hydrogenation of carbohydrates. US Patent 2,968,680 (1961).
66. Kruse WM, Wright LW. Polyhydric alcohol production using ruthenium zeolite catalyst. US Patent 3,963,788 (1976).
67. Van de Vyver S, Geboers J, Jacobs PA, Sels BF. Recent advances in the catalytic conversion of cellulose. ChemCatChem. 2011;3:82–94.
68. Yabushita M, Kobayashi H, Fukuoka A. Catalytic transformation of cellulose into platform chemicals. Appl Catal B Environ. 2014;145:1–9.
69. Li Y, Liao Y, Cao X, Wang T, Ma L, Long J, Liu Q, Xua Y. Advances in hexitol and ethylene glycol production by one-pot hydrolytic hydrogenation and hydrogenolysis of cellulose. Biomass Bioenergy. 2015;74:148–61.
70. Balandin AA, Vasyunina NA, Barysheva GS, Chepigo SV, Dubinin MM. Letters to the editor. Bull Acad Sci USSR Div Chem Sci. 1957;6:403–4.
71. Balandin AA, Vasyunina NA, Chepigo SV, Barysheva GS. Hydrolytic hydrogenation of cellulose. Dokl Akad Nauk SSSR. 1959;128:941–4.
72. Geboers J, Van de Vyver S, Carpentier K, Blochouse K, Jacobs P, Sels B. Efficient catalytic conversion of concentrated cellulose feeds to hexitols with heteropolyacids and Ru on carbon. Chem Commun. 2010;46:3577–9.
73. Palkovits R, Tajvidi K, Procelewska J, Rinaldi R, Ruppert A. Hydrogenolysis of cellulose combining mineral acids and hydrogenation catalysts. Green Chem. 2010;12:972–8.
74. Fukuoka A, Dhepe PL. Catalytic conversion of cellulose into sugar alcohols. Angew Chem. 2006;118:5285–7.
75. Dhepe PL, Fukuoka A. Cracking of cellulose over supported metal catalysts. Catal Surv Jpn. 2007;11:186–91.
76. Jollet V, Chambon F, Rataboul F, Cabiac A, Pinel C, Guillon E, Essayem N. Non-catalyzed and Pt/γ-$Al_2O_3$-catalyzed hydrothermal cellulose dissolution-conversion: influence of the reaction parameters and analysis of the unreacted cellulose. Green Chem. 2009;11:2052–60.
77. Han JW, Lee H. Direct conversion of cellulose into sorbitol using dual-functionalized catalysts in neutral aqueous solution. Catal Commun. 2012;19:115–8.
78. Luo C, Wang S, Liu HC. Cellulose conversion into polyols catalyzed by reversibly formed acids and supported ruthenium clusters in hot water. Angew Chem. 2007;119:7780–3.
79. Deng W, Tan X, Fang W, Zhang Q, Wang Y. Conversion of cellulose into sorbitol over carbon nanotube-supported ruthenium. Catal Lett. 2009;133:167–74.
80. Ribeiro LS, Órfão JJM, Pereira MFR. Enhanced direct production of sorbitol by cellulose ball-milling. Green Chem. 2015;17:2973–80.

81. Suzuki T, Nakagami J. Effect of crystallinity of microcrystalline cellulose on the compactability and dissolution of tablets. Eur J Pharm Biopharm. 1999;47:225–30.
82. Isik M, Sardon H, Mecerreyes D. Ionic liquids and cellulose: dissolution, chemical modification and preparation of new cellulosic materials. Int J Mol Sci. 2014;15:11922–40.
83. Ignatyev IA, Van Doorslaer C, Mertens PGN, Binnemans K, De Vos DE. Reductive splitting of cellulose in the ionic liquid 1-butyl-3-methylimidazolium chloride. ChemSusChem. 2010;3:91–6.
84. Creighton HJ, Hales RA. Electrolytic process for reducing sugars. US Patent 2,458,895 (1949).
85. Hefti HR, Kolb W. Electrolytic reduction of sugars. US Patent 2,507,973 (1950).
86. Parl K, Pintauro PN, Baizer MM, Nobe K. Flow reactor studies of the paired electro-oxidation and electroreduction of D-glucose. J Electrochem Soc Electrochem Sci Technol. 1985;132:1850–5.
87. Li H, Li W, Guo Z, Gu D, Cai S, Fujishima A. The paired electrochemical synthesis of gluconic acid and sorbitol. Collect Czechoslov Chem Commun. 1995;60:928–34.
88. Silveira MM, Jonas R. The biotechnological production of sorbitol. Appl Microbiol Biotechnol. 2002;59:400–8.
89. Sprenger GA. Carbohydrate metabolism in *Zymomonas mobilis*: a catabolic highway with some scenic routes. FEMS Microbiol Lett. 1996;145:301–7.
90. Viikari L. Formation of levan and sorbitol from sucrose by *Zymomonas mobilis*. Appl Microbiol Biotechnol. 1984;19:252–5.
91. Barrow KD, Collins JG, Leight DA, Rogers PL, Warr RG. Sorbitol production by *Zymomonas mobilis*. Appl Microbiol Biotechnol. 1984;20:225–32.
92. Leigh D, Scopes RK, Rogers PL. A proposed pathway for sorbitol production by *Zymomonas mobilis*. Appl Microbiol Biotechnol. 1984;20:413–5.
93. Zachariou M, Scopes RK. D-glucose-fructose oxidoreductase, a new enzyme isolated from *Zymomonas mobilis* that is responsible for sorbitol production. J Bacteriol. 1986;167:863–9.
94. Scopes RK, Rogers PL, Leigh DA. Method for the production of sorbitol and gluconate. US Patent 4,755,467 (1988).
95. Chun UH, Rogers PL. The simultaneous production of sorbitol from fructose and gluconic acid from D-glucose using an oxidoreductase of *Zymomonas mobilis*. Appl Microbiol Biotechnol. 1988;29:19–24.
96. Ichikawa Y, Kitamoto Y, Kato N, Mori N. Preparation of gluconic acid and sorbitol. European Patent Application 322,723 (1989).
97. Bringer-Meyer S, Sahm H. Process for obtaining sorbitol and gluconic acid by fermentation, and cell material suitable for this purpose. US Patent 5,017,485 (1991).
98. Rehr B, Wilhelm C, Sahm H. Production of sorbitol and gluconic acid by permeabilized cells of *Zymomonas mobilis*. Appl Microbiol Biotechnol. 1991;35:144–8.
99. Rehr B, Sahm H. Process for obtaining sorbitol and gluconic acid or gluconate. US Patent 5,102,795 (1992).
100. Rehr B, Sahm H. Process for obtaining sorbitol and gluconic acid or gluconate using *Zymomonas mobilis*. US Patent 5,190,869 (1993).
101. Jang KH, Jung SJ, Chang HS, Chun UH. Improvement of the process for sorbitol production with *Zymomonas mobilis* immobilized in κ-carrageenan. Process Biochem. 1996;31:485–92.
102. Ferraz HC, Alves TLM, Borges CP. Coupling of an electrodialysis unit to a hollow fiber bioreactor for separation of gluconic acid from sorbitol produced by *Zymomonas mobilis* permeabilized cells. J Membr Sci. 2001;191:43–51.
103. Wisbeck E, Silveira MM, Ninow J, Jonas R. Evaluation of the flocculent strain *Zymomonas mobilis* Z1–81 for the production of sorbitol and gluconic acid. J Basic Microbiol. 1997;6:445–9.

104. Silveira MM, Wisbeck E, Lemmel C, Erzinger GS, Lopes da Costa JP, Bertasso M, Jonas R. Bioconversion of D-glucose and fructose to sorbitol and gluconic acid by untreated cells of *Zymomonas mobilis*. J Biotechnol. 1999;75:99–103.
105. Ro H, Kim H. Continuous production of gluconic acid and sorbitol from sucrose using invertase and an oxidoreductase of *Zymomonas mobilis*. Enzym Microb Technol. 1991;13:920–4.
106. Kim DM, Kim HS. Continuous production of gluconic acid and sorbitol from Jerusalem artichoke and D-glucose using an oxidoreductase of *Zymomonas mobilis* and inulinase. Biotechnol Bioeng. 1992;39:336–42.
107. Cazetta ML, Celligoi MPC, Buzato JB, Scarmino IS, Da Silva RSF. Optimization study for sorbitol production by *Zymomonas mobilis* in sugar cane molasses. Process Biochem. 2005;40:747–51.
108. An K, Hu F, Bao J. Simultaneous saccharification of inulin and starch using commercial glucoamylase and the subsequent bioconversion to high titer sorbitol and gluconic acid. Appl Biochem Biotechnol. 2013;171:2093–104.
109. Liu C, Dong H, Zhong J, Ryu DD, Bao J. Sorbitol production using recombinant *Zymomonas mobilis* strain. J Biotechnol. 2010;148:105–12.
110. Pedruzzi I, da Silva EAB, Rodrigues AE. Production of lactobionic acid and sorbitol from lactose/fructose substrate using GFOR/GL enzymes from *Zymomonas mobilis* cells: a kinetic study. Enzym Microb Technol. 2011;49:183–91.
111. Gollhofer D, Nidetzky B, Fürlinger M, Kulbe KD. Efficient protection of D-glucose-fructose oxidoreductase from *Zymomonas mobilis* against irreversible inactivation during its catalytic action. Enzym Microb Technol. 1995;17:235–40.
112. Nidetzky B, Fürlinger M, Gollhofer D, Scopes RK, Haltrich D, Kulbe KD. Improved operational stability of cell-free D-glucose-fructose oxidoreductase from *Zymomonas mobilis* for the efficient synthesis of sorbitol and gluconic acid in a continuous ultrafiltration membrane reactor. Biotechnol Bioeng. 1997;53:623–9.
113. Silva-Martinez M, Haltrich D, Novalic S, Kulbe KD, Nidetzky B. Simultaneous enzymatic synthesis of gluconic acid and sorbitol: continuous process development using D-glucose-fructose oxidoreductase from *Zymomonas mobilis*. Appl Biochem Biotechnol. 1998;70–72:863–8.
114. Tani Y, Vongsuvanlert V. Sorbitol production by a methanol yeast, *Candida boidinii* (*Kloeckera* sp.) No. 2201. J Ferment Technol. 1987;65:405–11.
115. Vongsuvanlert V, Tani Y. Characterization of D-sorbitol dehydrogenase involved in D-sorbitol production of a methanol yeast, *Candida boidinii* (*Kloeckera* sp.) No. 2201. Agric Biol Chem. 1988;52:419–26.
116. Duvnjak Z, Turcotte G, Duan ZD. Production of sorbitol and ethanol from Jerusalem artichokes by *Saccharomyces cerevisiae* ATCC 36859. Appl Microbiol Biotechnol. 1991;35:711–5.
117. Nissen L, Perez-Martinez G, Yebra MJ. Sorbitol synthesis by an engineered *Lactobacillus casei* strain expressing a sorbitol-6-phosphate dehydrogenase gene within the lactose operon. FEMS Microbiol Lett. 2005;249:177–83.
118. De Boeck R, Sarmiento-Rubiano LA, Nadal I, Monedero V, Pérez-Martínez G, Yebra MJ. Sorbitol production from lactose by engineered *Lactobacillus casei* deficient in sorbitol transport system and mannitol-1-phosphate dehydrogenase. Appl Microbiol Biotechnol. 2010;85:1915–22.
119. Ladero V, Ramos A, Wiersma A, Goffin P, Schanck A, Kleerebezem M, Hugenholtz J, Smid EJ, Hols P. High-level production of the low-calorie sugar sorbitol by *Lactobacillus plantarum* through metabolic engineering. Appl Environ Microbiol. 2007;73:1864–72.
120. van Gorp K, Boerman E, Cavenaghi CV, Berben PH. Catalytic hydrogenation of fine chemicals: sorbitol production. 1999 52;349–61.
121. Puritech. Sorbitol/mannitol Separation. http://www.puritech.be/food-processing/sorbitol-mannitol (2016). Last accessed 15 July 2016.

122. Ferraz HC, Borges CP, Alves TLM. Produção de sorbitol e ácido glicônico por células permeabilizadas e imobilizadas de *Zymomonas mobilis* com separação simultânea dos produtos por eletrodiálise. In: Proceedings of the 13th national symposium on fermentation. Terezópolis, Brazil; (2000)
123. Silveira MM, Lopes da Costa JP, Jonas R. Processo de produção e recuperação de sorbitol e ácido glucônico ou gluconato. Brazilian Patent PI 9.403.981-0 (1994).
124. Jiang B, Li Z-G, Dai J-Y, Zhang D-J, Xiu ZL. Aqueous two-phase extraction of 2,3-butanediol from fermentation broths using an ethanol/phosphate system. Process Biochem. 2009;44:112–7.
125. Fang Z. Method for the dissolving and rapid hydrolyzing of lignocellulose biomass, device thereof and use of the same. US Patent 9,243,303 (2016).
126. Cantero DA, Bermejo D, Cocero MJ. High glucose selectivity in pressurized water hydrolysis of cellulose using ultra-fast reactors. Bioresour Technol. 2013;135:697–703.
127. Ochoa-Gómez JR, García-Luis A, Fernández-Carretero FJ, Lorenzo-Ibarreta L, Prieto S. Method for manufacturing 2,3-butanediol. European Patent Application EP14198812 (2014).
128. Pletcher D, Li X, Wang S. A comparison of cathodes for zero gap alkaline water electrolysers for hydrogen production. Int J Hydrog Energy. 2012;37:7429–35.

# Chapter 10
# Biotechnological Production of Xylitol from Biomass

**Felipe Antonio Fernandes Antunes, Júlio César dos Santos, Mário Antônio Alves da Cunha, Larissa Pereira Brumano, Thais Suzane dos Santos Milessi, Ruly Terán-Hilares, Guilherme Fernando Dias Peres, Kelly Johana Dussán Medina, Débora Danielle Virginio da Silva, Sai Swaroop Dalli, Swapnil Gaikwad, and Silvio Silvério da Silva**

**Abstract** Xylitol is a polyol of interest to food, dental and pharmaceutical industries because of its favorable characteristics such as sweetening capacity, insulin-independent metabolism effects and its lack of carcinogenic properties. It is usually produced by chemical processes that are expensive due to their high energy consumption and many purification steps. Biotechnological routes are promising because they can be carried out using mild conditions and have the possibility of using hydrolysates from renewable sources as raw materials without the need of extensive purification of xylose before the fermentation step. Different lignocellulosic materials have been studied as alternative raw materials in the fermentative process for xylitol production. However, the structure of lignocellulose is recalcitrant and a pretreatment step is necessary to release monomeric sugars that does not form compounds toxic to microorganisms. Another challenge for xylitol production by fermentation is the identification of efficient microorganisms for converting the pentose sugars present in hemicellulosic hydrolysates. Different strategies have also been investigated, aiming to optimize the biotechnological way, such as use of different configurations of bioreactors, process options and downstream steps. This

F.A.F. Antunes • J.C. dos Santos • L.P. Brumano • T.S.d.S. Milessi • R. Terán-Hilares
G.F.D. Peres • D.D.V. da Silva • S. Gaikwad • S.S. da Silva (✉)
Department of Biotechnology, Engineering School of Lorena, University of São Paulo, São Paulo, Brazil
e-mail: silviosilverio@usp.br

M.A.A. da Cunha
Department of Chemistry, Federal University of Technology of Paraná, Curitiba, Brazil

K.J.D. Medina
Department of Biochemistry and Technological Chemistry, Institute of Chemistry, State University of São Paulo – UNESP, São Paulo, Brazil

S.S. Dalli
Lakehead University, Thunder Bay, ON, Canada



Z. Fang et al. (eds.), *Production of Platform Chemicals from Sustainable Resources*, Biofuels and Biorefineries 7, DOI 10.1007/978-981-10-4172-3_10

chapter will explore biotechnological xylitol production from the selection and preparation of the raw material to fermentative process conditions, downstream strategies and future perspectives. These topics will be discussed to offer readers a better understanding of biotechnological routes to xylitol as well as their potential and future prospects.

**Keywords** Xylitol • Hemicellulosic hydrolysate • Lignocellulosic biomass • Pentose fermentation • Biotechnological process

## 10.1 Introduction

Xylitol is a sugar alcohol with different applications, largely used in food, odontological and pharmaceutical industries, with an increasing global market. It has a number of interesting characteristics, e.g. a sweetener whose power is similar to that of glucose and with antimicrobial and anti-cariogenic properties [1, 2]. Moreover, xylitol can be used in other applications, for example, for the synthesis of esters and polymers [3, 4].

Conventional production process of this polyalcohol uses purified xylose obtained by acid hydrolysis of lignocellulosic materials as substrate in a chemical route using metal as catalyst at severe conditions of pressure and temperature. Biotechnological alternative also use hemicellulosic hydrolysates from biomass as raw material with the conversion of pentose sugars by bacteria or yeast. Fermentation of the pentose sugars by suitable microorganisms is an environmental-friendly process that can be carried out under mild conditions such as atmospheric pressure and ambient temperature, avoiding also the need for purification of xylose, which is the major cost-intensive step in conventional catalytic processes [5, 6]. In addition to traditional fermentation, other alternatives have been evaluated for biological production of process. Xylitol production from xylose using enzyme technology can be an attractive alternative to both fermentation and chemical processes [7]. Future approaches can include techniques of metabolic engineering modification for filamentous fungus that can produce xylitol using the same metabolic pathway of yeasts [8].

Optimization of biotechnological xylitol production is another key issue addressed in the literature. For example, evaluation of the use of different biomass sources can be fundamental to decreasing the process cost and expanding the use of this compound. In this way, different techniques for biomass pretreatment and detoxification have been evaluated for enhanced yields in xylose produced from the hemicellulosic fraction of the material. Different strategies of fermentation for xylitol production have been carried out such as batch, fed-batch and continuous process, as well as the use of immobilized cells and bioreactors. In the downstream process, studies about the composition, purity, scale-up and operation costs have been performed for choosing suitable purification processes. In the sections below, different approaches regarding current and developed studies about xylitol biotechnological production from biomass will be contextualized with more details.

## 10.2 Xylitol-Producing Microorganisms and Metabolism

Xylitol is produced by different organisms such as plants, raspberry, strawberry, yellow plum, endive, lettuce and cauliflower. However, xylitol concentration in these vegetables is lower than 1% on dry weight mass, which makes the extraction of this biomolecule uneconomical [9]. In humans and animals, xylitol is produced as an intermediate of carbohydrate metabolism [10]. Regarding microorganisms, a considerable number of bacteria, fungi and yeasts can produce D-xylitol in their metabolism, secreting it extracellularly as a metabolic byproduct of ethanol production or as the mainly product from D-xylose. Among microorganisms, yeasts stands out, especially those from *Candida* genus [11]. However, studies reporting the use of filamentous fungi and bacteria for xylitol production are also found in the literature, as shown in Table 10.1.

Yeasts have been considered the best option for xylitol production process, due to its higher production capacity compared to other microorganisms. Among them, strains from genus *Candida*, as *C. guilliermondii*, *C. tropicalis*, *C. parapsilosis* and *C. boidinii*, have been quite investigated [9]. From previous studies, Guo et al. [19] investigated 274 yeast strains selected for xylitol. Among these screening, the five best xylitol producers were selected; all of them belong to the genus *Candida*. Studies that screened bacteria and filamentous fungi for xylitol production found relatively small quantities of xylitol produced by these microorganisms [33, 34]. For example, while the xylitol production using xylose as substrate by *Candida* yeasts as *Candida guilliermondii* Xu264 reached 26.07 g $L^{-1}$ [19], studies using bacteria and filamentous fungi (not genetically modified) reached much lower values, such as 4.6 and 0.52 g $L^{-1}$, by *Enterobacter* and *Penicillium crustosum*, respectively [29, 33].

An ideal microorganism for xylitol production is one that is easily cultivated, shows high capacity of xylitol production and has excellent tolerance to stress and toxic compounds [19, 35]. Within this context, researchers have been making efforts to discover new xylitol-producing microorganisms [19, 26] and to optimize production conditions [9]. Studies in the field of molecular engineering [36] have also been developed to increase xylitol biotechnological production and improve its economic viability.

**Table 10.1** Natural xylitol-producing microorganisms and some values of concentration, yield and productivity reported in literature

| Microorganism | Achieved xylitol concentration (g•L$^{-1}$) | Yield (g g$^{-1}$) | Productivity (g L$^{-1}$ h$^{-1}$) | Ref. |
|---|---|---|---|---|
| Yeast | | | | |
| *Candida boidinii* | 59.3 | 0.57–0.68 | 0.32–0.46 | [12] |
| *Candida mogii* NRRL Y-17032 | 30 | 0.65 | 0.40 | [13] |
| *Candida parapsilosis* | 210 | 0.7 | 3.18 | [14] |
| *Candida peltata* | n/a | 0.56 | n/a | [15] |
| *Hansenula polymorpha* | 58 | 0.62 | 0.60 | [16] |
| *Debaryomyces hansenii* UFV-170 | 76.6 | 0.73 | n/a | [17] |
| *Candida tropicalis* | 130 | 0.93 | n/a | [18] |
| *Candida maltosa* Xu316 | n/a | 0.49 | 0.97 | [19] |
| *Candida guilliermondii* | 54.6 | 0.78 | 0.58 | [20] |
| *Pichia stipitis* FPL-YS30 | n/a | 0.61 | 0.18 | [21] |
| *Candida amazonenses* | 25.2 | 0.59 | 0.52 | [22] |
| *Pachysolen tannophilus* | n/a | 0.80 | n/a | [23] |
| *Debaryomyces hansenii* | 56.23 | 0.95 | 2.34 | [24] |
| *Kluyveromyces marxianus* CCA 510 | 12.73 | 0.36 | n/a | [25] |
| *Cyberlindnera galapagoensis* f.a., sp. nov. | 24 | 0.64 | 0.33 | [26] |
| *Cyberlindnera xylosilytica* sp. nov. | 33.02 | 0.726 | 0.459 | [27] |
| Bacteria | | | | |
| *Corynebacterium* sp. | 69.0 | n/a | n/a | [28] |
| *Enterobacter liquefaciens* | 33.3 | n/a | n/a | [29] |
| *Mycobacterium smegmatis* | 11.3 | 0.56 | n/a | [30] |
| Filamentous fungi | | | | |
| *Fusarium oxysporum* | 1.0 | – | 0.02 | [31] |
| *Petromyces albertensis* | 39.8 | – | 0.16 | [32] |
| *Penicillium crustosum* | 0.52 | – | 0.005 | [33] |
| *Aspergillus niger* | 0.36 | – | 0.003 | [33] |
| *P. griseoroseum* | 0.16 | – | 0.006 | [33] |
| *P. italicum* | 0.16 | – | 0.003 | [33] |
| *P. roqueforti* | 0.27 | – | 0.003 | [33] |
| *P. citrinum* | 0.27 | – | 0.004 | [33] |
| *P. expansum* | 0.15 | – | 0.003 | [33] |
| *P. purpurogenum* | 0.14 | – | 0.003 | [33] |
| *P. janthinellum* | 0.29 | – | 0.006 | [33] |

*n/a* not available

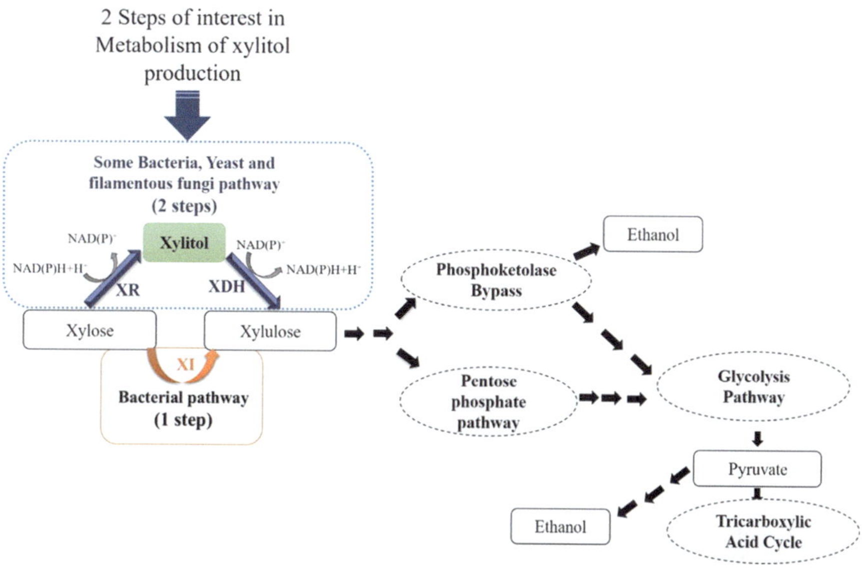


**Fig. 10.1** Metabolic pathway of xylose assimilation by bacteria and yeasts. *XI* xylose isomerase, *XK* xylulose kinase, *XR* Xylose reductase, *XDH* xylitol dehydrogenase, *Red-lined pathway*: xylose metabolism in bacteria; *Blue-lined pathway*: xylose metabolism in yeast

### *10.2.1 Metabolism of Xylitol Production*

Biotechnological xylitol production has been reported from D-xylose, sometimes using co-substrates such as D-glucose, ethanol, and glycerol [37]. Microorganisms assimilate and ferment glucose more readily than xylose. However, there are some bacteria, yeasts and fungi that can ferment pentoses as xylose to produce interesting compounds [38].

In the past, xylitol was considered only a by-product in the ethanol fermentation production from xylose [38]. Nevertheless, the use of xylose as a substrate for xylitol production has an interesting potential, considering that it can be obtained from agro-industrial waste and, unlike chemical synthesis, it does not require pure xylose, thus making the process more cost-effective [39].

Xylose metabolism depends on the type of microorganism: yeasts, bacteria or filamentous fungi. Some pathways are shown in the diagram below according to the different microorganisms (Fig. 10.1).

#### 10.2.1.1 Bacteria

In metabolism of xylose by bacteria, xylose is converted to xylulose in a single step (Fig. 10.1) by the enzyme xylose isomerase (EC 5.3.1.5), and after that, phosphorylated to xylulose-5-phosphate by xylulokinase (EC 2.7.1.17). Xylulose-5-phosphate can be converted into glyceraldehyde-3-phosphate and acetylphosphate by action of xylulose-5-phosphate phosphoketolase (EC 4.1.2.9) or enter in the pentose phosphate pathway (PPP). In addition, there are some bacteria that can present different flow, since they also have the oxidoreductase pathway in addition to/instead of xylose isomerase system (i.e., genes for xylose reductase (XR) and xylitol dehydrogenase (XDH)) [28, 30] and, thus, they can produce xylitol. For example, according to Yoshitake et al. [40], *E. liquefaciens* is able to use both enzymatic routes, such as XR and xylose isomerase, for initial xylose degradation. Other bacteria have been reported as xylitol producers, such as *Corynebacterium* sp. and *Mycobacterium smegmatis*.

#### 10.2.1.2 Filamentous Fungi

Unlike bacteria, fungi (filamentous and yeast) metabolize D-xylose into D-xylulose in two steps (Fig. 10.1). D-Xylose is firstly reduced to xylitol by either NADH- or NADPH dependent D-xylose reductase (XR) (EC 1.1.1.21). The resulting xylitol is either excreted or further oxidized to D-xylulose by $NAD^+$-dependent D-xylitol dehydrogenase (XDH) (EC 1.11.9). Details of these steps are discussed in the topic of yeast metabolism. There are few reports on xylitol production by fungi and its yield depends on the fungus in use and cultivation conditions. For example, as presented in the Table 10.1, the filamentous fungi *Penicillium crustosum* was found to produce 0.52 g $L^{-1}$ of xylitol, *Fusarium oxysporum* produced 1.0 g $L^{-1}$ and *Petromyces albertensis* produced 39.8 g $L^{-1}$ of xylitol. These differences can be explained not only by the difference among the strains but also due to cultivation conditions. In the work with *Penicillium crustosum*, initial xylose concentration was 11.50 g $L^{-1}$ of xylose, with consumption of 76% of this substrate in 96 h [33]. In the other works, *Fusarium oxysporum* was cultivated in a medium containing 50 g $L^{-1}$ of xylose for 48 h under aerobic conditions [31] and Petromyces albertensis, which produced the higher concentration of xylitol, was cultivated in a medium within 100 g $L^{-1}$ of xylose for 240 h [32].

#### 10.2.1.3 Yeasts

Among microorganisms, yeasts are considered to be more efficient producers of xylitol than bacteria or fungi. Xylitol production in yeasts occurs as a natural intermediate during D-xylose metabolism. Xylitol is accumulated between the two steps of bioconversion pathway from xylose to xylulose. The first step involves reduction of xylose to xylitol by XR (EC 1.1.1.21), whereas in the second step, intermediary xylitol is oxidized to xylulose XDH (EC 1.1.1.9). D-xylose reductase (XR) (EC 1.1.1.21) is a NADPH-dependent enzyme, while D-xylitol dehydrogenase (XDH) (EC 1.11.9) requires $NAD^+$ or $NADP^+$ (Fig. 10.1) [24].

Cofactor imbalance results in the secretion of xylitol as a by-product of D-xylose fermentation. Actually, availability of oxygen (aeration) is decisive for xylitol accumulation in yeasts, because it influences the activities XR and XDH, and thereby regulates xylose metabolism [37, 39]. In process with high aeration, oxidation of NADH to $NAD^+$ favors oxidation of xylitol to xylulose, and this compound is directed for biomass production pathway as well as for other cell activities. On the other hand, xylitol accumulation is favored under low oxygen availability, once an imbalance between XR and XDH activities occurs due to the reduction in the rate of oxidative pathway. In this way, xylitol oxidation to xylulose is limited and the polyalcohol can be accumulated and excreted [41].

Silva and Felipe [42] reported on the influence of the presence of other carbohydrates such as arabinose and glucose in the activities of xylose reductase and xylitol dehydrogenase in fermentations using sugarcane bagasse hemicellulose hydrolysate. Others factors as pH, temperature, agitation rate, oxygen transfer rate, percentage of inoculum, inoculum age, and initial xylose concentration are important and affect xylitol production [11]. The optimum conditions for these parameters depend on the microorganism. In addition, catabolite repression is known as one of the factors that affect xylitol production for influencing xylose assimilation.

#### 10.2.1.4 Strategies Relative to Microorganisms and Their Metabolism to Increase the Production of Xylitol

Seeking to improve biotechnological production of xylitol, some biological approaches have been developed by using natural microrganisms or genetically-modified strains. Among applied methodologies, there are traditional methods, such as screening for new strains or mutagenesis, metabolic engineering and enzymatic synthesis. Traditional methods are the main approach found in the literature, although these methodologies have a lower success rate and require high labor [19, 43].

Screening and characterization of new strains for xylitol production are mostly focused on microorganisms that are able to consume xylose as the carbon source. The strains can be obtained by isolating them from the environment or by mutating and selecting strains in the laboratory [22]. Some studies focus on the isolation of microorganisms from forests at lignocellulosic materials, insects and soil [19, 27, 29, 33, 34].

Investigations about engineering strains have been explored to improve the xylitol biotechnology process economically. After xylitol production by recombinant *Saccharomyces cerevisiae* was first reported by [36], various microorganisms have been developed to produce xylitol [37]. Most of the genetic modifications are focused on optimization of the metabolic pathway, simultaneous utilization of mixed sugars, disruption of xylitol uptake and secretion of xylitol (Table 10.2) [43].

Various strategies have been explored aiming to overcome the limitations of microbial conversion of xylitol from xylose at cellular and molecular level. Currently, approaches such as metabolic pathway engineering, increasing in xylose transport by heterologous expression of transport gene, changes in cofactor dependency and use of enzyme technology for increased level have been studied to

**Table 10.2** Characteristics of xylitol production by recombinant microorganisms using various metabolic engineering strategies

| Organism | Genetic modifications | Substrate | Productivity (g $L^{-1}$ $h^{-1}$) | Ref. |
|---|---|---|---|---|
| Yeast | | | | |
| *C. tropicalis* | Expressed At5g17010 | Xylose + glucose | 1.14 | [44] |
| *C. tropicalis* | Expressed *N. crassa* XR, optimized códon | Xylose + glucose | 1.44 | [45] |
| *C. tropicalis* | Overexpressed zwf and gnd | Xylose + glycerol | 1.25 | [46] |
| *C. tropicalis* | Expressed NADH-preferring *C. parapsilosis* XR | Xylose | 5.09 | [47] |
| *C. tropicalis* | Disrupted assimilation of Xylitol | Xylose + glucose/glycerol | 3.3 | [48] |
| *S. cerevisiae* | Expressed *P. stipitis* XYL1, cdt-1 and gh1- | Xylose + cellobiose | 0.55 | [49] |
| *S. cerevisiae* | Expressed *Pichia stipitis* XR gene XYL1 | Xylose + glucose | – | [37] |
| *Kluyveromyces marxianus YZJ74* | Expressed *Candida intermedia* glucose/xylose facilitator gene CiGXF1 | Xylose | 4.14 | [50] |
| *Debaryomyces hansenii* | XDH gene disruption | Xylose + glycerol | 2.34 | [24] |
| Bacteria | | | | |
| *E. coli* | Expressed *C. boidinii* XR replaced crp with crp*, deleted xylB | Xylose + glucose | 0.79<br>4.7 mol (mol glucose)$^{-1}$ | [51] |

| | | | | |
|---|---|---|---|---|
| *E. coli* | Expressed *N. crassa* XR, evolved the XR | Xylose + glucose + arabinose | Xylitol production reached near stoichiometric levels | [52] |
| *E. coli* | Expressed *C. boidinii* XR and *P. stipitis* Xyl3, deleted xylB | Xylose | 0.27 | [53] |
| *C. glutamicum* | Expressed *C. tenuis* XR and araE, deleted ldhA, ptsF, and3 xylB | Xylose + glucose | 7.9 | [54] |
| *C. glutamicum* | Expressed *R. mucilaginosa* NAD(P)H-dependent XR. *E. coli* araA, *A. tumefaciens* dpe and *M. smegmatis* lxr | Xylose + arabinose | 0.28 ± 0.05 g $g^{-1}$ cdw/h | [55] |
| *G. oxydans* | Disrupted adhB | Arabitol + ethanol | 0.98 | [56] |
| Fungi | | | | |
| *T. reesei* | Partial mutation of xdh1 | Xylose | 0.020 | [57] |

increase xylitol production. However, the use engineered strains present some challenges to be overcome as increase in their metabolic stability and tolerance to inhibitors presence [39].

Another possibility reported in literature is enzymatic method for xylitol production, that can result in high yield of xylose conversion. Compared to the fermentation process, the enzymatic method is expected to make a substantial increase in productivity considering no production of other metabolics. This method of production is an alternative biotechnological approach [58]. Studies in characterization, extraction and purification of XR from yeast cells have been carried out aiming to enhance the fermentative process or for direct use in conversion of xylose into xylitol [39, 59]. However, as observed in the metabolism pathway, for favorable xylitol production, coenzymes are a limiting factor in biocatalysis reactions. Thus, the addition of coenzymes, necessary for the use of isolated intracellular enzymes, makes this process expensive and, according to Branco et al. [59], an alternative could be the use of an enzymatic *in situ* regeneration system approach.

## 10.3 Use of Different Biomass for Xylitol Biotechnological Production

Because of their composition, the potential of lignocellulosic materials as raw materials for production of high value-added products, such as xylitol, has been investigated [60]. After a pretreatment step, that aims at releasing fermentable sugars from biomass, some carbon sources can be obtained, such as glucose and xylose. Thus, the conversion of these sugars, mainly xylose, into xylitol, has significant economic importance for the total use of biomass in biorefineries [61].

Although lignocellulosic materials are composed mainly of cellulose, hemicellulose and lignin, the composition proportion of these three main components is different according to each material, as exemplified in Table 10.3. This proportion directly influences the recalcitrant structure of biomass and consequently the sugar release profile and efficiency of the pretreatment step.

Availability of raw material is a key factor for choosing a profitable biomass for use in a bioprocess. For example, In Brazil, sugarcane is the most abundant agricultural crop, with an annual output higher than 600 million tons [68]. In this country,

**Table 10.3** Examples of the composition of lignocellulosic raw materials

| Raw material | Cellulose (%) | Hemicellulose (%) | Lignin (%) | Ref. |
|---|---|---|---|---|
| Sugarcane bagasse | 45.0 | 25.8 | 19.1 | [62] |
| Sugarcane straw | 33.6 | 28.9 | 31.8 | [63] |
| Rice straw | 43.4 | 22.9 | 17.2 | [64] |
| Corn stover | 34.4 | 22.8 | 18.0 | [65] |
| Corncob | 38.8 | 44.4 | 11.9 | [66] |
| Wheat straw | 40.1 | 32.8 | 14.1 | [67] |

the juice of milled sugarcane is largely used as carbon source for bioethanol generation as well as for sugar industry. However, large production of this vegetable generates high amounts of sugarcane bagasse (140 kg of bagasse are generated per ton of processed sugarcane [69]). This by-product, when generated in a high amount, can be potentially used as a source of carbon in biotechnological processes, such as xylitol production. As with sugarcane bagasse, a wide variety of lignocellulosic materials have been investigated for the production of xylitol such as rice straw, corncob, corn stover, sunflower stalks, and others (Table 10.4).

As shown in Table 10.4, the source of lignocellulosic material is one of the variables that directly influences xylitol yield and productivity. Canilha et al. [74], who used wheat straw as a raw material, obtained a xylitol yield of around 0.9 g $g^{-1}$, which was high value compared to Martínez et al. [75], who reported only 0.023 g $g^{-1}$ from sunflower stalks.

Regarding sugarcane bagasse, there are a number of studies in the literature. Silva et al. [70], investigated xylitol production from hemicellulosic hydrolysate of this material (obtained with dilute sulfuric acid hydrolysis) by using the wild yeast *Candida guilliermondii*. The authors observed a xylitol yield of 0.59 g $g^{-1}$ and productivity of 0.53 g $L^{-1}$ $h^{-1}$ from 46 g $L^{-1}$of initial concentration of xylose in the hydrolysate. In another study, Martini et al. [77], searching for new pentose fermenters microorganisms, evaluated xylitol production from sugarcane bagasse using a new yeast *Meyerozyma guilliermondii*, isolated from sugarcane juice. In that investigation, authors observed that this yeast was able to produce xylitol from acid hydrolysates, although it also produced ethanol as a byproduct.

Disruption of recalcitrance of lignocellulosic materials and release of sugars depend on the source of biomass as well as the type of pretreatment. Thus, the choice of the binomial pre-treatment + raw material is pivotal in the search for process viability. The next section contains further explanation about some techniques of biomass pre-treatment, aimed at the release of sugars, that may be used for xylitol production.

### 10.3.1 Pre-treatment of Biomass and Detoxification of Hemicellulosic Hydrolysates to Produce Xylitol

Maximum use of fermentable sugars present in carbohydrate polymeric fractions of biomass, such as glucose and pentoses, is necessary to make bioprocess more economically competitive, within the biorefinery concept of Silva et al. [78]. However, hydrolysis of polysaccharides contained in biomass into monomeric sugars requires a pretreatment method.

For xylitol production, dilute acid hydrolysis of hemicellulose is the most important technique used. Compounds as hydrochloric acid, phosphoric acid and sulfuric acid have been used to generate a hydrolysate rich in pentose sugars, such as xylose and arabinose, [79–81]. For example, in the work of Carvalho et al. [79], authors

**Table 10.4** Xylitol production from different lignocellulosic raw materials

| Raw material | Pretreatment process | Detoxification process | Microorganism | Concentration of produced Xylitol (g $L^{-1}$) | Yield (g $g^{-1}$) | Productivity (g $L^{-1}$ $h^{-1}$) | Ref. |
|---|---|---|---|---|---|---|---|
| Sugarcane Bagasse | Diluate-acid hydrolysis at 121 °C for 20 min with $H_2SO_4$ at 1:10 solid/liquid ratio (100 mg $H_2SO_4$/g dry matter) | pH adjusted 7.0 w/ CaO; pH to 2.5 w/ $H_3PO_4$, 1.0% w/v activ. charcoal (refin.), agitation (200 rpm, 60 °C) 30 min. Precipitates removed (filtr.) | *C. guilliermondii* FTI 20037 | 25 | 0.59 | 0.53 | [70] |
| Sugarcane straw | Dilute-acid hydrolysis with 1.0% (w/v) $H_2SO_4$ in 35-L steel reactor at 121 °C, 20 min at 1:10 solid/liquid ratio | pH adjusted 7.0 w/ CaO; pH to 2.5 w/ $H_3PO_4$, 1.0% w/v activ. charcoal (refin.) then, agitation (100 rpm, 60 °C) 30 min. Precipitates removed (filtr.) | *C. guilliermondii* FTI 20037 | 16.2 | 0.47 | 0.34 | [71] |
| Corncob | Dilute-acid hydrolysis in 1000 mL flask w/90 g corncobs loading 15% (w/w) w/$H_2SO_4$ (1%, w/w). Autoclaved, 125 °C, 1 h. Liquid separated. Solid residue washed w/180 mL water (60 °C) | Not detoxified | *Candida tropicalis* CCTCC M2012462 | 38.8 | 0.7 | 0.46 | [72] |

| | | | | | | | |
|---|---|---|---|---|---|---|---|
| Corn stover | Vapor-phase diethyl oxalate (DEO) pretreatment. Corn stover soaked in water 10 min. DEO (100 mL/0.75 kg stover) added to semi-pilot reactor. Hydrolysis at 156 °C, 30 min. Sugars water extracted, conc. by rev. osmosis | pH adjustment w/ $Ca(OH)_2$ added @RT w/mixing, increasing pH from 1.1 to 12; then pH reduc. to 5.6 w/85% $H_3PO_4$ or 45–50 wt% D-gluconic acid. Precipitates removed after each pH change (filt.) | *Pichia stipitis* FPL-YS30 | 13 | 0.61 | 0.18 | [21] |
| Corn straw | Steam explosion. Dried particles (100 g) impregnated 300 mL soln 1% (wt) $H_2SO_4$ or water @RT, 3 h. Wet solid in 1 L vessel, sealed, heated w/water/ steam 180–190 °C for 3–9 min | Not detoxified | *Candida tropicalis* | 35.6 | 0.71 | 0.94 | [73] |
| Wheat straw | Acid hydrolysis 145 °C 30 min w/2.5% (w/v) $H_2SO_4$ soln. Solid/liq ratio 1:17.5 (w/v), filtered, conc. 4x, 70 °C | pH adjusted 7.0 w/ CaO; pH to 5.5 w/ $H_3PO_4$, 10% w/v activ. charcoal then, agitation (200 rpm, 30 °C) 1 h. Precipitates removed (vac. filtr.) | *C. guilliermondii* FTI 20037 | 27.5 | 0.9 | 0.5 | [74] |

(continued)

**Table 10.4** (continued)

| Raw material | Pretreatment process | Detoxification process | Microorganism | Concentration of produced Xylitol (g $L^{-1}$) | Yield (g $g^{-1}$) | Productivity (g $L^{-1}$ $h^{-1}$) | Ref. |
|---|---|---|---|---|---|---|---|
| Rice straw | Hemicellulose hydrolysate in 25-L stain.-steel reactor, 145 °C, 20 min w/0.07 M $H_2SO_4$ 10:1 liq:solid ratio; centrifug. liq fract conc. w/heat, 70 °C | Initial pH 1.3 adjusted to 9.5 w/solid NaOH centrifug. (1000xg) 15 min to remove insolubles; pH supernatant lowered to 5.4 w/12 M $H_2SO_4$ | *C. mogii* NRRL Y-17032 | 29.8 | 0.65 | 0.4 | [13] |
| Sunflower stalks | Dilute-acid hydrolysis in 1-L stirred reactor, 50 g sunflower stalk, liq/sol ratio 20, 95 °C. $H_3PO_4$ conc. 0–2.7 M. Homogenized w/bladed shaft 200 rpm, reaction time, 240 min | Initial pH $\leq$ 1 to 4.5 w/KOH (10 mol/L) to avoid dilution in neutralization; Rot. evap. conc. w/ inhibitor removal to minimize deep color; hydrolysate bleached, activ. carbon pellets 8% weight, stirred 35 °C 1 h. Fines removal w/filtr | *Hansenula polymorpha* ATCC 34438 | 0.31 | 0.023 | – | [75] |

| | | | | | | | |
|---|---|---|---|---|---|---|---|
| Olive Stones | Pretreatment w/steam, hydrolysis; pressurized reactor, 1:5 (w/v) w/ pure water; heat 200 °C, 15 bar, fast cooling 40 °C; hydrolysis w/ HCl 0.4–5.7%, 1:10 (w/v), 90 °C, 180 min | pH adjusted to 4.0 w/ conc. KOH: centrifug. to remove precipitate, obtain clear solution to enable monitoring cell growth | *Pachysolen tannophilus* | 8.2 | 0.44 | 0.11 | [76] |
| Cashew apple bagasse | Dilute-acid hydrolysis; Dried bagasse, autoclaved 121 °C, 15 min in 0.6 M $H_2SO_4$, solid fraction, 20%. Liquid fraction collected by vac. filtr | Hydrolysate pH initially adjusted to 10 w/$Ca(OH)_2$; $H_2SO_4$ added to pH 6.0, detox w/gran. or powd. charcoals, 3 wv – 1% w/agitation 200 rpm, 35 °C, 2 h. Detox. conc. hemicellulosic hydrolysate recovered by vac. filtr | *Kluyveromyces marxianus* CCA 510 | 9.61 | 0.50 | 0.17 | [25] |

obtained sugarcane bagasse hemicellulosic hydrolysate with 17.1 $g.dm^{-3}$ of xylose, by using 70 mg of phosphoric acid/gram of bagasse. In addition, Nguyen et al. [81] recovered 50% for glucose and 70–98% of other sugars in hydrolysate obtained from diluted sulfuric mixed softwood forest thinnings pre-treatment. Moreover, Kumar et al. [82] studied xylitol and ethanol co-production from sugarcane bagasse hemicellulosic hydrolysates by a thermotolerant *Kluyveromyces* sp. The authors used two types of hydrolysates obtained through two-stage acid hydrolysis. In the first stage (8% sulfuric acid, 100 °C, 90 min and 1:4 solid liquid ratio), a hydrolysate rich in xylose was obtained for xylitol production, while a hydrolysate composed mainly by glucose was achieved in the second stage (40% sulfuric acid, 80 °C, 90 min and 1:4 solid liquid ratio) aiming at ethanol production. There were xylitol and ethanol yields of 0.61 and 0.43 $g\ g^{-1}$, respectively.

Physicochemical pretreatment is another category of methods known as hydrothermal, wet oxidation, ammonia recycle percolation, aqueous ammonia, organosolv, and others. For example, hydrothermal pretreatment is classified in physico-chemical methods commonly used in extraction of hemicellulosic biomaterial, noted by Ma et al. [83], in a process that offers hydrolysates rich in pentose sugars, mainly carbohydrates used as a source of carbon in xylitol production.

After pretreatment of lignocellulosic biomass, along with a high amount of monomeric sugars, some byproducts are produced, e.g., phenolics, other aromatics, aliphatic acids, furan derived compounds and inorganic ions [84]. These compounds can act as microbial growth inhibitors in the fermentation process. According to Rehman et al. [85], there are four ways to decrease the effect of inhibitors on the fermentation process to obtain better yields of desired products. They are: (i) efficient hydrolysis of biomass by reducing the byproduct formation; (ii) purification or detoxification of the hydrolysate using different techniques; (iii) use of metabolically engineered microorganisms; and (iv) chemical conversion of byproducts into non-toxic compounds [85].

Various detoxification methods have been reported in the literature such as enzymatic detoxification [86], alkali treatment [87], solvent extraction [88], anion and cation exchange resins, activated charcoal [89], heating and evaporation [90]. Among different detoxification methods, activated charcoal is generally one of the most often used to purify hemicellulosic hydrolysates because of its lower costs and lesser amount of labor involved. Various factors such as pH, temperature and incubation time have a significant effect on the efficiency of this method. According to Villarreal et al. [91], most of the phenolic compounds are removed in this treatment at acidic pH whereas organic acids are removed at alkaline pH [91]. In a study conducted by Kamal et al. [92], use of 2.5% of activated charcoal for the detoxification of sago trunk cortex hydrolysate for 60 min yielded maximum xylitol concentration of 19.53 $g\ L^{-1}$. Moreover, combining activated charcoal with other methods usually results in high efficiency processes. According to various works in the literature, detoxification methods using charcoal and ion exchange resin have been effective so far [91, 93]. For example, Prakash et al. [94] evaluated xylitol production from sugarcane bagasse using a thermotolerant strain of *Debaryomyces hansenii*. The hydrolysate was obtained through acid hydrolysisof steam-exploded sugarcane bagasse followed by activated charcoal and ion exchange detoxification, achieving xylitol yields and productivity of 0.69 g $g^{-1}$ and 0.28 g $L^{-1}$ $h^{-1}$ respectively. Actually, anion exchange resins have been widely studied for their efficiency in removing most of the inhibitors from the hydrolysate [91]. In a study conducted by Mancilha et al. [95], the use of ion-exchange resins A 103 S and A 860 S showed 99% and 100% removal of furfural and acetic acid, respectively, with 94–100% recovery of xylose [95]. The fermentation of the resulting hydrolysate gave high yields (0.41 g $g^{-1}$) of xylitol. Vithanage et al. [93] also studied the effect of Amberlite IRA 400 chloride form ion-exchange resin which resulted in a significant increase (348%) in the fermentative production of xylitol by *C. guilliermondii.* Moreover, treatment of hemicellulosic hydrolysate with alkaline solutions is believed to improve the fermentability of sugars in the production of high value added products [87]. A well studied alkaline solution is $Ca(OH)_2$ in a method called overliming. There are several studies on overliming of the hydrolysate before fermentation in the literature, aiming at xylitol production [96–98].

## 10.4 Different Fermentation Strategies for Xylitol Production

In the last few years, different fermentation strategies for xylitol production have been reported, for example, batch, fed-batch and continuous configuration. Moreover, the use of different kinds of bioreactors has been evaluated, including options such as: stirred tank reactor (STR), basket-type stirred tank reactor (BSTR), fluidized bed reactor (FBR) and airlift reactor. Some strategies for improving the efficiency of xylitol production are also related to the fermentation process using free or immobilized cells. In this section, strategies for xylitol production are reported, focusing the use of hydrolysates of lignocellulosic biomass.

### *10.4.1 Batch Fermentation of Xylitol*

This configuration is reported as one of the most often used for the production of xylitol and other bioproducts, because of advantages such as easy control of contamination. For xylitol production in batch configuration, stirred tank reactor (STR), basket-type stirred tank reactor (BSTR) and fluidized bed reactor (FBR) (Fig. 10.2) have been reported in the literature.

Zhang et al. [6], reported xylitol production from horticultural waste hemicellulosic hydrolysates (22.38 g $L^{-1}$ of glucose, 123.42 g $L^{-1}$ of xylose, 12.40 g $L^{-1}$ of mannose and 9.30 g $L^{-1}$ of arabinose) by *Candida athensensis* SB18. In that report, the fermentation process was carried out in a STR - BIOSAT® 2-L Microbial

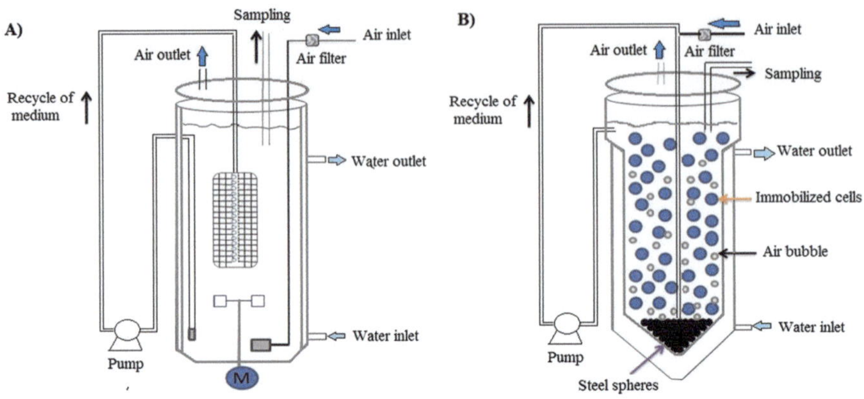


**Fig. 10.2** Some reactors used for xylitol production. (**a**) Immobilized cells in a basket-type stirred tank reactor (BSTR) and (**b**) Immobilized cells in a fluidized bed reactor (FBR)

Bioreactor (Sartorius Stedim Biotech, France) with agitation speed of 200 rpm, air flow rate of 0.7 L.min$^{-1}$ and temperature of 30 °C. After 102 h of batch fermentation, there were maximal xylitol production (100 g L$^{-1}$) and productivity (0.98 g L$^{-1}$ h$^{-1}$), with an efficiency of 89% of the theoretical yield (0.917 g of xylitol.g$^{-1}$ of xylose).

In another work, Wang et al. [99] evaluated xylitol production from corncob hydrolysates using immobilized *Candida tropicalis*. Polyurethane foam was used as a carrier for cells, and the fermentation was carried out in a 5 L STR bioreactor (B. Braun Biotech International, Germany) in the following established parameters: temperature of 30 °C; stirring of 200 rpm; dissolved oxygen (DO) value: first 24 h: 1.2 min$^{-1}$, then 0.6 min$^{-1}$; and initial pH value of 6. The initial concentration of sugars in corncob hydrolysates was 140 g L$^{-1}$ of xylose, 14.41 g L$^{-1}$ of glucose, 12.02 g L$^{-1}$ of arabinose and 6.98 g L$^{-1}$ of acetic acid. The authors found that the use of immobilized cells offered high yield of xylitol and stable cell performance. In multi-batch fermentation, the average values of volumetric productivity and xylitol yield using immobilized cells were 5–10% higher than the ones achieved with free cells, and the average batch period was 7–15% shorter than the one in fermentations using free cells.

Some authors reported use of techniques such as cell immobilization over free cells for enhanced xylitol process yields and productivity, given some advantages such as high cell density, easy biocatalyst recovery, comparatively easier reuse or continuous use of cells, reduction of the lag phase, higher conversion of substrate, reduction of fermentation time and lesser inhibition by substrate or product [100, 101]. By using immobilized cells, another kind of bioreactor evaluated for xylitol production is the fluidized bed reactor (FBR). The use of FBR is adequate when working with microorganisms sensible to high shear stress and for immobilized cells, once STR can break up the immobilized systems. A fluidized bed reactor (Bioengineering AG, Wald, Switzerland) was used in a batch configuration for xylitol production using immobilized cells [102]. In that study, hydrolyzed sugarcane bagasse (57.3 g $L^{-1}$ of xylose, 1.97 g $L^{-1}$ of glucose and 5.79 g $L^{-1}$ of arabinose) was fermented in xylitol using the yeast *Candida guilliermondii* FTI 20037 immobilized in porous glass beads (pore size <300 μm, porosity of 62% and absolute density of 2.42 g.$mL^{-1}$). The authors evaluated the influence of air flow rate and observed that the increase in air flow rate from 25 to 140 mL.$min^{-1}$ increased the volumetric xylitol productivity from 0.19 to 0.28 g $L^{-1}$ $h^{-1}$, but decreased the yield from 0.54 to 0.36 g of xylitol.$g^{-1}$ of xylose. As also shown in that study, aeration rate is an important parameter that directly influences xylose consumption and consequently xylitol production, therefore an optimization work is necessary for each specific reactor. Moreover, Sarrouh and da Silva [98] studied repeated batches using *Candida guilliermondii* FTI20037 cells immobilized in calcium alginate beads. Seven repeated batches were carried out in a fluidized bed reactor (FBR; Bioengineering AG, Wald, Switzerland). The average xylitol yield from sugarcane bagasse hydrolysates during six repeated batches was 0.7 $g_p.g_s^{-1}$ with a volumetric productivity of 0.42 g $L^{-1}$ $h^{-1}$, considering the total time (432 h). Xylitol yield decreased by 44% after six batches.

Immobilization of cells (*Candida guilliermondii*) for xylitol production using hydrolyzed sugarcane bagasse was reported by Carvalho et al. [96] and da Silva et al. [6]. In those studies, many different parameters (e.g. carrier and immobilization technique) that affect the efficiency of xylose conversion by immobilized cells were evaluated and some of the influence can be attributed to mass transfer conditions, considering the carrier used for immobilization. Using sugarcane bagasse hydrolysates, a comparative between BSTR and STR for xylitol production was reported by Carvalho et al. [97]. In that study, xylitol production and yield were higher in STR (23.5 g $L^{-1}$, 0.58 g of xylitol.$g^{-1}$ of xylose) than in BSTR (15 g $L^{-1}$, 0.46 g of xylitol.$g^{-1}$ of xylose).

One problem generally associated with batch fermentation is the susceptibility of microorganisms to high osmotic pressure when using high xylose concentration in the medium or because of high product concentration, which results in inhibition of cell growth. Nevertheless, during fermentation in batch configuration using xylose as substrate and the yeast *Candida athensensis* SB18, Zhang et al. [6] observed that an increase in xylose concentration from 100 to 250 g $L^{-1}$ did not result in cell growth inhibition, although the use of 300 g $L^{-1}$ of xylose resulted in a decrease in xylitol yield. In addition, by using another microbial species (*Candida tropicalis*), Tamburini et al. [9] observed that an increase in xylose concentration from 30 to 80 g•$L^{-1}$ increased xylitol productivity from 0.14 to 0.6 g $L^{-1}$ $h^{-1}$, but higher concentrations significantly decreased xylitol productivity due to a higher fermentation time.

### *10.4.2 Fed-Batch Fermentation of Xylitol*

Fed-batch fermentation presents some advantages as better yield and productivity in xylitol and other microbial metabolites [103]. In a studied conducted in the 1990s, Vandeska et al. [12] studied xylitol production from xylose by *Candida boidinii* in a fed-batch process using a stirred tank reactor (New Brunswick Scientific Co.) with a working volume of 1 L. In that work, the reactor was operated, in the first phase, in a batch process for up to 72 h at 30 °C, pH 5.5, using 100 g $L^{-1}$ of xylose, aeration of 0.3 $min^{-1}$ and agitation of 370 rpm. After 72 h, xylose feeding was started, with a flowrate of 0.8 g $L^{-1}$ $h^{-1}$ for 8 days, with a concomitant aeration decrease to 0.25 $min^{-1}$ and agitation of 250 rpm. In the first phase, xylose concentration dropped from 100 to 43.5 g $L^{-1}$ (55% of consumption), reaching 18.3 g $L^{-1}$. Maximum production of xylitol in the batch phase was about 20 g $L^{-1}$, a value that was increased to 59.3 g $L^{-1}$ during the fed-batch step. At the end of the fed-batch phase, xylose concentration was 110 g $L^{-1}$, because feeding flow exceeded the xylose consumption rate by the yeast.

Fed-batch systems have also been studied more recently for xylitol production. Tamburini et al. [9], for example, reported the use of *Candida tropicalis* in a fermentation process carried out in a stirred tank reactor with aeration of 0.5 min. The authors reported that this yeast is a promising strain for industrial application, given the possibility of working under non-sterile conditions. In that work, the fed-batch fermentation process was started when the biomass reached the exponential phase and xylose concentration dropped to 40 g $L^{-1}$. The xylose solution was fed to keep the concentration of this sugar at a medium level (about 80 g $L^{-1}$). The best condition used initial xylose concentration in the range of 60–80 g $L^{-1}$, at pH 5.5 and 37 °C, achieving 90% of theoretical xylitol yield. The optimized initial xylose concentration in that study was lower than the value of 100 g $L^{-1}$ previously established by Kim et al. [104] for *Candida tropicalis* ATCC 13803 in a fed-batch process. These authors observed, in the optimized condition, 187 g $L^{-1}$ of xylitol production, yield of 0.75 g of xylitol.$g^{-1}$ of xylose and 3.9 g $L^{-1}$ $h^{-1}$ of volumetric productivity.

Repeated fed-batch production of xylitol (three stages) by *Candida magnoliae* TISTR 5663 was reported by Sirisansaneeyakul et al. [105]. These authors evaluated a process carried out in a 2 L stirred-tank fermenter (Biostat B, B. Braun Biotech International, Germany) under oxygen limited condition (agitation speed of 300 rpm, aeration rate of 1.0 $min^{-1}$). When xylose was totally consumed in the first batch, part of the fermentation broth was removed and a new medium (xylose and nitrogen source) started to be fed in order to achieve 60 g $L^{-1}$ of xylose, 2.3 g $L^{-1}$ of urea and 1.0 g $L^{-1}$ of casamino acids. Xylitol production in the feed-batch I, II and III were 235 g $L^{-1}$, 284 g $L^{-1}$ and 280 g $L^{-1}$, respectively. As observed in the repeated feed-batch process, the quantity of biomass is totally viable and productive. Also, feed rate manipulation can be controlled by the concentration of the limiting nutrient in the medium and reduce inhibition by the substrate.

### *10.4.3 Continuous Fermentation of Xylitol*

Continuous production of xylitol or other metabolites has some advantages compared to batch processes: a significant reduction in total processing time by elimination of idle times (charge, discharge, cleaning and sterilization of vessel), reduction

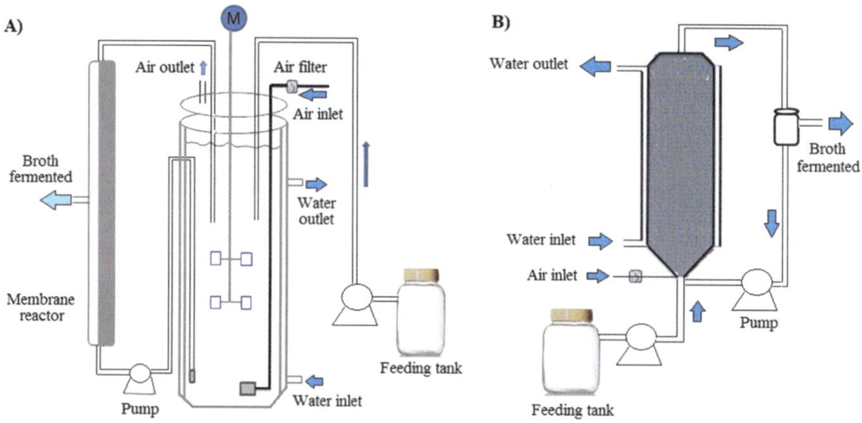


**Fig. 10.3** Bioreactors used for xylitol production. (**a**) Membrane reactor (MR) and (**b**) packed bed reactor (PBR)

in equipment size considering the same rate of production, in addition to easier instrumental control [106–108]. For continuous xylitol production, the use of different kinds of reactor has been reported, e.g., STR [107, 109], packed bed [110] and membrane systems [111, 112] (Fig. 10.3).

Continuous processes are also useful for evaluation of process parameters that influence fermentation, for example, aeration rate. This can be illustrated by the study about the effect of oxygen transfer rate carried out by Martínez et al. [107]. These authors evaluated continuous xylitol production from hemicellulosic hydrolysates containing 3 g $L^{-1}$ of glucose, 51 g $L^{-1}$ of xylose, 4.3 g $L^{-1}$ of arabinose and 4.4 g $L^{-1}$ of acetic acid in a 1.25-L STR (BIOFLO III; New Brunswick Scientific, NJ) using *C. guilliermondii* FTI 20037. In that study, maximum xylitol production rate was obtained by using lower $k_La$ values (10 and 20 $h^{-1}$). The best xylitol yield and volumetric productivity were 0.58 g of xylitol.$g^{-1}$ of xylose (63% of theoretical yield) and 0.69 g $L^{-1}$ $h^{-1}$, respectively. An increase of $k_La$ from 20 to 30 $h^{-1}$ significantly decreased xylitol yield from 0.58 to 0.24 g $g^{-1}$. A similar study was performed by Martínez et al. [109], using hemicellulosic hydrolysates and *C. guilliermondii* FTI 20037. In that study, the effect of pH and dilution rate (*D*) was evaluated. Highest xylitol production (28.7 g•$L^{-1}$) and yield (0.63 g $g^{-1}$) were obtained in the following condition: pH 4, $k_La$ of 30 $h^{-1}$, D of 0.01 $h^{-1}$.

Continuous conversion from xylose to xylitol by *C. guilliermondii* was evaluated by Faria et al. [111], using a membrane bioreactor. The best established conditions corresponded to a maximum pore diameter of membranes of 0.2 μm and permeability of 42.9 L.$m^{-2}$•$h^{-1}$ $bar^{-1}$. Initially, a batch process was performed and when xylose concentration dropped to 2 g $L^{-1}$ (after 60 h), a medium containing 50 g $L^{-1}$ of xylose was continuously fed and the fermented broth was continuously removed. The system was continuously operated for 140 h and the highest xylitol production was reached using dilution rate (*D*) of 0.03 $h^{-1}$ and $k_La$ of 42 $h^{-1}$, corresponding to 86% of conversion and volumetric productivity of 1.14 g $L^{-1}$ $h^{-1}$.

Another work using membrane bioreactor was reported for co-production of ethanol and xylitol from rice straw hydrolysates (75 g $L^{-1}$ glucose and 40 g $L^{-1}$ xylose) using *Saccharomyces cerevisiae* NCIM 3090 and *Candida tropicalis* NCIM 3119 [112]. The highest ethanol (37.8 g $L^{-1}$) and xylitol (24.2 g $L^{-1}$) production was achieved in an optimized dilution rate (*D*) of 0.03 L $h^{-1}$ after 150 h of process in a STR. Continuous fermentation for co-production of ethanol and xylitol was carried out by coupling a cross-flow microfiltration membrane with pore size of 0.45 μm and surface area of 0.1 $m^2$. The continuous process was started after complete consumption of glucose and xylose (72 h). Using this membrane bioreactor, 55 g $L^{-1}$ (yield of 0.70 g $g^{-1}$) of ethanol and 31 g $L^{-1}$ (yield of 0.78 g $g^{-1}$) of xylitol were obtained after 150 h, resulting in higher values than those obtained in single batch and co-culture conditions. A continuous process with co-cultures has some advantages in comparison with a batch process, such as reduction in the overall production cost, possibility of simultaneous production of valuable products such as xylitol and ethanol, and high concentration of the product.

Another strategy for xylitol production in a continuous configuration was proposed by Roca et al. [110]. In this case, an immobilized recombinant *Saccharomyces cerevisiae* S641 was used in a continuous packed-bed reactor (PBR), by using synthetic solutions of xylose and glucose as substrate. The authors evaluated the influence of hydraulic residence time (HRT), substrate/co-substrate ratio (SCSR), recycling ratio and aeration in the performance of the process. The highest overall xylitol yield (0.6 g $g^{-1}$) was achieved using a higher HRT (3.8 h), lower SCSR (0.5), recycling ratio of 10 and anaerobic condition.

## 10.5 Methods for Xylitol Separation and Purification

The downstream process of products obtained by fermentation plays an important role in industrial processes. The purification step is one of the bottlenecks of the fermentation production of xylitol due the complexity of fermented broth as well as the requirement to obtain a product with high purity that could be used for human consumption [113].

Although many studies have been reported regarding fermentation processes for xylitol production from hemicellulosic hydrolysates, just a few techniques of recovery and purification of polyols with the aim of obtaining uniform and high-purity crystals have been reported [114–116].

In some reports of the process for xylitol purification, crystallization was chosen as the main unit operation. Actually, crystallization creates thermodynamic conditions that lead molecules to approach and clump into highly organized structures, called crystals. However, in some cases, the operating conditions do not allow one to obtain pure crystals, and there are impurities that also have high affinity for the solute. According to Mullin [117], in general, there are three steps involving crystallization.

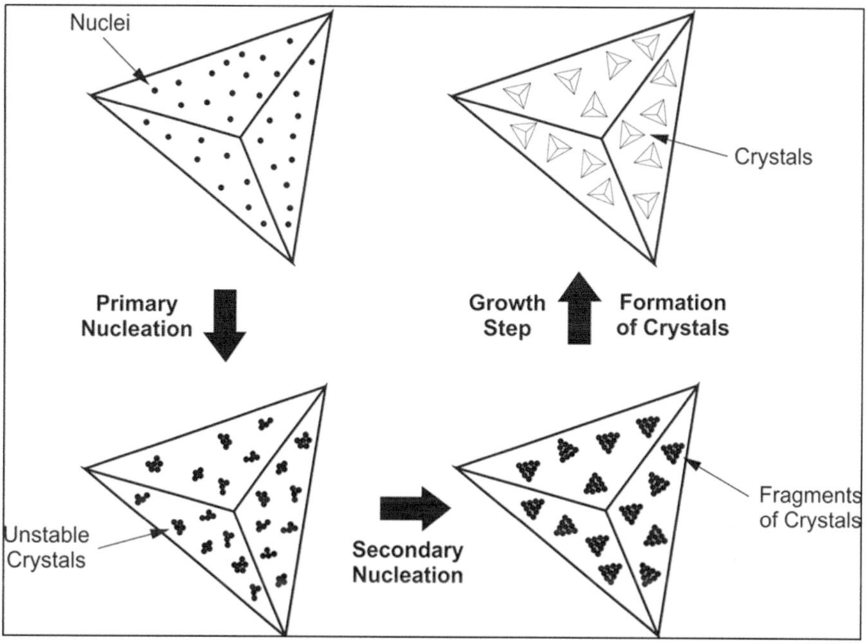


**Fig. 10.4** Main steps of crystallization process

Figure 10.4 shows the main steps of the crystallization process. This unit operation is based on mass and momentum transfer. The driving force for crystallization is the existence of supersaturation in the liquid mixture, e.g., the existence of a higher solute concentration in solution at saturation concentration (solubility limit).

The first step in the crystallization process is nucleation, required to create conditions for the molecules to approach and create the crystal. Agitation or circulation of the liquid mixture, which causes approach and collision between the molecules, occurring momentum transfer, is necessary for crystallization to occur. This mixing results in primary nucleation, in which the solid surface of the crystallizer may be nucleating agents [117]. Once the first crystals have been formed, small fragments of these crystals may be transformed into in the new nuclei. This step is called secondary nucleation. After forming, the crystals nuclei start to grow, entering the crystal growth step. The speed of agitation or circulation in the crystallizer, the degree of supersaturation, temperature, solution purity and cooling rate, are operative parameters that influence the growth speed of crystals and the characteristics of the final product [118]. Many nuclei are formed simultaneously and the final product is formed by small crystals [117].

Many parameters have been reported for xylitol purification, such as the type of techniques, solvent, time and temperature conditions using synthetic and lignocellulosic hydrolysates as fermented medium. Chromatography has been studied and used on an industrial scale for xylitol purification, particularly when xylitol production occurs through the chemical route. Nevertheless, for xylitol purification from fermented hemicellulosic hydrolysate, more efficient processes are required, since the fermented broth has impurities and low xylitol concentration [119].

Processes involving anion or/and cation exchange resin columns allow one to obtain solutions with high xylitol concentration before crystallization. In a study performed in the 1990s, Gurgel et al. [120] purified xylitol produced by fermentation on sugarcane bagasse hydrolysate by using column with anion and cation exchange resins and concluded that this technique resulted in low yield (46–57%) and was considered to be expensive on an industrial scale. However, Martínez et al. [114] tested a combined treatment with resins and crystallization as processes for xylitol purification and found better results. In that study, xylitol was obtained for fermentation of sugarcane hemicellulosic hydrolysate. Before the crystallization stage, the fermented broth was centrifuged, filtered, treated with cation and anion resins and concentrated. Crystallization was performed twice (initially 10 °C and after 2.5 °C) and pure crystals (92–94%) were obtained.

Liquid-liquid extraction and precipitation using activated charcoal followed by vacuum concentration and crystallization also offer an option to remove undesirable impurities. For example, Misra et al. [121] investigated various purification strategies aiming at maximum xylitol purification. Xylitol was produced from corncob hemicellulosic hydrolysates, and liquid-liquid extraction using ethyl acetate was evaluated, reaching 67.44% xylitol recovery. By combining activated charcoal treatment (15 g $L^{-1}$), vacuum concentration (10 times as much supersaturation as that of the initial concentration), and crystallization (first −20 °C and after 8 °C) techniques, they had 43.93% yield. This yield was improved by increasing crystallization cycles (4 cycles), obtaining 76.20% and 98.99% pure crystals.

Columns containing adsorbents have also been studied to purify the complex fermented broth, combined with crystallization procedures for crystal xylitol production. Wei et al. [116] purified and crystallized xylitol from fermented corncob hydrolysates by combining those techniques. The authors used activated carbon and two ion exchange resin columns for broth purification. The purified solution was concentrated up to supersaturation (750 g $L^{-1}$) and crystallized (−20 °C). Crystallization yield was 60.2% with 95% purity, and regular tetrahedral crystals were obtained. Mussatto et al. [122] studied adsorption using silica gel with different solvents, followed by the concentration stage (6.5-fold) and crystallization process for xylitol recovery, produced by fermentation of sugarcane bagasse hemicellulosic hydrolysates. The suitable condition in the adsorption stage that optimized xylitol recovery used a mixture of ethanol and ethyl acetate as solvents. The authors reached 60% crystallization yield and 33% efficiency in global xylitol recovery. Moreover, Sampaio et al. [123] studied the clarifying process as a step for xylitol recovery before crystallization, using synthetic fermented broth. For the clarifying step, activated charcoal was used in different concentrations and in the crystallization procedure, different temperatures and initial xylitol concentration were evaluated. The authors concluded that the clarifying treatment was efficient when 20 g $L^{-1}$of activated charcoal was used. The lower temperature (−10 °C) of the crystallization process allowed them to increase the yield (0.42); however, the crystal showed 11% of impurities and 85.3% xylitol concentration. The highest xylitol concentration (99.2%) and the lowest impurities (0.18%) were achieved by using a high temperature (15 °C). This study showed that they could operate crystallization at a low initial xylitol concentration and at a high temperature, indicating a positive effect when the residual xylose was present.

Moreover, some studies reported a process that combined low-pressure evaporation and cooling with crystallization. For example, De Faveri et al. [124] tested combined low-pressure evaporation and cooling for xylitol purification using a high concentration of fermented broth (xylose + xylitol). The results showed that the concentrated solution (730 g $L^{-1}$) and high cooled temperature (−5 °C) showed 50% yield and crystals with more than 0.90 purity degree.

Another approach was based on processes involving membrane combined with crystallization techniques, which were proposed for xylitol recovery from fermented broth by Affleck [119]. The author stated that this technology uses less energy and is feasible to obtain pure fermented broth before crystallization steps. The polysulfone membrane (10,000 nominal molecular weight) had a capacity to retain 49.2–53.6% impurities of fermented broth. After crystallization, crystals with 90.3% purity were obtained.

## 10.6 Market and Future Outlook

Due to the its favorable properties and benefits for human health, thousands of tons of xylitol are being produced per year. World xylitol production was around 160,000 metric tons in 2013, which makes the global market for xylitol to be around US$670 million per year. In 2020, market analysts have projected production levels of around 242,000 metric tons making the market to be approximately US$1bn [124].

Xylitol global market is driven mainly by application in food, dental and pharmaceuticals [1, 2, 125]. In food manufacturing, this polyalcohol is largely used due to its natural sweetener property, that can be used by diabetic patients since xylitol has metabolic pathways independent of insulin [126], as well as it has 40% less calories [127] compared with other sugars. Regarding health concerns, production of this polyalcohol for use in foods aiming to reduce diseases such as obesity, hypercholesterolemia, as well heart issues, represents significant portion of the xylitol market growth [128]. Other important sectors that has been leveraging xylitol production is the chewing gum industry, with projection of manufacturing to be 163,000 metric tons in 2020 for this single application [129]. Moreover, xylitol is largely produced for use in beverages and candies industry and in pharmaceutical applications, such as oral rinses, nasal decongestants, drugs for respiratory infections prevention, otitis, osteoporosis, and others [130]. Thus, the xylitol market is growing and there are many possibilities for taking advantage of the interesting properties of this molecule. For the next few years, perspectives includes enhancement of biotechnological process performance, including the production of xylitol from lignocellulosic materials in the context of biorefineries that will contribute to the integral use of carbohydrate fractions of agroindustrial and forest residues and by-products. Progress can includes strategies such as the use of different kinds of bioreactors, application of molecular genetic techniques to increase xylitol productivity by microorganisms and new developments in downstream steps.

**Acknowledgements** The authors would like to thank the Brazilian National Council for Scientific and Technological Development (CNPq) (Award Number 300127/2015-4), Brazilian Federal Agency for the Support and Evaluation of Graduate Education (CAPES), and the Research Council for the State of São Paulo (FAPESP) (Award Number 2014/27055-2) for financial support.

## References

1. Salli KM, Forssten SD, Lahtinen SJ, Ouwehand AC. Influence of sucrose and xylitol on an early *Streptococcus mutans* biofilm in a dental simulator. Arch Oral Biol. 2016;70:39–46.
2. Santi E, Facchin G, Faccio R, Barroso RP, Costa-Filho AJ, Borthagaray G, Torre MH. Antimicrobial evaluation of new metallic complexes with xylitol active against *P. aeruginosa* and *C. albicans*: MIC determination, post-agent effect and Zn-uptake. J Inorg Biochem. 2016;155:67–75.
3. Rufino AR, Biaggio FC, Santos JC, de Castro HF. Screening of lipases for the synthesis of xylitol monoesters by chemoenzymatic esterification and the potential of microwave and ultrasound irradiations to enhance the reaction rate. Int J Biol Macromol. 2010;47(1):5–9.
4. Albarrán-Preza E, Corona-Becerril D, Vigueras-Santiago E, Hernández-López S. Sweet polymers: synthesis and characterization of xylitol-based epoxidized linseed oil resins. Eur Polym J. 2016;75:539–51.
5. Prakasham RS, Sreenivas RR, Hobbs PJ. Current trends in biotechnology production of xylitol and future prospects. Curr Trends Biotechnol Pharm. 2009;3:8–36.
6. Zhang J, Geng A, Yao C, Lu Y, Li Q. Xylitol production from D-xylose and horticultural waste hemicellulosic hydrolysate by a new isolate of *Candida athensensis* SB18. Bioresour Technol. 2012;105:134–41.
7. Rafiqul ISM, Mimi Sakinah AM. Bioproduction of xylitol by enzyme technology and future prospects. Int Food Res J. 2012;19:405–8.
8. Chen X, Jiang Z, Chen S, Qin W. Microbial and bioconversion production of D-xylitol and its detection and application. Int J Biol Sci. 2010;6(7):834–44.
9. Tamburini E, Costa S, Marchetti SC, Pedrini P. Optimized production of xylitol from xylose using a hyper-acidophilic *Candida tropicalis*. Biomolecules. 2015;5:1979–89.
10. Pepper T, Olinger PM. Xylitol in sugar-free confections. Food Technol. 1988;42(10):98–106.
11. Ur-Rehman S, Mushtaq Z, Zahoor T, Jamil A, Murtaza MA. Xylitol: a review on bioproduction, application, health benefits, and related safety issues. Crit Rev Food Sci. 2015;55:1514–28.
12. Vandeska E, Amartey S, Kuzmanova S, Jeffries TW. Fed-batch culture for xylitol production by *Candida boidinii*. Process Biochem. 1996;31:265–70.
13. Mayerhoff ZDVL, Roberto IC, da Silva SS. Xylitol production from rice straw hemicellulose hydrolysate using different yeast strains. Biotechnol Lett. 1997;19(5):407–9.
14. Kim SY, Kim JH, Oh DK. Improvement of xylitol production by controlling oxygen supply in *Candida parapsilosis*. J Ferment Bioeng. 1997;83:267–70.
15. Saha BE, Bothast JR. Production of xylitol by *Candida peltata*. J Ind Microbiol Biot. 1999;22:633–6.

16. Suryadi H, Katsuragi T, Yoshida N, Suzuki S, Tani Y. Polyol production by culture of methanol-utilizing yeast. J Biosci Bioeng. 2000;89:236–40.
17. Sampaio FC, Torre P, Passos FM, Perego P, Passos FJ, Converti A. Xylose metabolism in *Debaryomyces hansenii* UFV-170: effect of the specific oxygen uptake rate. Biotechnol Prog. 2004;6:1641–50.
18. Lopez F, Delgado OD, Martinez MA, Spencer JF, Figueroa LI. Characterization of a new xylitol-producer *Candida tropicalis* strain. Antonie Van Leeuwenhoek. 2004;85:281–6.
19. Guo C, Zhao C, He P, Lu D, Shen A, Jiang N. Screening and characterization of yeasts for xylitol production. J Appl Microbiol. 2006;101:1096–104.
20. Mussatto SI, Roberto IC. Establishment of the optimum initial xylose concentration and nutritional supplementation of brewer's spent grain hydrolysate for xylitol production by *Candida guilliermondii*. Process Biochem. 2008;43:540–6.
21. Rodrigues RCLB, Kenealy WR, Jeffries TW. Xylitol production from DEO hydrolysate of corn stover by *Pichia stipitis* YS-30. J Ind Microbiol Biotechnol. 2011;38:1649–55.
22. Cadete RM, Melo MA, Dussán KJ, Rodrigues RC, Silva SS, Zilli JE, Vital MJ, Gomes FC, Lachance MA, Rosa CA. Diversity and physiological characterization of D-xylose-fermenting yeasts isolated from the Brazilian amazonian forest. PLoS One. 2012;7:431–5.
23. Ramesh S, Muthuvelayudham R, Kannan RR, Viruthagiri T. Enhanced production of xylitol from corncob by *Pachysolen tannophilus* using response surface methodology. Int J Food Sci. 2013;2013:Article ID 514676.
24. Pal S, Choudhary V, Kumar A, Biswas D, Mondal AK, Sahoo DK. Studies on xylitol production by metabolic pathway engineered *Debaryomyces hansenii*. Bioresour Technol. 2013;147:449–55.
25. Albuquerque TL, Gomes SDL, Marques Jr JE, Silva Jr IJ, Rocha MVP. Xylitol production from cashew apple bagasse by *Kluyveromyces marxianus* CCA510. Catal Today. 2015;255:33–40.
26. Guamán-Burneo MC, Dussán KJ, Cadete RM, Cheab MAM, Portero P, Carvajal-Barriga EJ, da Silva SS, Rosa CA. Xylitol production by yeasts isolated from rotting wood in the Galápagos Islands, Ecuador, and description of *Cyberlindnera galapagoensis* f.a., sp. nov. Antonie Van Leeuwenhoek. 2015;108(4):919–31.
27. Cadete RM, Cheab MA, Santos RO, Safar SV, Zilli JE, Basso LC, Lee CF, Kurtzman CZ. *Cyberlindnera xylosilytica* sp. nov., a xylitol-producing yeast species isolated from lignocellulosic materials. Int J Syst Evol Microbiol. 2015;65(9):2968–74.
28. Yoshitake J, Ohiwa H, Shimamura M, Imai T. Production of polyalcohol by a *Corynebacterium* sp Part I. Production of pentitol from aldopentose. Agric Biol Chem. 1971;35:905–11.
29. Yoshitake J, Ishizaki H, Shimamura M, Imai T. Xylitol production by an *Enterobacter* species. Agric Biol Chem. 1973;37:2261–6.
30. Izumori K, Tuzaki K. Production of xylitol from D-xylulose by *Mycobacterium smegmatis*. J Ferment Technol. 1988;66:33–6.
31. Suihko ML, Suomalainen I, Enari TM. D-xylose catabolism in *Fusarium oxysporum*. Biotechnol Lett. 1983;5:525–30.
32. Dahiya JS. Xylitol production by *Petromyces albertensis* grown on medium containing D-xylose. Can J Microbiol. 1991;37:14–31.
33. Sampaio FC, da Silveira WB, Chaves-Alves VM, Passos FML, Coelho JLC. Screening of filamentous fungi for production of xylitol from D-xylose. Braz J Microbiol. 2003;34:325–8.
34. Rangaswamy S, Agblevor FA. Screening of facultative anaerobic bacteria utilizing D-xylose for xylitol production. Appl Microbiol Biotechnol. 2002;60:88–93.
35. Kelly C, Jones O, Barnhart C, Lajoie C. Effect of furfural, vanillin and syringaldehyde on *Candida guilliermondii* growth and xylitol biosynthesis. Appl Biochem Biotechnol. 2008;148(1):97–108.

36. Hallborn J, Walfridsson M, Airaksinen U, Ojamo H, Hahn-Hag-erdahl B, Penttila M, Keranen S. Xylitol production by recombinant *Saccharomyces cerevisiae*. Biotechnology. 1991;9:1090–6.
37. Sazaki M, Inui M, Yukawa H. Microorganisms for xylitol production: focus on strain improvement. In: da Silva SS, Chandel AK, editors. D-xylitol fermentative production, application and commercialization. Heidelberg: Springer; 2012. p. 109–31.
38. Winkelhausen E, Kuzmanova S. Microbial conversion of D-xylose to xylitol. J Ferment Bioeng. 1998;86:1–14.
39. Pal S, Mondal AK, Sahoo DK. Molecular strategies for enhancing microbial production of xylitol. Process Biochem. 2016;51:809–19.
40. Yoshitake J, Shimamura K, Ishizaki H, Iris Y. Xylitol production by *Enterobacter liquefaciens*. Agric Biol Chem. 1976;40:1493–503.
41. van Dijken JP, Scheffers WA. Redox balances in the metabolism of sugars by yeasts. FEMS Microbiol Lett. 1986;32:199–224.
42. Silva DDV, Felipe MGA. Effect of glucose:xylose ratio on xylose reductase and xylitol dehydrogenase activities from *Candida guilliermondii* in sugarcane bagasse hydrolysate. J Chem Technol Biot. 2006;81:1294–300.
43. Su B, Wu M, Lin J, Yang L. Metabolic engineering strategies for improving xylitol production from hemicellulosic sugars. Biotechnol Lett. 2013;35:1781–9.
44. Jeon WY, Shim WY, Lee SH, Choi JH, Kim JH. Effect of heterologous xylose transporter expression in *Candida tropicalis* on xylitol production rate. Bioproc Biosyst Eng. 2013;36:809–17.
45. Jeon WY, Yoon BH, Ko BS, Shim WY, Kim JH. Xylitol production is increased by expression of codon-optimized *Neurospora crassa* xylose reductase gene in *Candida tropicalis*. Bioproc Biosyst Eng. 2012;35:191–8.
46. Ahmad I, Shim WY, Jeon WY, Yoon BH, Kim JH. Enhancement of xylitol production in *Candida tropicalis* by co-expression of two genes involved in pentose phosphate pathway. Bioprocess Biosyst Eng. 2012;35:199–204.
47. Lee JK, Koo BS, Kim SY. Cloning and characterization of the xyl1 gene, encoding an NADH-preferring xylose reductase from *Candida parapsilosis*, and its functional expression in *Candida tropicalis*. Appl Environ Microbiol. 2003;69:6179–88.
48. Jeon YJ, Shin HS, Rogers PL. Xylitol production from a mutant strain of *Candida tropicalis*. Lett Appl Microbiol. 2011;53:106–13.
49. Oh EJ, Ha SJ, Rin Kim S, Lee WH, Galazka JM, Cate JH, Jin YS. Enhanced xylitol production through simultaneous co-utilization of cellobiose and xylose by engineered *Saccharomyces cerevisiae*. Metab Eng. 2013;15:226–34.
50. Zhang J, Zhang B, Wang D, Gao X, Hong J. Improving xylitol production at elevated temperature with engineered *Kluyveromyces marxianus* through over-expressing transporters. Bioresour Technol. 2015;175:642–5.
51. Cirino PC, Chin JW, Ingram LO. Engineering *Escherichia coli* for xylitol production from glucose–xylose mixtures. Biotechnol Bioeng. 2006;95:1167–76.
52. Nair NU, Zhao H. Selective reduction of xylose to xylitol from a mixture of hemicellulosic sugars. Metab Eng. 2010;12:462–8.
53. Akinterinwa O, Cirino PC. Heterologous expression of D-xylulokinase from *Pichia stipitis* enables high levels of xylitol production by engineered *Escherichia coli* growing on xylose. Metab Eng. 2009;11:48–55.
54. Sasaki M, Jojima T, Inui M, Yukawa H. Xylitol production by recombinant *Corynebacterium glutamicum* under oxygen deprivation. Appl Microbiol Biotechnol. 2010;86:1057–66.
55. Dhar KS, Wendisch VF, Nampoothiri KM. Engineering of *Corynebacterium glutamicum* for xylitol production from lignocellulosic pentose sugars. J Biotechnol. 2016;230:63–71.
56. Suzuki S, Sugiyama M, Mihara Y, Hashiguchi K, Yokozeki K. Novel enzymatic method for the production of xylitol from D-arabitol by *Gluconobacter oxydans*. Biosci Biotechnol Biochem. 2002;66:2614–20.

57. Wang TH, Zhong YH, Huang W, Liu T, You YW. Antisense inhibition of xylitol dehydrogenase gene, xdh1 from *Trichoderma reesei*. Lett Appl Microbiol. 2005;40:424–9.
58. Nidetzky B, Neuhauser W, Haltrich D, Kulbe KD. Continuous enzymatic production of xylitol with simultaneous coenzyme regeneration in a charged membrane reactor. Biotechnol Bioeng. 1996;52:387–96.
59. Branco RF, Chandel AK, da Silva SS. Enzymatic production of xylitol: current status and future perspectives. In: Silva SS, Chandel AK, editors. D-Xylitol fermentative production, application and commercialization. Heidelberg: Springer; 2012. p. 193–204.
60. Santos JC, Converti A, Carvalho W, Mussatto SI, Silva SS. Influence of aeration rate and carrier concentration on xylitol production from sugarcane bagasse hydrolyzate in immobilized-cell fluidized bed reactor. Process Biochem. 2005;40:113–8.
61. Albuquerque TL, Silva Jr IJ, Macedo GR, Rocha MVP. Biotechnological production of xylitol from lignocellulosic wastes: a review. Process Biochem. 2014;49:1779–89.
62. Canilha L, Santos VTO, Rocha GJM, Silva JBA, Giulietti M, Silva SS, Felipe MGA, Ferraz A, Milagres AMF, Carvalho W. A study on the pretreatment of a sugarcane bagasse sample with dilute sulfuric acid. J Ind Microbiol Biotechnol. 2011;38:1467–75.
63. Silva AS, Inoue H, Endo T, Yano S, Bon EPS. Milling pretreatment of sugarcane bagasse and straw for enzymatic hydrolysis and ethanol fermentation. Bioresour Technol. 2010;101:7402–9.
64. Roberto IC, Mussatto SI, Rodrigues RCLB. Dilute-acid hydrolysis for optimization of xylose recovery from rice straw in a semi-pilot reactor. Ind Crop Prod. 2003;17:171–6.
65. Kumar P, Barrett DM, Delwiche MJ, Stroeve P. Methods for pretreatment of lignocellulosic biomass for efficient hydrolysis and biofuel production. Ind Eng Chem Res. 2009;48:3713–29.
66. Pointner M, Kutter P, Obrlik T, Kahr H. Composition of corncobs as substrate for fermentation of fuels. Agron Res. 2014;12:391–6.
67. Sun RC, Tomkinson J, Wang YX, Xiao B. Physico-chemical and structural characterization of hemicelluloses from wheat straw by alkaline peroxide extraction. Polymer. 2000;41:2647–56.
68. CONAB 2015/2016 – Companhia Nacional de Abastecimento. Acompanhamento da safra brasileira: cana-de-açúcar. Available on http://www.conab.gov.br/OlalaCMS/uploads/arquivos/15_12_17_09_03_29_boletim_cana_portugues_-_3o_lev_-_15-16.pdf. Accessed 21 Jan 2016.
69. Canilha L, Carvalgo W, Felipe MGA, Silva JBA, Giulietti M. Ethanol production from sugarcane bagasse hydrolysate using *Pichia stipitis*. Appl Biochem Biotechnol. 2010;161:84–92.
70. Silva DDV, Mancilha IM, Silva SS, Felipe MGA. Improvement of biotechnological xylitol production by glucose during cultive of *Candida guilliermondii* in sugarcane bagasse hydrolysate. Braz Arch Biol Technol. 2007;50:207–15.
71. Hernández-Pérez AF, Arruda PV, Felipe MGA. Sugarcane straw as a feedstock for xylitol production by *Candida guilliermondii* FTI 20037. Braz J Microbiol. 2016;47:489–96.
72. Ping Y, Ling H, Song G, Ge J. Xylitol production from non-detoxified corncob hemicellulose acid hydrolysate by *Candida tropicalis*. Biochem Eng J. 2013;75:86–91.
73. Wang W, Ling H, Zhao H. Steam explosion pretreatment of corn straw on xylose recovery and xylitol production using hydrolysate without detoxification. Process Biochem. 2015;50:1623–8.
74. Canilha L, Silva JBA, Felipe MGA, Carvalho W. Batch xylitol production from wheat straw hemicellulosic hydrolysate using *Candida guilliermondii* in a stirred tank reactor. Biotechnol Lett. 2003;25:1811–4.
75. Martínez ML, Sánchez S, Bravo V. Production of xylitol and ethanol by *Hansenula polymorpha* from hydrolysates of sunflower stalks with phosphoric acid. Ind Crop Prod. 2012;40:160–6.

76. Mateo S, Puentes JG, Moya AJ, Sánchez S. Ethanol and xylitol production by fermentation of acid hydrolysate from olive pruning with *Candida tropicalis* NBRC 0618. Bioresour Technol. 2015;190:1–6.
77. Martini C, Tauk-Tornisielo SM, Codato CB, Bastos RG, Ceccato-Antonini SR. A strain of *Meyerozyma guilliermondii* isolated from sugarcane juice is able to grow and ferment pentoses in synthetic and bagasse hydrolysate media. World J Microbiol Biotechnol. 2016;32:80.
78. da Silva SS, Chandel AK. Xylitol: fermentative production, application and commercialization. 1st ed. Berlin/New York: Springer; 2012. 348p.
79. Carvalho W, Batista MA, Canilha L, Santos JC, Converti A, Silva SS. Sugarcane bagasse hydrolysis with phosphoric and sulfuric acids and hydrolysate detoxification for xylitol production. J Chem Technol Biotechnol. 2004;79(11):1308–12.
80. Nguyen QA, Tucker MP, Keller FA, Eddy FP. Two-stage dilute-acid pretreatment of softwoods. ApplBiochemBiotechnol. 2000;84(86):561–76.
81. Nguyen QA, Tucker MP, Keller FA, Beaty DA, Connors KM, Eddy FP. Dilute acid hydrolysis of softwoods. Appl Biochem Biotechnol. 1999;77(79):133–42.
82. Kumar S, Dheeran P, Singh SP, Mishra IM, Adhikari DK. Bioprocessing of bagasse hydrolysate for ethanol and xylitol production using thermotolerant yeast. Bioprocess Biosyst Eng. 2015;38:39–47.
83. Ma XJ, Yang XF, Zheng X, Lin L, Chen LH, Haung LL, Cao SL. Degradation and dissolution of hemicellulose during bamboo hydrothermal pretreatment. Bioresour Technol. 2014;161:215–20.
84. Jönsson LJ, Alriksson B, Nilvebrant NO. Bioconversion of lignocellulose: inhibitors and detoxification. Biotechnol Biofuels. 2013;6:1–10.
85. Ur-Rehman S, Mushtaq Z, Zahoor T, Jamil A, Murtaza MA. Xylitol: a review on bioproduction, application, health benefits, and related safety issues. Crit Rev Food Sci Nutr. 2013;55:1514–28.
86. Jurado M, Prieto A, Martínez-Alcalá Á, Martínez ÁT, Martínez MJ. Laccase detoxification of steam-exploded wheat straw for second generation bioethanol. Bioresour Technol. 2009;100:6378–84.
87. Alriksson B, Sjöde A, Nilvebrant NO, Jönsson LJ. Optimal conditions for alkaline detoxification of dilute-acid lignocellulose hydrolysates. In: McMillan JD, Adney WS, Mielenz JR, Klasson KT, editors. Twenty-seventh symposium on biotechnology for fuels and chemicals. Totowa: Humana Press; 2006.
88. Cantarella M, Cantarella L, Gallifuoco A, Spera A, Alfani F. Comparison of different detoxification methods for steam-exploded poplar wood as a substrate for the bioproduction of ethanol in SHF and SSF. Process Biochem. 2004;39:1533–42.
89. Mateo S, Roberto IC, Sánchez S, Moya AJ. Detoxification of hemicellulosic hydrolyzate from olive tree pruning residue. Ind Crop Prod. 2013;49:196–203.
90. Larsson S, Reimann A, Nilvebrant NO, Jönsson LJ. Comparison of different methods for the detoxification of lignocellulose hydrolyzates of spruce. Appl Biochem Biotechnol. 1999;77:91–103.
91. Villarreal MLM, Prata AMR, Felipe MGA, Almeida E, Silva JB. Detoxification procedures of eucalyptus hemicellulose hydrolysate for xylitol production by *Candida guilliermondii*. Enzym Microb Technol. 2006;40:17–24.
92. Kamal SSM, Mohamad NL, Abdullah AGL, Abdullah N. Detoxification of sago trunk hydrolysate using activated charcoal for xylitol production. Procedia Food Sci. 2011;1:908–13.
93. Vithanage LNG, Barbosa AM, Kankanamge GRN, Rakshit SK, Dekker RFH. Valorization of hemicelluloses: production of bioxylitol from poplar wood prehydrolyzates by *Candida guilliermondii* FTI 20037. Bioenergy Res. 2015;9:181–97.
94. Prakasha G, Varmab AJ, Prabhunea A, Shouchec Y, Rao M. Microbial production of xylitol from D-xylose and sugarcane bagasse hemicellulose using newly isolated thermotolerant yeast *Debaryomyces hansenii*. Bioresour Technol. 2011;102(3):3304–8.

95. Maciel de Mancilha I, Karim MN. Evaluation of ion exchange resins for removal of inhibitory compounds from corn stover hydrolyzate for xylitol fermentation. Biotechnol Prog. 2003;19:1837–41.
96. Carvalho W, Silva SS, Converti A, Vitolo M, Felipe MG, Roberto IC, Silva MB, Mancilha IM. Use of immobilized Candida yeast cells for xylitol production from sugarcane bagasse hydrolysate: cell immobilization conditions. Appl Biochem Biotechnol. 2002;98(100):489–96.
97. Carvalho W, Silva SS, Santos JC, Converti A. Xylitol production by Ca-alginate entrapped cells: comparison of different fermentation systems. Enzym Microb Technol. 2003;32:553–9.
98. Sarrouh B, Silva SS. Repeated batch cell-immobilized system for the biotechnological production of xylitol as a renewable green sweetener. Appl Biochem Biotechnol. 2013;1:1–12.
99. Wang C, Li Y. Fungal pretreatment of lignocellulosic biomass. Biotechnol Adv. 2012;30(6):1447–57.
100. Wendhausen R. Estudo sobre utilização de crisotila como suporte de células de *Saccharomyces cerevisiae* para uso em processo contínuo de fermentação alcoólica e biorreduções. Brazil: University of Campinas – UNICAMP; 1998. 360p. (Dissertation).
101. Duarte JC, Rodrigues JAR, Moran PJS, Valença GP, Nunhez JR. Effect of immobilized cells in calcium alginate beads in alcoholic fermentation. AMB Express. 2013;3:31–9.
102. Santos JC, Carvalho W, Silva SS, Converti A. Xylitol production from sugarcane bagasse hydrolyzate in fluidized bed reactor. Effect of air flow rate. Biotechnol Prog. 2003;19:1210–5.
103. Gong C, Chen LF, Flickinger MC, Tsao GT. Conversion of hemicellulose carbohydrates. Adv Biochem Eng Biotechnol. 1981;20:93–118.
104. Kim JH, Han KC, Koh YH, Ryu YW, Seo JH. Optimization of fed-batch fermentation for xylitol production by *Candida tropicalis*. J Ind Microbiol Biotechnol. 2002;29:16–9.
105. Sirisansaneeyakul S, Wannawilai S, Chisti Y. Repeated fed-batch production of xylitol by *Candida magnoliae* TISTR 5663. J Chem Technol Biotechnol. 2013;88:1121–9.
106. Maxon WD. Continuous fermentation, a discussion of its principles and application. Microbiol Process Rep. 1954;28:110–21.
107. Martínez AE, Silva SS, Felipe MGA. Effect of the oxygen transfer coefficient on xylitol production from sugarcane bagasse hydrolysate by continuous stirred-tank reactor fermentation. Appl Biochem Biotechnol. 2000;84:633–41.
108. Rao VL, Goli JK, Gentela J, Koti S. Bioconversion of lignocellulosic biomass to xylitol: an overview. Bioresour Technol. 2016;213:299–310.
109. Martínez EA, Silva SS, Almeida E, Silva JB, Solenzal AIN, Felipe MGA. The influence of pH and dilution rate on continuous production of xylitol from sugarcane bagasse hemicellulosic hydrolysate by *C. guilliermondii*. Process Biochem. 2003;38:1677–83.
110. Roca E, Meinander N, Hahn-Hagerdal B. Xylitol production by immobilized recombinant *Saccharomyces cerevisiae* in a continuous packed-bed bioreactor. J Chem Inf Model. 1996;53:1689–99.
111. Faria LFF, Pereira N, Nobrega R. Xylitol production from D-xylose in a membrane bioreactor. Desalination. 2002;149:231–6.
112. Zahed O, Jouzani GS, Abbasalizadeh S, Khodaiyan F, Tabatabaei M. Continuous co-production of ethanol and xylitol from rice straw hydrolysate in a membrane bioreactor. Folia Microbiol. 2016;61:179–89.
113. De Faveri D, Torre P, Perego P, Converti A. Optimization of xylitol recovery by crystallization from synthetic solutions using response surface methodology. J Food Eng. 2004;61(3):407–12.
114. Martínez EA, de Almeida e Silva JB, Giulietti M, Solenzal AIN. Downstream process for xylitol produced from fermented hydrolysate. Enzym Microb Technol. 2007;40(5):1193–8.

115. Silva SS, Ramos RM, Rodrigues DC, Mancilha IM. Downstream processing for xylitol recovery from fermented sugar cane bagasse hydrolysate using aluminium polychloride. Z Naturforsch C. 2000;55(1–2):10–5.
116. Wei J, Yuan Q, Wang T, Wang L. Purification and crystallization of xylitol from fermentation broth of corncob hydrolysates. Front Chem Eng China. 2010;4(1):57–64.
117. Mullin JW. Crystallization. 4th ed. Oxford: Butterworth-Heinemann; 2001.
118. Martínez EA, Giulietti M, de Almeida e Silva JB, Derenzo S. Kinetics of the xylitol crystallization in hydro-alcoholic solution. Chem Eng Process Process Intensif. 2008;47(12):2157–62.
119. Affleck RP. Recovery of xylitol from fermentation of model hydrolysate using membrane technology. Vol. Master of Science in Biological Systems Engineering. Blacksburg: State University of Virginia; 2000.
120. Gurgel PV, Mancilha IM, Peçanha RP, Siqueira JFM. Xylitol recovery from fermented sugarcane bagasse hydrolyzate. Bioresour Technol. 1995;52(3):219–23.
121. Misra S, Gupta P, Raghuwanshi S, Dutt K, Saxena RK. Comparative study on different strategies involved for xylitol purification from culture media fermented by *Candida tropicalis*. Sep Purif Technol. 2011;78(3):266–73.
122. Mussatto SI, Santos JC, Ricardo Filho WC, Silva SS. A study on the recovery of xylitol by batch adsorption and crystallization from fermented sugarcane bagasse hydrolysate. J Chem Technol Biotechnol. 2006;81(11):1840–5.
123. Sampaio FC, Passos FML, Passos FJV, De Faveri D, Perego P, Converti A. Xylitol crystallization from culture media fermented by yeasts. Chem Eng Process Process Intensif. 2006;45(12):1041–6.
124. Shaw C. Global Xylitol demand to surge to US$1Bn by 2020. 2014. Companiesandmarkets.com – Tue, May 27, 2014. Available at https://uk.finance.yahoo.com/news/global-xylitol-demand-surge-us-000000759.html. Accessed Nov 2016.
125. Canilha L, Rodrigues RCLB, Antunes FAF, Chandel AK, Milessi TSS, Felipe MGA, da Silva SSS. Bioconversion of hemicellulose from sugarcane biomass into sustainable products. In: Chandel AK, da Silva SS, editors. Sustainable degradation of lignocellulosic biomass – techniques, applications and commercialization: InTech; 2013. doi:10.5772/53832. isbn:978-953-51-1119-1. Available from: http://www.intechopen.com/books/sustainable-degradation-of-lignocellulosic-biomass-techniques-applications-and-commercialization/bioconversion-of-hemicellulose-from-sugarcane-biomass-into-sustainable-products.
126. Islam MS, Indrajit M. Effects of xylitol on blood glucose, glucose tolerance, serum insulin and lipid profile in a type 2 diabetes model of rats. Ann Nutr Metab. 2012;61:57–64.
127. De Jong E, Higson A, Walsh P, Wellisch M. Value added products from biorefineries. IEA Bioenergy. 2012. Available: http://www.ieabioenergy.com/wp-content/uploads/2013/10/Task-42-Biobased-Chemicals-value-added-products-from-biorefineries.pdf. Accessed July 2016.
128. Research and Markets. Global Xylitol Market 2016–2020. 2015. http://www.researchandmarkets.com/research/dmslk6/global_xylitol. Accessed July 2016.
129. Research and Markets. Xylitol – A Global Market Overview. 2014. http://www.researchandmarkets.com/reports/2846975/xylitol-a-global-market-overview. May 2014. Accessed July 2016.
130. Branco RF, Santos JC, Silva SS. A novel use for sugarcane bagasse hemicellulosic fraction: xylitol enzymatic production. Biomass Bioenergy. 2011;35:3241–6.

# Chapter 11
# Production of Diols from Biomass

**Keiichi Tomishige, Yoshinao Nakagawa, and Masazumi Tamura**

**Abstract** Most building blocks in biomass refineries are oxygenates with high oxygen content. Substitution of petrochemicals with biomass-derived chemicals requires further transformations. One of the promising targets is the conversion into diols which have two OH groups in one molecule, especially since diols have been used for monomers of the production of polymers such as polyester and polyurethane. Since biomass-derived building blocks typically have OH, COOH and/or furan functional groups, the diols can be produced by removal of OH groups, hydrogenation of COOH to $CH_2OH$, and/or ring-opening C-O hydrogenolysis. This chapter summarizes development of heterogeneous catalysts for these reactions to produce diols.

**Keywords** Heterogeneous catalysis • Hydrogenolysis • Hydrogenation • Dehydration • Retro-aldol reaction • Deoxydehydration • Hydrodeoxygenation

## 11.1 Diols from Petroleum

Diols have been derived from petroleum-based building blocks such as ethylene, propylene and benzene. Ethylene glycol has been the most abundantly produced diol because it is one of the monomers for production of polyethylene terephthalate. Ethylene glycol has been synthesized by the oxidation of ethylene with $O_2$ to ethylene oxide and the subsequent hydration of ethylene oxide to ethylene glycol. Usually, ethylene is supplied from the thermal cracking of naphtha from petroleum refining. Ethylene can be supplied from dehydration of bioethanol and this enables production of biomass-derived ethylene glycol. Therefore, the production of ethylene glycol is not discussed here; recent researches for ethylene glycol synthesis from biomass are summarized in a review [1].

C3 diols are 1,2-propanediol (propylene glycol) and 1,3-propanediol. 1,2-Propanediol has been produced by the oxidation of propylene to propylene

K. Tomishige (✉) • Y. Nakagawa • M. Tamura
Department of Applied Chemistry, School of Engineering, Tohoku University, Sendai, Japan
e-mail: tomi@erec.che.tohoku.ac.jp



Z. Fang et al. (eds.), *Production of Platform Chemicals from Sustainable Resources*, Biofuels and Biorefineries 7, DOI 10.1007/978-981-10-4172-3_11

oxide and the hydration of propylene oxide to 1,2-propanediol. On the other hand, the production of 1,3-propanediol is not as easy as that of 1,2-propanediol. Two industrial chemical production routes to 1,3-propanediol have been used in the past. The first route is the oxidation of propylene to acrolein, the subsequent hydration of acrolein to 3-hydroxypropionaldehyde and then the hydrogenation to 1,3-propanediol. The second route is the hydroformylation of ethylene oxide to give 1,3-propanediol. Both methods are not so economical, and so the present practical method is the fermentation of glucose to 1,3-propanediol developed by DuPont [2].

C4 diols are 1,2-butanediol, 1,4-butanediol, 1,3-butanediol and 2,3-butanediol, in particular, 1,4- and 1,3-butanediols have been important in terms of the industrial field. The oxidation of *n*-butane to maleic anhydride and the subsequent hydrogenolysis of maleic anhydride produces 1,4-butanediol. 1,3-Butanediol is currently produced via 3-hydroxybutyraldehyde from aldol condensation of acetaldehyde, which is produced by the Wacker oxidation of ethylene.

Important C5 and C6 diols are 1,5-pentanediol and 1,6-hexanediol. These diols are produced by the hydrogenation of corresponding dicarboxylic acids, namely glutaric acid and adipic acid. Adipic acid has been produced by the oxidation with nitric acid of the mixture of cyclohexanol and cyclohexanone, which is supplied by the auto-oxidation with air of cyclohexane produced by the hydrogenation of benzene. Gultaric acid is a byproduct of adipic acid production process by the oxidation of cyclohexanol and cyclohexanone, and therefore, the yield of glutaric acid is not high. The production process of diols by the hydrogenation of dicarboxylic acids derived from the oxidation of petroleum-derived hydrocarbons is not energy-efficient because of the reduction reactions such as hydrogenation of the more deeply oxidized substrates. Therefore, energy-efficient production methods of diols from biomass-derived are needed.

## 11.2 Substrates for the Production of Biomass-Derived Diols

Generally speaking, chemical composition of the feedstock from biomass and biomass-based building blocks has much higher oxygen content than that from fossil resources including petroleum. The target of the present chapter is diols, which have been used as monomers for polymer synthesis. The building blocks proposed for biomass refinery include glycerol, lactic acid, succinic acid, levulinic acid, furans (furfural, 5-hydoxymethylfurfural), xylitol, and sorbitol [3]. In 2006, Schlaf described the importance of the development of the catalytic process for the production of $\alpha,\omega$-diols from biomass in the excellent and pioneering review article, which mainly focused on the homogeneous catalysis [4]. On the other hand, this review article also stimulated the field of heterogeneous catalysis and contributed to the progress in the field of development of heterogeneous catalysts. At present, the catalytic conversion of biomass to chemicals is one of the hot topics in chemical and engineering fields [5, 6]. In this chapter, recent progress in the field of heterogeneous catalysis for the production of biomass-derived diols is described.

## 11.3 Production of Biomass-Derived C3 Diols

### *11.3.1 Product 1,2-Propanediol from Glycerol and Lactic Acid*

Glycerol is one of the most promising building blocks in the biomass refinery because glycerol is a byproduct in the biodiesel production by the transesterfication of vegetable oils with methanol. Lactic acid is a typical biomass-based carboxylic acid that is produced practically by fermentation of sugars. Catalytic and chemical production of lactic acid from lignocellulose raw material, cellulose, sugars, and glycerol is nevertheless a research topic [7, 8].

Figure 11.1 shows the reaction network for C-O hydrogenolysis of glycerol. 1,2-Propanediol is one of the products. Many effective catalysts for the selective C-O hydrogenolysis of glycerol to 1,2-propanediol have been reported in this decade and are described in reviews [9–11]. The C-O hydrogenolysis of glycerol seems to proceed in a single step, however, in fact, the conventional reaction route is a multistep sequence. Fundamentally, it has been known that the glycerol hydrogenolysis to 1,2-propanediol proceeds in two routes: one is dehydration + hydrogenation route and the other is dehydrogenation + dehydration + hydrogenation route. Figure 11.2 shows the thermodynamic profiles of dehydration + hydrogenation route in the C-O hydrogenolysis of glycerol.

Regarding the dehydration + hydrogenation route (Fig. 11.2), glycerol dehydration gives acetol and 3-hydroxypropanal, and the dehydration proceeds more preferentially under more acidic conditions judging from the thermodynamic stability of acetol. The consecutive hydrogenation of acetol and 3-hydroxypropanal give 1,2- and 1,3-propanediols, respectively. As a result, the main product can be 1,2-propanediol in this route. For example, the combination of Ru/C as a hydrogenation catalyst with an ion exchange resin as a solid acid catalyst enhanced the formation rate of 1,2-propanediol in the hydrogenolysis of glycerol [12–16]. An interesting point is that the reactivity of 1,2-propanediol in the hydrogenolysis (dehydration + hydrogenation) is much lower than that of that of glycerol. This tendency is strongly related to high yields of 1,2-propanediol in the glycerol hydrogenolysis.

The dehydrogenation + dehydration + hydrogenation route is preferential to the dehydration + hydrogenation route, for example, under more basic conditions without the addition of acid catalysts. Figure 11.3 shows the thermodynamic profiles of dehydrogenation + dehydration + hydrogenation route in the C-O hydrogenolysis of glycerol. Glycerol dehydrogenation can give two products: glyceraldehyde and dihydroxyacetone. An important point is the formation of these two dehydrogenation products from glycerol is endothermic and the glyceraldehyde formation is more thermodynamically unfavorable than the dihydroxyacetone. Dehydration of glyceraldehyde can proceed because of the high acidity at C-H neighboring to formyl group, and this reaction is exothermic and thermodynamically favorable, which promotes the dehydrogenation reaction. However, dehydration of dihydroxyacetone is difficult to proceed, and this suppresses the dehydrogenation of glycerol to

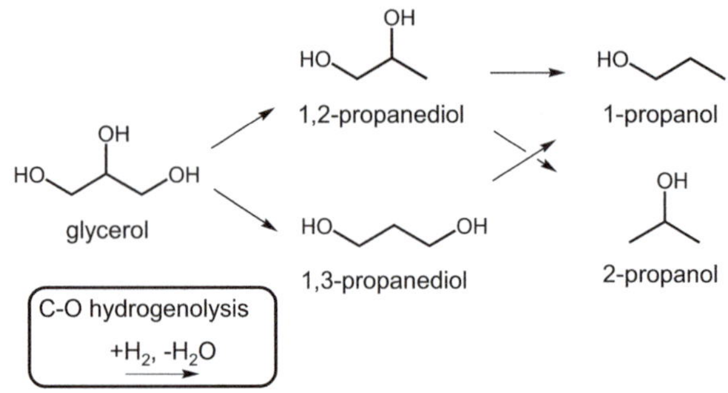


**Fig. 11.1** Reaction network for C-O hydrogenolysis of glycerol

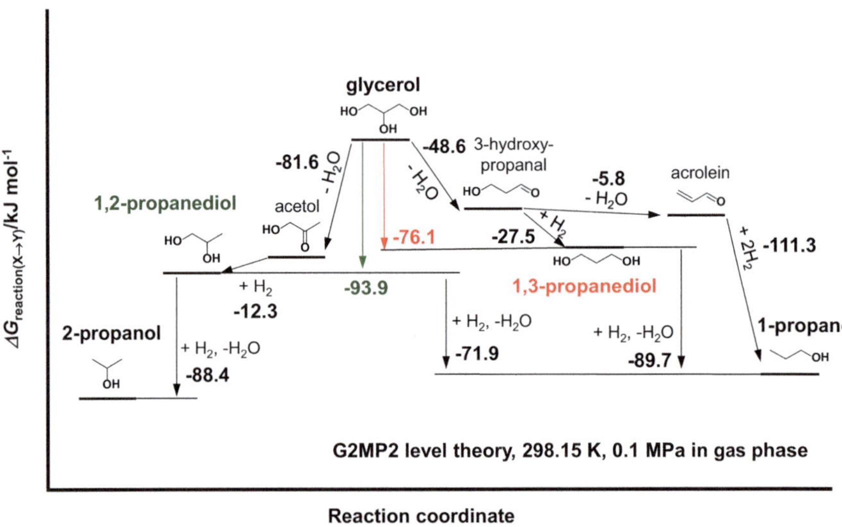


**Fig. 11.2** Thermodynamic profiles of dehydration + hydrogenation route in the C-O hydrogenolysis of glycerol

dihydroxyacetone. As a result, the glycerol is converted to 2-hydroxyacrolein or pyruvaldehyde via glyceraldehyde and subsequent hydrogen of 2-hydoxyacrolein or pyruvaldehyde gives 1,2-propanediol. It should be noted that retro-aldol reaction of glyceraldehyde can give the C-C cracking products such as ethylene glycol as a side reaction. Consecutive reaction of 1,2-propanediol is difficult to proceed because the dehydrogenation products of 1,2-propanediol such as acetol and 2-hydroxy-propanal cannot be dehydrated.

A very important point in both routes (Figs. 11.2 and 11.3) is that glycerol tends to be converted to 1,2-propanediol and the further reaction of 1,2-propanediol can be suppressed. Therefore, 1,2-propanediol can be produced from glycerol in high yield in principle. In fact, high yields of 1,2-propanediol above 90%, mainly over Cu catalysts, have been frequently reported [9–11].

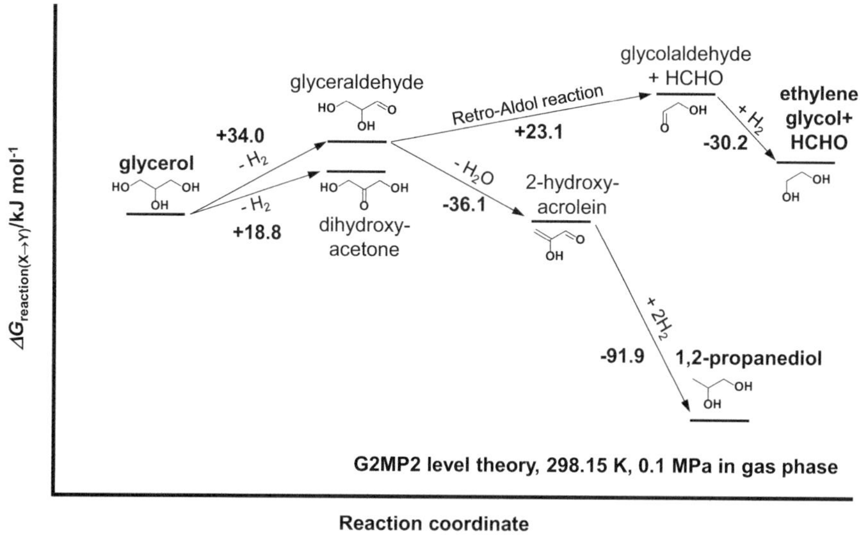


**Fig. 11.3** Thermodynamic profiles of dehydrogenation + dehydration + hydrogenation route in the C-O hydrogenolysis of glycerol

Another approach for the production of 1,2-propanediol is the hydrogenation of lactic acid (Fig. 11.4). Effective Ru heterogeneous catalysts such as Ru/$TiO_2$ [17], magnesia-supported poly-γ-aminopropylsiloxane–ruthenium complex (MgO-$NH_2$-Ru) [18], Ru-$MoO_x$ [19] catalysts have been reported, and high yields of 1,2-propanediol (>90%) are obtained. The use of Ru as an active species is due to its very low activity in the hydrogenolysis of 1,2-propanediol.

In the case of Ru-$MoO_x$, addition of a small amount of Mo to Ru/C and Ru/$SiO_2$ enhances the activity significantly [19]. The reaction conditions for the hydrogenolysis reaction are similar to those for the hydrogenation of carboxylic acids, and high activity in the hydrogenolysis of 1,2-propanediol can cause a decrease in the yield of 1,2-propanediol in the hydrogenation of lactic acid. Characterization results provide evidence that Ru is in a metallic state and Mo species are highly dispersed with about +4 valence, and these Mo species partially cover the surface of Ru metal by direct interaction between Mo and Ru. It is proposed that high activity of carboxylic acid hydrogenation can be due to the reactive hydride species from the heterolytic dissociation of $H_2$ on the Ru metal located near $MoO_x$ species [19]. The reusability of Ru-$MoO_x$/C catalyst without loss of activity and selectivity has been also confirmed.

Another important aspect of this reaction is the selective hydrogenolysis of L-lactic acid to (*S*)-1,2-propanediol. For example, Ru-$MoO_x$/C (Ru 5 wt%, Mo/Ru = 1/16) gave 98% ee (90% L-lactic acid conversion, 97% selectivity to 1,2-propanediol) at 353 K, 8 MPa $H_2$ in the aqueous phase. It is characteristic that the catalyst maintained high ee at higher reaction temperature (87% ee, 87% L-lactic acid conversion, 94% selectivity to 1,2-propanediol at 413 K) [19].

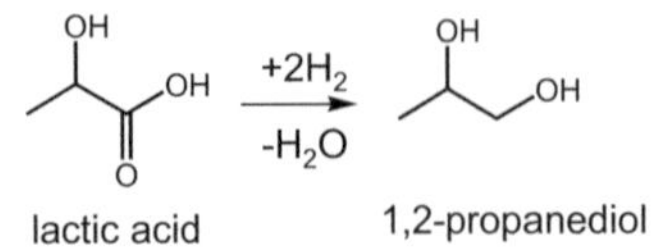


**Fig. 11.4** Hydrogenation of lactic acid to 1,2-propanediol

Generally speaking, higher reaction temperature can decrease ee through racemization of the substrate and the product with dehydrogenation + hydrogenation of C-OH groups and keto-enol tautomerism. It is verified that the Ru-$MoO_x$/C catalyst has comparable activity of (*S*)-1,2-propanediol racemization to Ru/C with modification with $MoO_x$ in spite of much higher lactic acid hydrogenation activity of Ru-$MoO_x$/C than Ru/C, suggesting that the addition of $MoO_x$ promotes the hydrogenation of lactic acid without the increase in the recemization activity. It has been also reported that the modification of $MoO_x$ on Rh catalysts is very effective for hydrogenation of amino acids [20, 21] and amides [22].

Another approach for the production of biomass-derived 1,2-propanediol is the catalytic conversion of saccharides and the derivatives such as cellulose, sorbitol and xylitol. Conversion of cellulose to 1,2-propanediol is catalysed by $WO_3$ + Ru/C catalyst system. The reaction route of cellulose to 1,2-propanediol is proposed as follows: (1) hydrolysis of cellulose to glucose catalysed by $WO_3$, (2) the isomerization of glucose to fructose catalysed by the carbon surface, (3) the retro-aldol reaction of fructose to glyceraldehyde or dihydroxyacetone, (4) the dehydration of glyceraldehyde to pyruvaldehyde (5) the hydrogenation of pyruvaldehyde to 1,2-propanediol catalysed by Ru. This reaction system has various kinds of side reactions. Therefore, the yields of 1,2-propanediol are about 30–40%. Various catalysts have been proposed to be effective: $WO_3$ + Ru/C [23], $Ca(OH)_2$ + supported noble metals [24], $Ca(OH)_2$ + Cu-$SiO_2$ [25, 26], $Ca(OH)_2$ (or $CeO_2$) + Ni/C [27], Pt-$SnO_x$/$Al_2O_3$ [28], Ni-$SnO_x$/$Al_2O_3$ [29], and Ni-$W_2C$/AC [30, 31].

## *11.3.2 Product 1,3-Propanediol from Glycerol*

Among the products in the glycerol hydrogenolysis such as 1,2- and 1,3-propanediols, ethylene glycol, 1- and 2-propanols and propane, the most valuable one is 1,3-propanediol. Therefore, the importance of conversion of glycerol to 1,3-propanediol has been recognized.

Effective heterogeneous catalysts for the selective hydrogenolysis of glycerol to 1,3-propanediol have been recently discovered [32], however, catalytic systems for this reaction are very limited to Ir-$ReO_x$ and the combination of Pt with acidic catalysts such as $WO_x$ or mordenite. The Ir-$ReO_x$/$SiO_2$ catalyst for the glycerol conversion was firstly reported in 2010, and showed the highest yield (38%) of 1,3-propanediol at that time [33, 34]. The Ir-$ReO_x$/$SiO_2$ catalyst shows the highest selectivity (~70% in initial stage of the reaction) to 1,3-propanediol among various combinations of noble metals and $MO_x$ (M = Re, Mo, W) on $SiO_2$ support. An important point is that monometallic Ir/$SiO_2$ and $ReO_x$/$SiO_2$ show almost no activity in the glycerol hydrogenolysis under the same reaction conditions, suggesting that the synergy exists between Ir and $ReO_x$ in the catalytic activity of Ir-$ReO_x$. The activity increases with increasing amount of Re up to Re/Ir = 2 at constant Ir loading (4 wt% Ir/$SiO_2$) [35]. Based on the characterization results, the surface of Ir metal nanoparticles is partially covered with three-dimensional $ReO_x$ clusters and the direct Ir-Re bond was formed [36, 37]. The synergy between Ir and Re suggests that the interface between Ir metal surface and $ReO_x$ clusters is the catalytically active site for the glycerol hydrogenolysis.

The reaction mechanism of the glycerol hydrogenolysis to 1,3-propanediol over Ir-$ReO_x$/$SiO_2$ was investigated in detail [33–39]. It is characteristic that the reaction order with respect to the glycerol concentration is almost zero, indicating that the adsorption of glycerol on the catalyst surface is almost saturated and the interaction between the glycerol and the active site is due to the formation of 1-glyceride on $ReO_x$ species. On the other hand, the reaction order with respect to $H_2$ pressure is first order, suggesting that one hydrogen molecule gives one active hydrogen species. From the reaction orders with respect to glycerol and $H_2$, the adsorption site of glycerol and $H_2$ can be different because the reaction order with respect to glycerol does not become negative and does not suppress $H_2$ adsorption. As a result, glycerol is adsorbed on $ReO_x$ by the formation of alkoxide, and $H_2$ is adsorbed on Ir metal. In particular, the first reaction order with respect to $H_2$ pressure indicates that the reaction rate can be controlled by the coverage of the adsorbed active hydrogen species, and this means that it is important to elucidate the active hydrogen species. To discuss the active hydrogen species, the reactivity of polyols in the hydrogenolysis was compared (Table 11.1) [39–41]. From the selectivity values, it is found that Ir-$ReO_x$ catalyzes the C-O dissociation neighboring to –$CH_2OH$ group in the reaction of polyols. An important point is that the reactivity of ethylene glycol is much higher than that of other polyols such as 1,2-butanediol, 1,2,4-butanetriol, glycerol and erythritol. This high reactivity of ethylene glycol is thought to exclude the acid-catalyzed C-O hydrogenolysis with a proton as the hydrogen species, considering the lower stability of primary carbocation than that of secondary carbocation. The reactivity of 1,2-butanediol is clearly higher than that of glycerol, erythritol and 1,2,4-butanetriol. From these tendencies, the reactivity can be concluded to be strongly influenced by the steric hindrance of the carbon atom in the dissociated C-O bond, suggesting that the C-O bond is dissociated with the $S_N2$-like attack and the active hydrogen species can be hydride. This $S_N2$-like attack by hydride is also supported by the first reaction order with respect to $H_2$ pressure. This mechanism is only applicable to the substrates that can have an anti-conformation of the dissociated C-O bond to the neighboring C-OH group. In fact, for 1,2-cyclohexanediols substrates which cannot have such an anti-conformation, the reaction order with respect to $H_2$ pressure is negative, suggesting a different mechanism [37]. The reaction mechanism proposed in the glycerol hydrogenolysis to 1,3-propanediol over the Ir-$ReO_x$ catalyst is novel and different from those in the glycerol hydrogenolysis to 1,2-propanediol (dehydration + hydrogenation and dehydrogenation + dehydration + hydrogenation), and it is regarded as "direct" $S_N2$-like attack C-O hydrogenolysis (Fig. 11.5). It should be noted that the Ir-$ReO_x$ catalysts with some modifications have been used by several research groups [42–47] for glycerol hydrogenolysis.

**Table 11.1** Reaction results of the hydrogenolysis of polyols on Ir-$ReO_x/SiO_2$ (Re/Ir = 1) (Drawn based on the data in Ref. [41])

| Substrate | Conv. / % | Products [selectivity / %] | | | | |
|---|---|---|---|---|---|---|
| Ethylene glycol | 94 | [98] | Others [2] | | | |
| 1,2-Butanediol | 51 | [88] | [10] | [2] | | |
| 1,2,4-Butanetriol | 22 | [73] | [12] | [7] | Others [7] | |
| Glycerol | 27 | [66] | [7] | [19] | [7] | Others [0] |
| Erythritol | 26 | [37] | [21] | [20] | [8] | Others [14] |

Reaction conditions: substrate 1 g, water 4 g, Ir-$ReO_x/SiO_2$ 0.3 g, $H_2SO_4$ ($H^+$/Ir=1), 373 K, $H_2$ 8 MPa, 4 h.

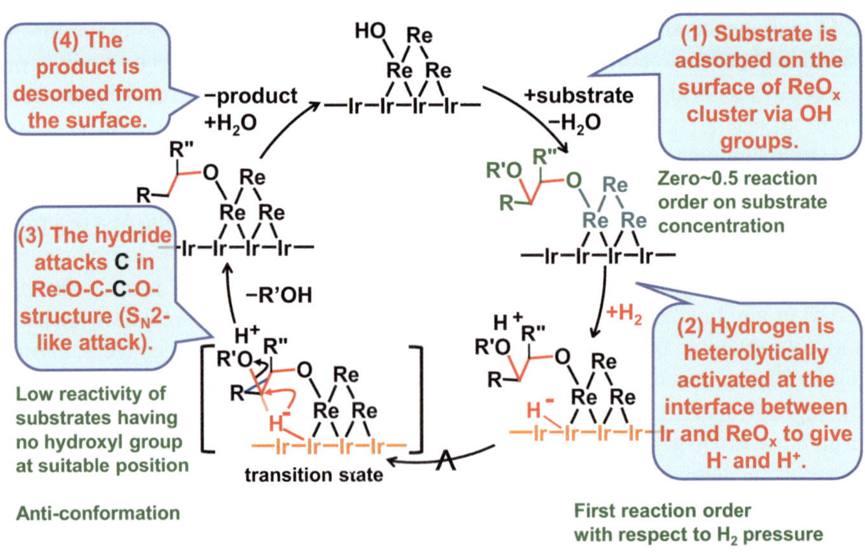


**Fig. 11.5** Catalytic cycle of the glycerol hydrogenolysis to 1,3-propanediol over Ir-$ReO_x$ catalyst

A very important point is that Ir-$ReO_x$ catalysts work for C-O hydrogenolysis at much lower reaction temperatures (for example, at 393 K) than many metal catalysts for the glycerol hydrogenolysis to 1,2-propanediol as mentioned above and than Pt-based catalysts for the glycerol hydrogenolysis to 1,3-propanediol as mentioned below. This low reaction temperature implies very high activity of Ir-$ReO_x$. The fact that significant amount of propanols are also formed from glycerol (Table 11.1) shows that Ir-$ReO_x$ also has some activity in the dissociation of C-O bonds without neighboring OH groups. The activity enables the application of Ir-$ReO_x$ to deeper C-O hydrogenolysis of saccharides and the derivatives such as cellulose, sorbitol and xylan into the corresponding alkanes (*n*-hexane or *n*-pentane) and monoalcohols [48–53]. This performance is also related to very low C-C hydrogenolysis of the Ir-$ReO_x$ catalyst. Moreover, Ir-$ReO_x$ shows excellent performance in the selective C=O hydrogenation of unsaturated aldehydes to unsaturated alcohols, which can be caused by the ability of heterolytic activation of $H_2$ molecule to hydride and proton [54, 55].

Other effective catalyst systems for the selective hydrogenolysis of glycerol to 1,3-propanediol are Pt-based catalyst systems. Since the report on Ir-$ReO_x$ catalysts in 2010 [33], various Pt-based catalysts giving high yields of 1,3-propanediol in the glycerol hydrogenolysis have been reported. This section is limited to the catalysts giving higher 1,3-propanediol yield than that of Ir-$ReO_x$ (38%).

Pt-sulfated zirconia gave 55% yield of 1,3-propanediol (66.5% conversion, 83.6% selectivity) at 443 K using 1,3-dimethyl-2-imidazolidinone (DMI) as a solvent [56]. In this reaction system, the DMI solvent was essential to high selectivity to 1,3-propanediol, and water solvent was found to be unsuitable to this reaction system, and the main product in water was 1,2-propanediol. Further investigation on Pt-sulfated zirconia catalysts has not been reported, and the field of this catalyst system is less-advanced.

Pt-$AlO_x$/$WO_3$ catalysts are effective for selective hydrogenolysis of glycerol to 1,3-propanediol and 40% yield of 1,3-propanediol (90% conversion) was reported at 453 K and 3 MPa $H_2$ in water [57]. It has been proposed that the aluminium oxide plays an important role on the formation of 1,3-propanediol which is promoted by the formation of primary aluminium alkoxide of glycerol. The secondary OH group in the adsorbed alkoxide of glycerol can be activated by the proton supplied from the surface of $WO_3$ support, and the hydrogenolysis reaction can proceed via secondary carbocation intermediate. Further development of this catalytic system brought some remarkable results. The same research group reported the catalyst with higher performance than Pt-$AlO_x$/$WO_3$. The improved catalyst, boehmite-supported Pt/W (Pt-$WO_x$/AlOOH), gives 66–69% yield of 1,3-propanediol (100% conversion) at 453 K and 5 MPa $H_2$ in water [58]. This yield of 1,3-propanediol is the highest at present. Moreover, the reusability of Pt-$WO_x$/AlOOH has been confirmed.

Regarding the results using fixed bed flow reaction systems, Pt-$WO_x$/$Al_2O_3$ (2 wt% Pt, 10 wt% $WO_x$) showed high yields of 1,3-propanediol (42.4% at 64.2% conversion) at 433 K and 5 MPa $H_2$ [59]. It is characteristic that the yield of 1,3-propanediol has a linear correlation with the number of Brønsted acid sites over

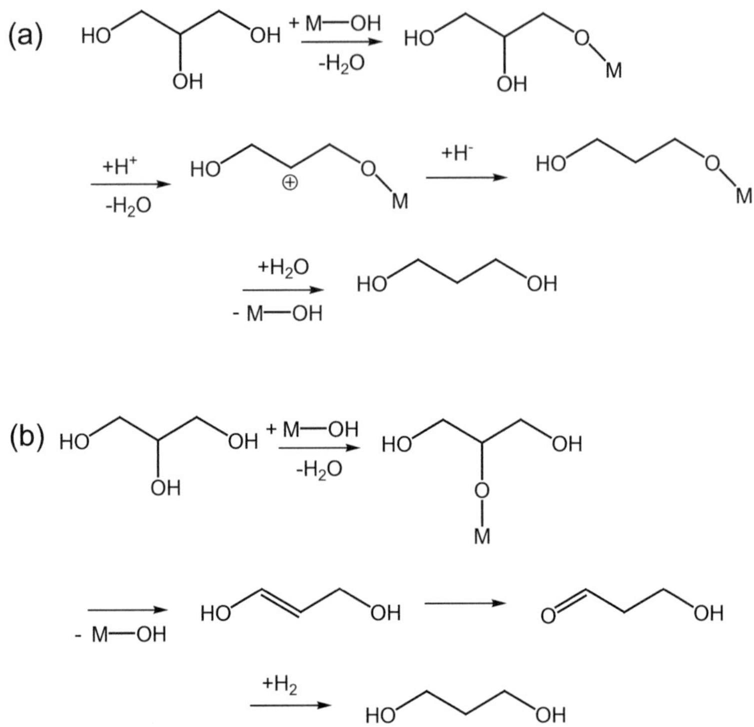


**Fig. 11.6** Reaction mechanism of glycerol hydrogenolysis to 1,3-propanediol catalyzed by Brønsted acid and Pt catalysts. (**a**) Primary alkoxide intermediate, (**b**) secondary alkoxide intermediate

Pt-$WO_x$/$Al_2O_3$ catalysts with 0–20 wt% $WO_x$. This strongly suggests that the Brønsted acid sites activate the secondary OH groups of glycerol.

Pt/H-Mordenite (Pt 2 wt%, $SiO_2$/$Al_2O_3$ = 20) has been recently reported to be effective for selective hydrogenolysis of glycerol to 1,3-propanediol in vapor phase, giving 46% 1,3-propanediol yield (94.9% conversion, 48.6% selectivity) at 498 K and atmospheric pressure of $H_2$ [60]. The average particle size of Pt was determined to be 4 nm, meaning that Pt metal particles were supported on the outer surface of the zeolite. In addition, Pt-Cu/H-Mordenite (2 wt% Pt, Cu 5 wt%) catalyst gave 53% 1,3-propanediol yield (90% conversion, 58.5% selectivity) at 488 K and atmospheric pressure of $H_2$ [61]. It is proposed that the key step is the formation of secondary carbocation by the attack of the proton on the Brønsted acid sites in the zeolite.

In both cases of Pt-$WO_x$ and Pt-zeolite catalysts, Brønsted acid sites play a crucial role in the activation of glycerol to the secondary carbocation, however, a different point is the alkoxide species (Fig. 11.6). In the case of Pt-$WO_x$, primary OH groups can be protected by the formation of primary alkoxide and the proton attacks the secondary carbon atom [57, 58]. In contrast, in the case of Pt-zeolite, the formation of a secondary alkoxide has been proposed [60]. In both cases, further investigation is necessary to elucidate the reaction intermediate.

## 11.4 Production of Biomass-Derived C4 Diols

Building blocks with four carbon atoms for biomass refineries are highly limited. Since succinic acid is produced by the fermentation of glucose in good yield, it is regarded as a C4 building block. On the other hand, various studies on the fermentation of glucose and glycerol for the production of erythritol have been carried out. The highest yield of erythritol from glucose reached 61% [62]. Erythritol production by fermentation of glucose has been already commercialized. In addition, it has been also reported that the yield of erythritol from glycerol reached 56% [63]. Therefore, it is thought that erythritol can be a candidate for C4 building blocks considering the limited available C4 building blocks. In particular, the merit of erythritol as a building block is that all of the carbon atoms in erythritol are functionalized. In the case of succinic acid, only the α and ω positions are functionalized, and so it is difficult to transform succinic acid to the products with functional groups at 2- and 3-positions such as 1,2-, 1,3-, and 2,3-butanediols. Here, the production of butanediols from succinic acid and erythritol is described. Production of C4 alcohols including diols from biomass-derived materials has been reviewed [64].

### *11.4.1 Product 1,4-Butanediol from Succinic Acid*

The hydrogenation of carboxyl group of succinic acid is used for the production of 1,4-butanediol (Fig. 11.7). In the case of hydrogenation of carboxylic acids to corresponding alcohols, the carboxylic acids are usually transformed to the esters by the reaction of alcohols before the hydrogenation reaction. On the other hand, the production process by fermentation gives succinic acid, rather than its ester. Therefore, hydrogenation of succinic acid without esterification has been investigated.

It is known that Re catalyst obtained by *in-situ* reduction of $Re_2O_7$ is effective for the reaction for a long time, providing high yields of 1,4-butanediol (94%; 483 K, $H_2$ 25 MPa, no solvent) [65]. Re(3.6 wt%)-Pd(2 wt%)/$TiO_2$ catalyst provides high yield of 1,4-butanediol (83%) at 433 K and 15 MPa in water [66]. Re(0.3 mol%)-Ru(0.3 mol%)/mesoporous carbon (MC) also shows 71% yield of 1,4-butanediol at 473 K and 8 MPa $H_2$ in 1,4-dioxane [67]. It has been reported that 89% yield of 1,4-butanediol at 413 K and 8 MPa $H_2$ in 1,4-dioxane is observed using Re-Pd/$SiO_2$ (Re 14 wt%, Re/Pd = 8) [68], which has been developed for the hydrogenation of higher carboxylic acids such as stearic acid [69]. This yield of 1,4-butanediol is

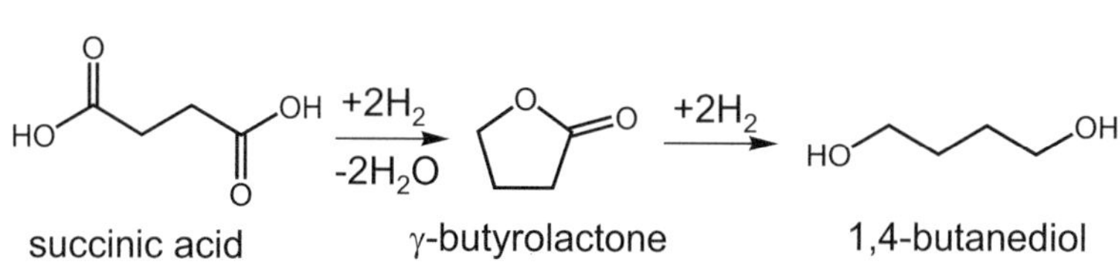


**Fig. 11.7** Hydrogenation of succinic acid to 1,4-butanediol via γ-butyrolactone

**Table 11.2** Comparison of the reactivity of succinic acid and γ-butyrolactone over $Re\text{-}Pd/SiO_2$ (Re 14 wt%, Re/Pd = 8)

| Substrate | Conv. (%) | Selectivity (%) | | | | |
|---|---|---|---|---|---|---|
| | | γ-Butyrolactone | 4-Hydroxybutyric acid | 1,4-Butanediol | 1-Butanol | Others |
| 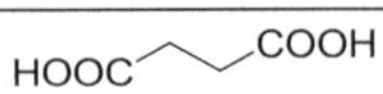 Succinic acid | 26 | 96 | - | 3.0 | 0.3 | 0.9 |
| 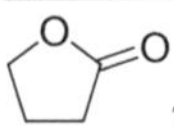 γ-Butyrolactone | 26 | - | - | 94 | 5.8 | 0.0 |

Reaction conditions: substrate 1 g, 1,4-dioxane 19 g, $Re\text{-}Pd/SiO_2$ 0.1 g, 413 K, $H_2$ 8 MPa, 4 h. Drawn based on the data in Ref. [68]

obtained at the reaction time of 96 h, and at shorter reaction time (24 h) high yield of γ-butyrolactone (85%) is obtained [68]. This indicates that the hydrogenation of succinic acid gives γ-butyrolactone first, and then the consecutive hydrogenation of γ-butyrolactone gives 1,4-butanediol. An important point is that the reactivity of γ-butyrolactone is comparable to that of succinic acid over $Re\text{-}Pd/SiO_2$ (Re 14 wt%, Re/Pd = 8) [68] (Table 11.2), meaning that the hydrogenation of γ-butyrolactone can be suppressed in the presence of succinic acid and as a result, high yields of γ-butyrolactone are obtained at shorter reaction time.

Generally speaking, it is difficult to determine the structure of Re species on Re-rich catalysts because Re species have various valence states. However, some Re-based catalysts for the hydrogenation of carboxylic acids have been characterized and the results suggest the structure of active Re species on $Re\text{-}Pd/SiO_2$ [68, 69] and $Re\text{-}Pd/TiO_2$ [70]. The $Re\text{-}Pd/SiO_2$ catalysts have $Pd^0$, $Re^0$(HCP), $Re^0$(FCC), $Re^{3+}$ and $Re^{4+}$, and the surface of $Pd^0$, $Re^0$(HCP) and $Re^0$(FCC) can be covered with $Re^{3+}$ and $Re^{4+}$ species. The reduction degree of Re species influences the distribution of $Re^0$, $Re^{3+}$, $Re^{4+}$, and $Re^{6+}$ species. High catalytic activity seems to be achieved when there is balanced distribution of these species. Therefore, both too high and too low degrees decrease the catalytic activity of the catalysts. This suggests that the presence of both $Re^0$ and $Re^{n+}$ species such as $Re^{3+}$ and $Re^{4+}$ can be connected to high catalytic activity [68, 69]. It also suggests that the role of Pd is to make the interaction between the substrate and the catalyst surface strong, and $Re^{n+}$ species on the metal surface can promote the heterolytic dissociation of $H_2$ to give active hydride species in the hydrogenation reaction.

The activation (reduction) method of the catalysts is important. On the $Re\text{-}Pd/SiO_2$ with *ex-situ* liquid-phase (in only 1,4-dioxane solvent) reduction with $H_2$, the catalyst shows much higher activity than that with *in-situ* liquid-phase (in the mixture of dicarboxylic acid and 1,4-dioxane) and gas-phase reduction with $H_2$ in the hydrogenation of succinic acid [68]. These tendencies suggest that the presence of succinic acid during the pre-reduction treatment suppresses the reduction of Re species to $Re^0$ on the calcined catalysts, which can be interpreted by the strong interaction of high valent (+7) Re species with succinic acid.

### 11.4.2 Butanediols from Erythritol and 1,4-Anhydroerythritol

Erythritol is a C4 sugar alcohol so that two-time C-O hydrogenolysis gives butanediols. However, a variety of the products can be formed in the hydrogenolysis of erythritol. Compared to the case of the glycerol hydrogenolysis, the reports on the erythritol hydrogenolysis are very limited. The reaction network of the C-O hydrogenolysis of erythritol is shown in Fig. 11.8. As mentioned for glycerol hydrogenolysis to 1,3-propanediol, Ir-$ReO_x$ shows very high catalytic activity of C-O hydrogenolysis, and this enables the C-O hydrogenolysis to occur at lower reaction temperatures like 393 K, which is suitable to suppress side reactions such as the dehydration to 1,4-anhydroerythritol. The result of the application of Ir-$ReO_x$ catalyst to the erythritol hydrogenolysis is shown in Fig. 11.9 [41]. The selective formation of diols such as 1,4-butanediol and 1,3-butanediol is not easy even using Ir-$ReO_x$ catalyst which has high selectivity in the glycerol hydrogenolysis to 1,3-propanediol.

Another approach for the production of butanediols from erythritol is the use of 1,4-anhydroerythritol as a starting material because 1,4-anhydroerythritol is obtained selectively from the acid-catalyzed dehydration of erythritol. The selective hydrodeoxygenation of 1,4-anhydroerythritol to 3-hydroxytetrahydrofuran using $WO_x$-Pd catalysts has been reported [71] (Fig. 11.10), where the yield of 3-hydroxytetrahydrofuran reaches 74% in 1,4-dioxane solvent. In addition, Ir-$ReO_x$/$SiO_2$ (Re/Ir = 2) catalyzes the selective C-O hydrogenolysis of 3-hydroxytetrahydrofuran to 1,3-butanediol in water to obtain 81% selectivity at 24% conversion [37] (Fig. 11.10). At present, higher conversions have not been

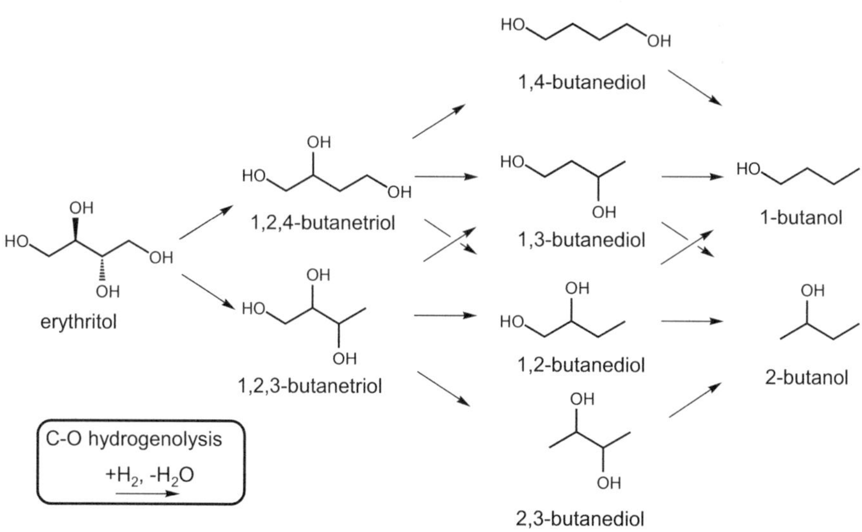


**Fig. 11.8** Reaction network for C-O hydrogenolysis of erythritol

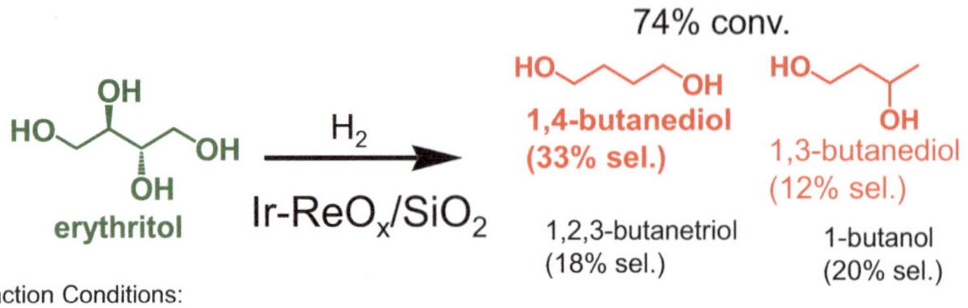


Reaction Conditions:

20 wt% erythritol aq. 5 g, $Ir-ReO_x/SiO_2$ (Re/Ir=1) 300 mg, $H_2SO_4$ ($H^+/Ir$ = 1), 393 K, 8 MPa $H_2$, 24 h.

**Fig. 11.9** Results of the hydrogenolysis of erythritol over $Ir-ReO_x/SiO_2$ catalyst (Drawn based on the data in Ref. [41])

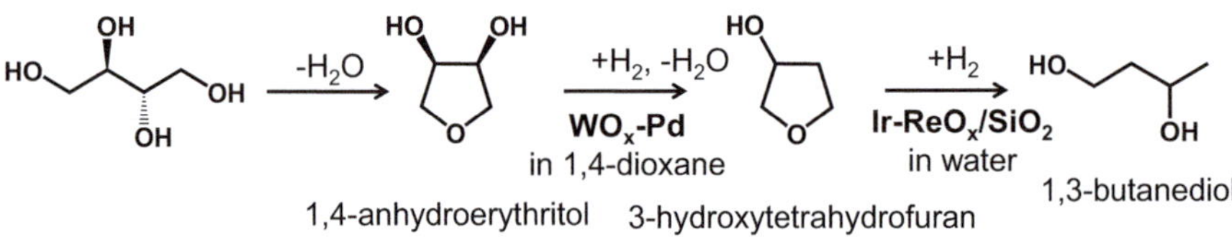


**Fig. 11.10** Reaction route for conversion of erythritol to 1,3-butanediol

reported, however, this reaction route can contribute to the development of 1,4-anhydroerythritol-to-1,3-butanediol catalytic system.

Another approach is the use of deoxydehydration reaction of vicinal diols. It has been known that Re, V, and Mo homogeneous catalysts, especially Re, are active in the deoxydehydration of vicinal OH groups to give corresponding alkenes. Deoxydehydration, which can be regarded as the reverse reaction of *cis*-dihydroxylation of alkenes, usually requires reducing agents such as secondary alcohols, $PPh_3$, sulfite, metal, and hydroaromatics [72, 73]. In these catalyst systems, $H_2$ is not a suitable reductant. On the contrary, it has been recently reported that heterogeneous $ReO_x$-Pd/$CeO_2$ catalyst shows excellent performance for simultaneous hydrodeoxygenation of vicinal OH groups using $H_2$ as a reducing agent, where hydrodeoxygenation consists of deoxydehydration of the substrates with vicinal OH groups to the compounds with C-C double bond and the subsequent hydrogenation of the C-C double bond [74–76]. Deoxydehydration is catalyzed by high valent Re species stabilized on $CeO_2$ surface, and the hydrogenation of C=C bond is catalyzed by very small Pd metal particles dispersed on $CeO_2$ with the modification of lower valent Re species. Simultaneous hydrodeoxygenation of vicinal OH groups using $H_2$ reducing agent over $ReO_x$-Pd/$CeO_2$ converts erythritol to 1,2-butanediol and 1,4-butanediol as the main product and the byproduct, respectively (Fig. 11.11) [74]. In the reaction mechanism, an important reaction intermediate is the diolate of Re species, and the much higher yield of 1,2-butanediol than that

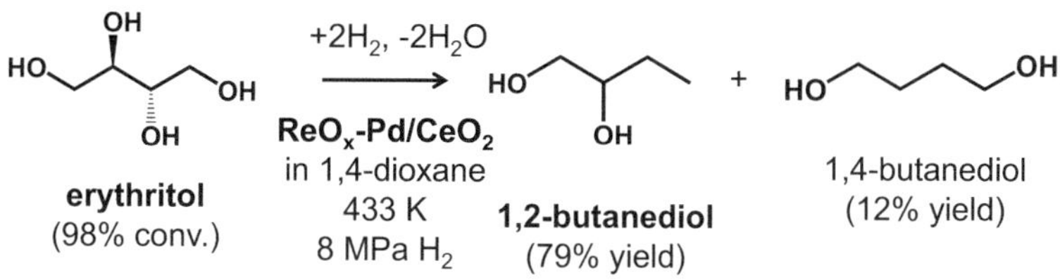


**Fig. 11.11** Results of the hydrodeoxygenation of erythritol over $ReO_x$-Pd/$CeO_2$ catalyst (Drawn based on the data in Ref. [74])

of 1,4-butanediol suggests that the 1,2-diolate from erythritol and Re species is more stable than the 2,3-diolate. In addition, the $ReO_x$-Pd/$CeO_2$ catalyst is very effective to hydrodeoxygenate vicinal OH groups in 1,4-anhydroerythritol to tetrahydrofuran with higher yield than 99%. The estimated turnover frequency per Re atom of the conversion of 1,4-anhydroerythritol to tetrahydrofuran in 1,4-dioxane at 443 K is about 300 $h^{-1}$. The stability of $ReO_x$-Pd/$CeO_2$ was verified by the recycle use of the catalyst, and the large total turnover number per Re atom (about 10,000) was demonstrated [74, 75].

Another reported route for 1,4-butanediol production from biomass consists of two consecutive reaction steps: selective oxidation of furfural to furanones and hydrogenation of the mixture of furanones to 1,4-butanediol. Pt/$TiO_2$-$ZrO_2$ (Ti/Zr = 1/1) is a very effective catalyst for both oxidation and hydrogenation reactions [77]. The oxidation of furfural with $H_2O_2$ proceeds at 298 K in HCOOH/$H_2O$/MeOH (v/v/v = 10/10/80). After the reactor is purged with nitrogen, the hydrogenation of furanones proceeds at 393 K and 3.5 MPa $H_2$. The yield of 1,4-butanediol reached 85%.

## 11.5 Production of Biomass-Derived C5 Diols

As described above, one strategy for production of C4 diols is twice C-O dissociations of erythritol. When the same strategy is applied to the production of C5 diols from C5 sugars and sugar alcohols, three C-O dissociations are required. However, this means that highly selective production of one specific C5 diol is more difficult than the case for C4 diol. Therefore, a different strategy is necessary.

The C5 building blocks include furfural and levulinic acid. Furfural is commercially produced by the acid-catalyzed dehydration of hemicellulose, and levulinic acid can be produced by the decomposition of 5-hydroxymethylfurfural along with formic acid. Selective production of one specific diol has been attempted from furfural and levulinic acid [78–80]. It is known that the ring-opening reaction of furfuryl alcohol, which is commercially produced from furfural by hydrogenation, to pentanediols is catalyzed by $CuCr_2O_4$ catalyst (Fig. 11.12) [81]. $CuCr_2O_4$ catalyst is

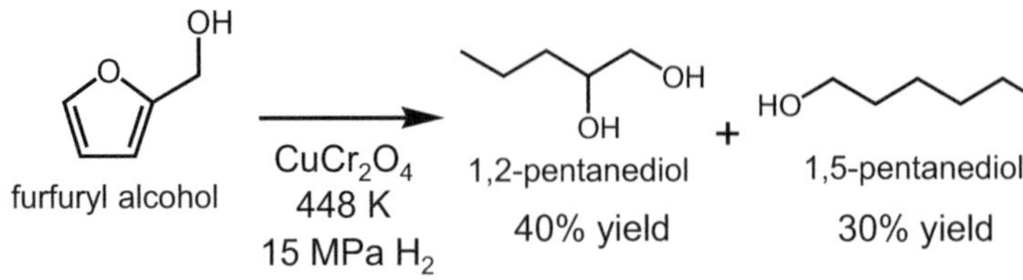


**Fig. 11.12** Conversion of furfuryl alcohol using Cu catalyst in the presence of hydrogen (Drawn based on the data in Ref. [81])

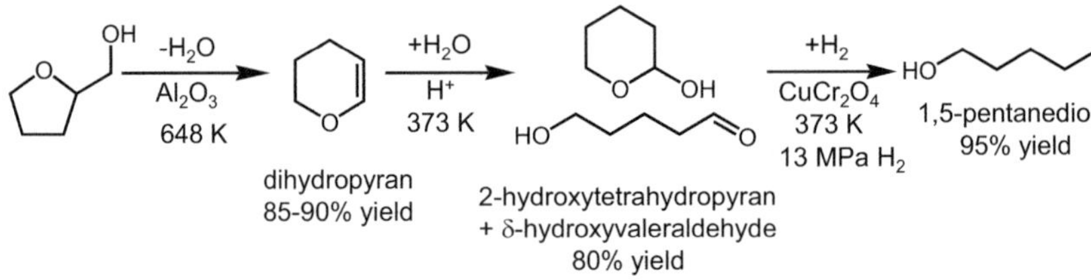


**Fig. 11.13** Three-step process for the transformation of tetrahydrofurfuryl alcohol to 1,5-pentanediol (Drawn based on the data in Ref. [82])

a conventional hydrogenation and hydrogenolysis catalyst, indicating that the reaction is not selective at all. Therefore, a selective production method of one pentanediol such as 1,5-pentanediol is needed. A possible starting substrate is tetrahydrofurfuryl alcohol, which is given from the total hydrogenation of furfural, and is dehydrated to dihydropyran. The dihydropyran is hydrated to the mixture of 2-hydroxytetrahydropyran and δ-hydroxyvaleraldehyde, and the subsequent hydrogenation using $CuCr_2O_4$ gives 1,5-pentanediol [82]. In Fig. 11.13, the values of the yield were obtained from the results using pure substrates in each step, and the yield of 1,5-pentanediol was estimated to be about 70%, meaning that the purification of the products in each step is necessary [82]. These works were reported in 1931 and 1946, and in 2006 Schlaf described in a review article as follows [4]: "a process that would allow a direct hydrogenation and hydrogenolysis of furfural to 1,5-pentanediol with higher selectivity and yield (possibly using homogeneous catalyst) would however still be desirable." In 2009, a selective hydrogenolysis of tetrahydrofurfuryl alcohol to 1,5-pentanediol using heterogeneous Rh-$ReO_x$ and Rh-$MoO_x$ catalysts has been reported [83, 84], and the related works promoted the catalysis field on the selective conversion of various biomass-derived substrates. Progress in the field of the production of C5 alcohols including diols is available in a review article [64].

### 11.5.1 Product 1,5-Pentanediol from Furfural and Tetrahydrofurfuryl Alcohol

Furfural is a very important building block, however, it is highly reactive so that it readily polymerizes to humins at many reaction conditions [85]. This is the reason why the saturated tetrahydrofurfuryl alcohol was selected as a starting compound for the multistep production of 1,5-pentanediol (Fig. 11.13). The selective hydrogenation of furfural to tetrahydrofurfuryl alcohol is therefore needed. Selective catalysts for the partial hydrogenation of furfural to furfuryl alcohol have been reported, whereas, in contrast, reports on the total hydrogenation of furfural to tetrahydrofurfuryl alcohol are very limited, especially in water solvent [86–89]. Figure 11.14 shows the reaction network of the total hydrogenation of furfural. Ni-Pd/$SiO_2$ (Ni/Pd = 7/1) gives 96% yield of tetrahydrofurfuryl alcohol under the conditions of furfural (0.5 M aq.; 10 ml), acetic acid (5.7 μl, 0.1 mmol), 0.1 g catalyst, $H_2$ (8 MPa) and 313 K [86]. Here, the leaching of Ni species was observed in the aqueous phase and this was connected to the catalyst deactivation in the repeated use of the catalyst. In contrast, the leaching of Pd was not observed. To overcome the leaching problem two different approaches have been taken.

The first approach is the gas phase catalytic hydrogenation of furfural using the Ni catalysts. Gas-phase hydrogenation of furfural to tetrahydrofurfuryl alcohol proceeds selectively over Ni/$SiO_2$ with <4 nm Ni metal particle size prepared by the reduction of $Ni(NO_3)_2$ on $SiO_2$, and the yields of tetrahydrofurfuryl alcohol reach 94% [87]. The reaction proceeds in two separated steps: hydrogenation of formyl group in furfural and subsequent hydrogenation of furan ring.

The second approach is the development of non-Ni catalysts. It has been reported that Pd-Ir/$SiO_2$ is effective for total hydrogenation of furfural in water [89]. In particular, the Pd-Ir/$SiO_2$ shows higher catalytic performance in the activity and selectivity than monometallic Pd/$SiO_2$ and Ir/$SiO_2$, which means that its high performance is due to synergistic effects between Pd and Ir by the formation of the alloy. High activity allows the reaction temperature to be decreased to values as low as 275 K, and this low reaction temperature suppresses the side reactions. As a result, the yield of tetrahydrofurfuryl alcohol reaches 94%. Further, the catalyst is recyclable.

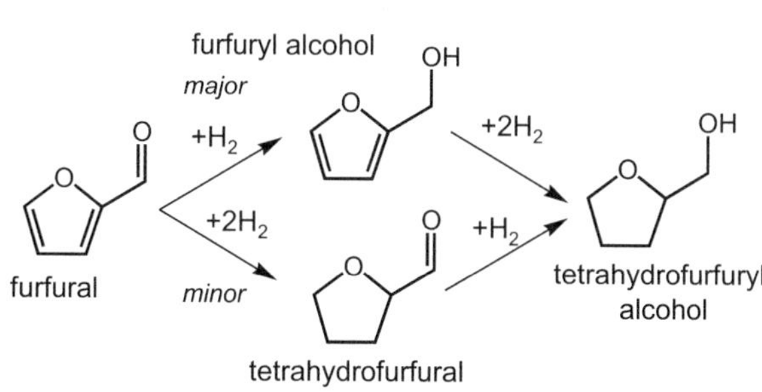


**Fig. 11.14** Reaction network for total hydrogenation of furfural

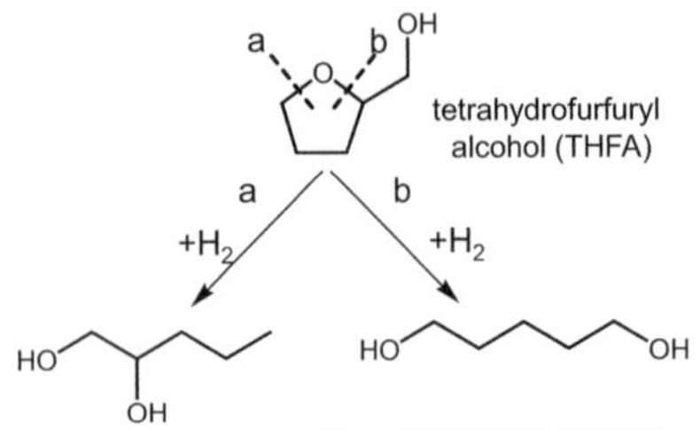


| | THFA conv. (%) | 1,2-pentanediol | 1,5-pentanediol |
|---|---|---|---|
| | | Selectivity (%) | |
| $Rh/SiO_2$ | 5.7 | 62 | 18 |
| $Rh\text{-}ReO_x/SiO_2$ (Re/Rh=0.5) | 57 | 0 | 94 |
| $Rh\text{-}MoO_x/SiO_2$ (Mo/Rh=0.13) | 50 | 0 | 96 |

**Fig. 11.15** Reaction scheme of the ring-opening C-O hydrogenolysis of tetrahydrofurfuryl alcohol (THFA) and the results of $Rh/SiO_2$ and $Rh\text{-}MO_x/SiO_2$ (M = Re, Mo) in this reaction. Reaction conditions: 8.0 MPa $H_2$, 393 K, 5 wt% THFA aqueous solution 20 ml, 0.05 g-cat, Rh 4 wt%, 4 h (Drawn based on the data in Ref. [84])

As shown above, furfural can be converted to tetrahydrofurfuryl alcohol, and it can be regarded as a substrate for the production of biomass-derived C5 diols. The ring-opening C-O hydrogenolysis of tetrahydrofurfuryl alcohol gives 1,2-pentanediol and 1,5-pentanediol (Fig. 11.15). $Rh/SiO_2$ shows some activity in the hydrogenolysis of tetrahydrofurfuryl alcohol to 1,2-pentanediol. Although the selectivity is not so high, the formation of 1,2-pentanediol is clearly preferable to that of 1,5-pentanediol [83]. The steric hindrance of $-CH_2OH$ group can explain the preferable formation of 1,2-pentanediol; the mechanism of this reaction has been also investigated theoretically [90]. On the other hand, $Rh\text{-}ReO_x$ and $Rh\text{-}MoO_x$ catalysts show much higher activity than $Rh/SiO_2$ and very high selectivity to 1,5-pentanediol [83, 84, 91–94]. It should be noted that $ReO_x/SiO_2$ and $MoO_x/SiO_2$ have no activity in the hydrogenolysis of tetrahydrofurfuryl alcohol. Similar behavior is also observed in the hydrogenolysis of glycerol [95–97]. These behaviors indicate that synergy between Rh metal and $MO_x$ (M = Re, Mo) exists that remarkably promotes the selective hydrogenolysis. This synergy is probably due to the structure of $Rh\text{-}ReO_x$ and $Rh\text{-}MoO_x$. According to the EXAFS analyses of the catalysts, the interaction of the Re and Mo species with Rh metal surface is direct and the Re and Mo species are the small two-dimensional $ReO_x$ clusters and the isolated $MoO_x$ species on the Rh metal surface, respectively. The synergy suggests that the active site is formed at the interface between Rh metal and $MO_x$ species. An important point is that the ring-opening C-O hydrogenolysis of tetrahydrofurfuryl alcohol to 1,5-pentanediol is similar to the glycerol hydrogenolysis to 1,3-propanediol from the viewpoint of regioselectivity: the C-O bonds neighboring to $-CH_2OH$ groups are dissociated. $Ir\text{-}ReO_x$, which is an effective catalyst for glycerol hydrogenolysis to 1,3-propanediol, also catalyzes the selective hydrogenolysis of tetrahydrofurfuryl alcohol to 1,5-pentanediol. Therefore, a similar reaction mechanism has been proposed: the hydride species formed at the interface between Rh metal surface and $MO_x$ (M = Re, Mo) can attack the carbon atom neighboring the $-CH_2OH$ group adsorbed by the formation of the alkoxide on $MO_x$ species [37, 91]. Another reaction mechanism has been proposed: selective C-O hydrogenolysis of polyols and cyclic ethers proceeds by acid-catalyzed dehydration reactions with Re-OH species coupled with metal-catalyzed hydrogenation [98, 99].

Based on the findings of selective hydrogenolysis of tetrahydrofurfuryl alcohol to 1,5-pentanediol over Rh-$ReO_x$, Rh-$MoO_x$ and Ir-$ReO_x$, catalyst systems of noble metals combined with oxophilic metals have been applied to similar reactions [100–103]. The highest yield of 1,5-pentanediol was obtained as 94% on Rh-$ReO_x$/C (Rh 4 wt%, Re/Rh = 0.25) under the conditions of aqueous tetrahydrofurfuryl alcohol solution (5 wt %, 20 ml), catalyst (100 mg), 373 K, 8 MPa $H_2$, 24 h [94].

After selective catalysts for the hydrogenolysis of tetrahydrofurfuryl alcohol to 1,5-pentanediol were identified, one-pot conversion of furfural to 1,5-pentanediol was attempted by the addition of the function for the total hydrogenation of furfural to the selective catalysts [104, 105]. The results show that Pd-Ir-$ReO_x$ and Rh-Ir-$ReO_x$ catalysts are effective for one-pot conversion of furfural. The reaction temperatures should be set to be largely different for the total hydrogenation of furfural and for the hydrogenolysis of tetrahydrofurfuryl alcohol. A one-pot two-step controlled temperature reaction gives 71% yield of 1,5-pentanediol over Pd-Ir-$ReO_x$ from diluted aqueous furfural (10 wt%) at 313 K for 8 h and 373 K for 72 h [104]. Rh-Ir-$ReO_x$ catalyst gives a higher yield of 1,5-pentanediol (78%) from diluted aqueous furfural at 313 K for 8 h and 373 K for 32 h [105]. In addition, high yields of 1,5-pentanediol from highly concentrated aqueous furfural (50 wt%) are obtained over Rh-Ir-$ReO_x$ (71%; at 313 K for 8 h and 373 K for 24 h) [105].

### *11.5.2 Pentanediols from Furfural and Furfuryl Alcohol via Opening of Furan Ring*

The key step in the production of pentanediols from furfural and the derivatives is the ring-opening reaction. In the case above, pentanediols are formed via teterahydrofurfuryl alcohol, which is the product of the total hydrogenation of furfural, and the hydrogenolysis of the C-O bond in the tetrahydrofuran ring. Another method for the production of pentanediols from the substrates with furan ring such as furfural and furfuryl alcohol is the direct ring-opening reaction of furan moiety.

When furfural is used as a starting substrate, two kinds of effective catalysts have been reported. Li-modified Pt/$Co_2AlO_4$ gives 34.9% yield of 1,5-pentanediol, 31.3% yield of tetrahydrofurfuryl alcohol, 16.2% yield of 1,2-pentanediol at 413 K and 1.5 MPa in ethanol solvent [106]. The selectivity is not so high in this case. High yield of tetrahydrofurfuryl alcohol implies its low reactivity over Li-modified Pt/$Co_2AlO_4$, suggesting that the reaction proceeds by direct ring-opening reaction before hydrogenation of the furan ring. Another proposed catalyst is hydrotalcite-supported Pt nanoparticle (Pt/HT). Pt/HT shows 73% yield of 1,2-pentanediol, 14% yield of tetrahydrofurfuryl alcohol, and 8% yield of 1,5-pentanediol in 2-propanol solvent at 423 K and 3 MPa $H_2$ [107]. The catalyst is reusable. When Pt/HT is applied to the reaction of furfuryl alcohol, 80% yield of 1,2-pentanediol is obtained. In both cases of furfural and furfuryl alcohol, Pt/HT gives 1,2-pentanediol selectively.

When furfuryl alcohol is used as a starting substrate, two kinds of effective catalysts have been reported. Ru/$MnO_x$ catalyst gives 42.1% yield of 1,2-pentanediol and 37% yield of tetrahydrofurfuryl alcohol in water solvent at 423 K and 1.5 MPa $H_2$ [108]. The reactivity of tetrahydrofurfuryl alcohol and tetrahydrofuran is much lower than furfuryl alcohol over Ru/$MnO_x$, and high reactivity of furfuryl alcohol can be due to the furan ring. The preferable formation of 1,2-pentanediol to 1,5-pentanediol can be interpreted as being due to steric hindrance of $-CH_2OH$ group in furfuryl alcohol. Other reported catalysts are Cu-based ones such as Cu-$Al_2O_3$ and Cu-$Mg_3AlO_{4.5}$ [109, 110]. Cu-$Mg_3AlO_{4.5}$ catalyst gives 51.2% yield of 1,2-pentanediol and 28.8% yield of 1,5-pentanediol in ethanol solvent at 413 K and 6 MPa $H_2$ [110].

Regarding the proposed catalysts for the direct ring-opening reaction of furan, the structural analysis of the catalysts and the elucidation of the reaction mechanism are needed in future studies.

### *11.5.3 Product 1,4-Pentanediol from Levulinic Acid*

The reactivity of carbonyl group is higher than that of carboxyl group, therefore, hydrogenation of levulinic acid can give 4-hydroxypentanoic acid first, and the dehydration of 4-hydroxypentanoic acid easily proceeds to $\gamma$-valerolactone (Fig. 11.16). Hydrogenation of the carboxyl group in 4-hydroxypentanoic acid or hydrogenolysis + hydrogenation of $\gamma$-valerolactone gives 1,4-pentanediol. One by-product is 2-methyltetrahydrofuran which is formed by dehydration of 1,4-pentanediol or hydrogenation + hydrogenolysis of $\gamma$-valerolactone (Fig. 11.16). Various kinds of catalysts have been proposed, however, the reaction scheme is essentially the same.

It has been reported that Cu/$ZrO_2$ (Cu 30 wt%) is very effective for hydrogenolysis of $\gamma$-valerolactone to give 96% yield of 1,4-pentanediol in ethanol solvent at 473 K and 6 MPa $H_2$ [111]. As the next step, aqueous phase hydrogenation of levulinic acid has been developed. Rh-$MoO_x$/$SiO_2$ (4 wt% Rh, Mo/Rh = 0.13) shows high performance in the aqueous phase selective hydrogenation of levulinic acid to 1,4-pentanediol, and it gives 70% yield of 1,4-pentanediol at 353 K and 6 MPa $H_2$ [112]. This catalyst was developed for C-O hydrogenolysis of saturated compounds such as tetrahydrofurfuryl alcohol [84, 91, 113], and therefore the yield of 1,4-pentanediol can be limited by consecutive reaction of 1,4-pentanediol via C-O hydrogenolysis as a side reaction. The change of the active metal from Rh to Pt (Pt-$MoO_x$/$SiO_2$) decreases the performance of 1,4-pentanediol production: the main reaction of levulinic acid on Pt-$MoO_x$/$SiO_2$ is the hydrogenation + dehydration to

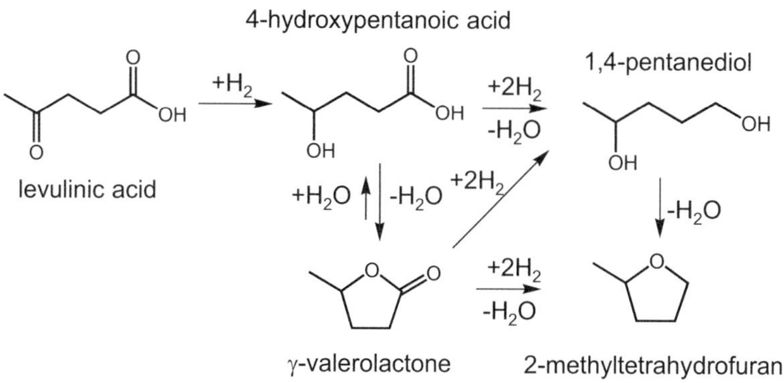


**Fig. 11.16** Reaction network for conversion of levulinic acid to 1,4-pentanediol and its side reactions

γ-valerolactone [112]. On the other hand, hydroxyapatite-supported Pt-Mo bimetallic nanoparticles (Pt-Mo/HAP) catalyzes the selective conversion of levulinic acid to 1,4-pentanadiol in water [114]. The yields of 1,4-pentanadiol reach 93% on Pt-Mo/HAP (Pt 2 mol%, Mo 0.5 mol%) at 403 K and 5 MPa $H_2$. The Pt–Mo/HAP catalyst is reusable. In this case, the formation of γ-valerolactone is not detected, and this can be interpreted as high activity of Pt-Mo/HAP in the hydrogenation of carboxyl group or ester group, where the reduced Mo species having oxygen vacancies can act as Lewis acid sites activating the carbonyl moieties of 4-hydroxypentanoic acid and γ-valerolactone.

## 11.6 Production of Biomass-Derived C6 Diols

The α,ω-C6 diol, 1,6-hexanediol, is an important monomer and has been supplied by the hydrogenation of adipic acid (or its ester). Adipic acid is also a very important monomer and produced in a large scale from petroleum. The C6 building blocks include 5-hydroxymethylfurfural and sorbitol. The selective conversion of 5-hydroxymethylfurfural to specific diols has been developed and will be discussed in Sect. 11.6.2. On the other hand, it is not so easy to use sorbitol as a precursor for the production of C6 diols because quadruple C-O hydrogenolysis is required. Another strategy for the synthesis of the C6 diols from biomass has been developed based on the derivation of pentanediols from furfural (Fig. 11.17). Tetrahydropyran-2-methanol is a key intermediate in this strategy. Tetrahydropyran-2-methanol is given from the total hydrogenation of acrolein dimer. While acrolein is currently supplied by oxidation of petroleum-derived propylene, acrolein can be produced by dehydration of glycerol, which means that tetrahydropyran-2-methanol can be an intermediate in the biomass refinery.

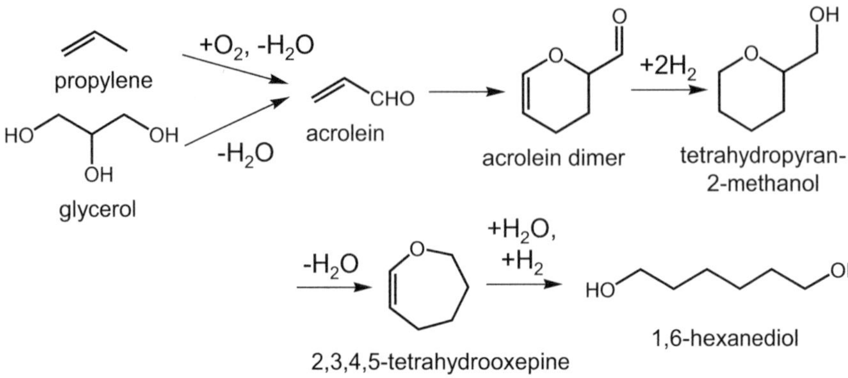


**Fig. 11.17** Production routes of 1,6-hexanediol via tetrahydropyran-2-methanol

## *11.6.1 Product 1,6-Hexanediol from Tetrahydropyran-2-Methanol*

It has been reported that tetrahydropyran-2-methanol can be converted to 1,6-hexanediol via 2,3,4,5-tetrahydrooxepine [115]. The 42% yield of 1,6-hexanediol from tetrahydropyran-2-methanol is obtained by the dehydration and subsequent hydrogenation.

The selective ring-opening C-O hydrogenolysis of tetrahydropyran-2-methanol was reported at almost the same time as that of tetrahydrofurfuryl alcohol [83], and the catalytic behavior of the hydrogenolysis of tetrahydropyran-2-methanol is similar to that of tetrahydrofurfuryl alcohol, however, the selectivity of 1,6-hexanediol is lower than that of 1,5-pentanediol. For example, Rh-$ReO_x$/$SiO_2$ (Re/Rh = 0.5) shows 94% 1,5-pentanediol selectivity at 57% conversion of tetrahydrofurfuryl alcohol, and it shows 90% 1,6-hexanediol selectivity at 32% conversion of tetrahydropyran-2-methanol (393 K, 8 MPa $H_2$) [94]. To obtain higher selectivity and higher yield of 1,6-hexanediol formation, Rh-$ReO_x$/C was developed [94]. Rh-$ReO_x$/C (Re/Rh = 0.25) gives 84% yield of 1,6-hexanediol in the hydrogenolysis of tetrahydropyran-2-methanol, which is clearly higher than that (42% yield) in the conversion via 2,3,4,5-tetrahydrooxepine mentioned above, while Rh-$ReO_x$/$SiO_2$ (Re/Rh = 0.50) gives 56% yield of 1,6-hexanediol as the highest yield obtained from experiments with different reaction times [94]. Later, Dumesic's group has reported the same reaction using Rh-Re, Rh-Mo, Pt-Mo catalysts [116–118]. Good to excellent selectivities (80–97%) are obtained at low conversion levels, although the maximum yields that would be obtained at longer reaction time have not been searched.

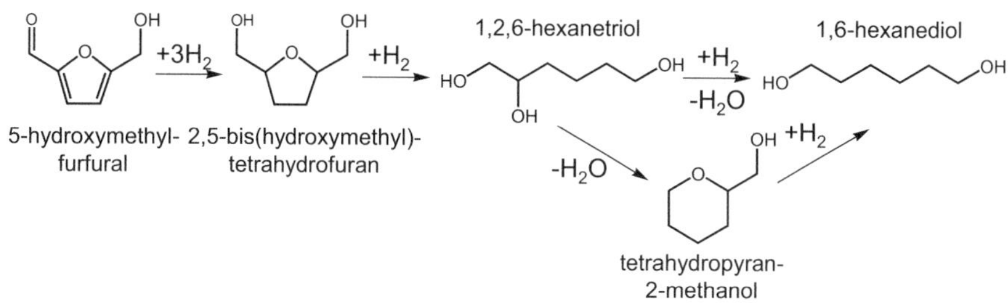


**Fig. 11.18** Production routes of 1,6-hexanediol from 5-hydroxymethylfurfural

## 11.6.2 Product 1,6-Hexanediol from 5-Hydroxymethylfurfural and Its Derivatives

The total hydrogenation of 5-hydroxymethylfurfural proceeds selectively by using the catalysts for the total hydrogenation of furfural to tetrahydrofurfuryl alcohol. Ni-Pd/$SiO_2$ (Ni/Pd = 7/1) gives 96% yield of 2,5-bis(hydroxymethyl)-tetrahydrofuran under the conditions of 5-hydroxymethylfurfural (0.5 M aq.; 10 ml), acetic acid (5.7 μl, 0.1 mmol), 0.1 g catalyst, 8 MPa $H_2$ and 313 K [86], and Pd-Ir/$SiO_2$ (Pd 2 wt%, Ir/Pd = 1/1) gives 95% yield of 2,5-bis(hydroxymethyl)-tetrahydrofuran under the conditions of 5-hydroxymethylfurfural (5 mmol) in water (9 g), 0.15 g catalyst, 8 MPa $H_2$, 275 K [89]. These works indicate that 2,5-bis(hydroxymethyl)-tetrahydrofuran can be synthesized from 5-hydroxymethylfurfural by catalytic total hydrogenation, and the researches on the conversion of 2,5-bis(hydroxymethyl)-tetrahydrofuran to 1,6-hexanediol have been carried out as mentioned below. Figure 11.18 shows the production routes of 1,6-hexanediol from 5-hydroxymethylfurfural via 2,5-bis(hydroxymethyl)-tetrahydrofuran.

One-pot conversion of 2,5-bis(hydroxymethyl)-tetrahydrofuran using Rh-Re/$SiO_2$ (6.5 wt% Rh, 6 wt% Re) and Nafion SAC-13 catalysts under the conditions of 0.9 mmol substrate, 0.010 g Rh-Re/$SiO_2$, 0.015 g Nafion SAC-13, 2 ml water, 8 MPa $H_2$ and 393 K gives 86% yield of 1,6-hexanediol and the 14% yield of 1,5-hexanediol [119–121]. Surprisingly, the reaction system does not give 1-hexanol, which is an expected by-product by the further reaction of 1,6-hexanediol, although the side reaction giving 1-hexanol and 1-pentanol has been often observed in the production of 1,6-hexanediol and 1,5-pentanediol, respectively. Two formation routes of 1,6-hexanediol have been proposed. One route is 2,5-bis(hydroxymethyl)-tetrahydrofuran → 1,2,6-hexanetriol → 1,6-hexanediol and the other route is 2,5-bis(hydroxymethyl)-tetrahydrofuran → 1,2,6-hexanetriol → tetrahydropyran-2-methanol → 1,6-hexanediol. It has been verified that the reaction of 1,2,6-hexanetriol to tetrahydropyran-2-methanol is easily catalyzed by acid [119]. The acid-catalyzed dehydration of 1,2,6-hexanetriol has been also investigated in detail [122], and it can be concluded that a ring closing dehydration reaction to a tetrahydropyran is the dominant reaction pathway.

One-step conversion of 5-hydroxymethylfurfural to 1,2,6-hexanetriol using Ni-Co-Al mixed oxide catalysts has been also reported [123]. Ni-Co-Al ((Ni + Co/Al = 3), Ni/Co = 0.2) catalyst gives 65% yield of 1,2,6-hexanetriol in the hydrogenolytic ring opening of 5-hydroxymethylfurfural in methanol solvent at 393 K and 4 MPa $H_2$, and the main by-product is 2,5-bis(hydroxymethyl)-tetrahydrofuran (about 20% yield) by the total hydrogenation reaction. The formation route of 1,2,6-hexanetriol is probably composed of the hydrogenolytic ring opening reaction of 2,5-bis(hydroxymethyl)-furan and the subsequent hydrogenation. Further investigation on the catalyst characterization and the elucidation of the role of Ni and Co are necessary.

Direct conversion of 5-hydroxymethylfurfural to 1,6-hexanediol has been attempted. Zirconium phosphate supported Pd catalyst (Pd/ZrP, Pd 7 wt%) shows 43% yield of 1,6-hexanediol from 5-hydroxymethylfurfural in ethanol solvent using formic acid as a reductant at 413 K [124]. The by-products were 2,5-hexanedione and 2,5-bis(hydroxymethyl)-tetrahydrofuran. The Brønsted acidity of zirconium phosphate support most likely promotes cleavage of the C-O bond in the furan ring.

Synthesis of 1,6-hexanediol from 5-hydroxymethylfurfural using the double-layered catalysts of $Pd/SiO_2$ + $Ir\text{-}ReO_x/SiO_2$ in a fixed-bed reactor has been reported [125]. This is the combination of 5-hydroxymethylfurfural hydrogenation to 2,5-bis(hydroxymethyl)-tetrahydrofuran catalyzed by $Pd/SiO_2$ and the C-O hydrogenolysis of 2,5-bis(hydroxymethyl)-tetrahydrofuran catalyzed by $Ir\text{-}ReO_x/SiO_2$. As the results of optimization of catalysts and reaction conditions including the solvent, 58% yield of 1,6-hexanediol is obtained using $Pd/SiO_2$ (Pd 0.6 wt%) and $Ir\text{-}ReO_x/SiO_2$ (5 wt% Ir, 5 wt% Re) at 373 K and 7 MPa $H_2$ in the mixed solvent of water + tetrahydrofuran (2:3). The by-products are 1,5-hexanediol, 1,2,6-hexanetriol and 1-hexanol. The stability of $Pd/SiO_2$ + $Ir\text{-}ReO_x/SiO_2$ has been verified for 24 h reaction time.

### *11.6.3 Products Hexanediols from Sorbitol*

Production of hexanediols from sorbitol requires removal of four OH groups. Like the case of the butanediol production from erythritol, the deoxydehydration reaction of vicinal diols can be used. When the $ReO_x\text{-}Pd/CeO_2$ catalyst was applied to the reaction of sorbitol, the obtained yields of 1,2-hexanediol, 1,4-hexanediol, 3,4-hexanediol, and 1,6-hexanediol are 36%, 32%, 11%, and 6%, respectively, at 433 K and 8 MPa $H_2$ in 1,4-dioxane solvent [74]. The selectivity of the specific hexanediol is not high, however, the sum of the yields of hexanediols is as high as 85%. The methods for the selective formation of specific diols are desired.

## 11.7 Conclusions and Future Outlook

The production of diols from biomass is one of important strategies for conversion of biomass as a renewable raw material to chemicals because of the relative high versatility and value of diols. At present, selective conversion to diols is becoming possible on the basis of the fundamental research introduced in this chapter. Catalysis technologies are essential for practical synthesis of biomass-derived diols. Improvement in selectivity to one specific diol is necessary for many target diols, especially for longer ones such as pentanediols and hexanediols. Because many systems use expensive active metals as catalysts, long catalyst life is also essential. Some systems show good reusability of catalysts for several runs; nevertheless industrialization requires verification of much longer catalyst life.

Some of reaction routes introduced in this chapter have been discovered recently, and therefore other new catalytic reactions and new types of catalysts will be discovered in the future. The new catalytic reactions as well as the reactions in this chapter with improved catalysts will have an expanding role in achieving practical biomass refineries in sustainable society.

## References

1. Pang J, Zheng M, Sun R, Wang A, Wang X, Zhang T. Synthesis of ethylene glycol and terephthalic acid from biomass for producing PET. Green Chem. 2016;18:342–59.
2. Hejna A, Kosmela P, Formela K, Piszczyk L, Haponiuk JT. Potential applications of crude glycerol in polymer technology–Current state and perspectives. Renew Sust Energy Rev. 2016;66:449–75.
3. Bozell JJ, Petersen GR. Technology development for the production of biobased products from biorefinery carbohydrates – the US Department of Energy's "Top 10" revisited. Green Chem. 2010;12:539–54.
4. Schlaf M. Selective deoxygenation of sugar polyols to α,ω-diols and other oxygen content reduced materials – a new challenge to homogeneous ionic hydrogenation and hydrogenolysis catalysis. Dalton Trans. 2006;4645–53
5. Besson M, Gallezot P, Pinel C. Conversion of biomass into chemicals over metal catalysts. Chem Rev. 2014;114:1827–70.
6. Delidovich I, Hausoul PJC, Deng L, Pfützenreuter R, Rose M, Palkovits R. Alternative monomers based on lignocellulose and their use for polymer production. Chem Rev. 2016;116:1540–99.
7. Mäki-Arvela P, Simakova IL, Salmi T, Murzin DY. Production of lactic acid/lactates from biomass and their catalytic transformations to commodities. Chem Rev. 2014;114:1909–71.
8. Dusselier M, Wouwe PV, Dewaele A, Makshina E, Sels BF. Lactic acid as a platform chemical in the biobased economy: the role of chemocatalysis. Energy Environ Sci. 2013;6:1415–42.
9. Sun D, Yamada Y, Sato S, Ueda W. Glycerol hydrogenolysis into useful C3 chemicals. Appl Catal A. 2016;193:75–92.
10. Nakagawa Y, Tomishige K. Heterogeneous catalysis of the glycerol hydrogenolysis. Cat Sci Technol. 2011;1:179–90.
11. Nanda MR, Yuan Z, Qin W, Xu C. Recent advancements in catalytic conversion of glycerol into propylene glycol: a review. Catal Rev Sci Eng. 2016;58:309–36.

12. Kusunoki Y, Miyazawa T, Kunimori K, Tomishige K. Highly active metal–acid bifunctional catalyst system for hydrogenolysis of glycerol under mild reaction conditions. Catal Commun. 2005;6:645–9.
13. Miyazawa T, Kusunoki Y, Kunimori K, Tomishige K. Glycerol conversion in the aqueous solution under hydrogen over Ru/C + an ion-exchange resin and its reaction mechanism. J Catal. 2006;240:213–21.
14. Miyazawa T, Koso S, Kunimori K, Tomishige K. Glycerol hydrogenolysis to 1,2-propanediol catalyzed by a heat-resistant ion-exchange resin combined with Ru/C. Appl Catal A. 2007;329:30–5.
15. Miyazawa T, Koso S, Kunimori K, Tomishige K. Development of a Ru/C catalyst for glycerol hydrogenolysis in combination with an ion-exchange resin. Appl Catal A. 2007;318:244–51.
16. Furikado I, Miyazawa T, Koso S, Shimao A, Kunimori K, Tomishige K. Catalytic performance of $Rh/SiO_2$ in glycerol reaction under hydrogen. Green Chem. 2007;9:582–8.
17. Primo A, Concepción P, Corma A. Synergy between the metal nanoparticles and the support for the hydrogenation of functionalized carboxylic acids to diols on $Ru/TiO_2$. Chem Commun. 2011;47:3613–5.
18. Mao BW, Cai ZZ, Huang MY, Jiang YY. Hydrogenation of carboxylic acids catalyzed by magnesia-supported poly-γ-aminopropylsiloxane–Ru complex. Polym Adv Technol. 2003;14:278–81.
19. Takeda K, Shoji T, Watanabe H, Tamura M, Nakagawa Y, Okumura K, Tomishige K. Selective hydrogenation of lactic acid to 1,2-propanediol over highly active ruthenium–molybdenum oxide catalysts. ChemSusChem. 2015;8:1170–8.
20. Tamura M, Tamura R, Takeda Y, Nakagawa Y, Tomishige K. Insight into the mechanism of hydrogenation of amino acids to amino alcohols catalyzed by a heterogeneous $MoO_x$-modified Rh catalyst. Chem Eur J. 2015;21:3097–107.
21. Tamura M, Tamura R, Takeda Y, Nakagawa Y, Tomishige K. Catalytic hydrogenation of amino acids to amino alcohols with complete retention of configuration. Chem Commun. 2014;50:6656–9.
22. Nakagawa Y, Tamura R, Tamura M, Tomishige K. Combination of supported bimetallic rhodium–molybdenum catalyst and cerium oxide for hydrogenation of amide. Sci Technol Adv Mater. 2015;16:014901. (7 pp)
23. Liu Y, Luo C, Liu H. Tungsten trioxide promoted selective conversion of cellulose into propylene glycol and ethylene glycol on a ruthenium catalyst. Angew Chem Int Ed. 2012;124:3303–7.
24. Sun J, Liu H. Selective hydrogenolysis of biomass-derived xylitol to ethylene glycol and propylene glycol on supported Ru catalysts. Green Chem. 2011;13:135–42.
25. Huang Z, Chen J, Jia Y, Liu H, Xia C, Liu H. Selective hydrogenolysis of xylitol to ethylene glycol and propylene glycol over copper catalysts. Appl Catal B. 2014;147:377–86.
26. Liu H, Huang Z, Xia C, Jia Y, Chen J, Liu H. Selective hydrogenolysis of xylitol to ethylene glycol and propylene glycol over silica dispersed copper catalysts prepared by a precipitation–gel method. ChemCatChem. 2014;6:2918–28.
27. Sun J, Liu H. Selective hydrogenolysis of biomass-derived xylitol to ethylene glycol and propylene glycol on Ni/C and basic oxide-promoted Ni/C catalysts. Catal Today. 2014;234:75–82.
28. Deng T, Liu H. Promoting effect of $SnO_x$ on selective conversion of cellulose to polyols over bimetallic $Pt–SnO_x/Al_2O_3$ catalysts. Green Chem. 2013;15:116–24.
29. Deng T, Liu H. Direct conversion of cellulose into acetol on bimetallic $Ni-SnO_x/Al_2O_3$ catalysts. J Mol Catal A. 2014;388–389:66–73.
30. Pang J, Zheng M, Wang A, Zhang T. Catalytic hydrogenation of corn stalk to ethylene glycol and 1,2-propylene glycol. Ind Eng Chem Res. 2011;50:6601–8.
31. Zhou L, Wang A, Li C, Zheng M, Zhang T. Selective production of 1,2-propylene glycol from Jerusalem artichoke tuber using $Ni–W_2C/AC$ catalysts. ChemSusChem. 2012;5:932–8.

32. Nakagawa Y, Tamura M, Tomishige K. Catalytic materials for the hydrogenolysis of glycerol to 1,3-propanediol. J Mater Chem A. 2014;2:6688–702.
33. Nakagawa Y, Shinmi Y, Koso S, Tomishige K. Direct hydrogenolysis of glycerol into 1,3-propanediol over rhenium-modified iridium catalyst. J Catal. 2010;272:191–4.
34. Amada Y, Shinmi Y, Koso S, Kubota T, Nakagawa Y, Tomishige K. Reaction mechanism of the glycerol hydrogenolysis to 1,3-propanediol over Ir-$ReO_x/SiO_2$ catalyst. Appl Catal B. 2011;105:117–27.
35. Amada Y, Watanabe H, Tamura M, Nakagawa Y, Okumura K, Tomishige K. Structure of $ReO_x$ clusters attached on the Ir metal surface in Ir-$ReO_x/SiO_2$ for the hydrogenolysis reaction. J Phys Chem C. 2012;116:23503–14.
36. Nakagawa Y, Ning X, Amada Y, Tomishige K. Solid acid co-catalyst for the hydrogenolysis of glycerol to 1,3-propanediol over Ir-$ReO_x/SiO_2$. Appl Catal A. 2012;433-434:128–34.
37. Chen K, Mori K, Watanbe H, Nakagawa Y, Tomishige K. C-O bond hydrogenolysis of cyclic ethers with OH groups over rhenium-modified supported iridium catalysts. J Catal. 2012;294:171–83.
38. Nakagawa Y, Mori K, Chen K, Amada Y, Tamura M, Tomishige K. Hydrogenolysis of C-O bond over Re-modified Ir catalyst in alkane solvent. Appl Catal A. 2013;468:418–25.
39. Tomishige K, Nakagawa Y, Tamura M. Role of Re species and acid cocatalyst on Ir-$ReO_x/SiO_2$ in the C-O hydrogenolysis of biomass-derived substrates. Chem Rec. 2014;14:1041–54.
40. Tomishige K, Nakagawa Y, Tamura M. Selective hydrogenolysis of C-O bonds using the interaction of the catalyst surface and OH groups. Top Curr Chem. 2014;353:127–62.
41. Amada Y, Watanabe H, Hirai Y, Kajikawa Y, Nakagawa Y, Tomishige K. Production of biobutanediols by the hydrogenolysis of erythritol. ChemSusChem. 2012;5:1991–9.
42. Deng C, Duan X, Zhou J, Zhou X, Yuan W, Scott SL. Ir–Re alloy as a highly active catalyst for the hydrogenolysis of glycerol to 1,3-propanediol. Cat Sci Technol. 2015;5:1540–7.
43. Deng C, Leng L, Duan X, Zhou J, Zhou X, Yuan W. Support effect on the bimetallic structure of Ir–Re catalysts and their performances in glycerol hydrogenolysis. J Mol Catal A. 2015;41:81–8.
44. Deng C, Leng L, Zhou J, Zhou X, Yuan W. Effects of pretreatment temperature on bimetallic Ir-Re catalysts for glycerol hydrogenolysis. Chin J Catal. 2015;36:1750–8.
45. Luo W, Lyu Y, Gong L, Du H, Wang T, Ding Y. Selective hydrogenolysis of glycerol to 1,3-propanediol over egg-shell type Ir–$ReO_x$ catalysts. RSC Adv. 2016;6:13600–8.
46. Luo W, Lyu Y, Gong L, Du H, Jiang M, Ding Y. The influence of impregnation sequence on glycerol hydrogenolysis over iridium-rhenium catalyst. React Kinet Catal Mech. 2016;118:481–96.
47. Tamura M, Amada Y, Liu S, Yuan Z, Nakagawa Y, Tomishige K. Promoting effect of Ru on Ir-$ReO_x/SiO_2$ catalyst in hydrogenolysis of glycerol. J Mol Catal A. 2014;388-389:177–87.
48. Chen K, Tamura M, Yuan Z, Nakagawa Y, Tomishige K. One-pot conversion of sugar and sugar polyols to *n*-alkanes without C-C dissociation over the Ir-$ReO_x/SiO_2$ catalyst combined with H-ZSM-5. ChemSusChem. 2013;6:613–21.
49. Liu S, Tamura M, Nakagawa Y, Tomishige K. One-pot conversion of cellulose into n-hexane over the Ir-$ReO_x/SiO_2$ catalyst combined with HZSM-5. ACS Sustainable Eng Chem. 2014;2:1819–27.
50. Liu S, Okuyama Y, Tamura M, Nakagawa Y, Imai A, Tomishige K. Production of renewable hexanols from mechanocatalytically depolymerized cellulose by using Ir-$ReO_x/SiO_2$ catalyst. ChemSusChem. 2015;8:628–35.
51. Nakagawa Y, Tamura M, Tomishige K. Catalytic total hydrodeoxygenation of biomass-derived polyfunctionalized substrates to alkanes. ChemSusChem. 2014;8:1114–32.
52. Liu S, Okuyama Y, Tamura M, Nakagawa Y, Imai A, Tomishige K. Selective transformation of hemicellulose (xylan) into n-pentane, pentanols or xylitol over a rhenium-modified iridium catalyst combined with acids. Green Chem. 2016;18:165–75.

53. Liu S, Okuyama Y, Tamura M, Nakagawa Y, Imai A, Tomishige K. Catalytic conversion of sorbitol to gasoline-ranged products without external hydrogen over Pt-modified Ir-$ReO_x$/$SiO_2$. Catal Today. 2016;269:122–31.
54. Tamura M, Tokonami K, Nakagawa Y, Tomishige K. Rapid synthesis of unsaturated alcohol in mild conditions by highly selective hydrogenation. Chem Commun. 2013;49:7034–6.
55. Tamura M, Tokonami K, Nakagawa Y, Tomishige K. Selective hydrogenation of crotonaldehyde to crotyl alcohol over metal oxide modified Ir catalysts and the mechanistic insight. ACS Catal. 2016;6:3600–9.
56. Oh J, Dash S, Lee H. Selective conversion of glycerol to 1,3-propanediol using Pt-sulfated zirconia. Green Chem. 2011;13:2004–7.
57. Mizugaki T, Yamakawa T, Arundhathi R, Mitsudome T, Jitsukawa K, Kaneda K. Selective hydrogenolysis of glycerol to 1,3-propanediol catalyzed by Pt nanoparticles–$AlO_x$/$WO_3$. Chem Lett. 2012;41:1720–2.
58. Arundhathi R, Mizugaki T, Mitsudome T, Jitsukawa K, Kaneda K. Highly selective hydrogenolysis of glycerol to 1,3-propanediol over a boehmite-supported platinum/tungsten catalyst. ChemSusChem. 2013;6:1345–7.
59. Zhu S, Gao X, Zhu Y, Li Y. Promoting effect of $WO_x$ on selective hydrogenolysis of glycerol to 1,3-propanediol over bifunctional Pt–$WO_x$/$Al_2O_3$ catalysts. J Mol Catal A. 2015;398:391–8.
60. Priya SS, Bhanuchander P, Kumar VP, Dumbre DK, Periasamy SR, Bhargava SK, Kantam ML, Chary KVR. Platinum supported on H-mordenite: a highly efficient catalyst for selective hydrogenolysis of glycerol to 1,3-propanediol. ACS Sustain Chem Eng. 2016;4:1212–22.
61. Priya SS, Bhanuchander P, Kumar VP, Bhargava SK, Chary KVR. Activity and selectivity of platinum–copper bimetallic catalysts supported on mordenite for glycerol hydrogenolysis to 1,3-propanediol. Ind Eng Chem Res. 2016;55:4461–72.
62. Jeya M, Lee KM, Tiwari MK, Kim JS, Gunasekaran P, Kim SY, Kim IW. Isolation of a novel high erythritol-producing *Pseudozyma tsukubaensis* and scale-up of erythritol fermentation to industrial level. Appl Microbiol Biotechnol. 2009;83:225–31.
63. Kobayashi Y, Iwata H, Mizushima D, Ogihara J, Kasumi T. Erythritol production by Moniliella megachiliensis using nonrefined glycerol waste as carbon source. Lett Appl Microbiol. 2015;60:475–80.
64. Sun D, Sato S, Ueda W, Primo A, Garcia H, Corma A. Production of C4 and C5 alcohols from biomass-derived materials. Green Chem. 2016;18:2579–97.
65. Broadbent HS, Campbell GC, Bartley WJ, Johnson JH. Rhenium and its compounds as hydrogenation catalysts. III. Rhenium heptoxide. J Org Chem. 1959;24:1847–54.
66. Ly BK, Minh DP, Pinel C, Besson M, Tapin B, Epron F. Effect of addition mode of Re in bimetallic Pd–Re/$TiO_2$ catalysts upon the selective aqueous-phase hydrogenation of succinic acid to 1,4-butanediol. Top Catal. 2012;55:466–73.
67. Kang KH, Hong UG, Bang Y, Choi JH, Kim JK, Lee JK, Han SJ, Song IK. Hydrogenation of succinic acid to 1,4-butanediol over Re–Ru bimetallic catalysts supported on mesoporous carbon. Appl Catal A. 2015;490:153–62.
68. Takeda Y, Tamura M, Nakagawa Y, Okumura K, Tomishige K. Hydrogenation of dicarboxylic acids to diols over Re-Pd catalysts. Cat Sci Technol. 2016;6:5668–83.
69. Takeda Y, Tamura M, Nakagawa Y, Okumura K, Tomishige K. Characterization of Re-Pd/$SiO_2$ catalysts for the hydrogenation of high fatty acid and its reaction mechanism. ACS Catal. 2015;5:7034–47.
70. Ly BK, Tapin B, Aouine M, Delichere P, Epron F, Pinel C, Especel C, Besson M. Insights into the oxidation state and location of rhenium in Re-Pd/$TiO_2$ catalysts for aqueous-phase selective hydrogenation of succinic acid to 1,4-butanediol as a function of palladium and rhenium deposition methods. ChemCatChem. 2015;7:2161–78.
71. Amada Y, Ota N, Tamura M, Nakagawa Y, Tomishige K. Selective hydrodeoxygenation of cyclic vicinal diols to cyclic alcohols over tungsten oxide–palladium catalysts. ChemSusChem. 2014;7:2185–92.

72. Raju S, Moret ME, Gebbink RJMK. Rhenium-catalyzed dehydration and deoxydehydration of alcohols and polyols: opportunities for the formation of olefins from biomass. ACS Catal. 2015;5:281–300.
73. Dethlefsen JR, Fristrup P. Rhenium-catalyzed deoxydehydration of diols and polyols. ChemSusChem. 2015;8:767–75.
74. Ota N, Tamura M, Nakagawa Y, Okumura K, Tomishige K. Hydrodeoxygenation of vicinal OH groups over heterogeneous rhenium catalyst promoted by palladium and ceria support. Angew Chem Int Ed. 2015;54:1897–900.
75. Ota N, Tamura M, Nakagawa Y, Okumura K, Tomishige K. Performance, structure, and mechanism of $ReO_x$–$Pd/CeO_2$ catalyst for simultaneous removal of vicinal OH groups with $H_2$. ACS Catal. 2016;6:3213–26.
76. Tazawa S, Ota N, Tamura M, Nakagawa Y, Okumura K, Tomishige K. Deoxydehydration with molecular hydrogen over ceria-supported rhenium catalyst with gold promoter. ACS Catal. 2016;6:6393–7.
77. Li F, Lu T, Chen B, Huang Z, Yuan G. Pt nanoparticles over $TiO_2$-$ZrO_2$ mixed oxide as multifunctional catalysts for an integrated conversion of furfural to 1,4-butanediol. Appl Catal A. 2014;478:252–8.
78. Nakagawa Y, Tomishige K. Catalyst development for the hydrogenolysis of biomass-derived chemicals to value-added ones. Catal Surv Jpn. 2011;15:111–6.
79. Nakagawa Y, Tomishige K. Production of 1,5-pentanediol from biomass via furfural and tetrahydrofurfuryl alcohol. Catal Today. 2012;195:136–43.
80. Nakagawa Y, Tamura M, Tomishige K. Catalytic conversions of furfural to pentanediols. Catal Surv Jpn. 2015;19:249–56.
81. Adkins H, Connor R. The catalytic hydrogenation of organic compounds over copper chromite. J Am Chem Soc. 1931;53:1091–5.
82. Schniepp LE, Geller HH. Preparation of dihydropyran, δ-hydroxyvaleraldehyde and 1,5-pentanediol from tetrahydrofurfuryl alcohol. J Am Chem Soc. 1946;68:1646–8.
83. Koso S, Furikado I, Shimao A, Miyazawa T, Kunimori K, Tomishige K. Chemoselective hydrogenolysis of tetrahydrofurfuryl alcohol to 1,5-pentanediol. Chem Commun. 2009:2035–7.
84. Koso S, Ueda N, Shinmi Y, Okumura K, Kizuka T, Tomishige K. Promoting effect of Mo on hydrogenolysis of teterhydrofurfuryl alcohol to 1,5-pentanediol over $Rh/SiO_2$. J Catal. 2009;267:89–92.
85. Mariscal R, Maireles-Torres P, Ojeda M, Sádaba I, Granados ML. Furfural: a renewable and versatile platform molecule for the synthesis of chemicals and fuels. Energy Environ Sci. 2016;9:1144–89.
86. Nakagawa Y, Tomishige K. Total hydrogenation of furan derivatives over silica-supported Ni-Pd alloy catalyst. Catal Commun. 2010;12:154–6.
87. Nakagawa Y, Nakazawa H, Watanabe H, Tomishige K. Total hydrogenation of furfural over a silica-supported nickel catalyst prepared by the reduction of a nickel nitrate precursor. ChemCatChem. 2012;4:1791–7.
88. Nakagawa Y, Tamura M, Tomishige K. Catalytic reduction of biomass-derived furanic compounds with hydrogen. ACS Catal. 2013;3:2655–68.
89. Nakagawa Y, Takada K, Tamura M, Tomishige K. Total hydrogenation of furfural and 5-hydroxymethylfurfural over supported Pd-Ir alloy catalyst. ACS Catal. 2014;4:2718–25.
90. Guan J, Li J, Yu Y, Mu X, Chen A. DFT studies of the selective C-O hydrogenolysis and ring-opening of biomass-derived tetrahydrofurfuryl alcohol over Rh(111) surface. J Phys Chem C. 2016;120:19124–34.
91. Koso S, Nakagawa Y, Tomishige K. Mechanism of the hydrogenolysis of ethers over silica-supported rhodium catalyst modified with rhenium oxide. J Catal. 2011;280:221–9.
92. Koso S, Watanabe H, Okumura K, Nakagawa Y, Tomishige K. Comparative study of Rh-$MoO_x$ and Rh-$ReO_x$ supported on $SiO_2$ for the hydrogenolysis of ethers and polyols. Appl Catal B. 2012;111–112:27–37.

93. Koso S, Watanabe H, Okumura K, Nakagawa Y, Tomishige K. Stable low-valence $ReO_x$ cluster attached on Rh metal particles formed by hydrogen reduction and its formation mechanism. J Phys Chem C. 2012;116:3079–90.
94. Chen K, Koso S, Kubota T, Nakagawa Y, Tomishige K. Chemoselective hydrogenolysis of tetrahydropyran-2-methanol to 1,6-hexanediol over rhenium-modified carbon-supported rhodium catalysts. ChemCatChem. 2010;2:547–55.
95. Shimao A, Koso S, Ueda N, Shinmi Y, Furikado I, Tomishige K. Promoting effect of Re addition to $Rh/SiO_2$ on glycerol hydrogenolysis. Chem Lett. 2009;38:540–1.
96. Shinmi Y, Koso S, Kubota T, Nakagawa Y, Tomishige K. Modification of $Rh/SiO_2$ catalyst for the hydrogenolysis of glycerol in water. Appl Catal B. 2010;94:318–26.
97. Amada Y, Koso S, Nakagawa Y, Tomishige K. Hydrogenolysis of 1,2-propanediol for the production of biopropanols from glycerol. ChemSusChem. 2011;3:728–36.
98. Chia M, Pagan-Torres YJ, Hibbits D, Tan Q, Pham HN, Datye AK, Neurock M, Davis RJ, Dumesic JA. Selective hydrogenolysis of polyols and cyclic ethers over bifunctional surface sites on rhodium–rhenium catalysts. J Am Chem Soc. 2011;133:12675–89.
99. Falcone DD, Hack JH, Klyushin AY, Knop-Gericke A, Schlögl R, Davis RJ. Evidence for the bifunctional nature of Pt-Re catalysts for selective glycerol hydrogenolysis. ACS Catal. 2015;5:5679–95.
100. Chatterjee M, Kawanami H, Ishizaka T, Sato M, Suzuki T, Suzuki A. An attempt to achieve the direct hydrogenolysis of tetrahydrofurfuryl alcohol in supercritical carbon dioxide. Cat Sci Technol. 2011;1:1466–147.
101. Wang Z, Phojaroen B, Li M, Dong W, Li N, Wang A, Wang X, Cong Y, Zhang T. Chemoselective hydrogenolysis of tetrahydrofurfuryl alcohol to 1,5-pentanediol over $Ir\text{-}MoO_x/SiO_2$ catalyst. J Energy Chem. 2014;23:427–34.
102. Guan J, Peng G, Cao Q, Mu X. Role of $MoO_3$ on a rhodium catalyst in the selective hydrogenolysis of biomass-derived tetrahydrofurfuryl alcohol into 1,5-pentanediol. J Phys Chem C. 2014;118:25555–66.
103. Pholjaroen B, Li N, Huang Y, Li L, Wang A, Zhang T. Selective hydrogenolysis of tetrahydrofurfuryl alcohol to 1,5-pentanediol over vanadium modified $Ir/SiO_2$ catalyst. Catal Today. 2015;245:93–9.
104. Liu S, Amada Y, Tamura M, Nakagawa Y, Tomishige K. One-pot selective conversion of furfural into 1,5-pentanediol over Pd-added $Ir\text{-}ReO_x/SiO_2$ bifunctional catalyst. Green Chem. 2014;16:617–26.
105. Liu S, Amada Y, Tamura M, Nakagawa Y, Tomishige K. Performance and characterization of rhenium-modified Rh-Ir alloy catalyst for one-pot conversion of furfural into 1,5-pentanediol. Cat Sci Technol. 2014;4:2535–49.
106. Xu W, Wang H, Liu X, Ren J, Wang Y, Lu G. Direct catalytic conversion of furfural to 1,5-pentanediol by hydrogenolysis of the furan ring under mild conditions over $Pt/Co_2AlO_4$ catalyst. Chem Commun. 2011;47:3924–6.
107. Mizugaki T, Yamakawa T, Nagatsu Y, Maeno Z, Mitsudome T, Jitsukawa K, Kaneda K. Direct transformation of furfural to 1,2-pentanediol using a hydrotalcite-supported platinum nanoparticle catalyst. ACS Sustain Chem Eng. 2014;2:2243–7.
108. Zhang B, Zhu Y, Ding G, Zheng H, Li Y. Selective conversion of furfuryl alcohol to 1,2-pentanediol over a $Ru/MnO_x$ catalyst in aqueous phase. Green Chem. 2012;14:3402–9.
109. Liu H, Huang Z, Kang H, Xia C, Chen J. Selective hydrogenolysis of biomass-derived furfuryl alcohol into 1,2- and 1,5-pentanediol over highly dispersed $Cu\text{-}Al_2O_3$ catalysts. Chin J Catal. 2016;37:700–10.
110. Liu H, Huang Z, Zhan F, Cui F, Li X, Xia C, Chen J. Efficient hydrogenolysis of biomass-derived furfuryl alcohol to 1,2- and 1,5-pentanediols over a non-precious $Cu\text{–}Mg_3AlO_{4.5}$ bifunctional catalyst. Cat Sci Technol. 2016;6:668–71.
111. Du XL, Bi QY, Liu YM, Cao Y, He HY, Fan KN. Tunable copper-catalyzed chemoselective hydrogenolysis of biomass-derived γ-valerolactone into 1,4-pentanediol or 2-methyltetrahydrofuran. Green Chem. 2012;14:935–9.

112. Li M, Li G, Li N, Wang A, Dong W, Wang X, Cong Y. Aqueous phase hydrogenation of levulinic acid to 1,4-pentanediol. Chem Commun. 2014;50:1414–6.
113. Arai T, Tamura M, Nakagawa Y, Tomishige K. Synthesis of 2-butanol by selective hydrogenolysis of 1,4-anhydroerythritol over molybdenum oxide-modified rhodium-supported silica. ChemSusChem. 2016;9:1680–8.
114. Mizugaki T, Nagatsu Y, Togo K, Maeno Z, Mitsudome T, Jitsukawa K, Kaneda K. Selective hydrogenation of levulinic acid to 1,4-pentanediol in water using a hydroxyapatite-supported Pt–Mo bimetallic catalyst. Green Chem. 2015;17:5136–9.
115. Misono A, Osa T, Sanami Y. The hydrogenation of pyran derivatives. IV. The skeletal rearrangement in the gas-phase dehydration of tetrahydropyran-2-methanol. Bull Chem Soc Jpn. 1968;41:2447–53.
116. Chia M, O'Neill BJ, Alamillo R, Dietrich PJ, Ribeiro FH, Miller JT, Dumesic JA. Bimetallic RhRe/C catalysts for the production of biomass-derived chemicals. J Catal. 2013;308:226–36.
117. Alba-Rubio AC, Sener C, Hakim SH, Gostanian TM, Dumesic JA. Synthesis of supported RhMo and PtMo bimetallic catalysts by controlled surface reactions. ChemCatChem. 2015;7:3881–6.
118. Hakim SH, Sener C, Alba-Rubio AC, Gostanian TM, O'Neill BJ, Ribeiro FH, Miller JT, Dumesic JA. Synthesis of supported bimetallic nanoparticles with controlled size and composition distributions for active site elucidation. J Catal. 2015;328:75–90.
119. Buntara T, Noel S, Phua PH, Melian-Cabrera I, de Vries JG, Heeres HJ. Caprolactam from renewable resources: catalytic conversion of 5-hydroxymethylfurfural into caprolactone. Angew Chem Int Ed. 2011;50:7083–7.
120. Buntara T, Noel S, Phua PH, Melian-Cabrera I, de Vries JG, Heeres HJ. From 5-hydroxymethylfurfural (HMF) to polymer precursors: catalyst screening studies on the conversion of 1,2,6-hexanetriol to 1,6-hexanediol. Top Catal. 2012;55:612–9.
121. Buntara T, Melian-Cabrera I, Tan Q, Fierro JLG, Neurock M, de Vries JG, Heeres HJ. Catalyst studies on the ring opening of tetrahydrofuran-dimethanol to 1,2,6-hexanetriol. Catal Today. 2013;210:106–16.
122. Nolan MR, Sun G, Shank BH. On the selective acid-catalysed dehydration of 1,2,6-hexanetriol. Cat Sci Technol. 2014;4:2260–6.
123. Yao S, Wang X, Jiang Y, Wu F, Chen X, Mu X. One-step conversion of biomass-derived 5-hydroxymethylfurfural to 1,2,6-hexanetriol over Ni–Co–Al mixed oxide catalysts under mild conditions. ACS Sustain Chem Eng. 2013;2:173–80.
124. Tuteja J, Choudhary H, Nishimura S, Ebitani K. Direct synthesis of 1,6-hexanediol from HMF over a heterogeneous Pd/ZrP catalyst using formic acid as hydrogen source. ChemSusChem. 2014;7:96–100.
125. Bin X, Zheng M, Li X, Pang J, Sun R, Wang H, Pang X, Wang A, Wang X, Zhang T. Synthesis of 1,6-hexanediol from HMF over double-layered catalysts of $Pd/SiO_2$ + $Ir–ReO_x/SiO_2$ in a fixed-bed reactor. Green Chem. 2016;18:2175–84.

# Chapter 12
# Production of Ethanol from Lignocellulosic Biomass

**Antonio D. Moreno, Pablo Alvira, David Ibarra, and Elia Tomás-Pejó**

**Abstract** Ethanol fuel is leading the transition towards a post-petrol era in the transport sector worldwide. Ethanol is produced via sugar fermentation processes by yeasts or bacteria. Although the current industrial production of ethanol mainly involves the use of starch- and sugar-based feedstocks, lignocellulosic biomass is expected to play a key role as renewable, carbohydrate-rich raw material. With the aim of placing lignocellulosic ethanol into the market, the scientific community has made great efforts to develop and implement efficient conversion technologies. Prior to fermentation, lignocellulosic biomass must be pretreated and hydrolysed to obtain the fermentable sugars. Biomass processing is, however, a major limiting step since it is hindered by the native structure of lignocellulose and generates different biomass-derived compounds that are inhibitors of the subsequent microbial conversion. In this context, different pretreatment, delignification and detoxification methods have been investigated to produce less inhibitory pretreated materials. Furthermore, several strategies such as working at high gravity conditions, high temperatures and/or different process configurations, have been shown to maximize ethanol production from lignocellulosic materials. The development of robust microbial strains tolerant to inhibitory compounds and capable of converting sugar mixtures is also needed for cost-effectiveness of the process. This chapter compiles

A.D. Moreno (✉)
Department of Energy, CIEMAT, Avda. Complutense 40, Madrid 28040, Spain
e-mail: david.moreno@ciemat.es

P. Alvira
LISBP, Université de Toulouse, CNRS, INRA, INSA, Toulouse, France
e-mail: pablo.alvira@insa-toulouse.fr

D. Ibarra
Forestry Products Department, INIA-CIFOR,
Ctra de La Coruña Km7.5, Madrid 28040, Spain
e-mail: ibarra.david@inia.es

E. Tomás-Pejó
Biotechnological Processes for Energy Production Unit, IMDEA Energía,
Móstoles 28935, Spain
e-mail: elia.tomas@imdea.org



Z. Fang et al. (eds.), *Production of Platform Chemicals from Sustainable Resources*, Biofuels and Biorefineries 7, DOI 10.1007/978-981-10-4172-3_12

recent advances in lignocellulosic ethanol production processes, from novel raw materials or fermenting microorganisms to new processing technologies addressed to commercialization.

**Keywords** Pretreatment • Lignocellulosic ethanol • Enzymatic hydrolysis • Detoxification • Delignification • Sugar fermentation • Process integration • Microbial robustness

## 12.1 Introduction

The implementation of a sustainable bio-based economy is considered a priority in today's society. To reach such a goal, lignocellulosic biomass – the major renewable organic matter in nature – has been recognized as a valuable raw material for the production of biofuels and several chemical building blocks within the biorefinery concept. Among lignocellulosic biofuels, bioalcohols are very attractive and promising alternatives for the transport sector, as they can share current fuel distribution systems and are easily stored and handled, in comparison to biogas and biodiesel [1].

With a long history, ethanol is the most widespread alcohol fuel. It has a low boiling point (78 °C), a high research octane number (RON; 107) and its energy content is comparable to that of gasoline (two thirds of the gasoline energy content) [1, 2]. Direct use of ethanol as fuel is possible in neat form (100% pure) or in blends with gasoline (e.g. E85: 85% ethanol and 15% gasoline). Ethanol can be converted to ethyl tert-butyl ether (ETBE), which is used as fuel additive.

Traditionally, bioethanol has been produced from sugar- and starch-based feedstocks such as sugarcane juice and molasses, and corn. Since January 2013, bioethanol also started to be produced from lignocellulosic feedstocks at commercial scale [3]. However, current prices for lignocellulosic ethanol are 0.57–1.20 USD/L, while conventional ethanol cost about 0.40–0.45 USD/L [4, 5]. To ensure a competitive lignocellulosic industry, some challenges both in biomass processing (such as having a good balance between biomass hydrolysability and biomass degradation) and microbial conversion processes (including the increase of the tolerance of fermenting microorganism to lignocellulose-derived compounds and the conversion of all lignocellulosic sugars into ethanol with high rates and yields) still need to be addressed. The present chapter reviews the current advances for a cost-effective lignocellulosic bioethanol production, from the use of novel raw materials and the development of new pretreatment technologies, to the investigation and engineering of fermentative microorganisms.

### 12.1.1 Lignocellulosic Bioethanol: A Process Overview

Ethanol production is based on sugar fermentation processes. With about 75% carbohydrate content on dry weight basis, lignocellulose represents an interesting feedstock for ethanol production [6]. However, in contrast to conventional bioethanol production technology, lignocellulosic ethanol production is very challenging due to the highly recalcitrant structure of lignocellulose. Lignocellulosic biomass is composed of three polymers: cellulose, hemicelluloses and lignin. The structural polymer, cellulose, is bonded with both hemicelluloses and lignin, forming a difficult to disrupt complex matrix.

Lignocellulosic ethanol production consists of pretreatment, enzymatic hydrolysis and fermentation steps. Pretreatment is needed to alter the structural characteristics of lignocellulose and increase the accessibility of cellulose and hemicelluloses to hydrolytic enzymes that are responsible for the hydrolysis of polysaccharides into fermentable sugars. Pretreatment is an important process since it has a great impact on final conversion yields and contributes to 30–40% to the overall process costs [7]. As it is further discussed in Sect. 12.3, there is no best pretreatment technology although dilute-acid pretreatment, steam explosion or certain ammonia-based technologies are effective methods that can be applied to a wide range of lignocellulosic feedstocks [8–10]. In pretreatment, high temperatures and pressures and/or the addition of solvents and chemical catalysts are required, which leads to biomass degradation, generating different enzymatic and microbial inhibitors that limit the subsequent saccharification and fermentation steps [11, 12]. Different physical, chemical, and biological detoxification methods have been evaluated with the aim of decreasing the inhibitory power of pretreated materials. Another important limiting factor is the residual lignin, which can unspecifically bind hydrolytic enzymes, decreasing saccharification yields. In a similar way to detoxification methods, different biological and chemical delignification processes have been also studied as complementary steps to enhance saccarification yields.

Considering the steps required for lignocellulosic ethanol production, different process configurations can be proposed as depicted in Fig. 12.1. There are three main process configurations, which are explained in detail in Sect. 12.4: separate hydrolysis and fermentation (SHF), simultaneous saccharification and (co)fermentation (SSF/SSCF) and consolidated bioprocessing (CBP) [13]. In these processes, different yeast, bacterial or fungi strains have been used for ethanol production as discussed in Sect. 12.5. However, microbial conversion processes are highly dependent on the composition of pretreated materials. The use and development of more robust fermentative microorganisms is therefore of the utmost importance and represents an interesting alternative to the aforementioned detoxification methods. Several metabolic and evolutionary engineering strategies have been used to obtain fermentative strains with increased capacity to convert and/or tolerate high concentrations of inhibitory compounds [14–16]. The ability of tolerating high temperatures and osmotic pressures, and converting the full range of sugars present in lignocellulosic feedstocks are important traits to take into consideration [17].

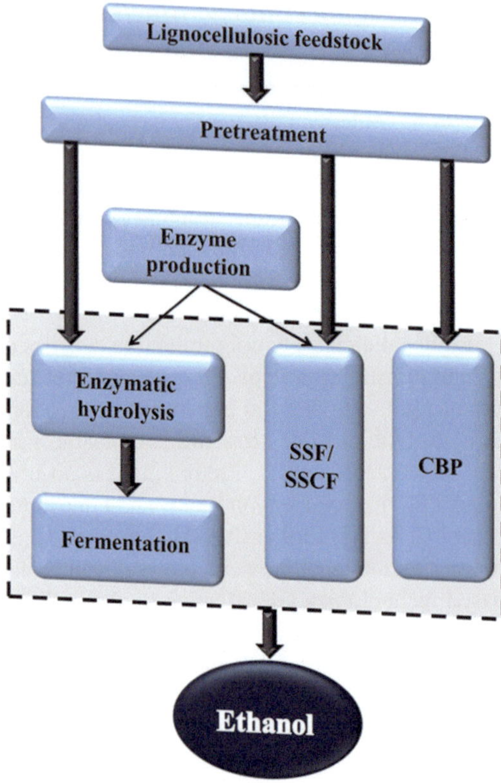


**Fig. 12.1** Process integration during lignocellulosic ethanol production. *SSF* simultaneous saccharification and fermentation, *CBP* consolidated bioprocessing (enzyme production, enzymatic hydrolysis and fermentation take place simultaneously)

## 12.2 Novel Promising Lignocellulosic Feedstocks

Lignocellulosic biomass has an estimated annual production of more than $10^{10}$ MT worldwide [18]. The most commonly used lignocellulosic sources include forestry woody feedstocks (spruce, eucalyptus, birch, etc.) and agricultural residues (cereal straw, sugarcane bagasse, corn stover, etc.). However, materials such as energy crops (poplar, switch grass, giant reed, elephantgrass, *Miscanthus giganteus*, etc.), industrial residues (brews' spent grains, paper mill residues, cheese whey, etc.), municipal solid wastes and marine algae (*Saccharina latissima*, *Laminaria* sp., *Gracilaria* sp., etc.) are considered attractive alternatives with high potential for biofuel production [19–21].

From the chemical point of view, lignocellulose is a highly heterogeneous material. Glucose, xylose, mannose or arabinose are the main sugar units in cellulose and hemicelluloses, while lignin is formed by guaiacyl, syringyl and *p*-hydroxyphenyl phenylpropanoid units. As no other class of material in nature, such a versatile composition of lignocellulose offers the possibility of its revalorization into a wide range of products with tremendous applicability.

The specific composition of the plant cell wall and its components varies depending on plant species, tissue type, and developmental state of the tissue [22]. Altering the qualitative traits of lignocellulosic biomass, including reducing recalcitrance to degradation or optimizing sugar composition for better fermentability is highly desired for optimal biofuel production. Specific genetic engineering approaches such as (1) reducing/altering the lignin polymer, (2) lowering the endogenous components that exert an inhibitory action on enzymes and fermentative microorganisms, and (3) increasing the abundance of sugars that are easily fermentable, have been investigated in several plants with the aim of enhancing the conversion of lignocellulose into biofuel [23].

Modifying the amount of lignin and its composition is one of the preferred approaches for reducing biomass recalcitrance. Yang et al. [24] were able to reduce about 20% lignin content in *Arabidopsis thaliana* by exchanging the promoter responsible for the expression of the cinnamate 4-hydroxylase with a vessel-specific promoter, without compromising plant cell growth. On the other hand, Wilkerson et al. [25] incorporated a feruloyl-CoA monolignol transferase in poplar for introducing monolignol ferulate esters that are more label towards a mild alkaline pretreatment.

Lignocellulose is also composed of acetyl esters, methyl esters and ethers. These groups of compounds act as inhibitors for the fermentation and enzymatic hydrolysis steps once they are released from biomass [11, 12]. Decreasing the concentration of those compounds in plant cell walls is therefore beneficial for the conversion of pretreated materials. For instance, the acetyl content or ferulate esters in biomass can be reduced by genetically interfering with the synthesis of the acetylated polysaccharides or reducing feruloylation, respectively [26, 27].

Another crucial engineering strategy that is being considered to enhance biomass conversion is to alter pentose content in lignocellulose. Pentoses such as xylose are fermented with lower efficiencies than hexoses. Thus, increasing the hexose:pentose ratio in raw materials would be of interest. Adjusting carbon partitioning from cell primary metabolism to wall glucan biosynthesis has been successful in poplar by the overexpression of a sucrose synthase [28]. Also, engineering glycan synthases and glycosyltransferases (which are directly involved in wall polymer synthesis) is a promising option. However, attempts to increase cellulose accumulation by overexpressing the multiple transmembrane spanning protein CesA in poplar results in a reduction of cellulose content instead [29].

Algal biomass is also considered a very interesting alternative since they do not require arable lands and can use seawater and wastewater instead of synthetic media [30]. In contrast, the low carbohydrate content and its complex composition (alginate, laminarin), and the low biomass density hamper its conversion into ethanol and makes imperative the discovery of new hydrolytic enzymes and fermenting microorganisms.

To make lignocellulosic biomass a suitable raw material for biofuel production, new engineering strategies and techno-economical modeling must be devised for genetically modifying plants/algae and obtaining the desired structural and chemical properties.

## 12.3 Important Aspects and Limitations of Biomass Processing

Different technologies have been evaluated for lignocellulose-to-ethanol conversion, with the biochemical platform being the most advantageous for scale-up [19]. However, biochemical processes are hindered by the high recalcitrant structure of lignocellulosic biomass that limits the accessibility of cellulose and hemicelluloses to the hydrolytic enzymes and prevent the release of sugars.

Pretreatment of biomass is essential to alter its composition and structure so that efficient and rapid enzymatic hydrolysis of carbohydrates can occur [31]. Since different lignocellulosic materials have different physicochemical characteristics, it is necessary to adopt suitable pretreatment technologies based on the properties of each raw material.

Another critical step in lignocellulosic ethanol production is the enzymatic hydrolysis of cellulose and hemicelluloses into fermentable sugars. This stage is affected by several factors including the composition and structure of feedstocks, the pretreatment technology applied, the type of enzymes used and the enzyme loadings [32]. Costs and catalytic efficiencies of enzymes represent a major bottleneck for improving the economy of the bioethanol industry. Optimization of enzyme cocktails for enzymatic hydrolysis is therefore crucial to make the lignocellulosic ethanol production economically viable.

### *12.3.1 Pretreatment and Hydrolysis of Lignocellulosic Biomass*

Cellulose, hemicelluloses and lignin are strongly intermeshed and bonded through non-covalent and covalent cross-linkages, forming a lignocellulosic matrix. These structural characteristics make lignocellulosic materials very recalcitrant to the action of hydrolytic enzymes. To alter the structure of lignocellulose and facilitate the production of fermentable sugars, a suitable pretreatment process is needed [31]. The main goal of pretreatment is to break down lignin polymer and disrupt the crystalline structure of cellulose. Factors such as lignin and hemicelluloses content [33], degree of polymerization of cellulose [34], and porosity level of lignocellulosic materials [35] are among the main parameters to consider when subjecting lignocellulose to pretreatment processes.

The effectiveness of pretreatment determines the overall efficiency of the ethanol production process, and there are factors both upstream and downstream of the pretreatment step that should be taken into consideration. In the upstream steps, pretreatment can be considered by the selection of the raw material, since biomass harvesting and storage may affect pretreatment conditions such as residence time, temperature and/or the addition of chemical/biological catalysts. On the downstream processing steps, pretreatment highly influences enzymatic hydrolysis and microbial fermentation by affecting enzyme loadings and enzymatic hydrolysis rates, and cell viability and final ethanol yields and productivities, respectively [36].

As mentioned previously, lignocellulose pretreatment has a high impact on the economy of biochemical ethanol production. In fact, Lynd [2] calculated pretreatment costs to be about one third of the total costs, while Mosier et al. [31] described it as one of the most expensive processing steps in the lignocellulose-to-ethanol conversion. Cost-effectiveness of pretreatment is dependent on factors such as sugar release, biomass degradation, byproduct formation, energy demand, addition of chemical and/or biological catalysts, feedstock particle size and moisture content [8].

In addition to pretreatment, enzymatic hydrolysis represents another technological and economical bottleneck in lignocellulosic ethanol processes [37]. During enzymatic hydrolysis, carbohydrates are depolymerized into soluble sugars. This process is influenced by both enzyme-related and substrate-related factors [38]. Enzyme-related factors include end product inhibition, thermal stability, synergism between different enzyme activities and the unspecific adsorption of enzymes to lignin. On the other hand, crucial substrate-related factors are cellulose crystallinity and its degree of polymerization, the available surface area of cellulose and the lignin and hemicelluloses content.

Complete hydrolysis of lignocellulose requires the action of different enzyme activities, grouped in cellulases, hemicellulases and ligninases (Fig. 12.2). Cellulases (endoglucanases, cellobiohydrolases and β-glucosidases) are needed to hydrolyze cellulose into glucose monomers, while hemicellulases (xylanases, β-xylosidases, α-L-arabinofuranosidases, esterases, etc.) and ligninases (laccases, ligninolytic per-

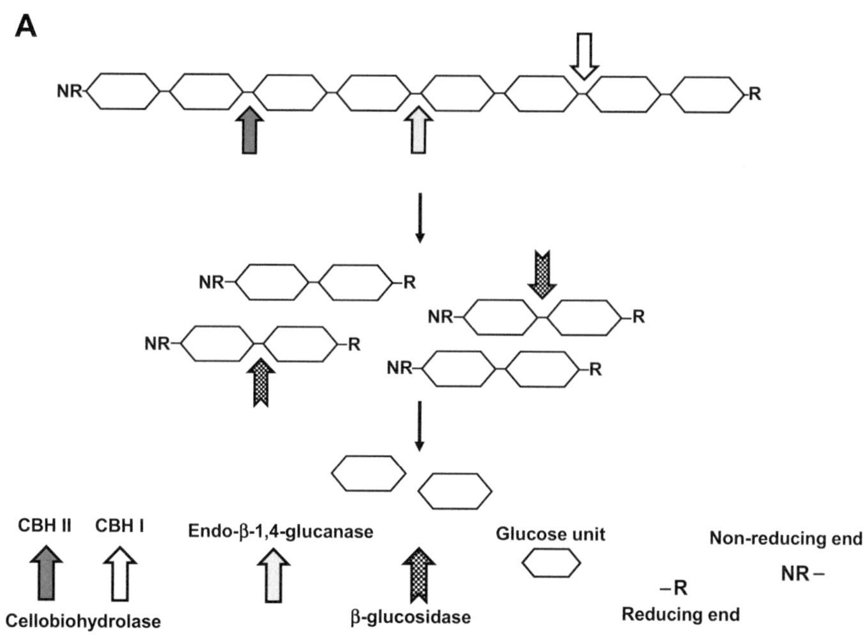


**Fig. 12.2** Schematic representation of the enzymes involved in cellulose (**a**), hemicellulose (**b**; adapted with permission from Ref. [38], Copyright © 2013 Springer), and lignin (**c**) biodegradation. *L* lignin

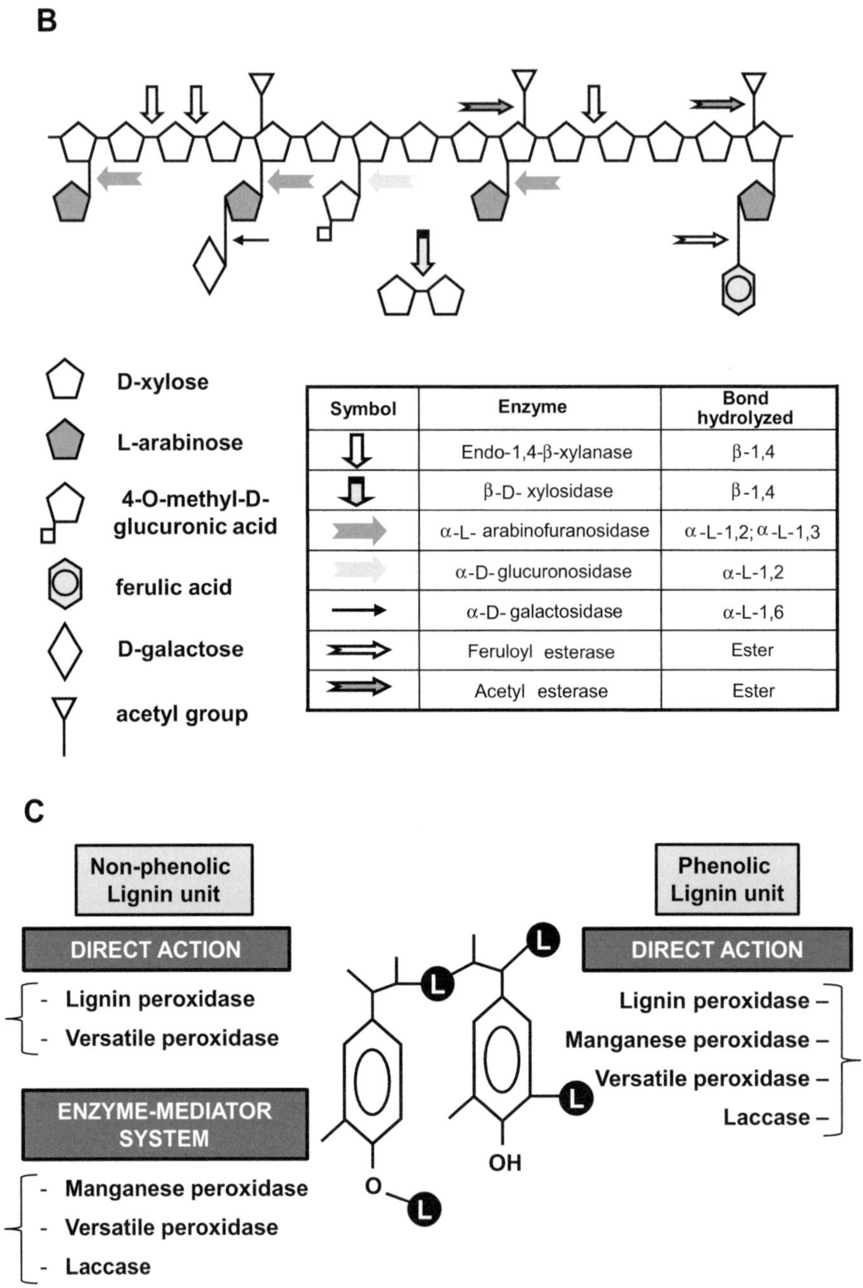


| Symbol | Enzyme | Bond hydrolyzed |
|---|---|---|
| | Endo-1,4-β-xylanase | β-1,4 |
| | β-D- xylosidase | β-1,4 |
| | α-L- arabinofuranosidase | α-L-1,2; α-L-1,3 |
| | α-D- glucuronosidase | α-L-1,2 |
| | α-D- galactosidase | α-L-1,6 |
| | Feruloyl esterase | Ester |
| | Acetyl esterase | Ester |

**Fig. 12.2** (continued)

oxidases, oxidases generating extracellular $H_2O_2$, reductases, etc.) are involved in breaking down hemicellulose and lignin polymers, respectively. A novel group of non-hydrolytic enzymes called polysaccharide monooxygenases (LPMOs) have been discovered that are highly promising enzymes for improving the efficiency of cellulases. Although their action mechanisms have not been completely elucidated, these LPMOs can oxidize crystalline regions of cellulose, creating new reducing and non-reducing ends for cellulases [39]. Finally, non-catalytic proteins such as swollenins and expansins have attracted some attention for enzymatic hydrolysis of lignocellulose. These proteins do not catalyze hydrolytic reactions but aid in disrupting the crystalline structure of cellulose, making it more accessible to hydrolytic enzymes [40].

Enzyme production costs have great impact on the overall process costs. Thus, different programs have funded studies to improve the enzyme production and make it affordable from the economic point of view. Also, combining and incorporating new enzyme activities in commercial preparations is required to provide an appropriate enzyme mixture. Significant advances have been achieved towards this aim. However, enzymatic mixtures still need to be optimized and adapted to different feedstocks and pretreatments [38]. Enzyme cocktail optimization involves different strategies such as engineering of cellulolytic microorganisms and/or their specific enzymes, screening of novel enzyme-producing microorganisms, enzyme recycling (enzyme immobilization, insoluble solids recycling, enzyme ultrafiltration, enzyme re-adsorption) and/or the addition of surfactants to avoid unproductive adsorption of hydrolytic enzymes to residual lignin.

#### 12.3.1.1 Recent Development in Pretreatment Technologies

Pretreatment research has been focused on identifying, evaluating, developing, and demonstrating promising approaches to enhance enzymatic hydrolysis of pretreated biomass, lowering enzyme dosages and shortening process times. Over the years, a large number of pretreatment technologies have been investigated on a wide variety of feedstocks [8, 32]. Pretreatment methods can be roughly divided into physical, chemical, physicochemical and biological processes. Some advantages and disadvantages of pretreatments processes are summarized in Table 12.1.

Combinations of pretreatment methods are usually needed to improve efficiencies. The complexity of lignocellulosic matrix depends very much on biomass feedstock. In this context, pretreatment effectiveness varies greatly among raw materials, and pretreatment optimization is therefore a must for each individual type of biomass. In the following sections, the most important pretreatment processes are summarized, and the most recent developments are highlighted. The impact of pretreatment methods on lignocellulosic materials and their effect on the subsequent enzymatic hydrolysis are discussed.

**Table 12.1** Advantages and disadvantages of the most important pretreatment processes

| Pretreatment | Advantages | Disadvantages | Ref. |
|---|---|---|---|
| Milling | Increases the available surface area of cellulose | High energy consumption | [41] |
| | Reduces cellulose crystallinity | | |
| Extrusion | Increases the available surface area of cellulose | High energy consumption | [9, 42] |
| | Higher efficiencies when combined with chemical/ biological catalysts | | |
| Diluted acid | Hemicellulose solubilisation | High formation of lignocellulose-derived inhibitors | [41, 43] |
| | Low cost | Difficulties for chemical recovery | |
| Alkaline | Reduces the content of lignin and hemicellulose | High pollution | [44, 45] |
| | Low cost | High costs of chemical recovery | |
| Oxidative | Efficient in reducing lignin content | High cost of reagents | [32] |
| Organosolv, COSLIF, CELF | Causes lignin and hemicellulose hydrolysis | High cost of organic solvents | [46–48] |
| | Recovery of lignin with high quality | Risk from operation with flammable solvents | |
| ILs pretreatment | Reduces cellulose crystallinity | High cost of ILs | [32, 49] |
| Steam explosion | Hemicellulose solubilisation and fibber destructuration | High formation of lignocellulose-derived inhibitors | [8, 32] |
| | Lignin redistribution and partial solubilization | High equipment cost | |
| | Cost-effective | | |
| AFEX | Increase accessible surface area | High equipment costs | [50] |
| | Low formation of lignocellulose-derived inhibitors | Lower efficiencies with lignin-rich materials | |
| EA pretreatment | Recovery of lignin with very high quality | High ammonia loadings | [10] |
| | Reduces cellulose crystallinity | High energy inputs | |
| Biological | Degrades lignin and hemicellulose | Low process rates | [12] |
| | Environmentally friendly and low energy consumption | | |
| | Low formation of lignocellulose-derived inhibitors | | |

### Physical Pretreatment

Physical pretreatment is aimed to increase the accessible surface area of lignocellulose to hydrolytic enzymes by reducing its particle size or disrupting the structure regularity. Such effects can be produced by a combination of shear stresses. Typical physical pretreatment includes chipping, grinding or milling processes [41]. However, the high energy input required in milling and refining processes makes them economically unfeasible.

Extrusion has been applied to improve the enzymatic hydrolysis of lignocellulosic materials. This technology is based on subjecting lignocellulosic biomass to shearing and heating stress factors, resulting in physical and chemical modifications (defibrillation, fibrillation and shortening of the fibers). In contrast to milling processes, extrusion represents a promising method for lignocellulosic biomass pretreatment due to its adaptability to process modifications (it can be combined with chemical and/or biological catalysts such as alkali or enzymes) and its versatility regarding the use of different raw materials [9, 42].

Irradiation with microwave, electron beam or gamma rays is also considered within physical pretreatments. These pretreatment methods are usually combined with alkalis, acids, ionic liquids (ILs) or salts for improving the digestibility of lignocellulosic materials [51, 52].

### Chemical Pretreatment

Chemical pretreatment uses different reagents to modify the structure and composition of lignocellulosic biomass. The main chemicals used are acids, alkalis, ILs, oxidant agents and organic solvents.

Of the chemical pretreatment methods, dilute-acid technology is very favorable for industrial applications. It has been studied to pretreat a wide range of lignocellulosic feedstocks including herbaceous crops, hardwoods and softwoods [41]. This technology is based mainly in hemicelluloses hydrolysis and solubilization, and can be performed at high or low temperatures with different residence times. Nonetheless, it promotes extensive biomass degradation, which generates several inhibitory compounds from both sugar degradation and partial lignin solubilization [43]. Mineral acids such as $H_2SO_4$ or HCl are the most used acid catalysts. However, organic acids including acetic acid, fumaric acid or maleic acid are appearing attractive alternatives [53].

In contrast to acid pretreatment, alkaline reagents increase cellulose digestibility by lignin removal [44]. Alkaline pretreatment can be performed at room temperature, with residence times ranging from seconds to days, and have shown to be more effective on agricultural residues than on woody materials [45]. Although some inhibitory compounds are generated during the process, alkaline pretreatments cause less sugar degradation than acid pretreatments. Among alkali catalysts, NaOH, KOH, $Ca(OH)_2$ and $NH_4OH$ and $NH_3$ are the most widely used [45]. Alkali-pretreated feedstocks usually show saccharification yields of about 50–70%. However, higher saccharification yields (up to 95%) can be obtained by combining alkali-based processes with other pretreatment methods (e.g. mechanical or biological pretreatments) or with oxidant agents such as $H_2O_2$ or copper-catalyzed alkaline hydrogen peroxide [12, 50, 54].

Oxidative delignification is a pretreatment method that uses ozone, oxygen, hydrogen peroxide, chlorine dioxide or elemental chlorine as oxidation reagents, causing a remarkable oxidative fragmentation and lignin removal [32]. Oxidative delignification is, in contrast, quite costly and it is normally used in combination with other traditional acid or alkaline pretreatments for removing residual lignin [32].

Organic solvents have also shown to be effective for pretreating lignocellulosic feedstocks. Pretreatment such as organosolv, COSLIF (cellulose and organic solvents-based lignocellulosic fractionation) or CELF (co-solvent enhanced lignocellulosic fractionation) are attractive technologies [46–48]. These processes can partially solubilize sugar components. However, the main action of these technologies is delignification, resulting in the recovery of a rather pure lignin fraction that can be used for further revalorization.

Pretreatment with ILs is a novel and promising alternative to improve ethanol production [32]. ILs are salts which exist as liquids at relatively low temperatures. Imidazolium salts are very common ILs [49]; however, tertiary amines derived from lignin and hemicellulose polymers (also called bioionic liquids) have also recently applied to pretreat lignocellulosic biomass [55]. One of the major advantages of ILs, is that their solvent properties can be adjusted in different ways for producing a simultaneous solubilization of sugars and lignin. Also, it should be highlighted that generation of inhibitory byproducts is avoided. In contrast, ILs can act as inhibitory compounds themselves, and an energy-efficient recycling method is required to compensate the high costs of these compounds.

### Physicochemical Pretreatment

Physicochemical pretreatment has been used for lignocellulosic ethanol production [8, 32, 56]. Of the developed types of physicochemical pretreatment, steam explosion, liquid hot water, ammonia fiber explosion/expansion (AFEX), extractive ammonia (EA), wet oxidation, and $CO_2$ pretreatment are the most important. Among them, steam explosion is the most widely used pretreatment technology and is one of the methods applied on the commercial scale. During steam explosion pretreatment, biomass is subjected to saturated steam at high temperatures and pressures, where acetyl groups are solubilized, promoting the autohydrolysis of hemicellulosic sugars [57]. Afterwards, there is a sudden depressurization that provokes a mechanical fiber deconstruction. Steam explosion has been successfully used for ethanol production from a wide range of agriculture residues [58, 59] and hardwoods [60]. In contrast, acid catalysts are needed for steam pretreating softwoods feedstocks, which have a lower content of acetyl groups [61]. The major drawback of steam explosion pretreatment is the extended biomass degradation, which limits the subsequent hydrolysis and fermentation steps [8, 32].

Ammonia-based pretreatments such as AFEX and EA are very promising pretreatment technologies, since ammonia is an inexpensive commodity chemical with easy recycling. AFEX technology is similar to steam explosion pretreatment [62]. It uses temperatures around 60–100 °C and high pressures (1.7–2.1 MPa). During AFEX pretreatment biomass composition is not very much altered and biomass accessibility is enhanced by a swollen effect that increases the water retention values in pretreated biomass [50]. EA is a novel pretreatment technology that uses liquid ammonia at elevated temperatures to solubilize lignin polymer [10]. EA can also alter crystallinity of cellulose, allowing the better hydrolyzability of pretreated materials. Although process conditions have to be optimized, EA pretreatment represents a very promising technology since it requires about 60% lower enzyme loadings to reach similar saccharification yields than those obtained with AFEX technology. A high purity lignin with a native-like structure is recovered after EA pretreatment, which offers a good possibility for lignin revalorization.

#### Biological Pretreatment

Several microorganisms and/or their enzymes have been used to pretreat lignocellulosic materials before enzymatic hydrolysis [63–68]. Some brown-, white-, and soft-rot fungi are capable of degrading lignin, hemicelluloses and small amount of cellulose. White-rot fungi are the most effective ligninolytic microorganisms due to their ability to produce enzymes (laccases and peroxidases), which can partially degrade lignin and/or modify its molecular structure [69]. White-rot basidiomycetes such as *Phanerochaete chrysosporium, Trametes versicolor, Panus tigrinus, Ceriporiopsis subvermispora, Pycnoporus cinnabarinus, Irpex lacteus, Ceriporia lacerata, Stereum hirsutum, Polyporus brumalis, Ganoderma austral* and *Pleurotus ostreatus* have been examined on different lignocellulosic feedstocks, showing high delignification efficiencies [12]. Although only white-rot basidiomycetes can degradade lignin extensively, some ascomycetes can also colonize lignocellulosic biomass and consequently improve saccharification yields [67]. Besides fungi, certain bacterial strains such as *Bacillus macerans, Cellulomonas cartae, Cellulomonas uda* and *Zymomonas mobilis* have also shown delignification abilities [12].

The use of ligninolytic enzymes (especially laccases) instead of microorganisms is another feasible alternative. Those enzymes show high reaction rates and are substrate specific, which offer the possibility to reduce the overall process time from weeks to hours without any loss in the sugar content [12].

The main advantages of biological pretreatment are their low capital costs, low energy demand and mild reaction conditions. Furthermore, these processes do not require the addition of chemical catalysts and do not release inhibitory byproducts. In contrast, major drawbacks include longer reaction times in comparison to other pretreatment technologies and the high enzyme production costs [8, 12].

Biological pretreatments can be combined with a mild acid/alkali pretreatment, organosolv, hydrogen peroxide, and thermal pretreatments to increase sugar recovery yields and reduce reaction times [67].

## 12.3.2 *Inhibitory Compounds and Residual Lignin*

Current pretreatment technologies still present several drawbacks that significantly influence saccharification and fermentation steps. Among them, residual lignin and biomass-degradation products are the most significant factors to consider for downstream processing.

Residual lignin promotes the unspecific adsorption of hydrolytic enzymes in pretreated materials, thus decreasing saccharification yields. It has been shown that the chemical and physical structure of residual lignin plays a large role in determining hydrolysis yields, which in turn, is heavily dependent on pretreatment conditions. Cellulases have been proposed to adsorb to residual lignin via hydrophobic, electrostatic and hydrogen bonding interactions [35]. Taking into account this negative role of residual lignin, a delignification process can be of benefit for increasing saccharification yields in pretreated materials.

On the other hand, lignocellulosic-derived compounds released during pretreatment process can act as inhibitors of the hydrolytic enzymes and fermentative microorganisms [11, 70–73]. Inhibitory compounds include furan derivatives (furfural and 5-hydroxymethylfurfural (5-HMF)), aliphatic acids (acetic acid, formic acid and levulinic acid), and phenolic compounds (4-hydroxybenzoic acid, 4-hydroxybenzaldehyde, vanillin, *p*-coumaric acid, ferulic acid, dihydroconiferyl alcohol, coniferyl aldehyde, syringaldehyde, syringic acid, and Hibber's cetones, etc.) [70]. Also, inorganic compounds and extractives (terpenes, fats, waxes and phenolics) can cause inhibitory effects. The nature and concentration of all these products is strongly dependent on the feedstock and pretreatment process [73]. Similar to delignification methods, several detoxification methods have been proposed to reduce the inhibitory power of pretreated substrates and increase saccharification and fermentation yields.

### 12.3.2.1 Delignification of Pretreated Materials

Different chemical and biological processes have been investigated to reduce the lignin content in pretreated materials. In comparison to the chemical delignification methods previously described as pretreatment processes (alkaline pretreatment, oxidative pretreatment, organosolv, EA, etc.), biological delignification is a promising technology due to the lower environmental impact and higher product yields.

Among the different biological strategies, the application of laccases has gained considerable attention in the last years [12]. Laccases are multicopper oxidases that catalyze the oxidation of substituted phenols, anilines and aromatic thiols to their corresponding radicals. The low redox potential of laccases only allows the direct oxidation of phenolic lignin units (Fig. 12.3), which represent only a small percentage of the whole lignin polymer [74]. However, in the presence of low molecular weight compounds (also called mediators) laccases can also oxidize non-phenolic lignin units (Fig. 12.3) [75]. In general, fungal laccases and laccases-mediator sys-

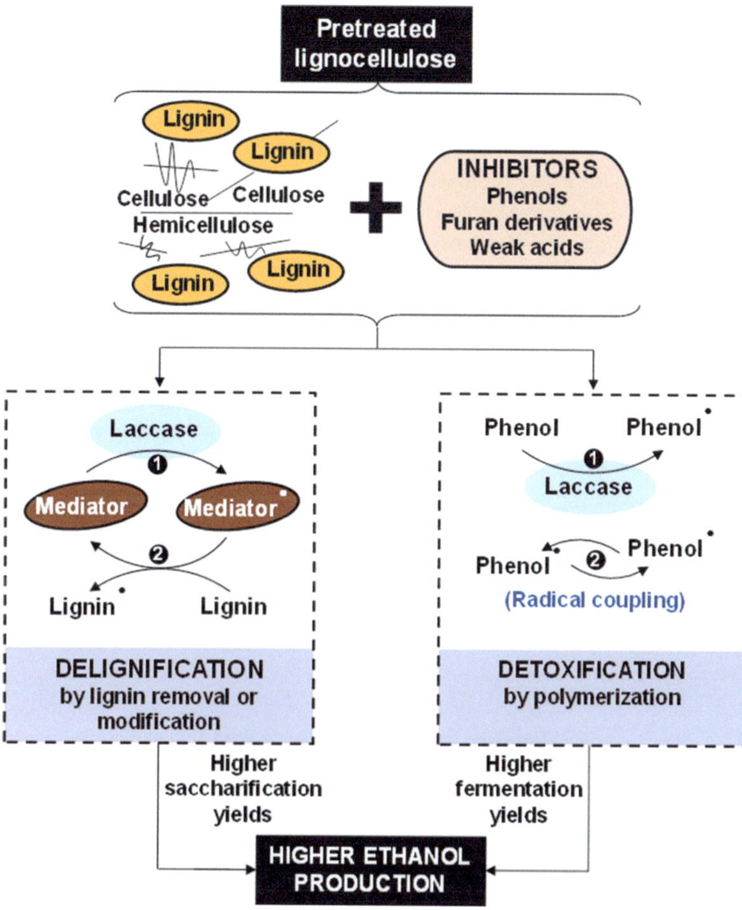


**Fig. 12.3** Role of laccase or laccase-mediator systems (LMS) in lignocellulosic ethanol production. Catalytic reactions include ether bond degradation, C–C degradation and aromatic ring cleavage as main delignification reactions, while oxidative polymerization is the main detoxification reaction

tems (LMS) have shown to be effective in modifying and/or partially removing lignin, increasing final hydrolysis yields of different pretreated materials [12, 65, 68]. Nevertheless, enzyme production costs and the use of costly synthetic mediators are still considered major drawbacks for the proper integration of these biological delignification methods in the current ethanol production process.

#### 12.3.2.2 Inhibitors Removal and Detoxification of Pretreated Materials

One way of reducing the amount of inhibitory compounds released from pretreatment is by using milder pretreatment conditions. However, mild pretreatment conditions usually lead to lower hydrolysis yields [73]. Therefore, a good balance between the accessibility of biomass to hydrolytic enzymes and biomass degradation is required during pretreatment processes. Another possibility to reduce the concentration of inhibitory compounds is to physically remove inhibitors or to perform a detoxification step prior to enzymatic hydrolysis and/or fermentation processes. Filtration and washing have been the most common methods employed to remove inhibitory compounds from pretreated materials. However, these processes involve the need of extra equipment, additional and expensive steps, loss of soluble sugars, waste of freshwater and production of wastewater [76].

Other methods developed to overcome the effects of inhibitory compounds include either physical, chemical or biological processes (Table 12.2). Vacuum evaporation for reducing volatile compounds such as furfural, acetic acid and vanillin is probably the most commonly applied physical method [77]. Ion-exchange resins or activated charcoal can adsorb different inhibitors [78, 79], while ethyl acetate and other solvents can extract them from hydrolysates [80]. Overliming processes with $Ca(OH)_2$ is among the most common chemical detoxification methods

**Table 12.2** Major considerations for the different methods used for inhibitor removal and/or detoxification of pretreated materials

| Method | Observations | Ref. |
|---|---|---|
| Evaporation | Removes volatile compounds (furfural, acetic acid and vanillin) | [77, 79] |
| | Concentration of non-volatile compounds (sugars, extractives and lignin derivatives) | |
| Overliming $Ca(OH)_2$ | Reduces the content of furfural, 5-HMF, and phenols | [77] |
| | Produces sugar loss | |
| Ion exchange resins | Reduces the content of lignin-derived compounds, acetic acid, and furfurals | [78, 79] |
| | High costs | |
| Activated charcoal | Reduces phenol content | [79] |
| | Cost-effective | |
| | No sugars loss | |
| Ethyl acetate extraction | Removes acetic acid, furfural, 5-HMF, vanillin, and hydroxybenzoic acid | [80] |
| Bio-abatement | Wide range of inhibitors profile modification | [12, 81, 82] |
| | Feasible and environmentally friendly | |
| | Few side-reactions and low energy requirements | |
| | Slow reaction time | |
| | Produces sugar loss | |

*5-HMF* 5-hydroxymethylfurfural

and can partially remove phenolic compounds, furfural and 5-HMF, improving hydrolysate fermentability [77].

Combination of different detoxification methods have also been used to detoxify pretreated materials [11]. For instance, chemical detoxification with $Ca(OH)_2$ has been combined with activated charcoal or ion-exchange resins, being one of the most efficient combinations [77].

Similar to biological delignification processes, biological detoxification methods with microorganisms or enzymes offer many advantages such as mild reaction conditions, no chemical addition, fewer side-reactions and less energy requirements [12, 81]. Among different microorganisms, fungi such as *Trichoderma reesei* have been mostly studied for this purpose [77, 82]. Moreover, this fungus can produce hydrolytic enzymes while detoxification take places [82].

Bacteria and yeasts have also been used for detoxification purposes [12]. For instance, most of *Saccharomyces cerevisiae* strains can convert furan derivatives into less inhibitory compounds (furfural to furfuryl alcohol and 5-HMF to 5-(hydroxymethyl)furfuryl alcohol). This yeast also possesses the capacity to metabolize several phenolic compounds and can convert some aromatic carboxylic acids to the corresponding vinyl derivatives [12, 81].

The most common enzymes used for enzymatic detoxification are laccases and peroxidases, which derive from diverse white-rot fungi (*T. versicolor, Trametes villosa, Coriolopsis rigida, P. cinnabarinus, Coltricia perennis, Cyathus stercoreus*) and bacteria [12, 81]. These enzymes act selectively to remove phenolic compounds and generate unstable phenoxy-radicals that polymerizes into less toxic aromatic compounds (Fig. 12.3) [83]. After laccase treatment, glucose consumption rates, ethanol volumetric productivity and ethanol yields are usually improved [77, 84–86]. Nevertheless, as discussed above, enzyme costs can limit the use of these biocatalysts.

For better economy in these processes, enzyme detoxification methods can be combined with robust fermenting microorganisms to reduce enzyme dosages. Also, the cloning of these enzymes into cellulase-producing microorganisms or their reutilization can reduce the costs of producing these detoxifying biological agents.

## 12.4 Process Integration

Pretreatment, enzymatic hydrolysis and fermentation steps need proper integration to achieve an efficient industrial ethanol production process. Strategies and conditions employed for each step have an important influence on the rest, making the process configuration crucial in each particular case.

The main configurations used have typically been SHF and SSF [6, 87]. Nevertheless, other configurations such as integrating pretreatment and hydrolysis in a bioextrusion process [9], or the combination of enzyme production, enzymatic hydrolysis and microbial fermentation in a single step (CBP) [88–90] have been developed.

### 12.4.1 Separate Hydrolysis and Fermentation (SHF)

SHF processes have been the strategy traditionally used in bioethanol production. In SHF, enzymatic hydrolysis and fermentation are performed in separate and sequential steps. The main advantage of this approach is the possibility of running both steps at optimal conditions of pH and temperature. Optimal temperature for fungal hydrolytic enzymes is in general higher (45–50 °C) than the optimal temperature for fermenting microorganisms (30–37 °C). Another advantage of SHF processes is the possibility of removing the insoluble solids after enzymatic hydrolysis. This allows performing liquid fermentation, facilitating the reutilization of the fermentative microorganisms [13].

The main disadvantage of SHF is the accumulation of sugars (free and oligomeric forms) during enzymatic hydrolysis. Sugar accumulation promotes end-product inhibition on hydrolytic enzymes and consequently reduces enzymatic hydrolysis yields [91, 92]. The *in situ* removal of sugars by dialysis has been proposed as an attractive option to solve this problem [93].

### 12.4.2 Simultaneous Saccharification and Fermentation (SSF)

In SSF, enzymatic hydrolysis (saccharification) and fermentation are performed simultaneously in the same vessel. This implies a cost reduction and a great advantage of an integrated process [94]. Another advantage of SSF in comparison to SHF is the reduction of end-product inhibition during the enzymatic hydrolysis step (sugars are simultaneously consumed by fermenting microorganisms). This results in higher hydrolysis yields, shorter reaction times and the possibility of reducing enzyme loadings [87, 95]. Furthermore, contamination risks are lower, since glucose is converted to ethanol as soon as it is released.

SSF processes require, in contrast to SHF processes, compatible fermentation and saccharification pH and temperature conditions. Thus, the main disadvantage of SSF process is the necessity of running enzymatic hydrolysis at suboptimal temperatures. The utilization of thermotolerant yeasts such as *Kluyveromyces marxianus* capable of fermenting at temperatures above 40 °C, involves a relevant advantage and has been evaluated in bioethanol production [96, 97]. Using high temperatures in SSF processes also implies other advantages such as less risk of contaminations and the reduction in cooling costs. In contrast, higher temperatures may involve fluidification of cell membranes (the physical protective barrier), reducing the tolerance of fermentative microorganisms towards inhibitory compounds [98].

When combining pentose-rich feedstocks and pentose-fermenting microorganisms, the process is called simultaneous saccharification and co-fermentation (SSCF) [15, 99]. As it is detailed later in this chapter, the use of native and engineered strains with the capacity of co-fermenting hexoses and pentoses represents a great advantage with higher potential for increasing ethanol production [96, 100].

In general, the choice between SHF and SSF approaches depends strongly on the process conditions: type of biomass, pretreatment method, solids loading, hydrolytic enzymes and fermentative microorganism. As an example of this variability, it has been recently reported that using pretreated wheat straw at high solids loading and applying an older generation of enzyme cocktails (Celluclast + Novozym188), a SSF process resulted in significant higher yields in comparison to SHF [101]. Interestingly, using new generation of improved enzymatic cocktails (Cellic CTec 2), SHF resulted in 20% higher final ethanol yield compared to SSF [101].

### *12.4.3 Consolidated Bioprocessing (CBP)*

CBP involves the integration of enzyme production, enzymatic hydrolysis and fermentation in a single bioprocessing system. The integration of all these biological processes needed for ethanol production entails great potential to save capital and operational costs [100, 102]. Different cellulolytic and non-cellulolytic microorganisms have been proposed as candidates for CBP processes, being bacteria from genus *Clostridium* (such as *Clostridium thermocellum* or *Clostridium phytofermentans*) the most studied [88–90]. These anaerobic bacteria display extracellular enzymatic systems or cellulosomes, which are attached to the cell surface and can degrade cellulose into soluble sugars [103]. Other bacteria including *Escherichia coli*, *Thermoanaerobacterium saccharolyticum* or *Caldicellulosiruptor bescii* have been also engineered and/or studied as CBP systems [104–106].

As an alternative to bacterial strains, genetically modified yeasts displaying cellulolytic mechanisms, can be also used in CBP processes. Examples of this strategy have been developed in *S. cerevisiae* [107, 108], *K. marxianus* [109] or *Yarrowia lipolytica* [110] strains.

CBP processes seem the logical endpoint in the evolution of ethanol production from lignocellulosic biomass. Application of CBP implies no operating cost or capital investment for dedicated enzyme production (or purchase), reduced diversion of substrate for enzyme production or compatible enzyme and fermentation systems. Nevertheless, most CBP organisms identified and developed up to date have limited overall performance when using real lignocellulosic substrates, either because they still need supplementation of exogenous enzymes, or because they show low ethanol titers, low tolerance to ethanol or limited growth [100, 111].

### *12.4.4 Operational Strategies*

In general, SHF, SSF and CBP processes can be performed in batch, fed-batch or in continuous mode. In addition, other operational strategies such as including a prehydrolysis step or the use of immobilized cells or membrane bioreactors have been studied for ethanol production.

Being a very simple method, batch processes (a closed culture system which contains an initial amount of substrate that is subjected to fermentation) are the most common fermentation strategies [111]. However, fed-batch processes (where substrate, culture medium and other required components and nutrients are loaded at specific time points) have shown to be effective for improving ethanol production when working at high substrates concentrations, mainly due to a better mixing and the presence of low concentration of inhibitor compounds, which facilitates their conversion by the fermenting microorganism. Combining substrate, enzyme and yeast feed in SSF processes, Koppram and Olsson [112] obtained 40 g/L of ethanol from a non-detoxified slurry (steam-exploded spruce) at a final substrate concentration of 20% (w/w) WIS (water insoluble solids), which showed to be totally inhibitory in a normal batch SSF process. Similarly, a multi-feed SSCF strategy was used by Wang et al. [113], increasing the final substrate (steam-exploded wheat straw) concentrations up to 22% (w/w) WIS (water insoluble solids) without observing mixing problems, both in a standard bioreactor and DEMO scale (10 $m^3$), and reaching ethanol concentrations as high as 57 g/L. Another variation of SSF/SSCF processes when using high solids loading is the introduction of a prehydrolysis or a presaccharification step (also known as Semi-Simultaneous Saccharification and Fermentation, SSSF). This presaccharification step allows a rapid liquefaction of the medium, offering more suitable conditions for fermentation and allowing higher ethanol yields [114, 115]. In the absence of cells, the prehydrolysis is performed at optimal conditions for enzymes for 1–80 h, after which the fermentative microorganism is inoculated. In a recent approach, a two stage hydrolysis configuration involving a quick liquefaction with strong mixing and a longer saccharification with no mixing was proposed to reduce the power required for mixing at high substrate loadings [116].

The use of immobilized cells by encapsulation has also been tested for lignocellulosic ethanol production. This strategy allows a better inhibitory tolerance of the fermenting strains and an optimal co-utilization of glucose and xylose [117]. A better inhibitory tolerance can be also obtained by working at high cell densities in membrane bioreactors. In this context, Ylitervo et al. [118] designed a membrane bioreactor with a cross-flow membrane to allow cell retention during a continuous process (substrate, culture medium and other required components and nutrients are loaded continuously), which enabled continuous ethanol production in the presence of high concentrations of acetic acid (up to 20 g/L).

## 12.5 Fermenting Microorganisms for Lignocellulosic Ethanol Production

In the biochemical conversion of lignocellulose, sugars contained in raw materials are converted to ethanol by microbial fermentation. Ethanologenic microorganisms ferment the glucose released during enzymatic hydrolysis into ethanol via glycolysis, with the consequent production of ATP and $CO_2$. The stoichiometric conversion of glucose to ethanol is 0.51 g/g. However, it is very difficult to reach such conversion yield, since microorganisms divert a certain part of the consumed carbon into cellular metabolism and growth.

In contrast to the conventional ethanol production processes from starchy/sugary feedstocks, lignocellulosic sugars are converted in a highly challenging environment. This fact, together with other inherent characteristics of the process, demands robust fermenting microorganisms with high tolerance to biomass-derived inhibitors, ethanol and/or mechanical and osmotic stress. Also, they must be able to ferment the wide range of lignocellulosic sugars and/or tolerate relatively high temperatures.

The yeast *S. cerevisiae* is the most commonly employed microorganism for industrial alcoholic fermentation. The most attractive characteristics of *S. cerevisiae*, are: (1) it is generally recognized as safe (GRAS), (2) it can consume all kinds of hexoses, (3) it reaches ethanol yields close to the theoretical and (4) it can produce ethanol at concentrations as high as 18% (v/v) [119]. Apart from monosaccharides, some strains of *S. cerevisiae* are also able to utilize disaccharides such as sucrose and maltose, or trisaccharides like maltotriose and raffinose [120]. Furthermore, the resistance of *S. cerevisiae* to lignocellulose-derived inhibitors is high and therefore it is the preferred microorganism for lignocellulosic ethanol production. In spite of showing all these interesting features, the main drawback of *S. cerevisiae* is its inability to ferment xylose, which is the second most important sugar after glucose in lignocellulose.

More than 2000 yeast species have been studied in the literature so far. Some of them show very interesting characteristics in overcoming the limitations inherent to *S. cerevisiae*. In this sense, researchers are exploring new alternatives in non-conventional yeasts or bacteria to boost the lignocellulosic ethanol production. Non-conventional yeasts are usually isolated from extreme environments in which they develop specific mechanisms to survive under harsh conditions. This makes these non-conventional yeasts incorporate relevant industrially attractive traits such as the ability to utilize complex substrates and/or having high tolerance to different stress factors [17].

Besides yeasts, other microorganisms like ethanologenic bacteria have shown promising results in terms of xylose fermentation and thermotolerance, which motivates researchers to explore the benefits of using these organisms in lignocellulosic ethanol production processes.

The following sections describe different fermentative microorganisms with interesting traits for lignocellulosic ethanol production and discuss advances in the development of novel strategies for strain engineering.

### 12.5.1 Hemicellulosic Sugars Fermentation

Efficient fermentation of all sugars present in lignocellulosic materials is crucial to increase the profitability of biological conversion of lignocellulose into ethanol. Since xylose is the second most abundant carbohydrate in nature, its commercial fermentation is essential to improve the global economy of the process.

A small number of native xylose-fermenting yeasts has been identified. *Candida tropicalis, Candida shehatae, Pachysolen tannophilus, Scheffersomyces stipitis* and *Spathaspora passalidarum* are among the xylose-utilizing yeast. These yeast species, however, present low ethanol yields, low tolerance to ethanol and inhibitors, require microaerophilic conditions and are very sensitive to pH changes, which limit their application on the industrial scale. Notwithstanding, advances have been made towards the improvement of these yeast species for their utilization in lignocellulosic ethanol production [121, 122].

As previously discussed, the yeast *S. cerevisiae* is one of the preferred microorganisms for lignocellulosic ethanol production as it has demonstrated to be highly tolerant towards lignocellulose-derived inhibitors and the end product ethanol. However, since wild-type *S. cerevisiae* strains are not able to ferment pentoses, hard efforts have been addresses to develop efficient engineered xylose-fermenting *S. cerevisiae* strains.

Among the three different metabolic pathways for xylose assimilation, only two of them have been introduced in *S. cerevisiae*: the oxidoreductive pathway, which involves xylose reductase (XR) and xylitol dehydrogenase (XDH) enzymes; and the xylose isomerase (XI) pathway. The product of both pathways, xylulose, is phosphorylated by the action of a xylulose kinase (XK) and metabolized via the pentose phosphate pathway (PPP). The *XYL1* and *XYL2* genes (mainly from *S. stipitis*) involved in the XR/XDH pathways have been cloned in several *S. cerevisiae* strains. However, these two enzymes have a different co-factor preference, which is translated in co-factor imbalance and xylitol accumulation. In this context, altering the co-enzyme preference of the XR and XDH has been one of the most effective approaches to decrease xylitol formation and enhance ethanol yields [123, 124]. The expression of XI from bacteria, which directly converts xylose into xylulose, does not present limitations regarding xylitol accumulation. In both pathways, the over expression of the endogenous *XKS1* gene encoding for XK enables *S. cerevisiae* to ferment xylose to ethanol more rapidly [125].

Besides xylitol accumulation, other constrains regarding xylose fermentation by recombinant *S. cerevisiae* strains have been ascribed to poor xylose uptake and limitation in the PPP and other metabolic fluxes. Thus, big efforts have been recently addressed to overcome these challenges by means of metabolic engineering techniques [126–128].

Directed adaptation or evolutionary engineering of yeast in the presence of xylose has been proven as one interesting strategy to develop more efficient xylose-fermenting strains. In this context, evolutionary engineering of xylose-fermenting *S. cerevisiae* strains to lignocellulosic hydrolysates not only led to an increase in xylose fermentation capacity but also in better results in terms of tolerance to inhibitors and ethanol yields, when compared with non-adapted cells [15, 129]. Furthermore, as it is detailed latter in this chapter, short-term adaptation of recombinant *S. cerevisiae* cells to hydrolysate rich in xylose and inhibitors during the propagation phase, also resulted in increased xylose fermentation in the subsequent ethanol production process [130, 131].

Apart from yeasts, xylose fermentation has also been reported with some fungi such as *Fusarium oxysporum*, *Mucor circinelloides* and *Rhizopus oryzae* [132, 133]. In comparison to xylose-fermenting yeasts, filamentous fungi are more tolerant to the inhibitors but they show slow fermentation rates for a competitive industrial process [134]. The ethanologenic xylose-fermenting bacteria that show the most promising characteristics for lignocellulosic ethanol production are *E. coli* and *Klebsiella oxytoca*. Both organisms utilize also a wide spectrum of sugars; however, they are inhibited at low sugar and ethanol concentrations, and the fermentation processes lead to considerable by-product formation.

### 12.5.2 Increased Tolerance to Inhibitors

The use of fermenting microorganisms that can cope with the inhibitory compounds released during pretreatment is crucial for achieving a cost-competitive production process. Several strategies to develop robust *S. cerevisiae* strains with improved tolerance to inhibitors have been described. The overexpression of genes that encode enzymes which confer resistance to specific inhibitors has improved the tolerance of *S. cerevisiae* to these degradation compounds [135, 136].

A significant increase in furfural tolerance has been observed by disrupting the *SIZ1* gene in *S. cerevisiae* [137]. The genome shuffling technique has also been used with success to improve the acid tolerance of *S. cerevisiae* [138]. Among non-metabolic engineering techniques, evolutionary engineering has been proposed as an effective method to improved tolerance to inhibitors in *S. cerevisiae* strains [129, 139]. During evolutionary engineering, microorganisms are subjected to high inhibitor concentrations over extended periods, which provoke random genetic changes that confer increased tolerance to the stress factors. Besides the long-term adaptation gained during evolutionary engineering, short-term adaptation of yeast to lignocellulosic hydrolysate during propagation has also been proven to increase detoxification rates of furfural and 5-HMF, reduce the lag phase of microorganisms and increase sugar consumption rates and ethanol yields in the subsequent fermentation step [130, 131, 140]. As introduced before, since tolerance to inhibitors is affecting sugar consumption rates, these strategies are very interesting for producing more efficient xylose-consuming strains, highly tolerant to lignocellulose-derived inhibitors.

In addition to all efforts addressed for increasing yeast tolerance to lignocellulose-derived inhibitors, little is known about the superior inhibitor tolerance detected in certain non-conventional yeasts. Among non-*Saccharomyces* species, *Pichia kudri-*

*avzevii* has been reported as an extremely robust microorganism, coping with up to 7 g/L of 5-HMF [141] and about 8 g/L of acetic acid [142]. *Zygosaccharomyces bailii* is the most acetic acid-tolerant specie described so far, showing the same growth rate reduction at 24 g/L of acetic acid than *S. cerevisiae* at 9 g/L [143].

Besides the described evolution and genetic engineering approaches, different alternatives such as cell retention, encapsulation, and flocculation of fermentative microorganisms have been developed to increase the intrinsic tolerance or the inherent detoxification capacity of some microorganisms [12, 81].

### 12.5.3 Ethanol Tolerance

*S. cerevisiae* has been described as the most ethanol-tolerant yeast species [141]. Similar to *S. cerevisiae*, *Dekkera bruxellensis* have shown to have similar ethanol tolerance levels. The common trait for the high ethanol tolerance in these two mentioned yeast species is reported to be the duplication of their alcohol dehydrogenase encoding genes. *D. bruxellensis* strains isolated from wine fermentation have demonstrated ethanol tolerance to concentrations ranging from 10 to 16% (v/v) [144]. In spite of the very promising features present in this yeast, genetic modification of this species is very difficult, which slows down the progress in using this organism for bioethanol production. Other highly ethanol-tolerant yeast species are *P. kudriavzevii*, and *Wickerhamomyces anomalus* which are able to tolerate 13% and 11% (v/v) ethanol concentrations, respectively [141].

### 12.5.4 Thermotolerance

The employment of thermotolerant yeasts in ethanol processes from lignocellulose is advantageous when applying SSF configurations, since the optimal conditions for enzymatic hydrolysis and fermentation are different. As mentioned before in this chapter, most hydrolytic enzymes perform better at 50 °C while most fermenting microorganisms have an optimum temperature of 30–37 °C. Thus, hydrolysis yields are increased when using thermotolerant microorganisms, owing to the higher temperature of SSF processes. There are other advantages that could be exploited when running SSF processes at higher temperatures, such as energy savings through a reduction in cooling costs, a significant reduction in contamination risks and the possibility of continuous ethanol removal by evaporation. Some studies conclude that an increase of 5 °C in the SSF temperature could considerably reduce the ethanol production costs [98].

Thermotolerant yeast strains of *Saccharomyces*, *Kluyveromyces* and *Fabospora* can grow at temperatures above 40 °C and produce ethanol at temperatures of 40 °C, 43 °C and 46 °C, respectively [98]. *K. marxianus* grow well at temperatures as high as 45–52 °C and can efficiently produce ethanol at temperatures of between 38

°C and 45 °C. Several strains belonging to the yeast species *K. marxianus* have been successfully employed in SSF processes from lignocellulose such as wheat straw [145], barley straw [146], eucalyptus [147], switchgrass [148] or even recycled paper sludge or waste [149, 150]. Apart from *K. marxianus*, some thermotolerant strains of *S. cerevisiae* and *K. fragilis* have also been successfully employed in SSF processes [151, 152].

*Ogataea polymorpha* is a yeast species able to grow at temperatures higher than 50 °C. This fact, together with its ability to ferment xylose and cellobiose to ethanol, makes *O. polymorpha* an interesting microorganism with high potential for use in SSF processes.

The use of thermophilic bacteria with broad substrate range and high yields may be another good option for ethanol production from lignocellulosic biomass. Highly ethanologen thermophilic bacteria are typically members of the genera *Thermoanaerobacterium*, *Thermoanaerobacter*, *Clostridium* or *Caldanaerobacter* [153]. Most thermophilic strains within the genus *Clostridium* have an optimum temperature in the range of 45–65 °C. On the other hand, *Thermoanaerobacterium* species have an optimum temperature ranging from 55 to 65 °C and have demonstrated good ethanol yields on lignocellulosic substrates. *Thermoanaerobacter* species have slightly higher optimum temperatures (65–75 °C) than *Thermoanaerobacterium*. Both *Thermoanaerobacter ethanolicus* and *Thermoanaerobacter* J1 have been proven to produce ethanol with high yields from lignocellulosic sugars [154, 155]. Other very promising ethanol producing thermophilic bacteria are the wild type strains of *Caloramator boliviensis* that have already been used in SSF processes from cassava starch [156].

### *12.5.5 Osmotolerance*

Running the bioethanol production process at high substrate concentrations (>15% (w/w)) results in improved overall productivity, reduced capital costs and lower energy input, when compared to processes at lower consistencies [157]. The high initial sugar concentrations in the broth may cause an increase in the osmolarity of the fermentation medium, which can lead to water loss and cell shrinking. *S. cerevisiae* can grow and ferment in media containing as much as 400 g/L glucose, which is one important advantage of this microorganism for producing ethanol at high substrate loading [158].

Two non-*Saccharomyces* yeast species isolated from sugar rich habitats (floral nectar and sugar beet juice) [141] show even higher osmotolerance than *S. cerevisiae*. *Candida bombi* grows well on rich media with 70% glucose. In the same study, *Starmerella bombicola* reached the same growth level as that of *C. bombi* on 60% glucose.

The application of extremely osmotolerant non-conventional yeast has recently appeared as interesting options in this kind of processes. *Zygosaccharomyces rouxii* is one of the most halotolerant and osmotolerant yeast species since it can cope with up to 90% (w/v) sugar concentration [159]. Studies have suggested that this extreme osmotolerance could rely on two plasma membrane transporters (ZrFfz1 and ZrFfz2) phylogenetically distant from any other fungal transporter [160].

Some attempts have been made to engineer new bacterial strains to cope with increased osmotolerance in ethanol production processes at high substrate loading. This is the case of recombinant *E. coli* strain FBR 5, which is able to ferment acid-treated wheat straw hydrolysates with 150 g/L total sugars. In this study, *E. coli* FBR5 was able to ferment both xylose and glucose with a final ethanol yield of 0.47 g/g [161].

## 12.6 Conclusions and Future Outlook

There is no doubt on the importance of developing and implementing a cost-effective lignocellulose-based industry, which will place into the market several renewable biofuels and other value-added products. Lignocellulosic ethanol is leading such scenario and quite a few commercial lignocellulosic ethanol plants have been opened in the last 3–4 years (BetaRenewables, Abengoa, Raizen, GranBio, Poet-DSM, and DuPont). However, there are still some challenges that should be addressed for establishing a competitive lignocellulosic ethanol industry.

The highly recalcitrant structure and the heterogeneous chemical composition of lignocellulosic materials hinder their utilization as sugar sources. Interesting approaches regarding the development of novel genetically modified feedstocks such as those having lower lignin content and/or a higher proportion of easily fermentable sugars have been reported.

During biomass processing, pretreatment is one of the most important steps in both process and economic terms. Different pretreatment processes have been developed to overcome the recalcitrant structure of lignocellulose and increase biomass digestibility. An optimal pretreatment process would be able to guarantee a proper balance between the increase in biomass digestibility and the extension of biomass degradation. Also, it would facilitate the recovery of a highly pure lignin fraction, thus offering possibilities for its revalorization. Methods such as steam explosion, dilute-acid pretreatment, extrusion, EA, CESF or ILs represent promising processes with high potential for industrial applications.

Enzymatic hydrolysis is another important process in biomass processing. Recent studies in saccharification processes have elucidated the need of incorporating new enzyme activities in the hydrolytic cocktails for the complete hydrolysis of lignocellulose. Some of these activities are LPMOs, hemicellulases, ligninases and other non-hydrolytic enzymes such as swollenins and expansins.

The development of more robust fermentative strains with abilities to cope with all the different stresses is also needed for the optimal conversion of the highly challenging hydrolysates. Different delignification and detoxification methods have been developed for increasing the hydrolyzability and fermentability of hydrolysates. However, promoting the *in situ* conversion and/or the tolerance of the inhibitory compounds present in hydrolysates would be economically beneficial. In addition, all lignocellulosic sugars must be converted during fermentation processes for increasing final ethanol concentrations and conversion yields. Strains with high pentose-conversion efficiencies are therefore very interesting options for lignocellulosic ethanol production.

Finally, although each process should be individually studied and optimized, proper process integration should be also evaluated in considering the selected pretreatment method, the enzyme cocktail and the fermentative microorganism. In this context, the development of an efficient CBP process where enzyme production, enzymatic hydrolysis and fermentation are integrated in one single step would be convenient for cost-effectiveness of lignocellulosic ethanol production.

In brief, it can be concluded that the breakthrough of key technologies both in biomass processing and fermentation processes, and the optimal integration of all the steps involved, are among the crucial aspects to overcome for the realization of a global bio-based economy, where lignocellulosic ethanol plays a key role.

**Acknowledgments** Authors thank the Spanish Ministry of Economy and Competitiveness for funding the present work via Projects ENE2014-54912-R and CTQ2013-47158-R. ETP acknowledges the People Programme (Marie Curie Actions) of the European Union's Seventh Framework Programme (FP7/2007-2013) under REA grant agreement n° 291803. ADM acknowledges the "Juan de la Cierva" Programme for contract FJCI-2014-22385.

## References

1. Imran A, Varman M, Masjuki HH, Kalam MA. Review on alcohol fumigation on diesel engine: a viable alternative dual fuel technology for satisfactory engine performance and reduction of environment concerning emission. Renew Sust Energ Rev. 2013;26:739–51.
2. Lynd LR. Overview and evaluation of fuel ethanol from cellulosic biomass: technology, economics, the environment, and policy. Annu Rev Energy Environ. 1996;21(1):403–65.
3. Balan V, Chiaramonti D, Kumar S. Review of US and EU initiatives toward development, demonstration, and commercialization of lignocellulosic biofuels. Biofuels Bioprod Biorefin. 2013;7(6):732–59.
4. Yu Y-S, Giles B, Oh V. State of the market report – uncovering the cost of cellulosic ethanol production. Luxresearch; 2016.
5. Statistics NEO. Fuel prices: Nebraska Energy Office; 2016. Available from: http://www.neo.ne.gov/statshtml/66.html. Access date: Nov 2016.
6. Sun Y, Cheng J. Hydrolysis of lignocellulosic materials for ethanol production: a review. Bioresour Technol. 2002;83(1):1–11.
7. Tomás-Pejó E, Alvira P, Ballesteros M, Negro MJ. Pretreatment technologies for lignocellulose-to-bioethanol conversion. In: Pandey A, Larroche C, Ricke SC, Dussap C-G, Gnansounou E, editors. Biofuels. Alternative feedstocks and conversion processes. Amsterdam: Academic Press; 2011. p. 149–76.
8. Alvira P, Tomás-Pejó E, Ballesteros M, Negro MJ. Pretreatment technologies for an efficient bioethanol production process based on enzymatic hydrolysis: a review. Bioresour Technol. 2010;101(13):4851–61.
9. Vandenbossche V, Brault J, Vilarem G, Hernández-Meléndez O, Vivaldo-Lima E, Hernández-Luna M, Barzana E, Duque A, Manzanares P, Ballesteros M, Mata J, Castellón E, Rigal L. A

new lignocellulosic biomass deconstruction process combining thermo-mechano chemical action and bio-catalytic enzymatic hydrolysis in a twin-screw extruder. Ind Crop Prod. 2014;55:258–66.

10. da Costa SL, Jin M, Chundawat SPS, Bokade V, Tang X, Azarpira A, Lu F, Avci U, Humpula J, Uppugundla N, Gunawan C, Pattathil S, Cheh AM, Kothari N, Kumar R, Ralph J, Hahn MG, Wyman CE, Singh S, Simmons BA, Dale BE, Balan V. Next-generation ammonia pretreatment enhances cellulosic biofuel production. Energy Environ Sci. 2016;9(4):1215–23.
11. Taherzadeh MJ, Karimi K. Fermentation inhibitors in ethanol processes and different strategies to reduce their effects. In: Pandey A, Larroche C, Ricke SC, Dussap C-G, Gnansounou E, editors. Biofuels. Alternative feedstocks and conversion processes. Amsterdam: Academic Press; 2011. p. 287–311.
12. Moreno AD, Ibarra D, Alvira P, Tomás-Pejó E, Ballesteros M. A review of biological delignification and detoxification methods for lignocellulosic bioethanol production. Crit Rev Biotechnol. 2015;35(3):342–54.
13. Olsson L, Jørgensen H, Krogh KBR, Roca CFA. Bioethanol production from lignocellulosic material. In: Dumitriu S, editor. Polysaccharides: structural diversity and functional versatility. New York: Marcel Dekker; 2005. p. 957–93.
14. Martín C, Marcet M, Almazan O, Jönsson LJ. Adaptation of a recombinant xylose-utilizing *Saccharomyces cerevisiae* strain to a sugarcane bagasse hydrolysate with high content of fermentation inhibitors. Bioresour Technol. 2007;98(9):1767–73.
15. Tomás-Pejó E, Ballesteros M, Oliva JM, Olsson L. Adaptation of the xylose fermenting yeast *Saccharomyces cerevisiae* F12 for improving ethanol production in different fed-batch SSF processes. J Ind Microbiol Biotechnol. 2010;37(11):1211–20.
16. Hasunuma T, Sanda T, Yamada R, Yoshimura K, Ishii J, Kondo A. Metabolic pathway engineering based on metabolomics confers acetic and formic acid tolerance to a recombinant xylose-fermenting strain of *Saccharomyces cerevisiae*. Microb Cell Factories. 2011;10(1):1–13.
17. Radecka D, Mukherjee V, Mateo RQ, Stojiljkovic M, Foulquie-Moreno MR, Thevelein JM. Looking beyond *Saccharomyces*: the potential of non-conventional yeast species for desirable traits in bioethanol fermentation. FEMS Yeast Res. 2015;15(6).
18. Sánchez OJ, Cardona CA. Trends in biotechnological production of fuel ethanol from different feedstocks. Bioresour Technol. 2008;99(13):5270–95.
19. Ballesteros M, Sáez F, Ballesteros I, Manzanares P, Negro MJ, Martínez JM, Castañeda R, Oliva JM. Ethanol production from the organic fraction obtained after thermal pretreatment of municipal solid waste. Appl Biochem Biotechnol. 2010;161(1–8):423–31.
20. Limayem A, Ricke SC. Lignocellulosic biomass for bioethanol production: current perspectives, potential issues and future prospects. Prog Energy Combust Sci. 2012;38(4):449–67.
21. Milledge J, Smith B, Dyer P, Harvey P. Macroalgae-derived biofuel: a review of methods of energy extraction from seaweed biomass. Energies. 2014;7(11):7194–222.
22. Pauly M, Keegstra K. Plant cell wall polymers as precursors for biofuels. Curr Opin Plant Biol. 2010;13(3):305–12.
23. Loque D, Scheller HV, Pauly M. Engineering of plant cell walls for enhanced biofuel production. Curr Opin Plant Biol. 2015;25:151–61.
24. Yang F, Mitra P, Zhang L, Prak L, Verhertbruggen Y, Kim JS, Sun L, Zheng K, Tang K, Auer M, Scheller HV, Loque D. Engineering secondary cell wall deposition in plants. Plant Biotechnol J. 2013;11(3):325–35.
25. Wilkerson CG, Mansfield SD, Lu F, Withers S, Park J-Y, Karlen SD, Gonzales-Vigil E, Padmakshan D, Unda F, Rencoret J, Ralph J. Monolignol ferulate transferase introduces chemically labile linkages into the lignin backbone. Science. 2014;344(6179):90–3.
26. Schultink A, Naylor D, Dama M, Pauly M. The role of the plant-specific ALTERED XYLOGLUCAN9 protein in Arabidopsis cell wall polysaccharide *O*-acetylation. Plant Physiol. 2015;167(4):1271–83.

27. Bartley LE, Peck ML, Kim SR, Ebert B, Manisseri C, Chiniquy DM, Sykes R, Gao L, Rautengarten C, Vega-Sanchez ME, Benke PI, Canlas PE, Cao P, Brewer S, Lin F, Smith WL, Zhang X, Keasling JD, Jentoff RE, Foster SB, Zhou J, Ziebell A, An G, Scheller HV, Ronald PC. Overexpression of a BAHD acyltransferase, *OsAt10*, alters rice cell wall hydroxycinnamic acid content and saccharification. Plant Physiol. 2013;161(4):1615–33.
28. Coleman HD, Yan J, Mansfield SD. Sucrose synthase affects carbon partitioning to increase cellulose production and altered cell wall ultrastructure. Proc Natl Acad Sci U S A. 2009;106(31):13118–23.
29. Joshi CP, Thammannagowda S, Fujino T, Gou JQ, Avci U, Haigler CH, McDonnell LM, Mansfield SD, Mengesha B, Carpita NC, Harris D, Debolt S, Peter GF. Perturbation of wood cellulose synthesis causes pleiotropic effects in transgenic aspen. Mol Plant. 2011;4(2):331–45.
30. Nguyen THM, Vu VH. Bioethanol production from marine algae biomass: prospect and troubles. J Viet Env. 2012;3(1):25–9.
31. Mosier N, Wyman C, Dale B, Elander R, Lee YY, Holtzapple M, Ladisch M. Features of promising technologies for pretreatment of lignocellulosic biomass. Bioresour Technol. 2005;96(6):673–86.
32. Sun S, Sun S, Cao X, Sun R. The role of pretreatment in improving the enzymatic hydrolysis of lignocellulosic materials. Bioresour Technol. 2016;199:49–58.
33. Pan X, Xie D, Gilkes N, Gregg DJ, Saddler JN. Strategies to enhance the enzymatic hydrolysis of pretreated softwood with high residual lignin content. Appl Biochem Biotechnol. 2005;121-124:1069–79.
34. Mansfield SD, Mooney C, Saddler JN. Substrate and enzyme characteristics that limit cellulose hydrolysis. Biotechnol Prog. 1999;15(5):804–16.
35. Chandra RP, Bura R, Mabee WE, Berlin A, Pan X, Saddler JN. Substrate pretreatment: the key to effective enzymatic hydrolysis of lignocellulosics? Adv Biochem Eng Biotechnol. 2007;108:67–93.
36. Mussatto SI, Dragone GM. Biomass pretreatment, biorefineries, and potential products for a bioeconomy development. In: Mussato SI, editor. Biomass fractionation technologies for a lignocellulosic feedstock based biorefinery. Amsterdam: Elsevier; 2016. p. 1–22.
37. Viikari L, Vehmaanperä J, Koivula A. Lignocellulosic ethanol: from science to industry. Biomass Bioenergy. 2012;46:13–24.
38. Alvira P, Ballesteros M, Negro MJ. Progress on enzymatic saccharification technologies for biofuels production. In: Gupta VK, Tuohy MG, editors. Biofuel technologies: recent developments. Berlin/Heidelberg: Springer; 2013. p. 145–69.
39. Martínez AT. How to break down crystalline cellulose. Science. 2016;352(6289):1050–1.
40. Jørgensen H, Kristensen JB, Felby C. Enzymatic conversion of lignocellulose into fermentable sugars: challenges and opportunities. Biofuels Bioprod Biorefin. 2007;1(2):119–34.
41. Taherzadeh MJ, Karimi K. Pretreatment of lignocellulosic wastes to improve ethanol and biogas production: a review. Int J Mol Sci. 2008;9(9):1621–51.
42. Duque A, Manzanares P, Ballesteros I, Negro MJ, Oliva JM, Gonzalez A, Ballesteros M. Sugar production from barley straw biomass pretreated by combined alkali and enzymatic extrusion. Bioresour Technol. 2014;158:262–8.
43. Saha BC, Iten LB, Cotta MA, Wu YV. Dilute acid pretreatment, enzymatic saccharification and fermentation of wheat straw to ethanol. Process Biochem. 2005;40(12):3693–700.
44. Carvalheiro F, Duarte LC, Gírio FM. Hemicellulose biorefineries: a review on biomass pretreatments. J Sci Ind Res. 2008;67(11):849–64.
45. Kumar P, Barrett DM, Delwiche MJ, Stroeve P. Methods for pretreatment of lignocellulosic biomass for efficient hydrolysis and biofuel production. Ind Eng Chem Res. 2009;48(8):3713–29.
46. Zhang YH, Ding SY, Mielenz JR, Cui JB, Elander RT, Laser M, Himmel ME, McMillan JR, Lynd LR. Fractionating recalcitrant lignocellulose at modest reaction conditions. Biotechnol Bioeng. 2007;97(2):214–23.

47. Zhao X, Cheng K, Liu D. Organosolv pretreatment of lignocellulosic biomass for enzymatic hydrolysis. Appl Microbiol Biotechnol. 2009;82(5):815–27.
48. Nguyen TY, Cai CM, Osman O, Kumar R, Wyman CE. CELF pretreatment of corn stover boosts ethanol titers and yields from high solids SSF with low enzyme loadings. Green Chem. 2016;18(6):1581–9.
49. Brandt A, Gräsvik J, Hallett JP, Welton T. Deconstruction of lignocellulosic biomass with ionic liquids. Green Chem. 2013;15(3):550.
50. Akhtar N, Gupta K, Goyal D, Goyal A. Recent advances in pretreatment technologies for efficient hydrolysis of lignocellulosic biomass. Environ Prog Sustain Energy. 2016;35(2):489–511.
51. Duarte CL, Ribeiro MA, Oikawa H, Mori MN, Napolitano CM, Galvão CA. Electron beam combined with hydrothermal treatment for enhancing the enzymatic convertibility of sugarcane bagasse. Radiat Phys Chem. 2012;81(8):1008–11.
52. Moretti MMS, Bocchini-Martins DA, Nunes CCC, Villena MA, Perrone OM, da Silva R, Boscolo M, Gomes E. Pretreatment of sugarcane bagasse with microwaves irradiation and its effects on the structure and on enzymatic hydrolysis. Appl Energy. 2014;122:189–95.
53. Kootstra AMJ, Beeftink HH, Scott EL, Sanders JPM. Comparison of dilute mineral and organic acid pretreatment for enzymatic hydrolysis of wheat straw. Biochem Eng J. 2009;46(2):126–31.
54. Bhalla A, Bansal N, Stoklosa RJ, Fountain M, Ralph J, Hodge DB, Hegg EL. Effective alkaline metal-catalyzed oxidative delignification of hybrid poplar. Biotechnol Biofuels. 2016;9:34.
55. Socha AM, Parthasarathi R, Shi J, Pattathil S, Whyte D, Bergeron M, George A, Tran K, Stavila V, Venkatachalam S, Hahn MG, Simmons BA, Singh S. Efficient biomass pretreatment using ionic liquids derived from lignin and hemicellulose. Proc Natl Acad Sci U S A. 2014;111(35):E3587–95.
56. Ruiz HA, Rodríguez-Jasso RM, Fernandes BD, Vicente AA, Teixeira JA. Hydrothermal processing, as an alternative for upgrading agriculture residues and marine biomass according to the biorefinery concept: a review. Renew Sust Energ Rev. 2013;21:35–51.
57. Pan X, Arato C, Gilkes N, Gregg D, Mabee W, Pye K, Xiao Z, Zhang X, Saddler J. Biorefining of softwoods using ethanol organosolv pulping: preliminary evaluation of process streams for manufacture of fuel-grade ethanol and co-products. Biotechnol Bioeng. 2005;90(4):473–81.
58. Ballesteros I, Negro MJ, Oliva JM, Cabañas A, Manzanares P, Ballesteros M. Ethanol production from steam-explosion pretreated wheat straw. Appl Biochem Biotechnol. 2006;129-132:496–508.
59. Cara C, Ruiz E, Ballesteros M, Manzanares P, Negro MJ, Castro E. Production of fuel ethanol from steam-explosion pretreated olive tree pruning. Fuel. 2008;87(6):692–700.
60. Oliva JM, Sáez F, Ballesteros I, González A, Negro MJ, Manzanares P, Ballesteros M. Effect of lignocellulosic degradation compounds from steam explosion pretreatment on ethanol fermentation by thermotolerant yeast *Kluyveromyces marxianus*. Appl Biochem Biotechnol. 2003;105-108:141–53.
61. Galbe M, Zacchi G. A review of the production of ethanol from softwood. Appl Microbiol Biotechnol. 2002;59(6):618–28.
62. Laureano-Perez L, Teymouri F, Alizadeh H, Dale BE. Understanding factors that limit enzymatic hydrolysis of biomass. Appl Biochem Biotechnol. 2005;124(1):1081–99.
63. Mendonça RT, Jara JF, González V, Elissetche JP, Freer J. Evaluation of the white-rot fungi *Ganoderma australe* and *Ceriporiopsis subvermispora* in biotechnological applications. J Ind Microbiol Biotechnol. 2008;35(11):1323–30.
64. Salvachúa D, Prieto A, López-Abelairas M, Lu-Chau T, Martínez AT, Martínez MJ. Fungal pretreatment: an alternative in second-generation ethanol from wheat straw. Bioresour Technol. 2011;102(16):7500–6.

65. Gutierrez A, Rencoret J, Cadena EM, Rico A, Barth D, del Rio JC, Martinez AT. Demonstration of laccase-based removal of lignin from wood and non-wood plant feedstocks. Bioresour Technol. 2012;119:114–22.
66. Martín-Sampedro R, Fillat U, Ibarra D, Eugenio ME. Towards the improvement of *Eucalyptus globulus* chemical and mechanical pulping using endophytic fungi. Int Biodeterior Biodegradation. 2015;105:120–6.
67. Martín-Sampedro R, Fillat U, Ibarra D, Eugenio ME. Use of new endophytic fungi as pretreatment to enhance enzymatic saccharification of *Eucalyptus globulus*. Bioresour Technol. 2015;196:383–90.
68. Rencoret J, Pereira A, del Río JC, Martínez AT, Gutiérrez A. Laccase-mediator pretreatment of wheat straw degrades lignin and improves saccharification. Bioenergy Res. 2016;9(3):917–30.
69. Sánchez C. Lignocellulosic residues: biodegradation and bioconversion by fungi. Biotechnol Adv. 2009;27(2):185–94.
70. Palmqvist E, Hahn-Hägerdal B. Fermentation of lignocellulosic hydrolysates. II: inhibitors and mechanisms of inhibition. Bioresour Technol. 2000;74(1):25–33.
71. Panagiotou G, Olsson L. Effect of compounds released during pretreatment of wheat straw on microbial growth and enzymatic hydrolysis rates. Biotechnol Bioeng. 2007;96(2):250–8.
72. Ximenes E, Kim Y, Mosier N, Dien B, Ladisch M. Deactivation of cellulases by phenols. Enzym Microb Technol. 2011;48(1):54–60.
73. Jonsson LJ, Martín C. Pretreatment of lignocellulose: formation of inhibitory by-products and strategies for minimizing their effects. Bioresour Technol. 2016;199:103–12.
74. Mayer AM, Staples RC. Laccase: new functions for an old enzyme. Phytochemistry. 2002;60(6):551–65.
75. Bourbonnais R, Paice MG. Oxidation of non-phenolic substrates. FEBS Lett. 1990;267(1):99–102.
76. García-Aparicio MP, Ballesteros I, González A, Oliva JM, Ballesteros M, Negro MJ. Effect of inhibitors released during steam-explosion pretreatment of barley straw on enzymatic hydrolysis. Appl Biochem Biotechnol. 2006;129(1):278–88.
77. Larsson S, Reimann A, Nilvebrant N-O, Jönsson LJ. Comparison of different methods for the detoxification of lignocellulose hydrolyzates of spruce. Appl Biochem Biotechnol. 1999;77(1):91–103.
78. Fargues C, Lewandowski R, Lameloise M-L. Evaluation of ion-exchange and adsorbent resins for the detoxification of beet distillery effluents. Ind Eng Chem Res. 2010;49(19):9248–57.
79. Rodrigues RCLB, Felipe MGA, Almeida e Silva JB, Vitolo M, Gómez PV. The influence of pH, temperature and hydrolyzate concentration on the removal of volatile and nonvolatile compounds from sugarcane bagasse hemicellulosic hydrolyzate treated with activated charcoal before or after vacuum evaporation. Braz J Chem Eng. 2001;18:299–311.
80. Wilson JJ, Deschatelets L, Nishikawa NK. Comparative fermentability of enzymatic and acid hydrolysates of steam-pretreated aspenwood hemicellulose by *Pichia stipitis* CBS 5776. Appl Microbiol Biotechnol. 1989;31(5):592–6.
81. Parawira W, Tekere M. Biotechnological strategies to overcome inhibitors in lignocellulose hydrolysates for ethanol production: review. Crit Rev Biotechnol. 2011;31(1):20–31.
82. Palmqvist E, Hahn-Hägerdal B, Szengyel Z, Zacchi G, Réczey K. Simultaneous detoxification and enzyme production of hemicellulose hydrolysates obtained after steam pretreatment. Enzym Microb Technol. 1997;20(4):286–93.
83. Alvira P, Moreno AD, Ibarra D, Sáez F, Ballesteros M. Improving the fermentation performance of *Saccharomyces cerevisiae* by laccase during ethanol production from steam-exploded wheat straw at high-substrate loadings. Biotechnol Prog. 2013;29(1):74–82.
84. Jurado M, Prieto A, Martínez-Alcalá A, Martínez AT, Martínez MJ. Laccase detoxification of steam-exploded wheat straw for second generation bioethanol. Bioresour Technol. 2009;100(24):6378–84.

85. Moreno AD, Ibarra D, Mialon A, Ballesteros M. A bacterial laccase for enhancing saccharification and ethanol fermentation of steam-pretreated biomass. Fermentation. 2016;2:11.
86. Moreno AD, Ibarra D, Alvira P, Tomás-Pejó E, Ballesteros M. Exploring laccase and mediators behavior during saccharification and fermentation of steam-exploded wheat straw for bioethanol production. J Chem Technol Biotechnol. 2016;91(6):1816–25.
87. Tomás-Pejó E, Oliva JM, Ballesteros M, Olsson L. Comparison of SHF and SSF processes from steam-exploded wheat straw for ethanol production by xylose-fermenting and robust glucose-fermenting *Saccharomyces cerevisiae* strains. Biotechnol Bioeng. 2008;100(6):1122–31.
88. Shao X, Jin M, Guseva A, Liu C, Balan V, Hogsett D, Dale BE, Lynd L. Conversion for avicel and AFEX pretreated corn stover by *Clostridium thermocellum* and simultaneous saccharification and fermentation: insights into microbial conversion of pretreated cellulosic biomass. Bioresour Technol. 2011;102(17):8040–5.
89. Jin M, Gunawan C, Balan V, Dale BE. Consolidated bioprocessing (CBP) of AFEX-pretreated corn stover for ethanol production using *Clostridium phytofermentans* at a high solids loading. Biotechnol Bioeng. 2012;109(8):1929–36.
90. Yee KL, Rodriguez Jr M, Thompson OA, Fu C, Wang Z-Y, Davison BH, Mielenz JR. Consolidated bioprocessing of transgenic switchgrass by an engineered and evolved *Clostridium thermocellum* strain. Biotechnol Biofuels. 2014;7:75.
91. Xiao Z, Zhang X, Gregg DJ, Saddler JN. Effects of sugar inhibition on cellulases and β-glucosidase during enzymatic hydrolysis of softwood substrates. Appl Biochem Biotechnol. 2004;115(1):1115–26.
92. Qing Q, Yang B, Wyman CE. Xylooligomers are strong inhibitors of cellulose hydrolysis by enzymes. Bioresour Technol. 2010;101(24):9624–30.
93. Andric P, Meyer AS, Jensen PA, Dam-Johansen K. Reactor design for minimizing product inhibition during enzymatic lignocellulose hydrolysis: I. Significance and mechanism of cellobiose and glucose inhibition on cellulolytic enzymes. Biotechnol Adv. 2010;28(3):308–24.
94. Wingren A, Galbe M, Zacchi G. Techno-Economic Evaluation of producing ethanol from softwood: comparison of SSF and SHF and identification of bottlenecks. Biotechnol Prog. 2003;19(4):1109–17.
95. Alfani F, Gallifuoco A, Saporosi A, Spera A, Cantarella M. Comparison of SHF and SSF processes for the bioconversion of steam-exploded wheat straw. J Ind Microbiol Biotechnol. 2000;25(4):184–92.
96. Zhang B, Zhang J, Wang D, Han R, Ding R, Gao X, Sun L, Hong J. Simultaneous fermentation of glucose and xylose at elevated temperatures co-produces ethanol and xylitol through overexpression of a xylose-specific transporter in engineered *Kluyveromyces marxianus*. Bioresour Technol. 2016;216:227–37.
97. Moreno AD, Ibarra D, Ballesteros I, Fernández JL, Ballesteros M. Ethanol from laccase-detoxified lignocellulose by the thermotolerant yeast *Kluyveromyces marxianus*-Effects of steam pretreatment conditions, process configurations and substrate loadings. Biochem Eng J. 2013;79:94–103.
98. Abdel-Banat BM, Hoshida H, Ano A, Nonklang S, Akada R. High-temperature fermentation: how can processes for ethanol production at high temperatures become superior to the traditional process using mesophilic yeast? Appl Microbiol Biotechnol. 2010;85(4):861–7.
99. Olsson L, Soerensen HR, Dam BP, Christensen H, Krogh KM, Meyer AS. Separate and simultaneous enzymatic hydrolysis and fermentation of wheat hemicellulose with recombinant xylose utilizing *Saccharomyces cerevisiae*. Appl Biochem Biotechnol. 2006;129-132:117–29.
100. Kumar R, Tabatabaei M, Karimi K, Sárvári Horváth I. Recent updates on lignocellulosic biomass derived ethanol - A review. Biofuel Research Journal. 2016;3(1):347–56.

101. Cannella D, Jørgensen H. Do new cellulolytic enzyme preparations affect the industrial strategies for high solids lignocellulosic ethanol production? Biotechnol Bioeng. 2014;111:59–68.
102. Lynd LR, van Zyl WH, McBride JE, Laser M. Consolidated bioprocessing of cellulosic biomass: an update. Curr Opin Biotechnol. 2005;16(5):577–83.
103. Bayer EA, Belaich JP, Shoham Y, Lamed R. The cellulosomes: multienzyme machines for degradation of plant cell wall polysaccharides. Annu Rev Microbiol. 2004;58:521–54.
104. Shaw AJ, Podkaminer KK, Desai SG, Bardsley JS, Rogers SR, Thorne PG, Hogsett DA, Lynd LR. Metabolic engineering of a thermophilic bacterium to produce ethanol at high yield. Proc Natl Acad Sci U S A. 2008;105(37):13769–74.
105. Bokinsky G, Peralta-Yahya PP, George A, Holmes BM, Steen EJ, Dietrich J, Lee TS, Tullman-Ercek D, Voigt CA, Simmons BA, Keasling JD. Synthesis of three advanced biofuels from ionic liquid-pretreated switchgrass using engineered *Escherichia coli*. Proc Natl Acad Sci U S A. 2011;108(50):19949–54.
106. Chung D, Cha M, Guss AM, Westpheling J. Direct conversion of plant biomass to ethanol by engineered *Caldicellulosiruptor bescii*. Proc Natl Acad Sci U S A. 2014;111(24):8931–6.
107. van Zyl WH, Lynd LR, den Haan R, McBride JE. Consolidated bioprocessing for bioethanol production using *Saccharomyces cerevisiae*. In: Olsson L, editor. Biofuels. Berlin/Heidelberg: Springer; 2007. p. 205–35.
108. Matano Y, Hasunuma T, Kondo A. Cell recycle batch fermentation of high-solid lignocellulose using a recombinant cellulase-displaying yeast strain for high yield ethanol production in consolidated bioprocessing. Bioresour Technol. 2013;135:403–9.
109. Yamada R, Hasunuma T, Kondo A. Endowing non-cellulolytic microorganisms with cellulolytic activity aiming for consolidated bioprocessing. Biotechnol Adv. 2013;31(6):754–63.
110. Guo Z, Duquesne S, Bozonnet S, Cioci G, Nicaud JM, Marty A, O'Donohue MJ. Development of cellobiose-degrading ability in *Yarrowia lipolytica* strain by overexpression of endogenous genes. Biotechnol Biofuels. 2015;8:109.
111. Balat M. Production of bioethanol from lignocellulosic materials via the biochemical pathway: a review. Energ Convers Manage. 2011;52(2):858–75.
112. Koppram R, Olsson L. Combined substrate, enzyme and yeast feed in simultaneous saccharification and fermentation allow bioethanol production from pretreated spruce biomass at high solids loadings. Biotechnol Biofuels. 2014;7:54.
113. Wang R, Unrean P, Franzén CJ. Model-based optimization and scale-up of multi-feed simultaneous saccharification and co-fermentation of steam pre-treated lignocellulose enables high gravity ethanol production. Biotechnol Biofuels. 2016;9:88.
114. Shen J, Agblevor FA. Ethanol production of semi-simultaneous saccharification and fermentation from mixture of cotton gin waste and recycled paper sludge. Bioprocess Biosyst Eng. 2011;34(1):33–43.
115. Gonçalves FA, Ruiz HA, Silvino dos Santos E, Teixeira JA, de Macedo GR. Bioethanol production by *Saccharomyces cerevisiae*, *Pichia stipitis* and *Zymomonas mobilis* from delignified coconut fibre mature and lignin extraction according to biorefinery concept. Renew Energy. 2016;94:353–65.
116. Liu K, Zhang J, Bao J. Two stage hydrolysis of corn stover at high solids content for mixing power saving and scale-up applications. Bioresour Technol. 2015;196:716–20.
117. Westman JO, Bonander N, Taherzadeh MJ, Franzén CJ. Improved sugar co-utilisation by encapsulation of a recombinant *Saccharomyces cerevisiae* strain in alginate-chitosan capsules. Biotechnol Biofuels. 2014;7:102.
118. Ylitervo P, Franzén CJ, Taherzadeh MJ. Continuous ethanol production with a membrane bioreactor at high acetic acid concentrations. Membranes. 2014;4(3):372–87.
119. Lin Y, Tanaka S. Ethanol fermentation from biomass resources: current state and prospects. Appl Microbiol Biotechnol. 2006;69(6):627–42.
120. Ostergaard S, Olsson L, Nielsen J. Metabolic engineering of *Saccharomyces cerevisiae*. Microbiol Mol Biol R. 2000;64(1):34–50.

121. Dussán KJ, Silva DDV, Perez VH, da Silva SS. Evaluation of oxygen availability on ethanol production from sugarcane bagasse hydrolysate in a batch bioreactor using two strains of xylose-fermenting yeast. Renew Energy. 2016;87:703–10.
122. Long TM, Su YK, Headman J, Higbee A, Willis LB, Jeffries TW. Cofermentation of glucose, xylose, and cellobiose by the beetle-associated yeast *Spathaspora passalidarum*. Appl Environ Microbiol. 2012;78(16):5492–500.
123. Krahulec S, Klimacek M, Nidetzky B. Analysis and prediction of the physiological effects of altered coenzyme specificity in xylose reductase and xylitol dehydrogenase during xylose fermentation by *Saccharomyces cerevisiae*. J Biotechnol. 2012;158(4):192–202.
124. Petschacher B, Nidetzky B. Altering the coenzyme preference of xylose reductase to favor utilization of NADH enhances ethanol yield from xylose in a metabolically engineered strain of *Saccharomyces cerevisiae*. Microb Cell Factories. 2008;7:9.
125. Karhumaa K, García Sánchez R, Hahn-Hägerdal B, Gorwa-Grauslund MF. Comparison of the xylose reductase-xylitol dehydrogenase and the xylose isomerase pathways for xylose fermentation by recombinant *Saccharomyces cerevisiae*. Microb Cell Factories. 2007;6:5.
126. Gonçalves DL, Matsushika A, de Sales BB, Goshima T, Bon EP, Stambuk BU. Xylose and xylose/glucose co-fermentation by recombinant *Saccharomyces cerevisiae* strains expressing individual hexose transporters. Enzym Microb Technol. 2014;63:13–20.
127. Kricka W, Fitzpatrick J, Bond U. Challenges for the production of bioethanol from biomass using recombinant yeasts. In: Sariaslani S, Gadd GM, editors. Advances in applied microbiology. Amsterdam: Academic Press; 2015. p. 89–125.
128. Runquist D, Hahn-Hägerdal B, Radstrom P. Comparison of heterologous xylose transporters in recombinant *Saccharomyces cerevisiae*. Biotechnol Biofuels. 2010;3:5.
129. Koppram R, Albers E, Olsson L. Evolutionary engineering strategies to enhance tolerance of xylose utilizing recombinant yeast to inhibitors derived from spruce biomass. Biotechnol Biofuels. 2012;5:32.
130. Nielsen F, Tomás-Pejó E, Olsson L, Wallberg O. Short-term adaptation during propagation improves the performance of xylose-fermenting *Saccharomyces cerevisiae* in simultaneous saccharification and co-fermentation. Biotechnol Biofuels. 2015;8:219.
131. Tomás-Pejó E, Olsson L. Influence of the propagation strategy for obtaining robust *Saccharomyces cerevisiae* cells that efficiently co-ferment xylose and glucose in lignocellulosic hydrolysates. Microb Biotechnol. 2015;8(6):999–1005.
132. Komeda H, Yamasaki-Yashiki S, Hoshino K, Asano Y. Identification and characterization of D-xylulokinase from the D-xylose-fermenting fungus, *Mucor circinelloides*. FEMS Microbiol Lett. 2014;360(1):51–61.
133. Millati R, Edebo L, Taherzadeh MJ. Performance of *Rhizopus*, *Rhizomucor*, and *Mucor* in ethanol production from glucose, xylose, and wood hydrolyzates. Enzym Microb Technol. 2005;36(2–3):294–300.
134. Hahn-Hägerdal B, Galbe M, Gorwa-Grauslund MF, Lidén G, Zacchi G. Bio-ethanol – the fuel of tomorrow from the residues of today. Trends Biotechnol. 2006;24(12):549–56.
135. Gorsich SW, Dien BS, Nichols NN, Slininger PJ, Liu ZL, Skory CD. Tolerance to furfural-induced stress is associated with pentose phosphate pathway genes *ZWF1*, *GND1*, *RPE1*, and *TKL1* in *Saccharomyces cerevisiae*. Appl Microbiol Biotechnol. 2006;71(3):339–49.
136. Pereira FB, Guimarães PMR, Gomes DG, Mira NP, Teixeira MC, Sá-Correia I, Domingues L. Identification of candidate genes for yeast engineering to improve bioethanol production in very high gravity and lignocellulosic biomass industrial fermentations. Biotechnol Biofuels. 2011;4:57.
137. Xiao H, Zhao H. Genome-wide RNAi screen reveals the E3 SUMO-protein ligase gene *SIZ1* as a novel determinant of furfural tolerance in *Saccharomyces cerevisiae*. Biotechnol Biofuels. 2014;7:78.
138. Lu Y, Cheng YF, He XP, Guo XN, Zhang BR. Improvement of robustness and ethanol production of ethanologenic *Saccharomyces cerevisiae* under co-stress of heat and inhibitors. J Ind Microbiol Biotechnol. 2012;39(1):73–80.

139. Çakar ZP, Turanli-Yildiz B, Alkim C, Yilmaz U. Evolutionary engineering of *Saccharomyces cerevisiae* for improved industrially important properties. FEMS Yeast Res. 2012;12(2):171–82.
140. Alkasrawi M, Rudolf A, Lidén G, Zacchi G. Influence of strain and cultivation procedure on the performance of simultaneous saccharification and fermentation of steam pretreated spruce. Enzym Microb Technol. 2006;38(1–2):279–86.
141. Ruyters S, Mukherjee V, Verstrepen KJ, Thevelein JM, Willems KA, Lievens B. Assessing the potential of wild yeasts for bioethanol production. J Ind Microbiol Biotechnol. 2015;42(1):39–48.
142. Dandi ND, Dandi BN, Chaudhari AB. Bioprospecting of thermo- and osmo-tolerant fungi from mango pulp-peel compost for bioethanol production. Antonie Van Leeuwenhoek. 2013;103(4):723–36.
143. Lindberg L, Santos AX, Riezman H, Olsson L, Bettiga M. Lipidomic profiling of *Saccharomyces cerevisiae* and *Zygosaccharomyces bailii* reveals critical changes in lipid composition in response to acetic acid stress. PLoS One. 2013;8(9):e73936.
144. Echeverrigaray S, Randon M, da Silva K, Zacaria J, Delamare AP. Identification and characterization of non-*Saccharomyces* spoilage yeasts isolated from Brazilian wines. World J Microbiol Biotechnol. 2013;29(6):1019–27.
145. Tomás-Pejó E, Oliva JM, González A, Ballesteros I, Ballesteros M. Bioethanol production from wheat straw by the thermotolerant yeast *Kluyveromyces marxianus* CECT 10875 in a simultaneous saccharification and fermentation fed-batch process. Fuel. 2009;88(11):2142–7.
146. García-Aparicio MP, Oliva JM, Manzanares P, Ballesteros M, Ballesteros I, González A, Negro MJ. Second-generation ethanol production from steam exploded barley straw by *Kluyveromyces marxianus* CECT 10875. Fuel. 2011;90(4):1624–30.
147. Ballesteros M, Oliva JM, Negro MJ, Manzanares P, Ballesteros I. Ethanol from lignocellulosic materials by a simultaneous saccharification and fermentation process (SFS) with *Kluyveromyces marxianus* CECT 10875. Process Biochem. 2004;39(12):1843–8.
148. Faga BA, Wilkins MR, Banat IM. Ethanol production through simultaneous saccharification and fermentation of switchgrass using *Saccharomyces cerevisiae* $D_5A$ and thermotolerant *Kluyveromyces marxianus* IMB strains. Bioresour Technol. 2010;101(7):2273–9.
149. Ballesteros M, Oliva JM, Manzanares P, Negro MJ, Ballesteros I. Ethanol production from paper material using a simultaneous saccharification and fermentation system in a fed-batch basis. World J Microbiol Biotechnol. 2002;18(6):559–61.
150. Kádár Z, Szengyel Z, Réczey K. Simultaneous saccharification and fermentation (SSF) of industrial wastes for the production of ethanol. Ind Crop Prod. 2004;20(1):103–10.
151. Edgardo A, Carolina P, Manuel R, Juanita F, Baeza J. Selection of thermotolerant yeast strains *Saccharomyces cerevisiae* for bioethanol production. Enzym Microb Technol. 2008;43(2):120–3.
152. Hari Krishna S, Janardhan Reddy T, Chowdary GV. Simultaneous saccharification and fermentation of lignocellulosic wastes to ethanol using a thermotolerant yeast. Bioresour Technol. 2001;77(2):193–6.
153. Scully S, Orlygsson J. Recent advances in second generation ethanol production by thermophilic bacteria. Energies. 2014;8(1):1–30.
154. Hild HM, Stuckey DC, Leak DJ. Effect of nutrient limitation on product formation during continuous fermentation of xylose with *Thermoanaerobacter ethanolicus* JW200 Fe(7). Appl Microbiol Biotechnol. 2003;60(6):679–86.
155. Jessen JE, Orlygsson J. Production of ethanol from sugars and lignocellulosic biomass by *Thermoanaerobacter* J1 isolated from a hot spring in Iceland. J Biomed Biotechnol. 2012;2012:186982.
156. Moshi AP, Hosea KM, Elisante E, Mamo G, Mattiasson B. High temperature simultaneous saccharification and fermentation of starch from inedible wild cassava (*Manihot glaziovii*) to bioethanol using *Caloramator boliviensis*. Bioresour Technol. 2015;180:128–36.

157. Koppram R, Tomás-Pejó E, Xiros C, Olsson L. Lignocellulosic ethanol production at high-gravity: challenges and perspectives. Trends Biotechnol. 2014;32(1):46–53.
158. Saito H, Posas F. Response to hyperosmotic stress. Genetics. 2012;192(2):289–318.
159. Martorell P, Stratford M, Steels H, Fernandez-Espinar MT, Querol A. Physiological characterization of spoilage strains of *Zygosaccharomyces bailii* and *Zygosaccharomyces rouxii* isolated from high sugar environments. Int J Food Microbiol. 2007;114(2):234–42.
160. Leandro MJ, Sychrova H, Prista C, Loureiro-Dias MC. The osmotolerant fructophilic yeast *Zygosaccharomyces rouxii* employs two plasma-membrane fructose uptake systems belonging to a new family of yeast sugar transporters. Microbiology. 2011;157(Pt 2):601–8.
161. Saha BC, Nichols NN, Cotta MA. Ethanol production from wheat straw by recombinant *Escherichia coli* strain FBR5 at high solid loading. Bioresour Technol. 2011;102(23):10892–7.

# Part V
# Production of Lactones and Amino Acids

# Chapter 13
# Production of γ-Valerolactone from Biomass

**Kai Yan and Huixia Luo**

**Abstract** This chapter surveys the methodology and recent advances in the production of γ-valerolactone (GVL) from different renewable biomass-derived sources, from the pioneering studies to the present state of the art. The mechanism for the production of GVL is discussed. An overview of the different methods and advances in the synthesis of GVL are then analyzed and compared. Different advanced catalysts (e.g., homogeneous as well as heterogeneous catalysts) and catalytic systems (e.g., different hydrogen sources) are highlighted. More emphasis is placed on a comparative analysis of recently developed methods for GVL production in terms of efficiency, selectivity and cost-effectiveness. Challenges and areas that need improvement are also given. Specific examples are reviewed with emphasis on different synthetic systems, comparing the behavior of different metal catalysts and reaction parameters. The main prospects and constraints related to the several conversion routes are presented including the technical barriers, promise of scale-up, and potentially environmental issues.

**Keywords** Biomass • Gamma-valerolactone • Production • Catalysts • Catalytic system • Activity

## 13.1 Introduction

Biomass often refers to the organic material of renewable sources [1, 2], it is the only renewable source of carbon neutral which can be used for the production of fuels and value-added chemicals as an alternative to reduce the use of fossil resources [3–6]. Figure 13.1 describes the typically chemical structures of the major component of biomass, hemicellulose, cellulose and lignin [2, 7]. Lignocellulosic biomass mainly contributes to the structural integrity of plants (as shown in Fig. 13.2), which consists of three polymeric components, the percentages have

K. Yan (✉)
School of Engineering, Brown University, 182 Hope Street, Providence, RI 02912, USA
e-mail: kai_yan@brown.edu

H. Luo (✉)
Department of Chemistry, Princeton University, Princeton, NJ 08544, USA
e-mail: huixial@princeton.edu



Z. Fang et al. (eds.), *Production of Platform Chemicals from Sustainable Resources*, Biofuels and Biorefineries 7, DOI 10.1007/978-981-10-4172-3_13

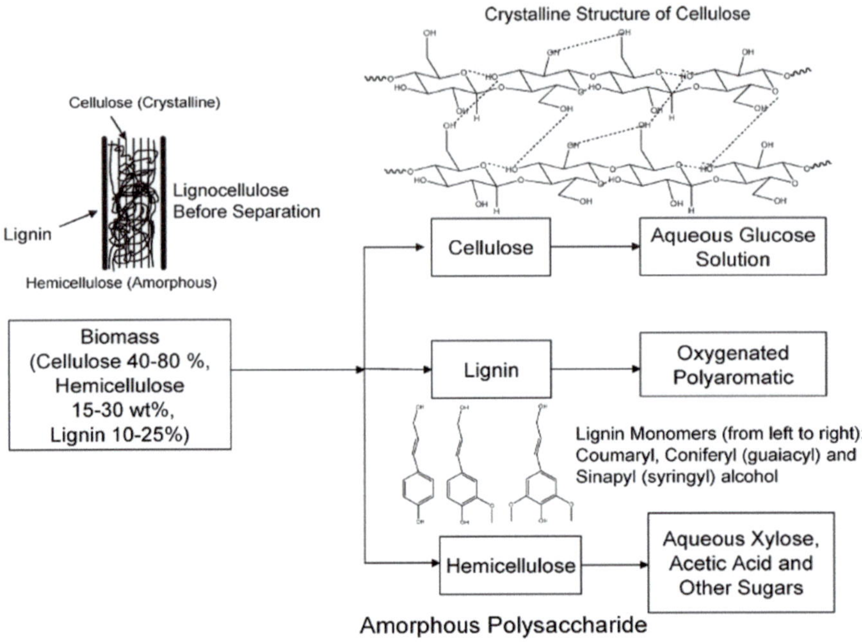


**Fig. 13.1** The chemical structure of the major components of biomass and their typically upgraded products (Reprinted with permission from Ref. [2]. Copyright © 2006, American Chemical Society)

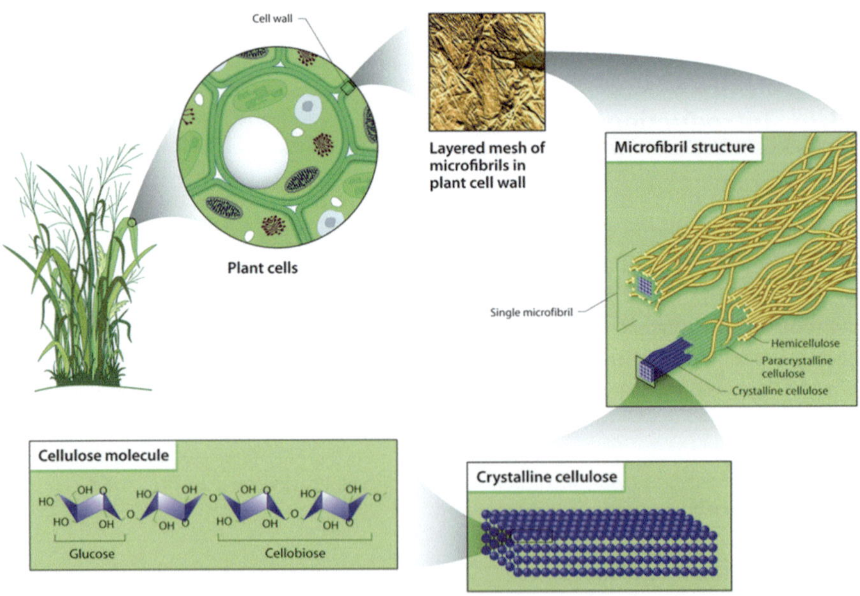


**Fig. 13.2** The structural organization of the plant cell wall (Source from the Office of Biological and Environmental Research of the U.S. Department of Energy Office of Science. science.energy.gov/ber/)

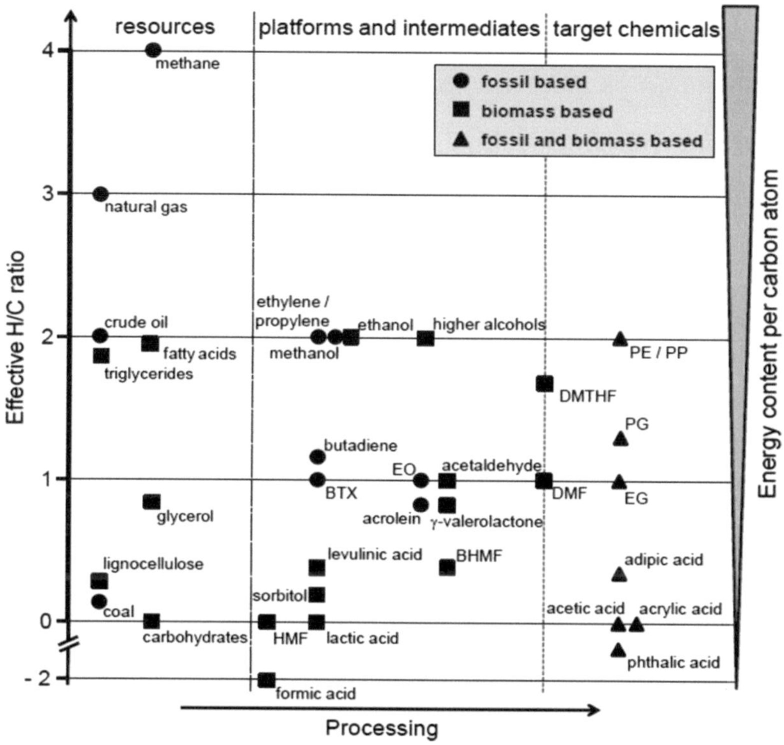


**Fig. 13.3** Effective H/C ratio vs the degree of processing. Note: *HMF* 5-hydroxymethylfurfural, *BTX* benzene, toluene, xylene, *EO* ethylene oxide, *BHMF* 2,5-bis-(hydroxymethyl)furfural, *DMF* 2,5-dimethylfuran, *DMTHF* 2,5-dimethyltetrahydrofuran, *EG* ethylene glycol, *PG* propylene glycol, *PE* polyethylene, *PP* polypropylene (Reprinted with permission from Ref. [18]. Copyright © 2015, Royal Society of Chemistry)

difference based on the species of origin and the distributed locations [8, 9]. Cellulose (up to 40–80%) is the main component in the plant cell walls (Fig. 13.2), consisting of the glucose unit linked unbranched chains [10, 11]. The parallel unbranched D-glucopyranose units linked by β-1,4-glycosidic bonds would form the crystalline and highly organized microfibrils through numerous internal and intramolecular hydrogen bonds and Van der Waals forces. Hemicellulose (between 15% and 30%) is made of C5 and C6-derived sugars through the hydrogen bonds [12, 13]. Lignin is formed by substituted phenolic units and accounts for 15–25% of biomass composition [14–17]. The investigation of energy- and chemical processes is to efficiently transform biomass into fuel candidates, value-added chemicals, and materials. Similar to the petrochemical industry, the fully integrated biorefinery produces a variety of products upon the advanced catalysts and efficient catalytic system.

Considering the transformation of renewable biomass into the highly attractive biofuels, a high effective H/C ratio has the direct relationship to high energy content per carbon (i.e. heating value or combustion enthalpy). As shown in Fig. 13.3,

**Table 13.1** Main physical properties of γ-valerolactone in comparison with ethanol fuel

| Terms | Ethanol | GVL |
|---|---|---|
| M (g $mol^{-1}$) | 46.07 | 100.12 |
| Carbon (wt %) | 52.2 | 60 |
| Hydrogen (wt %) | 13.1 | 8 |
| Oxygen (wt %) | 34.7 | 32 |
| Boiling point (°C) | 78 | 207 |
| Melting point (°C) | −114 | −31 |
| Flash Point (°C) | 13 | 96.1 |
| Density (g $mL^{-1}$) | 0.789 | 1.0485 |
| Solubility in water/ (mg/ml) | Miscible | ≥100 |
| Octane number | 108.6 | – |
| Cetane number | 5 | – |
| ΔHvap (kJ $mol^{-1}$) | 42.590 | −54.8 |
| $\Delta cH^{\circ}_{liquid}$ (kJ $mol^{-1}$) | −1367.6 ± 0.3 | −2649.6 |



different resources including renewable and non-renewable, platform intermediates, and target chemicals have been classified based on their effective H/C ratios.

Along the line of these value-added candidates, GVL is one of the most promising biofuel and valuable chemical. Due to its attractive physical-chemical properties and unique fuel characteristics as depicted in Table 13.1, it has attracted a voluminous body of work and intensive efforts over the last several decades. GVL is renewable, has a low melting point of −31 °C, a high boiling point of 207 °C and flash point of 96 °C, a definitive but acceptable smell for easy recognition of leaks and spills, low toxicity, and high solubility in water to assist biodegradation. Besides, GVL has similar combustion energy to ethanol (29.7 MJ $kg^{-1}$) and a higher energy density in comparison with ethanol fuel. A number of researchers have focused on the developing efficient catalytic system for the production of GVL from different lignocellulosic biomass-derived sources [5, 7, 20–25]. Horvath et al. [26, 27] have demonstrated that GVL neither hydrolyzed under neutral conditions nor formed measurable amounts of peroxides in a glass flask under air in weeks, making it a safe and promising material for the industrial applications. Using the criteria H/C ratio to compare the variously value-added chemicals, the effective H/C ratio of the hydrodeoxygenation of carbohydrates was 0, the H/C ratio of levulinic acid (LA) was 0.4, and the ratio of GVL was 0.8. The combustion enthalpies per carbon atom for methane and octane are 890.4 and 683.8 kJ/mol, respectively, compared to 485.4 kJ/mol for LA and 463.5 kJ/mol for 5-hydroxymethylfurfural (HMF) [18]. Horvath et al. [26, 27] further studied the mixture of 90% (v/v) conventional gasoline with 10% GVL and 10% ethanol, having particularly the same octane number and similar physical properties [26–28], which has been suggested as an attractive liquid fuel. GVL, an ideal sustainable green solvent and chemical intermediate, can be further hydrogenated to produce fuel additives (e.g., 2-methyltetrahydrofuran

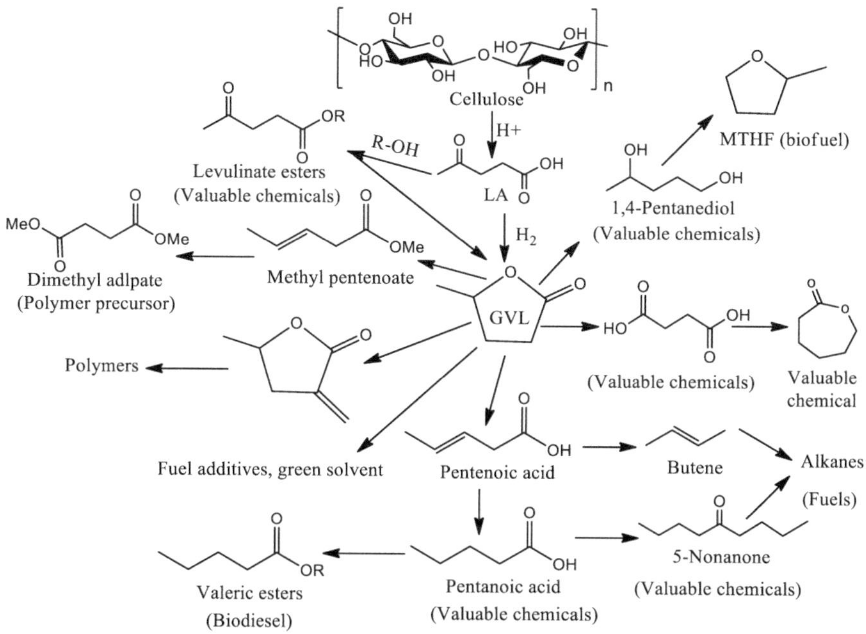


**Fig. 13.4** Catalytic upgrading of γ-valerolactone to fuels, fuel additives and value-added chemicals (Reprinted with permission from Ref. [20]. Copyright © 2015, Elsevier)

(MTHF)), or value-added chemicals (e.g., 1,4-pentanediol) as shown in Fig. 13.4. Biomass-derived polymers can be produced from the intermediate 1,4-pentanediol, while MTHF can be used as solvent and fuel additive [22, 29].

## 13.2 Mechanism Studies on the Production of GVL

Among different methodologies and catalytic system to produce GVL, the most efficient synthesis of GVL is the direct hydrogenation of LA [30–37], where LA is often produced from biomass as shown in Step I of Fig. 13.5. Although many issues (e.g., mechanisms) are associated with the production of LA are not clear yet, the production of LA have obtained many progress. Detailed reviews have been commented on the production of LA [5, 8, 43–47]. It was proposed that the first step during the hydrogenation reaction is chemisorption of molecular hydrogen and liquid LA on the metal support [34, 48]. Three endergonic reaction pathways are discussed in literature for GVL formation then, which depend on the order of the hydrogenation and dehydration step (Fig. 13.5) [49]. The first mechanism assumes the formation of pseudolevulinic acid by an intramolecular addition of the carboxyl on the carbonyl group in an equilibrium reaction. Pseudolevulinic acid is then dehydrated to α-angelica lactone which is finally hydrogenated to GVL. The second

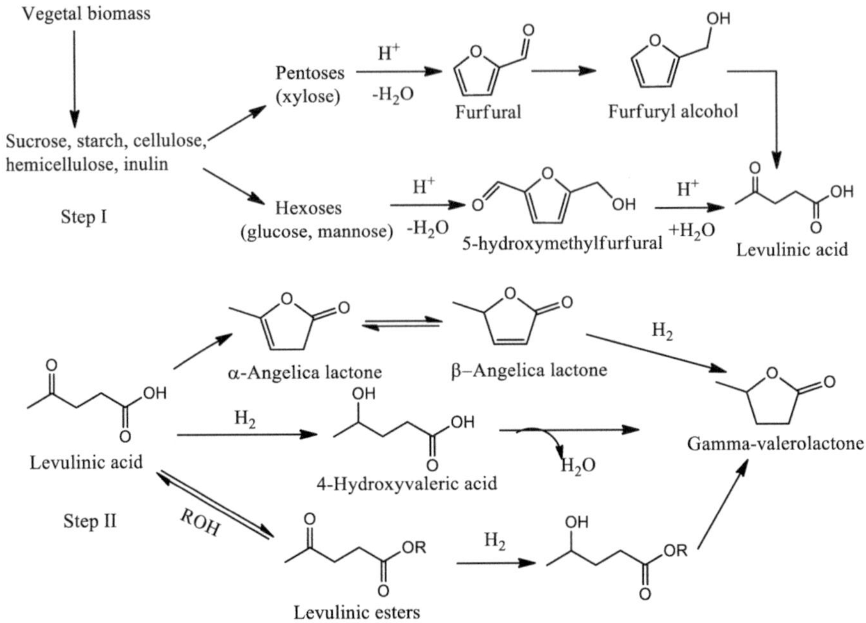


**Fig. 13.5** Different reaction pathways to produce γ-valerolactone [7, 8, 38–42]

pathway starts with the hydrogenation of LA to 4-hydroxypentanoic acid. Then simple lactonization will give GVL. Which of these pathways occurs preferentially is certainly a function of the solvent (water free conditions favor route 1) and the hydrogen pressure (high values more prefer route 2). As a side reaction, water elimination of 4-hydroxypentanoic acid to penteonic acid occurs which tends to form pentanoic acid under these hydrogen rich reaction conditions [8, 19, 29]. GVL can also be generated from the hydrogenation of esters of LA. In the first step, a hydroxyl levulinic ester is produced from the hydrogenation of levulinate esters. In the second step, GVL and the corresponding alcohol will be produced through an intramolecular transesterification.

In 2004, Manzer et al. [50] studied the reaction of LA in an autoclave reactor at a temperature of 150 °C for 4 h at 3.4 MPa $H_2$, dioxane solvent and 5% Ru/C catalyst, whereas nearly perfect conversion and a high GVL selectivity of 97% were reported. Manzer and Hutchenson [51] further described a continuous flow hydrogenation of LA in supercritical $CO_2$ ($scCO_2$), which was carried out in a tubular reactor under the reaction conditions of 200 °C, 20 MPa and 1% $Ru/Al_2O_3$ catalyst. Ninety-eight percent conversion of LA and perfect selectivity of GVL were reported [51, 52]. In 2007, Bourne et al. [53] studied another method for the continuous hydrogenation of LA in $scCO_2$, where water was used as a co-solvent and 10 MPa $H_2$ was used in a 3 molar excess to LA at 200 °C and $Ru/SiO_2$ catalyst. Almost perfect conversion of LA and full selectivity of GVL were reached. The main advantage of the catalytic system was the solubility of GVL in the combined water/$scCO_2$ mixture, allowing continuous separation of GVL without extra energy input. In

2009, formic acid was reported as the hydrogen source for the hydrogenation of LA since it was a co-product of LA in the dehydration of biomass [54]. The formic acid will supply the hydrogen source, avoiding the external sources of $H_2$ and separating of LA from formic acid in the aqueous reaction mixture. In 2010, Lange et al. [55] screened a number of ~50 catalysts for the hydrogenation of LA. The best performances were reported using Pt on $TiO_2$ or $ZrO_2$ at 200 °C and 40 bar $H_2$, where a high selectivity of GVL (> 95 mol%) was produced and a marginal deactivation of catalyst was found after 100 h. In 2011, Dumesic and coworker [56] did a pioneering work on the catalytic production of GVL from LA and its ester through Meerwein–Ponndorf–Verley (MPV) reduction using alcohol as hydrogen donor, where the MVP reaction is also often called as catalytic transfer hydrogenation (CTH) process. They found that reduction of LA and its esters to GVL can be accomplished over various metal oxide catalysts (e.g., $ZrO_2$, $MgO/Al_2O_3$) using the secondary alcohols (e.g., isopropanol, ethanol, butanol) as the hydrogen donor, where the $ZrO_2$ was demonstrated to be the most active catalyst in both batch and continuous flow reactor. In 2012, Dumesic and coworkers further studied the production of GVL along with LA and furfural from the integrated conversion of hemicellulose and cellulose [57]. The catalytic system was either a monophasic medium composed of GVL and an inorganic acid (e.g., HCl), or a biphasic system comprising an organic layer of GVL and an immiscible aqueous layer. During the process, LA was selectively converted into GVL at the mild conditions of 180 °C and 35 bar $H_2$ with Ru or Sn-based catalysts. The GVL acted as a product, also worked as a solvent, offering several clear advantages in the system: (1) GVL was a product of the system, it reduced the steps in the separation; (2) GVL solubilized the cellulose and hemicellulose, avoiding the formation of deposits on catalyst; (3) As they reported, the GVL decreased the rate of furfural degradation and enhanced the conversion of cellulose.

Researchers have more focused on the developing of new supports to improve the catalytic performances, catalyst stability, recycle and regeneration. Different metal oxides [45, 58, 59], mesoporous supports [60–62], carbon support [33, 63, 64], polymers [65] have been designed for this purpose. Catalyst recycle and regeneration in the hydrogenation of LA still need more efforts for future studies.

## 13.3 Production of GVL Using Homogeneous Catalysts

In homogeneous catalytic reaction, the reactants, products and catalyst are all dissolved in the same solvent and in one phase. Water is the most often solvent used due to environmental considerations, thus the transition metal ion catalyst is often designed and developed [66–70]. In contrast, homogeneous catalysts have a higher activity and are more selective under mild conditions. Burtoloso and coworkers [71] have reported that the $Fe_3(CO)_{12}$ catalyst was effective in the hydrogenation of LA using formic acid as hydrogen source and 92% yield of GVL was produced at 180 °C. While the crude liquor from the acid hydrolysis of sugarcane biomass was

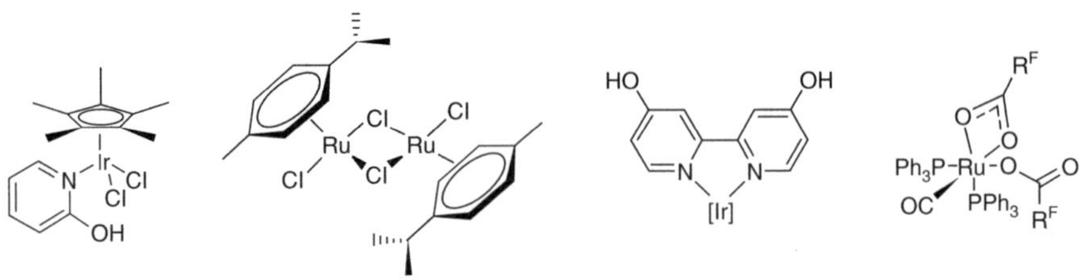


**Fig. 13.6** Typical Ru- and Ir-complex for the hydrogenation of levulinic acid [68, 70, 73–76]

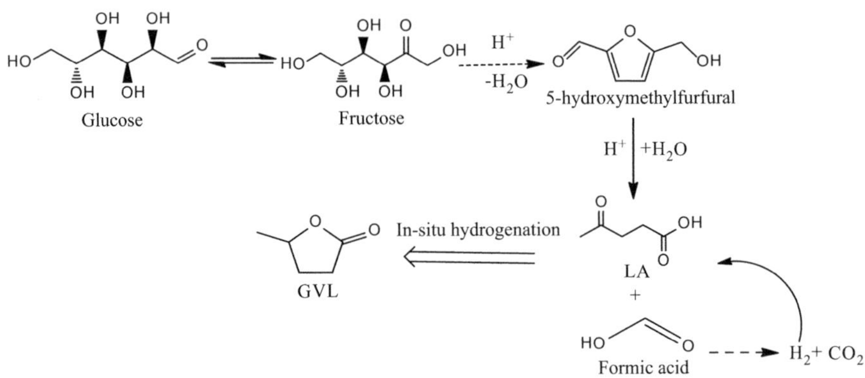


**Fig. 13.7** Production of γ-valerolactone from biomass-derived units using the generated formic acid in the process

directly converted into 50% yield GVL without the need for isolating LA. Transitional metal Pd-complex such as (dtbpe)$PdMe_2$ and (dtbpe)$PdCl_2$ (dtbpe = 1,2-(bis-di-tert-butylphosphino)ethane) were also reported to be efficient catalysts, where perfect yield of GVL was obtained at 100 °C in water solvent and formic acid as the hydrogen source [72]. Ru- or Ir-based homogeneous complex such as $RuCl_2(PPh_3)_3$, $Ru(acac)_3/PBu_3$, $Ru(acac)_3/P(n\text{-}Oct)_3$ and $[Ir(COE)_2Cl]_2$ catalysts have been designed and synthesized. The typical Ru- and Ir-complex as shown in Fig. 13.6 for the hydrogenation of LA has been frequently reported with good conversion of LA and high selectivity of GVL [68, 70, 73–76].

Horvath et al. did some pioneering works and studied the addition of the Shvo catalyst in the dehydration of fructose, where a high yield of 55% GVL was produced in one-step [77, 78]. In 2012, Dumesic and coworkers further reported the production of GVL along with LA and furfural from C6- carbohydrate or C5-carbohydrate [57]. The direct conversion of biobased carbohydrates to GVL would be more promising. Besides, if the produced side product formic acid can be further utilized as hydrogen source for the *in-situ* hydrogenation of LA (as shown in Fig. 13.7), the whole process will be more economic and attractive, albeit many reactions occurred simultaneously. More typical works of the advanced catalysts and catalytic system for the hydrogenation of LA to GVL have been compared in Table 13.2.

**Table 13.2** Production of γ-valerolactone from levulinic acid and carbohydrates using homogeneous catalysts[a]

| No. | Substrate | Catalyst | Solvent | $T$ (°C) | $H_2$ source | $t$ (h) | $Y_{GVL}$(%) | Ref. |
|---|---|---|---|---|---|---|---|---|
| 1 | LA | $RuCl_2(PPh_3)_3$ | $ScCO_2$ | 150 | 40 bar | 6 | 72.5 | [79] |
| 2 | LA | $(PPh_2)_3NC_3H_6$ + RuHCl $(PPh_3)_3$ | THF | 160 | 70 bar | 18 | 78 | [80] |
| 3 | LA | $Ru(acac)_3$ | THF | 160 | 70 bar | 18 | 99 | [80] |
| 4 | LA | $RuH_2(PPh_3)_4$ | THF | 160 | 70 bar | 18 | 95 | [80] |
| 5 | LA | $RuCl_2(PPh_3)_3 + (PPh_2)_3NC_3H_6$ | THF | 160 | 70 bar | 18 | 95 | [80] |
| 6 | LA | $Ru(acac)_3 + (PPh_2)_3$ $NC_3H_6$ | THF | 160 | 70 bar | 18 | 95 | [80] |
| 7 | LA | $Ru(acac)_3$ + TPPTS | $H_2O$ | 140 | 69 bar $H_2$ | 12 | 95 | [78] |
| 8 | LA | $[(\eta^6\text{-}C_6Me_6)$ $Ru(bpy)(H_2O)][SO_4]$ | $H_2O$ | 70 | HCOONa | 18 | 25 | [78] |
| 9 | LA | $Ru(acac)_3 + PBu_3$ | – | 200 | 82.8 bar $H_2$ | 6 | 37 | [78] |
| 10 | LA | $Ru(acac)_3 + PnOct_3 + NH_4PF_6$ | – | 160 | 100 bar $H_2$ | 18 | >99 | [69] |
| 11 | LA | $RuCl_2(PPh_3)$ | – | 180 | 12 kg/cm$^2$ $H_2$ | 24 | 99 | [81] |
| 12 | LA | $RuH_2(PPh_3)_4$ | – | 180 | 12 kg/cm$^2$ $H_2$ | 24 | 58 | [81] |
| 13 | LA | $Ru(acac)_3$/ $RP(C_6H_4\text{-m-}SO_3Na)_2$[b] | – | 140 | 10 bar $H_2$ | 4.5 | 39–100 | [82, 83] |
| 14 | LA | $RuCl_3{\bullet}H_2O/PPh_3$ | – | 150 | HCOOH | 12 | 93 | [54] |
| 15 | Glucose | $Ru(CO)_4I_2$ + HI | $H_2O$ | 200 | 30 bar $H_2$ | 8 | 39.1 | [74] |
| 16 | Fructose | $HCO_2\text{-}CF_3$, Ru/C | $H_2O + HCO_2\text{-}CF_3$ | 180 | 94 bar $H_2$ | 8 | 62 | [84] |
| 17 | Glucose | $HCO_2\text{-}CF_3$, Ru/C | $H_2O + HCO_2\text{-}CF_3$ | 180 | 94 bar $H_2$ | 8 | 38 | [84] |
| 18 | Sucrose | $HCO_2\text{-}CF_3$, Ru/TPPTS | $H_2O + HCO_2\text{-}CF_3$ | 180 | 94 bar $H_2$ | 8 | 52 | [84] |



[a]Note: *T: temperature, t: time, Y: yield of GVL*

## 13.4 Production of GVL Using Heterogeneous Catalysts

### *13.4.1 Supported Noble-Metal Catalysts*

The supported Noble-metal catalysts often present good crystallinity, mechanical stability, fine particle sizes, large surface area and many exposed active sites on the surface. These parameters have important influence on their physical and chemical properties, influencing their catalytic performances in a significant way. Besides, through tuning the interaction of the support and metal nanoparticles, specific functions can be introduced for particular utilizations. The synergic effect between metal nanoparticle and support can also make great contribution to the high activity and stability. The most commonly supported Noble metal catalysts are Ru, Rh, Pd, Pt and Au particles that are often used for the production of GVL from various substrates. Manzer et al. studied various metals (e.g., Ir, Rh, Ru, Pt, Ni and Pd) supported on carbon for GVL synthesis at 150 °C and 34.5 bar $H_2$ in dioxane, where the Ru/C catalyst display the highest activity with 97% yield [50]. Chang et al. [37] further investigated Noble metal (e.g., Ru, Pt and Pd) on carbon for the selective vapour phase hydrogenation of LA using a continuous flow fixed-bed reactor system, where 5% Ru/C catalyst presented 100% selectivity of GVL. Recently, Venugopal et al. [85] reported the vapor phase hydrogenation of aqueous LA over Pd, Pt, Ru, Cu and Ni on hydroxyapatite, where the supported Ru catalyst display much better performance in term of GVL yield and TOF value. Besides, except the change of metal type, it was found the support has also crucial influence on the yield of GVL. Du et al. [86, 87] have reported the various Noble metals (e.g., Pd, Pt, Ru, Au) supported on different types of supports (e.g., $TiO_2$, $SiO_2$, C and $ZrO_2$), whereby ~97% yield of GVL was generated on Au/$ZrO_2$ catalysts.

The green solvent $CO_2$ has been used for the LA hydrogenation over the Ru/$Al_2O_3$ and Ru/$SiO_2$ catalysts showing 99% conversion of LA with complete selectivity to GVL under conditions of 200 °C and 200 bar $H_2$ [53]. $CO_2$ is the most often used fluid because it is cheap, nonflammable, nontoxic and exhibits easily accessible critical parameters ($P_c$ = 73.8 bar, $T_c$ = 30.98 °C). $CO_2$ has been proven to offer a number of interesting opportunities as a medium for performing various catalytic reactions [88]. We have synthesized a series of Pd nanoparticles on the $SiO_2$ and $Al_2O_3$ supports in $CO_2$ (liquid phase) for the selective hydrogenation of LA, where these catalysts displayed good catalytic performance as well as high durability [33–35].

Ebitani et al. [36] found the Au nanoparticle catalyst was efficient for the synthesis of GVL through the comparison of Ru/C, Ru/SBA, Au/ZrC and Au/$ZrO_2$ catalysts in the hydrogenation of LA, whereas 90% yield of GVL was obtained in water solvent. The Au/$ZrO_2$ catalyst have presented good stability and it can be used over five runs [36]. Wettstein et al. [89] reported a biphasic reaction system for the efficient synthesis of LA and GVL from the depolymerization of cellulose using GVL as a solvent in the biphasic reaction system. Yang et al. [64] synthesized Ru nanoparticle (NP) on the porous carbon nanofiber through the transformation of Ru-functionalized metal organic framework. The authors utilized different proportional $RuCl_3$, $Zn(Ac)_2 \cdot 2H_2O$, trimesic acid in N,N-dimethylformamide solvent to construct the fiber precursor. The fiber precursor went through the high-temperature pyrolysis to form the uniform Ru NPs of ca. 12–16 nm on the hierarchically porous carbon fiber. The resulting Ru-catalysts display good performances in the liquid-phase hydrogenation of LA to GVL with 96.0% yield obtained [64]. More typical Noble metal catalysts used for the synthesis of GVL from various biomass-derived substrates was depicted in Table 13.3. In general, the vapor phase

**Table 13.3** Production of γ-valerolactone using the supported Noble-metal catalysts[a]

| No. | Substrate | Catalyst | Other conditions | $Y_{GVL}$ (%) | Ref. |
|---|---|---|---|---|---|
| 1 | LA | Ru/C | Dioxane, 150 °C, 34.5 bar $H_2$, 4 h | 97 | [50] |
| 2 | LA | $Ru/SiO_2$ | $H_2O + CO_2$, 200 °C, 100 bar $H_2$ | >99 | [53] |
| 3 | LA | $Ru\text{-}P/SiO_2$ | $H_2O$, 150 °C, HCOOH, 6 h | 30 | [90] |
| 4 | LA | Ru/C | $H_2O$, 150 °C, 40 bar $H_2$, 1 h | 30 | [90] |
| 5 | LA | $Ru/TiO_2$ | $H_2O$, 150 °C, 1 h, HCOOH +40 bars $H_2$ | 63 | [90] |
| 6 | LA | Ru/C | $CH_3OH$, 130 °C, 12 bar $H_2$, 3 h | 91 | [48] |
| 7 | LA | $Pd/Al_2O_3$ | 220 °C, HCOOH, 12 h | 29 | [91] |
| 8 | LA | 5% Ru/C | Dioxane, 265 °C, 1–25 bar $H_2$, 50 h | 98.6 | [37] |
| 9 | LA | 5%Ru/C | Dioxane, 201 °C, $H_2 + CO_2$ (201 bar) | 73.2 | [92] |
| 10 | LA | 5% $Ru/Al_2O_3$ | Dioxane, 201 °C, $H_2 + CO_2$ (200 bar) | 75.3 | [92] |
| 11 | LA | 5% Rh/C | Dioxane, 141 °C, $H_2 + CO_2$ (247 bar) | 98.9 | [92] |
| 12 | LA | 5% Ir/C | Dioxane, 141 °C, $H_2 + CO_2$ (250 bar) | 43.0 | [92] |
| 13 | LA | $Au/TiO_2$ | $H_2O$, 150 °C, formic acid, 6 h | 55 | [86] |
| 14 | LA[b] | $Au/ZrO_2$-VS | $H_2O$, 180 °C, formic acid, 3 h | 99 | [86] |
| 15 | BL[c] | $Au/ZrO_2$ | $H_2O$, 170 °C, 6 h | 95 | [90] |
| 16 | LA | 5%Ru/C | $H_2O$, 150 °C, formic acid, 5 h | 90 | [36] |
| 17 | LA | 5%$Au/ZrO_2$ | $H_2O$, 150 °C, formic acid, 5 h | 97 | [36] |
| 18 | Fructose | $Au/ZrO_2$ | $H_2O$, 150 °C, formic acid, 5 h | 48 | [36] |
| 19 | Fructose | Au/ZrC | $H_2O$, 150 °C, formic acid, 5 h | 47 | [36] |
| 20 | Fructose | Ru/SBA-15 | $H_2O$, 150 °C, formic acid, 5 h | 26 | [36] |
| 21 | Fructose | Ru/C | $H_2O$, 150 °C, formic acid, 5 h | 21 | [36] |
| 22 | LA | 5%$Au/ZrO_2$ | $H_2O$, 150 °C, formic acid, 5 h | 66 | [36] |
| 23 | LA | Ru–Ni/Meso-C | 150 °C, 45 bar $H_2$, 2 h | 96 | [93] |
| 24 | LA | 5% Ru/hydroxyapatite | $H_2O$, 70 °C, 5 bar $H_2$, 4 h | 99 | [94] |


[a]Note: *Y: yield of GVL*; [b]*$ZrO_2$-VS: acid-tolerant $ZrO_2$*; [c]*BL: butyl levulinate*

hydrogenation often displays good activity, while it is more energy sensitive with the vaporization of LA [95]. In comparison, the liquid-phase hydrogenation is simpler and more economical, albeit the activity is relatively lower than the vapor phase [95, 96].

## 13.5 Non-Noble Metal Catalysts

Albeit good catalytic performances have been frequently reported, the high cost and limited existence have hindered the application of Noble-metals in the production of GVL. As an alternative, Non-noble metal catalysts have obtained much attention [48]. $Al_2O_3$-supported Ni-Cu bimetallic catalysts have been studied for the LA hydrogenation, where high yield of GVL up to 96% was obtained [97]. It was found the decline of activity at 265 °C and 25 bars $H_2$ over silica supported copper catalysts during the time on stream. To reduce the coke formation during the time on stream, the addition of Ni will promote the activity and stability. Haan et al. [98] reported 71% yield of GVL over a Ni catalyst. Rao et al. [99, 100] reported Ni catalysts over the supports of $Al_2O_3$, $SiO_2$, ZnO, $ZrO_2$, $TiO_2$ and MgO for the vapour phase hydrogenation of LA to GVL without using any organic additives. The characterizations results by pyridine adsorbed IR showed the Lewis and Brønsted acid sites play an important role to generate GVL in the dehydration of the intermediate 4-hydroxy pentanoic acid.

The addition of another metal to the pristine metal as a bimetallic catalysts will change the electronic structure, which will change the catalytic activity some of extent [101]. For example, $Al_2O_3$ supported Ni-Cu bimetallic catalysts for the hydrogenation of LA, where the highest yield of GVL with 96% was obtained at 250 °C, 6.5 MPa and 2 h [102]. Recently, we have shown bimetallic Cu-derived catalysts were facilely synthesized from Cu-hydrotalcite (i.e., CuCr-, CuAl-, CuFe-) precursors highly efficient for the hydrogenation of LA as well as furfural [11, 103–106]. Hengne et al. [107] reported nanocomposites of Cu–$ZrO_2$ and Cu–$Al_2O_3$ quantitatively catalyzed the hydrogenation of LA and its methyl ester to give 90–100% selectivity to GVL in methanol and water respectively. More typical works are compared in Table 13.4. The future investigations on various catalytic systems, further optimization of reaction parameters still require more efforts.

## 13.6 Different Hydrogen Sources Utilized for the Hydrogenation of LA

### *13.6.1 Formic Acid as Hydrogen Source*

Formic acid (HCOOH), a colorless and non-toxic liquid has a highly pungent, penetrating odor at room temperature [114]. It is miscible with water and most polar organic solvents, thus it is often sued as an important intermediate in chemical

**Table 13.4.** Production of γ-valerolactone from levulinic acid, levulinates esters and furfural using Non-noble metal catalysts[a]

| No. | Substrate | Catalyst | Other reaction conditions | $Y_{GVL}$ (%) | Ref. |
|---|---|---|---|---|---|
| 1 | LA | Raney Ni | ~200 °C, 48.3 bars $H_2$ | 94 | [108] |
| 2 | LA | Raney Ni | 220 °C, 48 bars $H_2$, 3 h | 94 | [109] |
| 3 | LA | Ni-MoOx/C | 140 °C, 8 bars $H_2$, 5 h | 97 | [101] |
| 4 | LA | $Ni/TiO_2$ | 140 °C, 8 bars $H_2$, 5 h | 38 | [101] |
| 5 | LA | Ni-MoOx/C | Toluene solvent, 140 °C, 8 bar $H_2$, 5 h | 21 | [101] |
| 6 | LA | $Cu/SiO_2$ | 265 °C, 10 bars $H_2$ | 99.9 | [97] |
| 7 | LA | $Cu\text{-}ZrO_2$ | $H_2O$ solvent, 200 °C, 34.5 bar $H_2$, 5 h | 100 | [107] |
| 8 | LA | $Cu\text{-}ZrO_2$ | Methanol solvent, 200 °C, 34.5 bar $H_2$, 5 h | 90 | [107] |
| 9 | LA | $Cu\text{-}Al_2O_3$ | $H_2O$ solvent, 200 °C, 34.5 bar $H_2$, 5 h | 100 | [107] |
| 10 | LA | $Cu\text{-}Al_2O_3$ | Methanol solvent, 200 °C, 34.5 bar $H_2$, 5 h | 86 | [107] |
| 11 | LA | Cu-Cr oxide | $H_2O$ solvent, 200 °C, 70 bar $H_2$, 10 h | 90.7 | [103] |
| 12 | LA | Cu-Fe oxide | $H_2O$ solvent, 200 °C, 70 bar $H_2$, 10 h | 88.9 | [105] |
| 13 | ML[b] | $Cu\text{-}ZrO_2$ | Methanol solvent, 200 °C, 34.5 bar $H_2$, 5 h | 92 | [107] |
| 14 | ML | $Cu\text{-}Al_2O_3$ | Methanol solvent, 200 °C, 34.5 bar $H_2$, 5 h | 88 | [107] |
| 15 | EL[c] | 10% Ni/Si | 200 °C, HCOOH | 40 | [98] |
| 16 | EL | 10% Ni/Si | 250 °C, HCOOH | 73 | [98] |
| 17 | LA | $PtO_2$ | Ethyl ether solvent, 250 °C, 44 h, 3 bar $H_2$ | 87 | [110] |
| 18 | LA | CuCr-oxide | 190 °C, 1.3 h, 200 bars $H_2$ | 94 | [109] |
| 19 | LA | Zr-Beta-100 | 2-propanol, 250 °C, 1 bar pressure | >99 | [111] |
| 20 | LA | Zr-Beta-100 | 2-pentanol solvent, 118 °C, 10 h | 96 | [111] |
| 21 | LA | Zr-Beta-100 | Cyclohexanol solvent, 150 °C, 6 h | 82 | [111] |
| 22 | LA | ZrAl-Beta-100 | 2-pentanol solvent, 118 °C, 6 h | 72 | [111] |
| 23 | LA | $ZrO(OH)_n$-400 | 2-butanol solvent, 150 °C, 16 h | 41 | [111] |
| 24 | BL[d] | $MgO/Al_2O_3$ | 2-butanol solvent, 150 °C, 16 h | 14.6 | [56] |
| 25 | BL | CeZrOx | 2-butanol solvent, 150 °C, 16 h | 15.8 | [56] |
| 26 | BL | $ZrO_2$ | 2-butanol solvent, 150 °C, 16 h | 84.7 | [56] |
| 27 | EL | $ZrO_2$ | Isopropanol solvent, 150 °C, 4 h | 62.4 | [56] |
| 28 | LA | Zr-beta | 2-butanol solvent, 120 °C, 11 h | >98 | [112] |
| 29 | ML | Zr-beta | 2-butanol solvent, 120 °C, 5 h | >97 | [112] |
| 30 | ML | Al-beta | Butanol solvent, 120 °C, 5 h | 10.8 | [112] |

(continued)

**Table 13.4.** (continued)

| No. | Substrate | Catalyst | Other reaction conditions | $Y_{GVL}$ (%) | Ref. |
|---|---|---|---|---|---|
| 31 | Furfural | Zr-beta +Amberlyst-70 | 2-butanol solvent, 120 °C, 24 h | 66 | [112] |
| 32 | EL | $Zr(OH)_4$ | 2-butanol solvent, 200 °C, 1 h | 81.3 | [13] |
| 33 | LA | β-$Mo_2C$ | Water, dioxane, 180 °C, 30 bar $H_2$ | >81 | [113] |


[a]Note: *Y*: yield of γ-valerolactone (GVL); [b]*ML: methyl levulinate*; [c]*EL: ethyl levulinate*; [d]*BL: butyl levulinate*

synthesis [115]. Besides, formic acid was often used to promote transfer hydrogenation reactions [116] due to the generation of $H_2$ and $CO_2$ from the decomposition based on a simple equation 13.1 [117]:

$$\mathrm{HCOOH} \rightarrow CO_2 + \mathrm{H}_2 \tag{13.1}$$

The selectivity of the product depends on the types of catalyst used, CO and water as side-products were often produced simultaneously following the formula (Eq. 13.2):

$$\mathrm{HCOOH} \rightarrow CO + \mathrm{H_2O} \tag{13.2}$$

recent investigations reported that several transition-metal complexes or Noble metal catalysts could act as catalysts to improve the selectivity of $H_2$ in the decomposition of formic acid. Laurenczy et al. [118] studied the decomposition of formic acid into $H_2$ and $CO_2$ in aqueous solutions catalyzed by $[Ru(H_2O)_6]^{2+}$, $[Ru(H_2O)_6]^{3+}$, and $RuCl_3$ $xH_2O$ in the presence of TPPTS (TPPTS = meta-trisulfonatedtriphenylp hosphine). The decomposition of formic acid is slow, unless sodium formate is present as an initiator. Beller and co-workers studied the promotion effect through the addition of different amines (i.e. triethylamine, N,N-dimethylhexylamine, N,N-dimethylaminoethanol) in the catalytic decomposition of formic acid in the aqueous mixture of HCOOH and HCOONa [119]. Under such conditions, the reaction was observed even at room temperature over Ru-based catalysts including $[RuCl_2(p\text{-}cymene)]_2$ and $[RuCl_2(Benzene)]_2$ /$(PPh_3)_3$. Ogo et al. studied a homogeneous Ru catalyst $[(\eta_6\text{-}C_6Me_6)Ru(bpy)(H_2O)][SO_4]$ able to promote the transfer hydrogenation of LA with HCOONa in liquid water solutions and the possible pathway was proposed in Fig. 13.8, where the GVL and 1,4-pentanediol were obtained in 25% yield [120]. Heeres et al. [84] studied a one-pot synthesis of GVL from the conversion of C6-sugars using trifluoroacetic acid, Ru/C catalyst, and formic acid as the hydrogen donor.

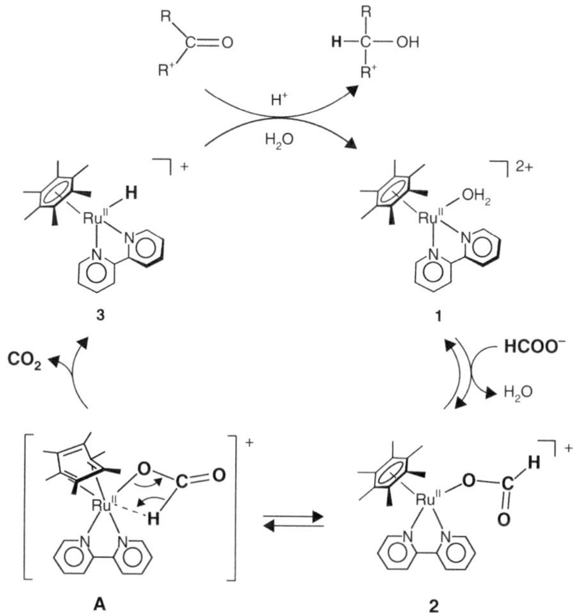


**Fig. 13.8** Mechanism for the transfer hydrogenation of levulinic acid (LA) using $[(\eta_6\text{-}C_6Me_6)Ru(bpy)(H_2O)][SO_4]$ as a catalyst precursor in water (Reprinted with permission from Ref. [120]. Copyright © 2002, American Chemical Society)

### 13.6.2 Hydrogen

Hydrogen probably is the most used sources for the hydrogenation. Among the different catalytic system, batch autoclaves and continuous flow reactor are the most often utilized. As previously discussed, the pioneered works reported for the hydrogenation of LA to GVL on Raney-nickel catalyst and platinum oxide with good yield of GVL [109, 121]. The supported Ru, Pd and Pt nanoparticles have been frequently used for the hydrogenation of LA to GVL using hydrogen has received much attention in both continuous and batch reactor [82, 122–124]. Al-Shaal et al. [125] studied the Ru on the activated carbon, $Al_2O_3$, $TiO_2$ and $SiO_2$ as catalysts for the GVL synthesis in batch autoclaves using different solvents, where the best activity was obtained at 20 bar $H_2$ and 130 °C. It has been reported that the bimetallic systems can enhance the catalyst stability, especially for Ru-derived system. For example, Yang et al. [93] investigated Ru–Ni bimetallics in ordered mesoporous carbons for the hydrogenation of LA to GVL. A high TOF (>2000 $h^{-1}$) was reported, and the $Ru_{0.9}Ni_{0.1}$–OMC catalyst can be used at least 15 times without obvious loss of its catalytic performance. Ftouni et al. [126] investigated the Ru-based catalysts supported on $TiO_2$, $ZrO_2$, and C for the hydrogenation of LA to GVL at 30 bar of $H_2$ and 423 K in dioxane solvent. All catalysts display good activity for the synthesis of GVL. However, high yields over several runs were only obtained on the $Ru/ZrO_2$ catalyst. The authors further utilized different analysis tools of XPS, CO/FT-IR, TGA, STEM, and physisorption-IR to explore the deactivation and found the partial deactivation was due to the reduction of the titania support and the partial coverage on the catalyst surface. Besides, the addition of water was reported to promote the selective hydrogenation reaction. The groups of Manzer [50] and Poliakoff [53] have performed the continuous flow reactor for the hydrogenation of LA, where high selectivity of GVL has been reported. Continuous production of GVL has many advantages compared to batch processes and it may deserve more attention for future studies.

### 13.6.3 Meerwein-Ponndorf-Verley Reaction

The conversion of LA and its esters to GVL via transfer hydrogenation was pioneered by a study of Chia and Dumesic [56]. GVL production through MPV reduction is beneficial due to the use of inexpensive metal oxides or zeolites without external hydrogen source. $ZrO_2$ was reported to be the most active metal oxide for the transfer hydrogenation, where GVL yields up to 85% under the conditions of 2-butanol solvent using butyllevulinate reactant at 150 °C and 16 h [111]. Hwang et al. [127] reported a series of zirconium-based metal–organic frameworks (Zr-MOFs) using isopropanol solvent for catalytic transfer hydrogenation of ethyl levulinate (EL) to GVL as shown in Fig. 13.9. 92.7% yield of GVL was achieved in 2 h at 200 °C using the UiO-66(Zr) catalyst and it can be recycled over five times without a notable change in catalytic activity and product selectivity.

Luo et al. [128] investigated the reaction kinetics of the MPV reduction of methyl levulinate (ML) to GVL catalyzed by Lewis acid zeolites. Reaction kinetics studies show the hydride shift is the rate-limiting step. Secondary alcohols exhibit higher reaction rates than primary alcohols with lower apparent activation energies. Besides, the authors also found the increasing polarity of the hydrogen donor would cause the decrease of reaction rates. Han et al. [129] reported the catalytic transfer hydrogenation of EL to GVL using isopropanol solvent over a porous Zr-containing catalyst with a phenate group in its structure that was prepared by the coprecipitation of 4-hydroxybenzoic acid dipotassium salt and $ZrOCl_2$ (Zr-HBA) in water. Zr-HBA was very active for the hydrogenation and 94.4% yield of GVL was reported. The existence of a phenate in the structure of Zr-HBA was found to be favorable for the CTH of EL. Lin et al. [13] studied a series of metal hydroxides and $ZrO(OH)_2{\cdot}xH_2O$ was found to be most active, whereas 93.6% conversion of EL and 94.5% selectivity of GVL were reported when 2-propanol was used as a hydrogen

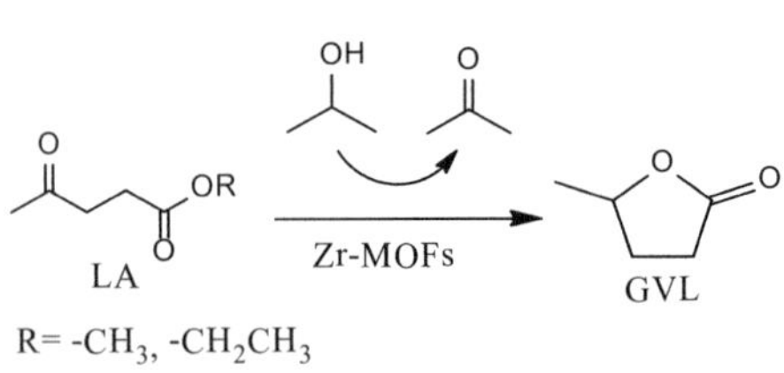


**Fig. 13.9** Meerwein-Ponndorf-Verley (MPV) reduction of levulinic acid (LA) to γ-valerolactone

**Table 13.5** Typical works on the transfer hydrogenation to produce γ-valerolactone[a]

| No. | Substrate | Catalyst | Alcohol | Other conditions | $Y_{GVL}$ (%) | Ref. |
|---|---|---|---|---|---|---|
| 1 | ML[b] | $ZrO_2$/SBA-15 | 2-propanol | 150 °C, 3 h, 10 bar Ar | 95 | [130] |
| 2 | LA | $ZrO_2$/SBA-15 | 2-propanol | 150 °C, 3 h, 10 bar Ar | >90 | [130] |
| 3 | ML | $ZrO_2$/SBA-15 | Ethanol | 150 °C, 3 h, 10 bar Ar | 41 | [130] |
| 4 | EL[c] | $ZrO_2$/SBA-15 | 2-propanol | 150 °C, 3 h, 10 bar Ar | 91 | [130] |
| 5 | BL | $ZrO_2$/SBA-15 | 2-propanol | 150 °C, 3 h, 10 bar Ar | 71 | [130] |
| 6 | EL | Al-Zr mixed oxides | 2-propanol | 220 °C, 4 h | 83.2 | [131] |
| 7 | BL[d] | $MgO/Al_2O_3$ | 2-butanol | 150 °C, 16 h | 14.6 | [56] |
| 8 | BL | CeZrOx | 2-butanol | 150 °C, 16 h | 15.8 | [56] |
| 9 | BL | $ZrO_2$ | 2-butanol | 150 °C, 16 h | 84.7 | [56] |
| 10 | EL | $ZrO_2$ | Isopropanol | 150 °C, 4 h | 62.4 | [56] |
| 11 | LA | Zr-beta | 2-butanol | 120 °C, 11 h | >98 | [112] |
| 12 | ML | Zr-beta | 2-butanol | 120 °C, 5 h | >97 | [112] |
| 13 | ML | Al-beta | 2-butanol | 120 °C, 5 h | 10.8 | [112] |
| 14 | Furfural | Zr-beta + Amberlyst-70 | 2-butanol | 120 °C, 24 h | 66 | [112] |

[a]Note: *Y*: yield of γ-valerolactone (GVL); [b]*ML: methyl levulinate*; [c]*EL: ethyl levulinate*; [d]*BL: butyl levulinate*

donor at 473 K with a reaction time of 1 h. The $ZrO(OH)_2{\cdot}xH_2O$ catalyst exhibited good stability over ten times. Typical works on the transfer hydrogenation to produce GVL have been depicted in Table 13.5.

## 13.7 Conclusions and Future Outlook

This chapter summarizes recent research works on the synthesis of GVL using different catalytic system ranged from different advanced catalysts and hydrogen sources. Many progress on the homogeneous, heterogeneous catalysts and the reaction systems utilizing them for the efficient conversion of biorenewable feedstocks to GVL have been commented in this work. The future direction may require more attention to the following direction:

(i) The hydrogenation of platform chemical LA is still the main route for the production of GVL. A more promising for GVL is more directly from lignocellulosic biomass-derived monomers, then more pristine biomass in a large scale. Lignocellulosic biomass conversion is still performed under harsh environment (e.g., acidic medium, high temperature and pressure). Development of a robust catalyst with ecofriendly property is more attractive for future studies.

(ii) Heterogeneous catalytic production of GVL still faces the issues of operating conditions and product purification. The use of heterogeneous catalysts in continuous flow reactor process would be more attractive for industrial application due to the simpler processes and more economical.
(iii) Transfer hydrogenation still faces the problems of product separation and purification. It requires more efforts from the engineering and reactor design. Besides, the hydrogen source produced by gasification of biomass or other sustainable processes (e.g., water splitting using electrocatalysis [132] or photocatalysis [133]) for the production of GVL would be more sustainable.

## References


1. Li H, Fang Z, Smith Jr RL, Yang S. Efficient valorization of biomass to biofuels with bifunctional solid catalytic materials. Prog Energy Combust Sci. 2016;55:98–194.
2. Huber GW, Iborra S, Corma A. Synthesis of transportation fuels from biomass: chemistry, catalysts, and engineering. Chem Rev. 2006;106:4044–98.
3. Dutta S, De S, Saha B, Alam MI. Advances in conversion of hemicellulosic biomass to furfural and upgrading to biofuels. Cat Sci Technol. 2012;2:2025–36.
4. Kikas T, Tutt M, Raud M, Alaru M, Lauk R, Olt J. Basis of energy crop selection for biofuel production: cellulose vs. lignin. Int J Green Energy. 2016;13:49–54.
5. Alonso DM, Bond JQ, Dumesic JA. Catalytic conversion of biomass to biofuels. Green Chem. 2010;12:1493–513.
6. Gallezot P. Catalytic conversion of biomass: challenges and issues. ChemSusChem. 2008;1:734–7.
7. Climent MJ, Corma A, Iborra S. Conversion of biomass platform molecules into fuel additives and liquid hydrocarbon fuels. Green Chem. 2014;16:516–47.
8. Yan K, Jarvis C, Gu J, Yan Y. Production and catalytic transformation of levulinic acid: a platform for speciality chemicals and fuels. Renew Sust Energ Rev. 2015;51:986–97.
9. Li C, Zhao X, Wang X, Huber GW, Zhang T. Catalytic transformation of lignin for the production of chemicals and fuels. Chem Rev. 2015;115:11559–624.
10. Luo J, Fang Z, Smith Jr RL. Ultrasound-enhanced conversion of biomass to biofuels. Prog Energy Combust Sci. 2014;41:56–93.
11. Yan K, Wu G, Lafleur T, Jarvis C. Production, properties and catalytic hydrogenation of furfural to fuel additives and value-added chemicals. Renew Sust Energ Rev. 2014;38:663–76.
12. Noordermeer LPAMA. Biomass to biofuels, a chemical perspective. Green Chem. 2006;8:861–7.
13. Tang X, Chen HW, Hu L, Hao WW, Sun Y, Zeng XH, Lin L, Liu SJ. Conversion of biomass to gamma-valerolactone by catalytic transfer hydrogenation of ethyl levulinate over metal hydroxides. Appl Catal B Environ. 2014;147:827–34.
14. Ragauskas AJ, Beckham GT, Biddy MJ, Chandra R, Chen F, Davis MF, Davison BH, Dixon RA, Gilna P, Keller M, Langan P, Naskar AK, Saddler JN, Tschaplinski TJ, Tuskan GA, Wyman CE. Lignin valorization: improving lignin processing in the biorefinery. Science. 2014;344:1246843.
15. Upton BM, Kasko AM. Strategies for the conversion of lignin to high-value polymeric materials: review and perspective. Chem Rev. 2016;116:2275–306.
16. Saidi M, Samimi F, Karimipourfard D, Nimmanwudipong T, Gates BC, Rahimpour MR. Upgrading of lignin-derived bio-oils by catalytic hydrodeoxygenation. Energy Environ Sci. 2014;7:103–29.

17. Rinaldi R, Jastrzebski R, Clough MT, Ralph J, Kennema M, Bruijnincx PCA, Weckhuysen BM. Paving the way for lignin valorisation: recent advances in bioengineering, biorefining and catalysis. Angew Chem Int Ed. 2016;55:8164–215.
18. Hengst K, Schubert M, Kleist W, Grunwaldt JD. Hydrodeoxygenation of lignocellulose-derived platform molecules. In: Catalytic hydrogenation for biomass valorization. Cambridge: The Royal Society of Chemistry; 2015. p. 125–50.
19. Yan K, Yang Y, Chai J, Lu Y. Catalytic reactions of gamma-valerolactone: a platform to fuels and value-added chemicals. Appl Catal B Environ. 2015;179:292–304.
20. Lange JP. Lignocellulose conversion: an introduction to chemistry, process and economics. Biofuels Bioprod Biorefin. 2007;1:39–48.
21. Werpy T, Petersen G, Aden A, Bozell J, Holladay J, White J, Manheim A, Eliot D, Lasure L, Jones S. Top value added chemicals from biomass. Volume 1-Results of screening for potential candidates from sugars and synthesis gas, www.nrel.gov/docs/fy04osti/35523.pdf (2004).
22. Gallezot P. Conversion of biomass to selected chemical products. Chemical Society Review. 2012;41:1538–58.
23. Bond JQ, Wang D, Alonso DM, Dumesic JA. Interconversion between γ-valerolactone and pentenoic acid combined with decarboxylation to form butene over silica/alumina. J Catal. 2011;281:290–9.
24. Bond JQ, Upadhye AA, Olcay H, Tompsett GA, Jae J, Xing R, Alonso DM, Wang D, Zhang T, Kumar R, Foster A, Sen SM, Maravelias CT, Malina R, Barrett SRH, Lobo R, Wyman CE, Dumesic JA, Huber GW. Production of renewable jet fuel range alkanes and commodity chemicals from integrated catalytic processing of biomass. Energy Environ Sci. 2014;7:1500–23.
25. Bond JQ, Martin Alonso D, West RM, Dumesic JA. γ-Valerolactone ring-opening and decarboxylation over $SiO_2/Al_2O_3$ in the presence of water. Langmuir. 2010;26:16291–8.
26. Fabos V, Koczo G, Mehdi H, Boda L, Horvath IT. Bio-oxygenates and the peroxide number: a safety issue alert. Energy Environ Sci. 2009;2:767–9.
27. Horvath IT, Mehdi H, Fabos V, Boda L, Mika LT. [gamma]-valerolactone-a sustainable liquid for energy and carbon-based chemicals. Green Chem. 2008;10:238–42.
28. Fegyverneki D, Orha L, Láng G, Horváth IT. Gamma-valerolactone-based solvents. Tetrahedron. 2010;66:1078–81.
29. Yan K, Lafleur T, Wu X, Chai J, Wu G, Xie X. Cascade upgrading of [gamma]-valerolactone to biofuels. Chem Commun. 2015;51:6984–7.
30. Abdelrahman OA, Heyden A, Bond JQ. Analysis of kinetics and reaction pathways in the aqueous-phase hydrogenation of levulinic acid to form γ-valerolactone over Ru/C. ACS Catal. 2014;4:1171–81.
31. Tang X, Hu L, Sun Y, Zeng XH, Lin L. Conversion of biomass to novel platform chemical gamma-valerolactone by selective reduction of levulinic acid. Prog Chem. 2013;25:1906–14.
32. van de Graaf WD, Lange JP, Haan RJ. Conversion of furfuryl alcohol into ethyl levulinate using solid acid catalysts. ChemSusChem. 2009;2:437–41.
33. Yan K, Lafleur T, Liao J. Facile synthesis of palladium nanoparticles supported on multi-walled carbon nanotube for efficient hydrogenation of biomass-derived levulinic acid. J Nanopart Res. 2013;15:1–7.
34. Yan K, Lafleur T, Wu G, Liao J, Ceng C, Xie X. Highly selective production of value-added γ-valerolactone from biomass-derived levulinic acid using the robust Pd nanoparticles. Appl Catal A Gen. 2013;468:52–8.
35. Yan K, Jarvis C, Lafleur T, Qiao Y, Xie X. Novel synthesis of Pd nanoparticles for hydrogenation of biomass-derived platform chemicals showing enhanced catalytic performance. RSC Adv. 2013;3:25865–71.
36. Son PA, Nishimura S, Ebitani K. Production of [gamma]-valerolactone from biomass-derived compounds using formic acid as a hydrogen source over supported metal catalysts in water solvent. RSC Adv. 2014;4:10525–30.

37. Upare PP, Lee JM, Hwang DW, Halligudi SB, Hwang YK, Chang JS. Selective hydrogenation of levulinic acid to [gamma]-valerolactone over carbon-supported noble metal catalysts. J Ind Eng Chem. 2011;17:287–92.
38. Lange JP, van de Graaf WD, Haan RJ. Conversion of furfuryl alcohol into ethyl levulinate using solid acid catalysts. ChemSusChem. 2009;2:437–41.
39. Raspolli Galletti AM, Antonetti C, Ribechini E, Colombini MP, Nassi o Di Nasso N, Bonari E. From giant reed to levulinic acid and gamma-valerolactone: a high yield catalytic route to valeric biofuels. Appl Energy. 2013;102:157–62.
40. Serrano-Ruiz JC, West RM, Dumesic JA. Catalytic conversion of renewable biomass resources to fuels and chemicals. Annu Rev Chem Biomol Eng. 2010;1:179–00.
41. Yan K, Wu G, Wen J, Chen A. One-step synthesis of mesoporous $H_4SiW_{12}O_{40}$-$SiO_2$ catalysts for the production of methyl and ethyl levulinate biodiesel. Catal Commun. 2013;34:58–63.
42. Chen B, Li F, Huang Z, Lu T, Yuan Y, Yuan G. Integrated catalytic process to directly convert furfural to levulinate ester with high selectivity. ChemSusChem. 2014;7:202–9.
43. Yan K, Luo H. Recent development of metal nanoparticles catalysts and their use for efficient hydrogenation of biomass-derived levulinic acid. Green Processes for Nanotechnology: Springer; 2015. p. 75–98.
44. Pileidis FD, Titirici MM. Levulinic acid biorefineries: new challenges for efficient utilization of biomass. ChemSusChem. 2016;9:562–82.
45. Tong X, Ma Y, Li Y. Biomass into chemicals: conversion of sugars to furan derivatives by catalytic processes. Appl Catal A Gen. 2010;385:1–13.
46. West RM, Kunkes EL, Simonetti DA, Dumesic JA. Catalytic conversion of biomass-derived carbohydrates to fuels and chemicals by formation and upgrading of mono-functional hydrocarbon intermediates. Catal Today. 2009;147:115–25.
47. Zhou CH, Xia X, Lin CX, Tong DS, Beltramini J. Catalytic conversion of lignocellulosic biomass to fine chemicals and fuels. Chem Soc Rev. 2011;40:5588–617.
48. Yan ZP, Lin L, Liu S. Synthesis of γ-Valerolactone by hydrogenation of biomass-derived levulinic acid over Ru/C catalyst. Energy Fuel. 2009;23:3853–8.
49. Bond JQ, Wang D, West RM, Dumesic JA. Integrated catalytic conversion of γ-Valerolactone to liquid alkenes for transportation fuels. Science. 2010;327:1110–4.
50. Manzer LE. Catalytic synthesis of [alpha]-methylene-[gamma]-valerolactone: a biomass-derived acrylic monomer. Appl Catal A Gen. 2004;272:249–56.
51. Manzer LEW, Hutchenson KWLU, Process for the production of y-methyl-a-methylene-y--butyrolactone from reaction of levulinic acid and hydrogen with recycle of unreacted levulinic acid followed by reaction of crude y-valerolactone and formaldehyde, both reactions being carried out in the supercritical or near-critical fluid phase. United States Patent Application 20060100447.
52. Manzer LEW, Production of 5-methyl-N-(methyl aryl)-2-pyrrolidone, 5-methyl-N-(methyl cycloalkyl)-2-pyrrolidone and 5-methyl-N-alkyl-2-pyrrolidone by reductive amination of levulinic acid with cyano compounds United States Patent Application 20040204592 .
53. Bourne RA, JaStevens JG, Ke J, Poliakoff M. Maximising opportunities in supercritical chemistry: the continuous conversion of levulinic acid to γ-valerolactone in $CO_2$. Chem Commun. 2007:4632–4.
54. Deng L, Li J, Lai DM, Fu Y, Guo QX. Catalytic conversion of biomass-derived carbohydrates into gamma-Valerolactone without using an external $H_2$ supply. Angew Chem Int Ed. 2009;48:6529–32.
55. Lange JP, Price R, Ayoub PM, Louis J, Petrus L, Clarke L, Gosselink H. Valeric biofuels: a platform of cellulosic transportation fuels. Angew Chem Int Ed. 2010;49:4479–83.
56. Chia M, Dumesic JA. Liquid-phase catalytic transfer hydrogenation and cyclization of levulinic acid and its esters to [gamma]-valerolactone over metal oxide catalysts. Chem Commun. 2011;47:12233–5.
57. Alonso DM, Wettstein SG, Mellmer MA, Gurbuz EI, Dumesic JA. Integrated conversion of hemicellulose and cellulose from lignocellulosic biomass. Energy Environ Sci. 2013;6:76–80.

58. Li H, Fang Z, Yang S. Direct catalytic transformation of biomass derivatives into biofuel component gamma-valerolactone with magnetic nickel-zirconium nanoparticles. ChemPlusChem. 2016;81:135–42.
59. Li H, Fang Z, Yang S. Direct conversion of sugars and ethyl levulinate into γ-valerolactone with superparamagnetic acid–base bifunctional ZrFeOx nanocatalysts. ACS Sustai Chem Eng. 2016;4:236–46.
60. Yan K, Lafleur T, Liao J, Xie X. Facile green synthesis of palladium nanoparticles for efficient liquid-phase hydrogenation of biomass-derived furfural. Sci Adv Mater. 2014;6:135–40.
61. Yan K, Wu X, An X, Xie X. Novel preparation of nano-composite $CuO-Cr_2O_3$ using CTAB-template method and efficient for hydrogenation of biomass-derived furfural. Funct Mater Lett. 2013;6:1350007.
62. Yan K, Lafleur T, Jarvis C, Wu G. Clean and selective production of γ-valerolactone from biomass-derived levulinic acid catalyzed by recyclable Pd nanoparticle catalyst. J Clean Prod. 2014;72:230–2.
63. Upare PP, Lee M, Lee SK, Yoon JW, Bae J, Hwang DW, Lee UH, Chang JS, Hwang YK. Ru nanoparticles supported graphene oxide catalyst for hydrogenation of bio-based levulinic acid to cyclic ethers. Catal Today. 2016;265:174–83.
64. Yang Y, Sun CJ, Brown DE, Zhang L, Yang F, Zhao H, Wang F, Ma X, Zhang X, Ren Y. A smart strategy to fabricate Ru nanoparticle inserted porous carbon nanofibers as highly efficient levulinic acid hydrogenation catalysts. Green Chem. 2016;18:3558–66.
65. Xue Z, Jiang J, Li G, Zhao W, Wang J, Mu T. Zirconium-cyanuric acid coordination polymer: highly efficient catalyst for conversion of levulinic acid to [gamma]-valerolactone. Cat Sci Technol. 2016;6:5374–9.
66. Deuss PJ, Barta K, de Vries JG. Homogeneous catalysis for the conversion of biomass and biomass-derived platform chemicals. Cat Sci Technol. 2014;4:1174–96.
67. FÁBOS V. Gamma-valerolactone, and its synthesis by catalytic transfer hydrogenation of levulinic acid, Ph. D. Thesis, Eötvös Loránd University (Hungary), (2009).
68. Geilen FMA, Engendahl B, Hölscher M, Klankermayer JR, Leitner W. Selective homogeneous hydrogenation of biogenic carboxylic acids with [Ru(TriPhos)H]+: a mechanistic study. J Am Chem Soc. 2011;133:14349–58.
69. Geilen F, Engendahl B, Harwardt A, Marquardt W, Klankermayer J, Leitner W. Selective and flexible transformation of biomass-derived platform chemicals by a multifunctional catalytic system. Angew Chem Int Ed. 2010;49:5510–4.
70. Braca G, Raspolli Galletti AM, Sbrana G. Anionic ruthenium iodorcarbonyl complexes as selective dehydroxylation catalysts in aqueous solution. J Organomet Chem. 1991;417:41–9.
71. Metzker G, Burtoloso ACB. Conversion of levulinic acid into [gamma]-valerolactone using $Fe_3(CO)_{12}$: mimicking a biorefinery setting by exploiting crude liquors from biomass acid hydrolysis. Chem Commun. 2015;51:14199–202.
72. Ortiz-Cervantes C, Flores-Alamo M, García JJ. Hydrogenation of biomass-derived levulinic acid into γ-valerolactone catalyzed by palladium complexes. ACS Catal. 2015;5:1424–31.
73. Joó F, Tóth Z, Beck MT. Homogeneous hydrogenations in aqueous solutions catalyzed by transition metal phosphine complexes. Inorg Chim Acta. 1977;25:L61–2.
74. Li W, Xie JH, Lin H, Zhou QL. Highly efficient hydrogenation of biomass-derived levulinic acid to [gamma]-valerolactone catalyzed by iridium pincer complexes. Green Chem. 2012;14:2388–90.
75. Brewster TP, Miller AJM, Heinekey DM, Goldberg KI. Hydrogenation of carboxylic acids catalyzed by half-sandwich complexes of iridium and rhodium. J Am Chem Soc. 2013;135:16022–5.
76. Delhomme C, Schaper LA, Zhang-Preße M, Raudaschl-Sieber G, Weuster-Botz D, Kühn FE. Catalytic hydrogenation of levulinic acid in aqueous phase. J Organomet Chem. 2013;724:297–9.

77. Qi L, Horváth IT. Catalytic conversion of fructose to γ-valerolactone in γ-valerolactone. ACS Catal. 2012;2:2247–9.
78. Mehdi H, Fabos V, Tuba R, Bodor A, Mika LT, Horvath IT. Integration of homogeneous and heterogeneous catalytic processes for a multi-step conversion of biomass: from sucrose to levulinic acid, gamma-valerolactone, 1,4-pentanediol, 2-methyl-tetrahydrofuran, and alkanes. Top Catal. 2008;48:49–54.
79. Yang W, Cheng H, Zhang B, Li Y, Liu T, Lan M, Yu Y, Zhang C, Lin W, Fujita SI, Arai M, Zhao F. Hydrogenation of levulinic acid by $RuCl_2(PPh_3)_3$ in supercritical $CO_2$: the significance of structural changes of Ru complexes via interaction with $CO_2$. Green Chem. 2016;18:3370–7.
80. Deng L, Kang B, Englert U, Klankermayer J, Palkovits R. Direct hydrogenation of biobased carboxylic acids mediated by a nitrogen-centered tridentate phosphine ligand. ChemSusChem. 2016;9:177–80.
81. Osakada K, Ikariya T, Yoshikawa D. Preparation and properties of hydride triphenylphosphine ruthenium complexes with 3-formyl (or acyl) propionate [RuH(ocochrchrcor′)$(PPh_3)_3$] (R --- H, $CH_3$, $C_2H_5$; R --- H, $CH_3$, $C_6H_5$) and with 2-formyl (or acyl) benzoate [RuH(o-$OCCOC_6H_4$COR′)$(PPh_3)_3$] (R′ --- H, $CH_3$). J Organomet Chem. 1982;231:79–90.
82. Tukacs JM, Kiraly D, Stradi A, Novodarszki G, Eke Z, Dibo G, Kegl T, Mika LT. Efficient catalytic hydrogenation of levulinic acid: a key step in biomass conversion. Green Chem. 2012;14:2057–65.
83. Tukacs JM, Jones RV, Darvas F, Dibo G, Lezsak G, Mika LT. Synthesis of [gamma]-valerolactone using a continuous-flow reactor. RSC Adv. 2013;3:16283–7.
84. Heeres H, Handana R, Chunai D, Rasrendra CB, Girisuta B, Heeres HJ. Combined dehydration/(transfer)-hydrogenation of C6-sugars (D-glucose and D-fructose) to γ-valerolactone using ruthenium catalysts. Green Chem. 2009;11:1247–55.
85. Sudhakar M, Kumar VV, Naresh G, Kantam ML, Bhargava SK, Venugopal A. Vapor phase hydrogenation of aqueous levulinic acid over hydroxyapatite supported metal (M = Pd, Pt, Ru, Cu, Ni) catalysts. Appl Catal B Environ. 2016;180:113–20.
86. Du XL, Bi QY, Liu YM, Cao Y, Fan KN. Conversion of biomass-derived levulinate and formate esters into γ-valerolactone over supported gold catalysts. ChemSusChem. 2011;4:1838–43.
87. Du XL, He L, Zhao S, Liu YM, Cao Y, He HY, Fan KN. Hydrogen-independent reductive transformation of carbohydrate biomass into γ-valerolactone and pyrrolidone derivatives with supported gold catalysts. Angew Chem Int Ed. 2011;50:7815–9.
88. Leitner W. Supercritical carbon dioxide as a green reaction cedium for catalysis. Acc Chem Res. 2002;35:746–56.
89. Wettstein SG, Alonso DM, Chong Y, Dumesic JA. Production of levulinic acid and gamma-valerolactone (GVL) from cellulose using GVL as a solvent in biphasic systems. Energy Environ Sci. 2012;5:8199–203.
90. Deng L, Zhao Y, Li J, Fu Y, Liao B, Guo QX. Conversion of levulinic acid and formic acid into γ-Valerolactone over heterogeneous catalysts. ChemSusChem. 2010;3:1172–5.
91. Kopetzki D, Antonietti M. Transfer hydrogenation of levulinic acid under hydrothermal conditions catalyzed by sulfate as a temperature-switchable base. Green Chem. 2010;12:656–60.
92. Manzer LE. Production of 5-methylbutyrolactone from levulinic acid, US Patent 6,617,464, 2003.
93. Yang Y, Gao G, Zhang X, Li F. Facile fabrication of composition-tuned Ru–Ni bimetallics in ordered mesoporous carbon for levulinic acid hydrogenation. ACS Catal. 2014;4:1419–25.
94. Sudhakar M, Lakshmi Kantam M, Swarna Jaya V, Kishore R, Ramanujachary KV, Venugopal A. Hydroxyapatite as a novel support for Ru in the hydrogenation of levulinic acid to γ-valerolactone. Catal Commun. 2014;50:101–4.
95. Wright WRH, Palkovits R. Development of heterogeneous catalysts for the conversion of levulinic acid to γ-valerolactone. ChemSusChem. 2012;5:1657–67.

96. Zhang J, Wu S, Li B, Zhang H. Advances in the catalytic production of valuable levulinic acid derivatives. ChemCatChem. 2012;4:1230–7.
97. Upare PP, Lee JM, Hwang YK, Hwang DW, Lee JH, Halligudi SB, Hwang JS, Chang JS. Direct hydrocyclization of biomass-derived levulinic acid to 2-methyltetrahydrofuran over nanocomposite copper/silica catalysts. ChemSusChem. 2011;4:1749–52.
98. Haan R, Lange JP, Petrus L, Petrus-Hoogenbosch C. Hydrogenation process for the conversion of a carboxylic acid or an ester having a carbonyl group, US. Patent Application 11/680,437, 2007.
99. Mohan V, Raghavendra C, Pramod CV, Raju BD, Rama Rao KS. Ni/H-ZSM-5 as a promising catalyst for vapour phase hydrogenation of levulinic acid at atmospheric pressure. RSC Adv. 2014;4:9660–8.
100. Mohan V, Venkateshwarlu V, Pramod CV, Raju BD, Rao KSR. Vapour phase hydrocyclisation of levulinic acid to [gamma]-valerolactone over supported Ni catalysts. Cat Sci Technol. 2014;4:1253–9.
101. Shimizu KI, Kanno S, Kon K. Hydrogenation of levulinic acid to [gamma]-valerolactone by Ni and MoOx co-loaded carbon catalysts. Green Chem. 2014;16:3899–903.
102. Obregón I, Corro E, Izquierdo U, Requies J, Arias PL. Levulinic acid hydrogenolysis on $Al_2O_3$-based Ni-Cu bimetallic catalysts. Chin J Catal. 2014;35:656–62.
103. Yan K, Chen A. Efficient hydrogenation of biomass-derived furfural and levulinic acid on the facilely synthesized noble-metal-free Cu–Cr catalyst. Energy. 2013;58:357–63.
104. Yan K, Liao J, Wu X, Xie X. A noble-metal free Cu-catalyst derived from hydrotalcite for highly efficient hydrogenation of biomass-derived furfural and levulinic acid. RSC Adv. 2013;3:3853–6.
105. Yan K, Chen A. Selective hydrogenation of furfural and levulinic acid to biofuels on the ecofriendly Cu–Fe catalyst. Fuel. 2014;115:101–8.
106. Yan K, Wu G, Jin W. Recent advances in the synthesis of layered double-hydroxide-based materials and their applications in hydrogen and oxygen evolution. Energ Technol. 2016;4:354–68.
107. Hengne AM, Rode CV. Cu-$ZrO_2$ nanocomposite catalyst for selective hydrogenation of levulinic acid and its ester to [gamma]-valerolactone. Green Chem. 2012;14:1064–72.
108. Kyrides LP, Craver JK. Process for the production of lactones, US Patent 2,368,366, 1945.
109. Christian RV, Brown HD, Hixon RM. Derivatives of γ-Valerolactone, 1,4-Pentanediol and 1,4-di-(β-cyanoethoxy)-pentane. J Am Chem Soc. 1947;69:1961–3.
110. Schuette HA, Thomas RW. Nomral valerolactone. III Its preparation by catalytic reduction of levulinic acid with hydrogen in the prsence of platinum oxide. J Am Chem Soc. 1930;52:3010–2.
111. Wang J, Jaenicke S, Chuah JK. Zirconium-Beta zeolite as a robust catalyst for the transformation of levulinic acid to [gamma]-valerolactone via Meerwein-Ponndorf-Verley reduction. RSC Adv. 2014;4:13481–9.
112. Bui L, Luo H, Gunther WR, Román-Leshkov Y. Domino reaction catalyzed by zeolites with Brønsted and Lewis acid sites for the production of γ-Valerolactone from furfural. Angew Chem Int Ed. 2013;52:8022–5.
113. Quiroz J, Mai EF, Teixeira da Silva V. Synthesis of nanostructured molybdenum carbide as catalyst for the hydrogenation of levulinic acid to γ-Valerolactone. Top Catal. 2016;59:148–58.
114. Youn DH, Bae G, Ham DJ, Lee JS. Electrocatalysts for electrooxidation of methyl formate. Appl Catal A Gen. 2011;393:309–16.
115. Ferenc J. Breakthroughs in hydrogen storage – formic acid as a sustainable storage material for hydrogen. ChemSusChem. 2008;1:805–8.
116. Zhang J, Blazecka PG, Bruendl MM, Huang Y. Ru-TsDPEN with formic acid/Hünig's base for asymmetric transfer hydrogenation, a practical synthesis of optically enriched N-propyl pantolactam. J Org Chem. 2009;74:1411–4.

117. Zhou X, Huang Y, Xing W, Liu C, Liao J, Lu T. High-quality hydrogen from the catalyzed decomposition of formic acid by Pd-Au/C and Pd-Ag/C. Chem Commun. 2008;14(30):3540–2.
118. Fellay C, Yan N, Dyson PJ, Laurenczy G. Selective formic acid decomposition for high-pressure hydrogen generation: a mechanistic study. Chem Eur J. 2009;15:3752–60.
119. Boddien A, Loges B, Junge H, Gärtner F, Noyes JR, Beller M. Continuous hydrogen generation from formic acid: highly active and stable ruthenium catalysts. Adv Synth Catal. 2009;351:2517–20.
120. Ogo S, Abura T, Watanabe Y. pH-dependent transfer hydrogenation of ketones with HCOONa as a hydrogen donor promoted by (η6-$C_6Me_6$)Ru complexes. Organometallics. 2002;21:2964–9.
121. Thomas RW, Schuette HA. Studies on levulinic acid I: its preparaton from carbohydrate by digestion with hydrochloric acid under pressure. J Am Chem Soc. 1931;53:2324–8.
122. Adams RD, Boswell EM, Captain B, Hungria AB, Midgley PA, Raja R, Thomas JM. Bimetallic Ru–Sn nanoparticle catalysts for the solvent-free selective hydrogenation of 1,5,9-Cyclododecatriene to cyclododecene. Angew Chem Int Ed. 2007;46:8182–5.
123. Al-Naji M, Yepez A, Balu AM, Romero AA, Chen Z, Wilde N, Li H, Shih K, Gläser R, Luqueb R. Insights into the selective hydrogenation of levulinic acid to γ-valerolactone using supported mono- and bimetallic catalysts. J Mol Catal A Chem. 2016;417:145–52.
124. Hengst K, Schubert M, Carvalho HWP, Lu C, Kleist W, Grunwaldt JD. Synthesis of γ-valerolactone by hydrogenation of levulinic acid over supported nickel catalysts. Appl Catal A Gen. 2015;502:18–26.
125. Al-Shaal MG, Dzierbinski A, Palkovits R. Solvent-free [gamma]-valerolactone hydrogenation to 2-methyltetrahydrofuran catalysed by Ru/C: a reaction network analysis. Green Chem. 2014;16:1358–64.
126. Ftouni J, Muñoz-Murillo A, Goryachev A, Hofmann J, Hensen EJM, Lu L, Kiely CJ, Bruijnincx PCA, Weckhuysen B. $ZrO_2$ is preferred over $TiO_2$ as support for the Ru-catalyzed hydrogenation of levulinic acid to γ-valerolactone. ACS Catal. 2016;6:5462–72.
127. Valekar AH, Cho KH, Chitale SK, Hong DY, Cha GY, Lee UH, Hwang D, Serre C, Chang JS, Hwang YK. Catalytic transfer hydrogenation of ethyl levulinate to γ-Valerolactone over zirconium-based metal-organic frameworks. Green Chem. 2016;18:4542–52.
128. Luo HY, Consoli DF, Gunther WR, Román-Leshkov Y. Investigation of the reaction kinetics of isolated Lewis acid sites in Beta zeolites for the Meerwein–Ponndorf–Verley reduction of methyl levulinate to γ-valerolactone. J Catal. 2014;320:198–207.
129. Song J, Wu L, Zhou B, Zhou H, Fan H, Yang Y, Meng Q, Han B. A new porous Zr-containing catalyst with a phenate group: an efficient catalyst for the catalytic transfer hydrogenation of ethyl levulinate to [gamma]-valerolactone. Green Chem. 2015;17:1626–32.
130. Kuwahara Y, Kaburagi W, Osada Y, Fujitani T, Yamashita H. Catalytic transfer hydrogenation of biomass-derived levulinic acid and its esters to γ-valerolactone over $ZrO_2$ catalyst supported on SBA-15 silica. Catal Today. 2016; doi:10.1016/j.cattod.2016.05.016.
131. He J, Li H, Lu Y-M, Liu YX, Wu ZB, Hu DY, Yang S. Cascade catalytic transfer hydrogenation–cyclization of ethyl levulinate to γ-valerolactone with Al–Zr mixed oxides. Appl Catal A Gen. 2016;510:11–9.
132. Yan K, Maark TA, Khorshidi A, Sethuraman VA, Peterson A, Guduru PR. The influence of elastic strain on catalytic activity in the hydrogen evolution reaction. Angew Chem Int Ed. 2016;128:6283–9.
133. Yan K, Wu G, Jarvis C, Wen J, Chen A. Facile synthesis of porous microspheres composed of $TiO_2$ nanorods with high photocatalytic activity for hydrogen production. Appl Catal B Environ. 2014;148:281–7.

# Chapter 14
# Production of Amino Acids (L-Glutamic Acid and L-Lysine) from Biomass

**Yota Tsuge and Akihiko Kondo**

**Abstract** Since the discovery of Gram-positive bacterium *Corynebacterium glutamicum* over a half century ago, amino acid fermentation has been one of the largest microbial-based industries, with L-glutamic acid and L-lysine accounting for the majority of amino acids produced. After considerable effort at generating over-producing mutants by random mutagenesis, advances in metabolic and genetic engineering techniques and sequence analysis have greatly facilitated rational design of over-producing strains with expanded carbon utilization. Recent enthusiasm on the utilization of biomass to produce fuels and chemicals has reached the field of amino acid fermentation, and many studies have reported methods to produce L-glutamic acid and L-lysine from biomass. Although L-glutamic acid and L-lysine are predominantly used as flavor enhancers and animal feed additives, respectively, these two amino acids have received considerable attention as platform chemicals for synthesizing building blocks of polymers and carbon fibers. This chapter provides an overview of the microbial production of L-glutamic acid and L-lysine as platform chemicals from biomass.

**Keywords** L-Glutamic acid • L-Lysine • Amino acid fermentation • Biomass • Lignocellulose • *Corynebacterium glutamicum*

Y. Tsuge
Institute for Frontier Science Initiative, Kanazawa University,
Kakuma-machi, Kanazawa, Ishikawa 920-1192, Japan

A. Kondo (✉)
Graduate School of Science, Technology and Innovation, Kobe University,
1-1 Rokkodaicho, Nada-ku, Kobe, Hyogo 657-8501, Japan

Department of Chemical Science and Engineering, Graduate School of Engineering,
Kobe University, 1-1 Rokkodai, Nada, Kobe 657-8501, Japan

RIKEN Center for Sustainable Resource Science,
1-7-22 Suehiro-cho, Tsurumi-ku, Yokohama, Kanagawa 230-0045, Japan
e-mail: akondo@kobe-u.ac.jp



Z. Fang et al. (eds.), *Production of Platform Chemicals from Sustainable Resources*, Biofuels and Biorefineries 7, DOI 10.1007/978-981-10-4172-3_14

## 14.1 Introduction

L-Glutamic acid and L-lysine are essential amino acids that are used as additives in the food and animal feed industries. Currently, almost all amino acids, including L-glutamic acid and L-lysine, are produced through microbial fermentation processes. As amino acids are the essential building blocks of proteins, and concentrations of amino acids in microbial cells were long thought to be strongly limited. However, in 1956, the bacterium *Corynebacterium glutamicum* was isolated and was found to accumulate and excrete L-glutamic acid [1, 2]. This finding allowed for the bulk production of L-glutamic acid through microbial fermentation, as prior to the isolation of *C. glutamicum*, L-glutamic acid was produced through extraction from kelp or protein hydrolysis. In this bacterium, the production of L-glutamic acid is induced by biotin limitation, addition of fatty acid ester surfactants such as Tween 40, or $\beta$-lactam antibiotics, such as penicillin [3–5]. To date, several amino acids, including L-glutamic acid and L-lysine, have been produced industrially using *C. glutamicum* [6].

Worldwide production of amino acids has doubled to nearly 5,000,000 tons within the past decade [7]. L-Glutamic acid, which is exclusively used as a flavor enhancer and represents the largest share of the amino acid market, and L-lysine, which is primarily used as an additive in animal feed, have attracted attention as economically feasible sources of nitrogen-containing building block chemicals for the synthesis of polyamides and carbon fibers, as large amounts of energy are required to incorporate nitrogen into hydrocarbons by chemical-based processes [8–11]. In industry, the amino acids are produced from sugar molasses or starch hydrolysates that contain sucrose and glucose, respectively, as the main components [12]. Because the demand for amino acids continues to increase, the use of inedible biomass, such as lignocellulose and chitin, as a feedstock for amino acid production is highly desirable [13–17].

In this chapter, metabolic engineering strategies are introduced that have been applied for L-lysine overproduction. Then, advances are summarized on expanding the carbon availability of *C. glutamicum* for allowing the production of L-glutamic acid and L-lysine from biomass. Future utilization of L-glutamic acid and L-lysine as building block chemicals for synthesizing commodity chemicals and non-protein amino acids, particularly cadaverine, 5-aminovaleric acid, gamma-aminobutyric acid, 5-aminolevulinic acid, and putrescine is discussed. The chemical conversion of L-glutamic acid into succinonitrile and acrylonitrile is also described.

## 14.2 Metabolic Engineering of *C. glutamicum* for Improving L-Lysine Production

L-Glutamic acid is synthesized in a one-step reaction catalyzed by glutamate dehydrogenase from the tricarboxylic acid [TCA] cycle intermediate 2-oxloglutarate. Genetic engineering is not necessary for L-glutamic acid production in *C. glutamicum*; however, the decreased activity of 2-oxoglutarate dehydrogenase [ODH] has been shown to lead to L-glutamic acid overproduction [18, 19]. Regulatory mechanisms for controlling ODH activity have been found. A small protein OdhI binds to the E1o subunit of ODH [OdhA] depending on its phosphorylation state [20]. OdhI synthesis is induced by addition of Tween 40 or penicillin, and most synthesized OdhI is unphosphorylated, resulting in a decrease in ODH activity. The decreased ODH activity triggers the L-glutamic acid overproduction [21].

L-lysine is synthesized from oxaloacetate through the L-lysine biosynthesis pathway (Fig. 14.1). In contrast to L-glutamic acid, L-lysine biosynthesis is tightly regulated by feedback resistance involving aspartokinase [22]. In addition, four moles of the reducing equivalent NADPH are required to produce one mole of L-lysine (Fig. 14.1). Therefore, deregulation of feedback resistance and increased NADPH availability are necessary for L-lysine overproduction [6]. In the pre-genomic era, classical strategies based on mutagenesis and screening were used to generate L-lysine-overproducing strains [23]. However, with the completion of the genomic sequences of *C. glutamicum* strains [24–26], knowledge on the metabolism and pathway regulation in *C. glutamicum* could be coupled with metabolic engineering for rational design of strains with high productivity and yields of target compounds.

To increase L-lysine productivity in *C. glutamicum*, the following strategies have been used: [i] release of feedback regulation in the key enzyme for L-lysine biosynthesis; [ii] increase of the supply of NADPH and [iii] increase of carbon flux into the L-lysine biosynthesis pathway. The metabolic engineering approaches in *C. glutamicum* for improving L-lysine production are summarized in Fig. 14.1. Prior to the genomic era, deregulation of feedback resistance was typically achieved through the screening of mutants resistant to an L-lysine analogue [27]. Feedback resistant aspartokinase [encoded by the *lysC* gene] was demonstrated to be critical for L-lysine overproduction in *C. glutamicum* [28]. Although several mutations that lead to the deregulation of feedback resistance of aspartokinase were identified, the substitution of threonine at residue 311 with isoleucine was found to be highly effective for increasing L-lysine production [28, 29]. Strains with limited L-threonine synthesizing capability also exhibit elevated L-lysine production, because aspartokinase is inhibited through the concerted feedback by L-lysine and threonine. The substitution of valine with alanine at residue 59 in homoserine dehydrogenase [encoded by the *hom* gene] contributed to this effect [28].

Because NADPH is mainly produced in the pentose phosphate pathway [PPP], increasing the cellular carbon flux toward the PPP is required for increasing L-lysine production by *C. glutamicum*. The mutation of serine at residue 361 to proline in 6-phosphogluconate dehydrogenase [encoded by the *gnd* gene] effectively increases the availability of NADPH [28]. As a more direct approach to increase the carbon flux toward the PPP, overexpression of the *tkt* operon, which is comprised of the *tkt*, *tal*, *zwf*, and *pgl* genes in the PPP has also been shown to markedly increase L-lysine production [30]. Overexpression of the *fbp* gene, which encodes the gluconeogenic enzyme fructose 1,6-bisphosphatase, also improves L-lysine yields by increasing the availability of NADPH [31, 32]. As an alternative approach to promote NADPH production by *C. glutamicum*, the native NADH-generating glyceraldehyde

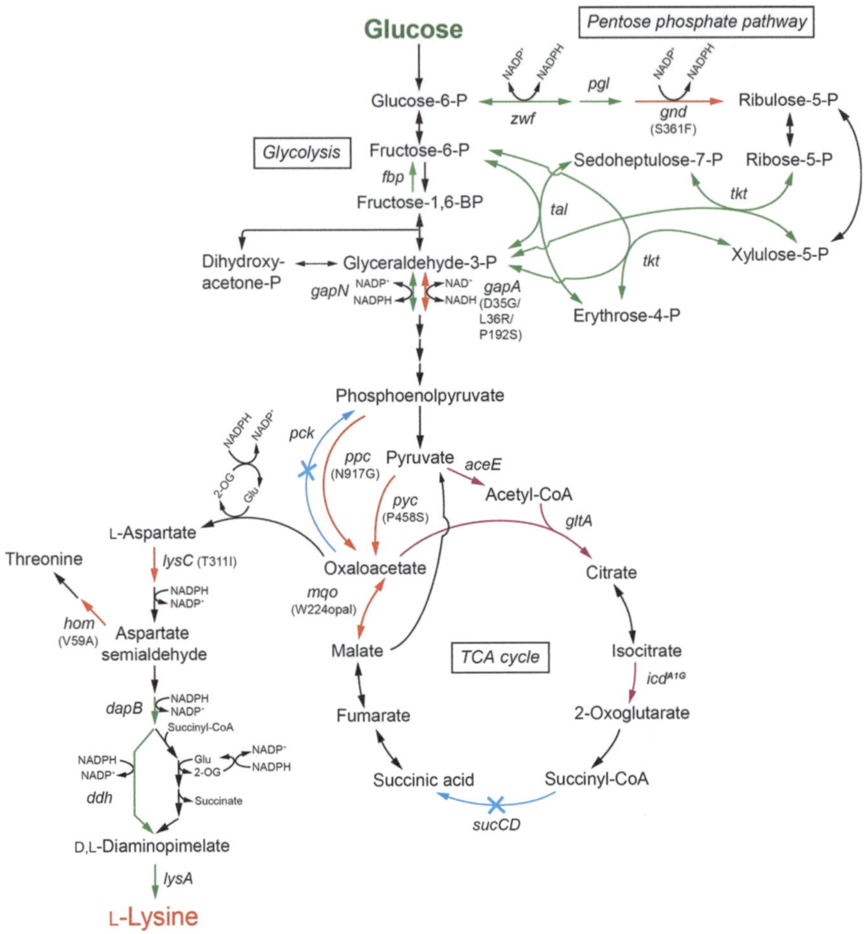


**Fig. 14.1** Metabolic engineering strategies used for improving L-lysine production in *C. glutamicum*. *Green arrows* indicate genes that have been amplified; *red arrows* indicate proteins with amino acid substitutions shown in *brackets*; *purple arrows* indicate genes reduced expression level by either promoter exchange or nucleotide exchange; *blue arrows* indicate genes that have been deleted

3-phosphate dehydrogenase [GAPDH; encoded by the *gapA* gene] can be replaced with a NADPH-generating GAPDH from *Streptococcus mutans* [encoded by the *gapN* gene] harboring a suppressor mutation with improved growth [33–35]. In addition, the co-factor specificity of the endogenous GAPDH can be changed from $NAD^+$ to $NADP^+$ using a rational protein design strategy by targeted mutagenesis [36]. Targeted mutagenesis has also been used to deregulate the feedback inhibition of phosphoenolpyruvate carboxylase [PEPC; encoded by the *ppc* gene] and results in the increased flux toward the PPP, leading to the increased supply of NADPH and acceleration of L-lysine production [37].

To increase the carbon flux into the L-lysine biosynthesis pathway, several genes in this pathway have been overexpressed in *C. glutamicum*. For example, overexpression of the genes encoding deregulated aspartokinase and diaminopimelate decarboxylase [*lysA*], together with those encoding NADPH-consuming dihydrodipicolinate reductase [*dapB*] and diaminopimelate dehydrogenase [*ddh*] that converts L-2,3-dihydrodipicolinate into D,L-diaminopimelate through L-tetrahydrodipicolinate, results in the overproduction of L-lysine [30]. To promote the accumulation of oxaloacetate, which is an important precursor for L-lysine biosynthesis, a mutated pyruvate carboxylase [encoded by the *pyc* gene] with increased activity due to a substitution of proline at residue 458 with serine has been overexpressed [38, 39]. Deletion of the *pck* gene [encoding phosphoenolpyruvate carboxykinase] also contributes to oxaloacetate accumulation and results in increased L-lysine production [40].

Reducing the metabolic flux into the TCA cycle also increases the carbon flux toward the L-lysine biosynthesis pathway, because oxaloacetate can be synthesized from pyruvate via anaplerotic reaction as described above. To this end, decreasing the expression of the *icd* gene [encoding isocitrate dehydrogenase] by replacing adenine in the start codon with guanine increases L-lysine production by *C. glutamicum* [38]. The level of citrate synthase [encoded by the *gltA gene*] activity has also been found to correlate with L-lysine production, with decreased enzyme activity resulting in increased L-lysine production [40]. Replacing the native promoter of the *aceE* gene [encoding the E1p subunit of the pyruvate dehydrogenase complex [PDHC]] with a weaker promoter results in a decrease in growth rate and PDHC activity, but an increase in L-lysine production [41]. In the succinylase branch of the L-lysine biosynthesis pathway, succinyl-CoA is utilized in a reaction catalyzed by tetrahydrodipicolinate *N*-succinyltransferase [encoded by the *dapD* gene], and succinate is then released by the activity of succinyl-diaminopimelate desuccinylase [encoded by the *dapE* gene]. Therefore, the succinylase branch may function as a bridging reaction to convert succinyl-CoA to succinate in the TCA cycle. Indeed, deletion of the *sucCD* gene [encoding the subunits of succinyl-CoA synthetase] increases L-lysine production [42]. A nonsense mutation of tryptophan at residue 224 in the *mqo* gene [encoding malate:quinone oxidoreductase] has been also demonstrated to improve L-lysine production [23]. As the strain expressing the mutated *mqo* exhibits down-regulation of TCA cycle genes in DNA microarray experiments, it is likely that *mqo* gene deletion down-regulates the flux of the TCA cycle to maintain redox balance and results in redirection of the metabolic flux from oxaloacetate towards L-lysine biosynthesis [23].

## 14.3 Expanding the Carbon Availability of *C. glutamicum* for L-Glutamic Acid and L-Lysine Production

The ability to use a wide range of carbon substrates is key for bio-based production of chemicals from biomass that contains many types of complex polysaccharides and sugars. Although wild-type *C. glutamicum* has a relatively narrow substrate

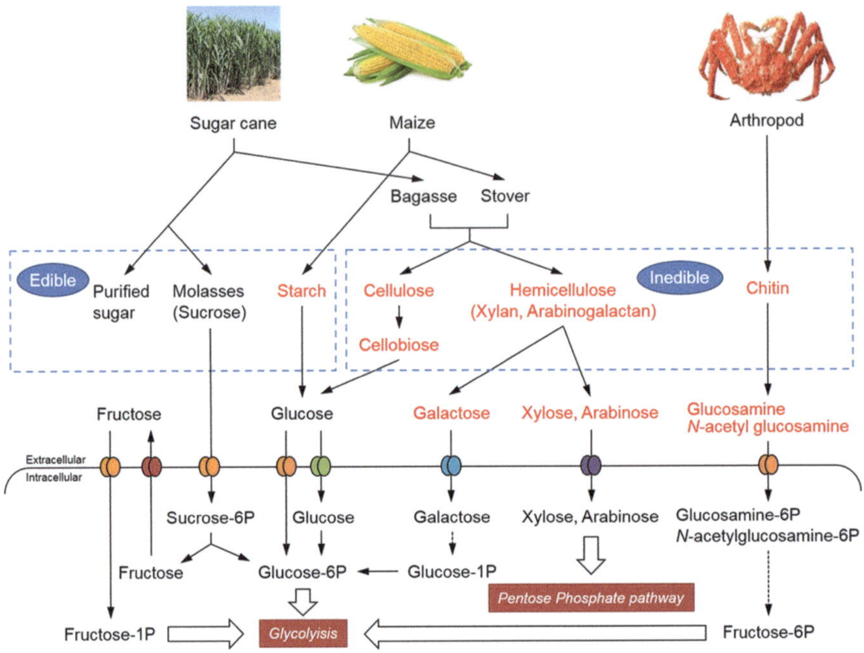


**Fig. 14.2** Schematic representation of biomass utilization in *C. glutamicum. Red text* indicates poly-, di- and monosaccharides that the wild type *C. glutamicum* strain is unable to utilize. *Dotted lines* indicate multi-step enzymatic reactions

range compared with *Escherichia coli*, *C. glutamicum* is capable of utilizing several monosaccharides and disaccharides, including glucose, fructose, sucrose and maltose, for growth [43]. However, *C. glutamicum* is unable to utilize a number of polymeric carbon sources, such as starch, cellulose, hemicellulose, and chitin, or many of their monomeric constituents, including xylose, arabinose, and *N*-acetylglucosamine (Fig. 14.2).

## *14.3.1 Utilization of Edible Biomass*

Starch, in addition to molasses, is a major feedstock for industrial amino acid production. However, prior to utilization by *C. glutamicum*, starch must be enzymatically hydrolyzed into glucose due to lack of starch-degrading enzymes in this organism (Fig. 14.2). The heterologous expression and secretion of α-amylase from *Streptomyces griseus* in *C. glutamicum* enables cell growth on starch as a sole carbon source [44]. Surface display of α-amylase from *Streptococcus bovis* fused with the anchoring protein PgsA from *Bacillus subtilis* also enables the direct utilization of starch by *C. glutamicum* [45]. Another anchoring protein, Ncgl1221, has also been shown to be effective for displaying α-amylase on the surface of *C.*

*glutamicum* cells [46]. This surface display system has been confirmed to be stable for repeated use of the cells [47]. Direct production of L-glutamic acid and L-lysine from starch by *C. glutamicum* cells with surface-displayed or secreted α-amylase has been demonstrated by several groups [44–46].

### 14.3.2 Utilization of Inedible Biomass and Derived Sugars

Lignocellulose is the most abundant renewable resource on Earth. The use of inedible biomass as a substrate for the microbial production of building block chemicals is considered ideal to avoid competition with the food supply. For example, bagasse and corn stover, which are inedible lignocellulosic materials obtained from sugarcane and maize after sugar purification and grain removal, respectively, contain cellulose [40–50%], hemicellulose [20–30%], and lignin [10–20%] [48], and cannot be directly utilized by most microorganisms, including *C. glutamicum*. However, cellulose, which is a linear chain of glucose molecules, can be utilized as a suitable carbon source following depolymerization by chemical or enzymatic processes. Similarly, hemicellulose, which consists primarily of glucose and the C5 sugars xylose, arabinose and galactose, also requires pretreatment by acidic or enzymatic hydrolysis. However, because *C. glutamicum* is only able to utilize glucose among the monosaccharides derived from lignocellulose, the engineering of metabolic pathways is required to use other hemicellulose sugars for target compound production.

Xylose is converted to xylulose-5-phosphate in the PPP in a two-step reaction catalyzed by xylose isomerase [encoded by the *xylA* gene] and xylulokinase [encoded by the *xylB* gene] (Fig. 14.3). Although *C. glutamicum* possesses an endogenous *xylB* gene, xylose utilization in *C. glutamicum* was first demonstrated in strain simultaneously overexpressing the *xylA* and *xylB* genes from *E. coli* [49]. Production of L-lysine and L-glutamic acid from xylose was demonstrated in a strain expressing *xylA* from *Xanthomonas campestris* and endogenous *xylB*, which shows fast consumption of xylose [50]. Arabinose utilization has also been achieved by heterologous expression of the *E. coli araBAD* operon, encoding arabinoisomerase, ribulokinase, and ribulose 5-phosphate 4-epimerase, respectively, which generates xylulose-5-phosphate as in Fig. 14.3 [51]. Production of L-lysine and L-glutamic acid from arabinose has been demonstrated by introducing the same *araBAD* operon [52]. The hemicellulose sugar galactose could also be assimilated by *C. glutamicum* through the heterologous expression of the *galK* [encoding galactokinase], *galT* [encoding UDP-glucose-1-phosphate uridylyltransferase], *galE* [encoding UDP-galactose-4-epimerase], and *galM* genes [encoding aldose-1-epimerase] from *Lactococcus lactis* subsp. cremoris MG1363 as in Fig. 14.3 [53].

The amino sugars glucosamine and *N*-acetylglucosamine, which have the potential to serve as a combined carbon and nitrogen source for the production of L-lysine and L-glutamic acid, are the monomeric components of chitin (Fig. 14.2). Chitin, which is found in the exoskeletons of crustaceans and insects, is the second most

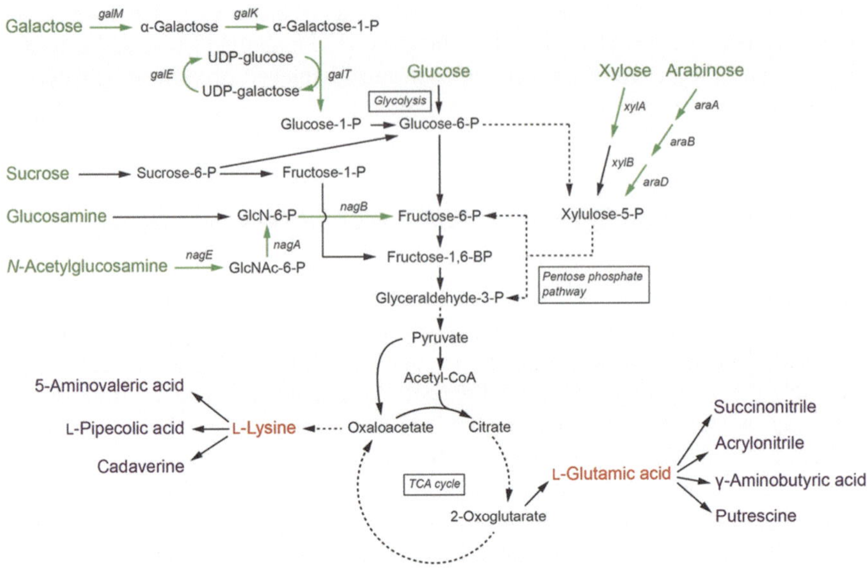


**Fig. 14.3** Overview of the metabolic pathways for the production of L-lysine and L-glutamic acid in *C. glutamicum* from various sugars. *Dotted lines* indicate multi-step enzymatic reactions. Substrate sugars are written in *green*, and target compounds are written in *blue*. *Green arrows* indicate genes that were heterologously or endogenously overexpressed to assimilate the corresponding sugars

abundant organic compound in nature after cellulose. Although *C. glutamicum* possesses genes for glucosamine utilization, the rapid growth on glucosamine as a sole carbon and nitrogen source requires overproduction of endogenous glucosamine-6-phosphate deaminase [encoded by the *nagB* gene] as in Fig. 14.3 [54]. To enable the growth of *C. glutamicum* on *N*-acetylglucosamine, the *N*-acetylglucosamine-specific phosphotransferase system [PTS] gene *nagE* from *Corynebacterium glycinophilum*, in addition to the endogenous *nagB* and *nagA* genes [encoding cytoplasmic *N*-acetylglucosamine-6-phosphate deacetylase], have been overexpressed as in Fig. 14.3 [55]. Using these two metabolic engineering approaches, L-lysine production from glucosamine and *N*-acetylglucosamine has been achieved [54, 55].

*C. glutamicum* has been genetically modified to allow the direct utilization of multi-saccharides present in lignocellulosic biomass and eliminate the need for enzymatic hydrolysis procedures. For example, a strain of *C. glutamicum* ATCC13032 that is capable of utilizing the disaccharide cellobiose, which is formed during enzymatic hydrolysis of cellulose by endoglucanases and exoglucanases, was established by anchoring β-glucosidase [Sed1394] from *Saccharophagus degradans* on the cell surface. This allowed production of L-lysine directly from cellobiose without the requirement for enzymatic hydrolysis by β-glucosidase prior to fermentation [56]. Although *C. glutamicum* ATCC13032 requires genetic engineering to use cellobiose, *C. glutamicum* R is able to degrade cellobiose without genetic engineering because of a single amino acid substitution in the β-glucoside-specific enzyme IIBCA component [BglF] of the phosphotransferase system [57]. *C. glutamicum* has also been genetically modified to allow it to use xylan, which is a polymer of xylose. In addition to optimizing the expression of *xylA*, *xylB*, and xylose transporter *xylE* genes, an endoxylanase gene [*xlnA*] from *Streptomyces coelicolor* A3 [2] and a xylosidase gene [*xynB*] from *Bacillus pumilus* have been expressed with a signal peptide under the control of an optimized promoter to efficiently secrete both enzymes into the culture medium for degrading xylan to xylose [58]. In this modified strain, L-lysine is directly produced from xylan.

## 14.4 L-Glutamic Acid and L-Lysine as Platform Chemicals

### *14.4.1 L-Lysine-Derived Chemicals*

Environmental concerns and continued depletion of oil reserves have prompted many governments to fund research into the development of environmentally benign and sustainable processes for the production of bio-based chemicals and fuels from renewable resources. In particular, L-glutamic acid and L-lysine have attracted attention as an economically feasible source of nitrogen-containing platform chemicals for the synthesis of polymer building blocks that are currently generated from petroleum.

Cadaverine [1,5-diaminopentane], which is a five-carbon diamine, is a promising candidate for the synthesis of "green" nylon [59] and is generated by the one-step decarboxylation of L-lysine (Fig. 14.4). The microbial production of cadaverine was first demonstrated by introducing the lysine decarboxylase gene [*cadA*] from *E. coli* into *C. glutamicum*, in combination with the deletion of the endogenous *hom* gene [encoding homoserine dehydrogenase] [60]. Cadaverine was also directly produced from starch with an α-amylase secreting *C. glutamicum* strain [61]. The Wittmann group extensively examined cadaverine production by *C. glutamicum*. They used another lysine decarboxylase-encoding gene, *ldcC*, from *E. coli* instead of *cadA*, because the LdcC protein prefers neutral pH [62]. However, approximately 20% of the intracellular cadaverine produced by the recombinant strain was acetylated [63]. The group later identified the gene responsible for cadaverine acetylation [*Ncgl1469*, encoding diaminopentane acetyltransferase] in *C. glutamicum* [64]. Genome-wide transcriptional analysis led to the identification of a cadaverine exporter gene [*cg2893*], the product of which was later determined to additionally transport putrescine as described in the following section. The overexpression of *cg2893* in *C. glutamicum* increases cadaverine secretion by 20% [65]. The resultant strain produces 88 g/L cadaverine, corresponding to a molar yield of 50% [66]. The generated cadaverine polymerizes with a bio-based C10-dicarboxylic acid [sebacic acid] to allow synthesize of 100% bio-polyamide [nylon 5,10] that has a comparable melting point [215°C] and glass transition temperature [50°C], but a higher

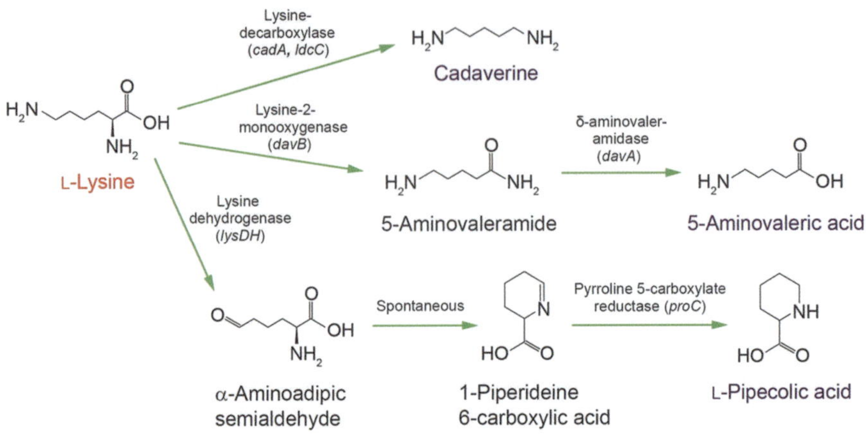


**Fig. 14.4** Reactions for converting L-lysine to cadaverine and 5-aminovaleric acid. Target compounds are written in *blue*

transparency than that of the petrochemical-based polymers nylon 6 and nylon 6,6 [66]. The same research group also produced cadaverine from xylose at a high titer [103 g/L] and product yield [32%] by introducing the *E. coli xylA* and *xylB* genes into *C. glutamicum* [67, 68].

The non-protein amino 5-aminovaleric acid [5-AVA] is a potential C5 building block chemical for synthesizing glutarate, 5-hydroxyvalerate, and 1,5-pentanediol [69]. With increasing demand for nitrogen-containing bulk chemicals, 5-AVA is also considered to be useful for the production of bio-polyamides, such as nylon 5. The 5-AVA is naturally produced by *Pseudomonas putida* as an intermediate in the L-lysine degradation pathway via 5-aminovaleramide in reactions catalyzed by lysine 2-monooxygenase [encoded by the *davB* gene] and delta-aminovaler amidase [encoded by the *davA* gene] (Fig. 14.4). *E. coli* cells expressing the *davAB* genes from *P. putida* have been used for 5-AVA production in both batch and fed-batch fermentations from glucose, and by the bio-conversion of L-lysine as a precursor [70, 71].

The non-protein amino acid L-pipecolic acid [L-PA], a precursor of immunosuppressants, peptide antibiotics, or piperidine alkaloids, can also be synthesized from L-lysine (Fig. 14.4). To enable production of L-PA by a L-lysine-producing strain, the L-lysine 6-dehydrogenase gene *lysDH* from *Silicibacter pomeroyi* and the endogenous pyrroline 5-carboxylate reductase gene *proC*, an enzyme of L-proline biosynthesis in many bacteria including *C. glutamicum*, were overexpressed as synthetic operon [72].

### 14.4.2 L-Glutamic Acid-Derived Chemicals

The non-protein amino acid 5-aminolevulinic acid [5-ALA] is in increasing demand in many fields, including medicine, cosmetics, and agriculture. The 5-ALA is synthesized from L-glutamic acid in three steps in the C5 pathway (Fig. 14.5). In this pathway, the five-carbon skeleton of L-glutamate is first converted to glutamyl adenylate, which is then transferred to a specific tRNA to form glutamyl-tRNA by the action of glutamyl-tRNA ligase [encoded by the *gltX* gene]. The resulting glutamyl-tRNA is converted to glutamate-1-semialdehyde by glutamyl-tRNA reductase [encoded by the *hemA* gene] and is then transaminated in a reaction catalyzed by glutamate-1-semialdehyde aminotransferase [encoded by the *hemL* gene] to generate 5-ALA [73]. Because *C. glutamicum* possesses an endogenous *gltX* gene, the *hemA* gene from *Salmonella typhimurium* or *S. arizona*, and the *hemL* gene from *E. coli* is introduced into *C. glutamicum* to generate a strain capable of producing 5-ALA [74–76].

Putrescine [1,4-diaminobutane] is a four-carbon diamine that is industrially produced by chemical synthesis involving the addition of hydrogen cyanide to acrylonitrile, followed by the hydrogenation of the resulting succinonitrile. Putrescine, in combination with adipic acid, is used for the synthesis of nylon 4,6 [distributed by DSM as Stanyl], which possesses mechanical and physical properties, including melting point, glass transition temperature, tensile strength, solvent resistance, and crystallization rate that are comparable, or even superior, to those of nylon 6,6 [77]. Putrescine can be biologically synthesized from L-ornithine through a single decarboxylation reaction catalyzed by ornithine decarboxylase [encoded by *speC*/*speF*] (Fig. 14.5). To date, the highest titer of microbial-produced putrescine has been obtained using an engineered strain of *E. coli* that overexpresses the *potE* gene [encoding putrescine/ornithine antiporter] and contains deletions of *puuP* [encoding putrescine importer] and the genes encoding enzymes of competitive and degradation routes for putrescine [78]. The native promoters of the key putrescine biosynthetic genes, which consist of the *argECB* operon, encodes *N*-acetylornithine decarboxylase, *N*-acetyl-γ-glutamyl-phosphate reductase, and *N*-acetylglutamate kinase, respectively, *argD* [encoding *N*-acetylornithine aminotransferase] and *speC*, are replaced with a stronger promoter (Fig. 14.5). The resultant strain is capable of producing 24.2 g/L putrescine [78]. *C. glutamicum* has higher tolerance to putrescine compared with *E. coli* and *S. cerevisiae* [79]. Although the putrescine metabolic pathway has not been identified in *C. glutamicum*, introduction of the *speC* gene from *E. coli* enables *C. glutamicum* to synthesize putrescine [79]. Deletion of the *argF* gene [encoding ornithine transcarbamylase] effectively increases putrescine production by increasing the supply of ornithine [80]. Furthermore, the gene responsible for putrescine acetylation, *snaA*, was identified through a systematic gene deletion approach of 18 *N*-acetyltransferase genes. The gene is the same as the one responsible for cadaverine acetylation, and the deletion of the gene minimizes the generation of acetylputrescine as by-product [81]. As mentioned above, the

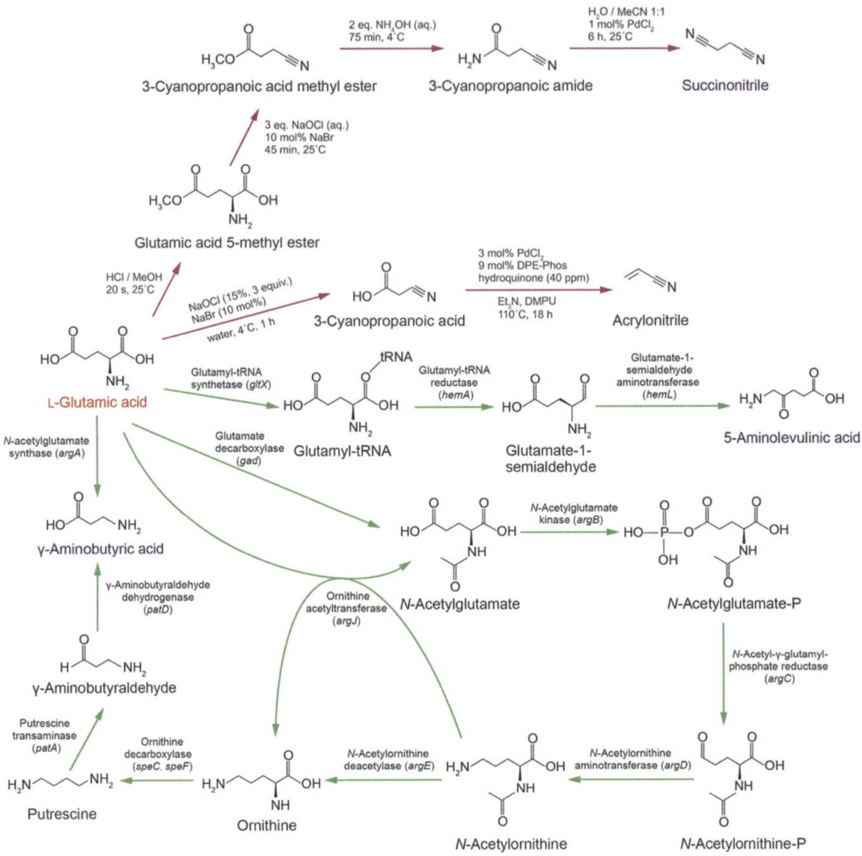


**Fig. 14.5** Reactions for converting L-glutamic acid to succinonitrile, acrylonitrile gamma-aminobutyric acid, 5-aminolevulinic acid and putrescine. Target compounds are written in *blue*. *Purple arrows* indicate chemical reactions; *green arrows* indicate biological reactions. Gene names encoding the corresponding enzymes are written in *parentheses*

decreased activity of ODH in *C. glutamicum* is associated with glutamate overproduction. When ODH activity was made to decrease five-fold by exchanging the translational start codon of a subunit gene *odhA* from GTG to TTG and exchanging the threonine residue 15 of OdhI, which inhibits the ODH complex, putrescine production by *C. glutamicum* increased by 28% [82].

Gamma-aminobutyric acid [GABA] is a non-protein amino acid that is used in the pharmaceutical and food industries. GABA is readily cyclized to form the highly stable lactam 2-pyrrolidone, which can be chemically converted to bio-based polyamide 4 [83, 84]. GABA is synthesized in a one-step decarboxylation reaction catalyzed by glutamate decarboxylase [encoded by the *gad* gene] ([Fig. 14.5). Heterologous expression of the glutamate decarboxylase gene from *E. coli* or *Lactobacillus brevis* in *C. glutamicum* results in GABA production under culture conditions that stimulate L-glutamic acid production, as described above in Sect. 14.1 [85–88]. Moreover, a protein specific for importing GABA [GabP[Cg]] encoded by *Ncgl0464* was identified, and deletion of the gene improved extracellar GABA concentration by 12.5% [89]. Very recently, a new metabolic route for the production of GABA through putrescine by two-step reactions was identified (Fig. 14.5). Putrescine is firstly converted to γ-aminobutyraldehyde by putrescine transaminase encoded by *patA*. GABA is then synthesized by γ-aminobutyraldehyde dehydrogenase encoded by *patD* [90].

L-Glutamic acid is a useful starting chemical for the synthesis of nitrogen-containing compounds, such as acrylonitrile and succinonitrile. In particular, the demand for acrylonitrile has largely increased in the past decade for use as a building block material for carbon fibers. Although the target compounds are synthesized by chemical conversion of L-glutamic acid, use of L-glutamic acid as a starting chemical would alleviate the need for fossil fuels in these processes [91]. Acrylonitrile can be chemically produced from L-glutamic acid in a two-step procedure without the use of additional ammonia (Fig. 14.5). In the first step, 3-cyanopropanoic acid is generated through the oxidative decarboxylation of L-glutamic acid in water, followed by a decarbonylation-elimination reaction using a palladium catalyst to produce acrylonitrile [92]. L-glutamic acid can also be converted to succinonitrile, which is an industrial precursor of putrescine, in a four-step reaction (Fig. 14.5). L-Glutamic acid is first esterified using methanol to form glutamic acid 5-methyl ester, which is then oxidatively decarboxylated to generate 3-cyanopropanoic acid methyl ester. This reaction is followed by amidation of the ester in aqueous ammonia to form 3-cyanopropanoic amide, which is then dehydrated to generate succinonitrile [93].

## 14.5 Conclusions and Future Outlook

Microbial strains capable of producing high yields of L-glutamic acid and L-lysine from biomass have been generated and optimized using metabolic engineering approaches. In particular, expanding the carbon source availability for *C. glutamicum* through the use of metabolic engineering technology is a desirable goal for large scale bio-production of nitrogen-containing platform chemicals for synthesizing polyamide and carbon fiber materials. Utilization of L-glutamic acid and L-lysine as platform chemicals is expected to help reduce environmental burden associated with the use of petroleum-based materials. Further challenges remain in giving *C. glutamicum* tolerance to various stresses such as low pH and elevated temperatures. Being able to culture the bacterium at low pH and at elevated temperatures would reduce a significant amount of downstream purification and cooling costs and thus would improve process economics. High tolerance of bacterium to fermentation inhibitors that are produced during pretreatment of lignocellulosic biomass is also required for the practical processing of lignocellulose. Further optimization of the enzymes and transporters involved in the production and extracellular secretion of these platform chemicals and target compounds through protein engineering, evolutionary engineering, and synthetic biology is expected to enable more efficient redirection and acceleration of carbon flux toward the target chemicals and to further increase their yields.

## References


1. Kinoshita S, Udaka S, Shimono M. Studies on the amino acid fermentation part 1. Production of L-glutamic acid by various microorganisms. J Gen Appl Microbiol. 1957;3:193–205.
2. Udaka S. Screening method for microorganisms accumulating metabolites and its use in the isolation of *Hicrococcus alatamicus*. J Bactariol. 1960;79:754–5.
3. Nara T, Samejima H, Kinoshita S. Effect of penicillin on amino acid fermentation. Agric Biol Chem. 1964;28:120–4.
4. Shiio I, Otsuka SI, Takahashi M. Effect of biotin on the bacterial formation of glutamic acid. I. Glutamate formation and cellular permeability of amino acids. J Biochem. 1962;51:56–62.
5. Takinami K, Yoshii H, Tsuri H, Okada H. Biochemical effects of fatty acid and its derivatives on L-glutamic acid fermentation. Part III. Biotin-Tween 60 relationship in the accumulation of L-glutamic acid and the growth of *Brevibacterium lactofermentum*. Agric Biol Chem. 1965;29:351–9.
6. Hirasawa T, Shimizu H. Recent advances in amino acid production by microbial cells. Curr Opin Biotechnol. 2016;42:133–46.
7. Eggeling L, Bott M. A giant market and a powerful metabolism: L-lysine provided by *Corynebacterium glutamicum*. Appl Microbiol Biotechnol. 2015;9:3387–94.
8. Becker J, Wittmann C. Advanced biotechnology: metabolically engineered cells for the bio-based production of chemicals and fuels, materials, and health-care products. Angew Chem Int Ed Engl. 2015;54:3328–50.
9. Scott E, Peter F, Sanders J. Biomass in the manufacture of industrial products--the use of proteins and amino acids. Appl Microbiol Biotechnol. 2007;75:751–62.
10. Tsuge Y, Kawaguchi H, Sasaki K, Kondo A. Engineering cell factories for producing building block chemicals for bio-polymer synthesis. Microb Cell Factories. 2016;15:19.
11. Wendisch VF. Microbial production of amino acids and derived chemicals: synthetic biology approaches to strain development. Curr Opin Biotechnol. 2014;30:51–8.
12. Hermann T. Industrial production of amino acids by coryneform bacteria. J Biotechnol. 2003;104:155–72.
13. Gopinath V, Murali A, Dhar KS, Nampoothiri KM. *Corynebacterium glutamicum* as a potent biocatalyst for the bioconversion of pentose sugars to value-added products. Appl Microbiol Biotechnol. 2012;93:95–106.
14. Jojima T, Omumasaba CA, Inui M, Yukawa H. Sugar transporters in efficient utilization of mixed sugar substrates: current knowledge and outlook. Appl Microbiol Biotechnol. 2010;85:471–80.
15. Tsuge Y, Hasunuma T, Kondo A. Recent advances in the metabolic engineering of *Corynebacterium glutamicum* for the production of lactate and succinate from renewable resources. J Ind Microbiol Biotechnol. 2015;42:375–89.
16. Zahoor A, Lindner SN, Wendisch VF. Metabolic engineering of *Corynebacterium glutamicum* aimed at alternative carbon sources and new products. Comput Struct Biotechnol J. 2012;3:e201210004.
17. Watanabe A, Hiraga K, Suda M, Yukawa H, Inui M. Functional Characterization of *Corynebacterium alkanolyticum* β-Xylosidase and Xyloside ABC Transporter in *Corynebacterium glutamicum*. Appl Environ Microbiol. 2015;81:4173–83.

18. Asakura Y, Kimura E, Usuda Y, Kawahara Y, Matsui K, Osumi T, Nakamatsu T. Altered metabolic flux due to deletion of *odhA* causes L-glutamate overproduction in *Corynebacterium glutamicum*. Appl Environ Microbiol. 2007;73:1308–19.
19. Kawahara Y, Takahashi-Fuke K, Shimizu E, Nakamatsu T, Nakamori S. Relationship between the glutamate production and the activity of 2-oxoglutarate dehydrogenase in *Brevibacterium lactofermentum*. Biosci Biotechnol Biochem. 1997;61:1109–12.
20. Niebisch A, Kabus A, Schultz C, Weil B, Bott M. Corynebacterial protein kinase G controls 2-oxoglutarate dehydrogenase activity via the phosphorylation status of the OdhI protein. J Biol Chem. 2006;281:12300–7.
21. Kim J, Hirasawa T, Saito M, Furusawa C, Shimizu H. Investigation of phosphorylation status of OdhI protein during penicillin- and Tween 40-triggered glutamate overproduction by *Corynebacterium glutamicum*. Appl Microbiol Biotechnol. 2011;91:143–51.
22. Yoshida A, Tomita T, Kurihara T, Fushinobu S, Kuzuyama T, Nishiyama M. Structural insight into concerted inhibition of alpha 2 beta 2-type aspartate kinase from *Corynebacterium glutamicum*. J Mol Biol. 2007;368:521–36.
23. Ikeda M, Ohnishi J, Hayashi M, Mitsuhashi S. A genome-based approach to create a minimally mutated *Corynebacterium glutamicum* strain for efficient L-lysine production. J Ind Microbiol Biotechnol. 2006;33:610–5.
24. Ikeda M, Nakagawa S. The *Corynebacterium glutamicum* genome: features and impacts on biotechnological processes. Appl Microbiol Biotechnol. 2003;62:99–109.
25. Kalinowski J, Bathe B, Bartels D, Bischoff N, Bott M, Burkovski A, Dusch N, Eggeling L, Eikmanns BJ, Gaigalat L, Goesmann A, Hartmann M, Huthmacher K, Krämer R, Linke B, McHardy AC, Meyer F, Möckel B, Pfefferle W, Pühler A, Rey DA, Rückert C, Rupp O, Sahm H, Wendisch VF, Wiegräbe I, Tauch A. The complete *Corynebacterium glutamicum* ATCC 13032 genome sequence and its impact on the production of L-aspartate-derived amino acids and vitamins. J Biotechnol. 2003;104:5–25.
26. Yukawa H, Omumasaba CA, Nonaka H, Kos P, Okai N, Suzuki N, Suda M, Tsuge Y, Watanabe J, Ikeda Y, Vertès AA, Inui M. Comparative analysis of the *Corynebacterium glutamicum* group and complete genome sequence of strain R. Microbiology. 2007;153:1042–58.
27. Sano K, Shiio I. Microbial production of L-lysine III. Production by mutants resistant to *S*-(2-aminoethyl)- L-cysteine. J Gen Appl Microbiol. 1970;16:373–91.
28. Ohnishi J, Mitsuhashi S, Hayashi M, Ando S, Yokoi H, Ochiai K, Ikeda M. A novel methodology employing *Corynebacterium glutamicum* genome information to generate a new L-lysine-producing mutant. Appl Microbiol Biotechnol. 2002;58:217–23.
29. Kelle R, Hermann T, Bathe B. L-lysine production: Handbook of *Corynebacterium glutamicum*. Boca Raton: CRC Press; 2005. p. 465–88.
30. Becker J, Zelder O, Häfner S, Schröder H, Wittmann C. From zero to hero--design-based systems metabolic engineering of *Corynebacterium glutamicum* for L-lysine production. Metab Eng. 2011;13:159–68.
31. Becker J, Klopprogge C, Zelder O, Heinzle E, Wittmann C. Amplified expression of fructose 1,6-bisphosphatase in *Corynebacterium glutamicum* increases in vivo flux through the pentose phosphate pathway and lysine production on different carbon sources. Appl Environ Microbiol. 2005;71:8587–96.
32. Georgi T, Rittmann D, Wendisch VF. Lysine and glutamate production by *Corynebacterium glutamicum* on glucose, fructose and sucrose: roles of malic enzyme and fructose-1,6-bisphosphatase. Metab Eng. 2005;7:291–301.
33. Takeno S, Murata R, Kobayashi R, Mitsuhashi S, Ikeda M. Engineering of *Corynebacterium glutamicum* with an NADPH-generating glycolytic pathway for L-lysine production. Appl Environ Microbiol. 2010;76:7154–60.
34. Komati Reddy G, Lindner SN, Wendisch VF. Metabolic engineering of an ATP-neutral Embden-Meyerhof-Parnas pathway in *Corynebacterium glutamicum*: growth restoration by an adaptive point mutation in NADH dehydrogenase. Appl Environ Microbiol. 2015;81:1996–2005.

35. Takeno S, Hori K, Ohtani S, Mimura A, Mitsuhashi S, Ikeda M. L-Lysine production independent of the oxidative pentose phosphate pathway by *Corynebacterium glutamicum* with the *Streptococcus mutans gapN* gene. Metab Eng. 2016;37:1–10.
36. Bommareddy RR, Chen Z, Rappert S, Zeng AP. A *de novo* NADPH generation pathway for improving lysine production of *Corynebacterium glutamicum* by rational design of the coenzyme specificity of glyceraldehyde 3-phosphate dehydrogenase. Metab Eng. 2014;25:30–7.
37. Chen Z, Bommareddy RR, Frank D, Rappert S, Zeng AP. Deregulation of feedback inhibition of phosphoenolpyruvate carboxylase for improved lysine production in *Corynebacterium glutamicum*. Appl Environ Microbiol. 2014;80:1388–93.
38. Becker J, Klopprogge C, Schroder H, Wittmann C. Metabolic engineering of the tricarboxylic acid cycle for improved lysine production by *Corynebacterium glutamicum*. Appl Environ Microbiol. 2009;75:7866–9.
39. Riedel C, Rittmann D, Dangel P, Möckel B, Petersen S, Sahm H, Eikmanns BJ. Characterization of the phosphoenolpyruvate carboxykinase gene from *Corynebacterium glutamicum* and significance of the enzyme for growth and amino acid production. J Mol Microbiol Biotechnol. 2001;3:573–83.
40. van Ooyen J, Noack S, Bott M, Reth A, Eggeling L. Improved L-lysine production with *Corynebacterium glutamicum* and systemic insight into citrate synthase flux and activity. Biotechnol Bioeng. 2012;109:2070–81.
41. Buchholz J, Schwentner A, Brunnenkan B, Gabris C, Grimm S, Gerstmeir R, Takors R, Eikmanns BJ, Blombach B. Platform engineering of *Corynebacterium glutamicum* with reduced pyruvate dehydrogenase complex activity for improved production of L-lysine, L-valine, and 2-ketoisovalerate. Appl Environ Microbiol. 2013;79:5566–75.
42. Kind S, Becker J, Wittmann C. Increased lysine production by flux coupling of the tricarboxylic acid cycle and the lysine biosynthetic pathway-metabolic engineering of the availability of succinyl-CoA in *Corynebacterium glutamicum*. Metab Eng. 2013;15:184–95.
43. Blombach B, Seibold GM. Carbohydrate metabolism in *Corynebacterium glutamicum* and applications for the metabolic engineering of L-lysine production strains. Appl Microbiol Biotechnol. 2010;86:1313–22.
44. Seibold G, Auchter M, Berens S, Kalinowski J, Eikmanns BJ. Utilization of soluble starch by a recombinant *Corynebacterium glutamicum* strain: growth and lysine production. J Biotechnol. 2006;124:381–91.
45. Tateno T, Fukuda H, Kondo A. Direct production of L-lysine from raw corn starch by *Corynebacterium glutamicum* secreting *Streptococcus bovis* alpha-amylase using *cspB* promoter and signal sequence. Appl Microbiol Biotechnol. 2007;77:533–41.
46. Yao W, Chu C, Deng X, Zhang Y, Liu M, Zheng P, Sun Z. Display of alpha-amylase on the surface of *Corynebacterium glutamicum* cells by using NCgl1221 as the anchoring protein, and production of glutamate from starch. Arch Microbiol. 2009;191:751–9.
47. Tsuge Y, Tateno T, Sasaki K, Hasunuma T, Tanaka T, Kondo A. Direct production of organic acids from starch by cell surface-engineered *Corynebacterium glutamicum* in anaerobic conditions. AMB Express. 2013;3:72.
48. Aristidou A, Penttila M. Metabolic engineering applications to renewable resource utilization. Curr Opin Biotechnol. 2000;11:187–98.
49. Kawaguchi H, Vertès AA, Okino S, Inui M, Yukawa H. Engineering of a xylose metabolic pathway in *Corynebacterium glutamicum*. Appl Environ Microbiol. 2006;72:3418–28.
50. Meiswinkel TM, Gopinath V, Lindner SN, Nampoothiri KM, Wendisch VF. Accelerated pentose utilization by *Corynebacterium glutamicum* for accelerated production of lysine, glutamate, ornithine and putrescine. Microb Biotechnol. 2013;6:131–40.
51. Kawaguchi H, Sasaki M, Vertes AA, Inui M, Yukawa H. Engineering of an L-arabinose metabolic pathway in *Corynebacterium glutamicum*. Appl Microbiol Biotechnol. 2008;77:1053–62.
52. Schneider J, Niermann K, Wendisch VF. Production of the amino acids L-glutamate, L-lysine, L-ornithine and L-arginine from arabinose by recombinant *Corynebacterium glutamicum*. J Biotechnol. 2011;154:191–8.

53. Barrett E, Stanton C, Zelder O, Fitzgerald G, Ross RP. Heterologous expression of lactose- and galactose-utilizing pathways from lactic acid bacteria in *Corynebacterium glutamicum* for production of lysine in whey. Appl Environ Microbiol. 2004;70:2861–6.
54. Uhde A, Youn JW, Maeda T, Clermont L, Matano C, Kramer R, Wendisch VF, Seibold GM, Marin K. Glucosamine as carbon source for amino acid-producing *Corynebacterium glutamicum*. Appl Microbiol Biotechnol. 2013;97:1679–87.
55. Matano C, Uhde A, Youn JW, Maeda T, Clermont L, Marin K, Kramer R, Wendisch VF, Seibold GM. Engineering of *Corynebacterium glutamicum* for growth and L-lysine and lycopene production from *N*-acetylglucosamine. Appl Microbiol Biotechnol. 2014;98:5633–43.
56. Adachi N, Takahashi C, Ono-Murota N, Yamaguchi R, Tanaka T, Kondo A. Direct L-lysine production from cellobiose by *Corynebacterium glutamicum* displaying β-glucosidase on its cell surface. Appl Microbiol Biotechnol. 2013;97:7165–72.
57. Kotrba P, Inui M, Yukawa H. A single V317A or V317M substitution in Enzyme II of a newly identified β-glucoside phosphotransferase and utilization system of *Corynebacterium glutamicum* R extends its specificity towards cellobiose. Microbiology. 2003;149:1569–80.
58. Yim SS, Choi JW, Lee SH, Jeong KJ. Modular optimization of a hemicellulose-utilizing pathway in *Corynebacterium glutamicum* for consolidated bioprocessing of hemicellulosic biomass. ACS Synth Biol. 2016;15:334–43.
59. Kind S, Wittmann C. Bio-based production of the platform chemical 1, 5-diaminopentane. Appl Microbiol Biotechnol. 2011;91:1287–96.
60. Mimitsuka T, Sawai H, Hatsu M, Yamada K. Metabolic engineering of *Corynebacterium glutamicum* for cadaverine fermentation. Biosci Biotechnol Biochem. 2007;71:2130–5.
61. Tateno T, Okada Y, Tsuchidate T, Tanaka T, Fukuda H, Kondo A. Direct production of cadaverine from soluble starch using *Corynebacterium glutamicum* coexpressing α-amylase and lysine decarboxylase. Appl Microbiol Biotechnol. 2009;82:115–21.
62. Yamamoto Y, Miwa Y, Miyoshi K, Furuyama J, Ohmori H. The *Escherichia coli ldcC* gene encodes another lysine decarboxylase, probably a constitutive enzyme. Genes Genet Syst. 1997;72:167–72.
63. Kind S, Jeong WK, Schröder H, Wittmann C. Systems-wide metabolic pathway engineering in *Corynebacterium glutamicum* for bio-based production of diaminopentane. Metab Eng. 2010;12:341–51.
64. Kind S, Jeong WK, Schröder H, Zelder O, Wittmann C. Identification and elimination of the competing *N*-acetyldiaminopentane pathway for improved production of diaminopentane by *Corynebacterium glutamicum*. Appl Environ Microbiol. 2010;76:5175–80.
65. Kind S, Kreye S, Wittmann C. Metabolic engineering of cellular transport for overproduction of the platform chemical 1,5-diaminopentane in *Corynebacterium glutamicum*. Metab Eng. 2011;13:617–27.
66. Kind S, Neubauer S, Becker J, Yamamoto M, Völkert M, Gv A, Zelder O, Wittmann C. From zero to hero – Production of bio-based nylon from renewable resources using engineered *Corynebacterium glutamicum*. Metab Eng. 2014;25:113–23.
67. Buschke N, Schröder H, Wittmann C. Metabolic engineering of *Corynebacterium glutamicum* for production of 1,5-diaminopentane from hemicellulose. Biotechnol J. 2011;6:306–17.
68. Buschke N, Becker J, Schäfer R, Kiefer P, Biedendieck R, Wittmann C. Systems metabolic engineering of xylose-utilizing *Corynebacterium glutamicum* for production of 1,5-diaminopentane. Biotechnol J. 2013;8:557–70.
69. Park SJ, Oh YH, Noh W, Kim HY, Shin JH, Lee EG, Lee S, David Y, Baylon MG, Song BK, Jegal J, Lee SY, Lee SH. High-level conversion of L-lysine into 5-aminovalerate that can be used for nylon 6,5 synthesis. Biotechnol J. 2014;9:1322–8.
70. Adkins J, Jordan J, Nielsen DR. Engineering *Escherichia coli* for renewable production of the 5-carbon polyamide building-blocks 5-aminovalerate and glutarate. Biotechnol Bioeng. 2013;110:1726–34.
71. Park SJ, Kim EY, Noh W, Park HM, Oh YH, Lee SH, Song BK, Jegal J, Lee SY. Metabolic engineering of *Escherichia coli* for the production of 5-aminovalerate and glutarate as C5 platform chemicals. Metab Eng. 2013;16:42–7.

72. Pérez-García F, Peters-Wendisch P, Wendisch VF. Engineering Corynebacterium glutamicum for fast production of L-lysine and L-pipecolic acid. Appl Microbiol Biotechnol. 2016;100:8075–90.
73. Liu L, Zhang G, Li X, Zhang J. Microbial production and applications of 5- aminolevulinic acid. Appl Microbiol Biotechnol. 2014;98:7349–57.
74. Feng L, Zhang Y, Fu J, Mao Y, Chen T, Zhao X, Wang Z. Metabolic engineering of *Corynebacterium glutamicum* for efficient production of 5-aminolevulinic acid. Biotechnol Bioeng. 2016;113:1284–93.
75. Ramzi AB, Hyeon JE, Kim SW, Park C, Han SO. 5-Aminolevulinic acid production in engineered *Corynebacterium glutamicum* via C5 biosynthesis pathway. Enzym Microb Technol. 2015;81:1–7.
76. Yu X, Jin H, Liu W, Wang Q, Qi Q. Engineering *Corynebacterium glutamicum* to produce 5-aminolevulinic acid from glucose. Microb Cell Factories. 2015;14:183.
77. Roerdink E, Warnier JMM. Preparation and properties of high molar mass nylon-4, 6: a new development in nylon polymers. Polymer. 1985;26:1582–8.
78. Qian ZG, Xia XX, Lee SY. Metabolic engineering of *Escherichia coli* for the production of putrescine: a four carbon diamine. Biotechnol Bioeng. 2009;104:651–62.
79. Schneider J, Wendisch VF. Putrescine production by engineered *Corynebacterium glutamicum*. Appl Microbiol Biotechnol. 2010;88:859–68.
80. Schneider J, Eberhardt D, Wendisch VF. Improving putrescine production by *Corynebacterium glutamicum* by fine-tuning ornithine transcarbamoylase activity using a plasmid addiction system. Appl Microbiol Biotechnol. 2012;95:169–78.
81. Nguyen AQ, Schneider J, Wendisch VF. Elimination of polyamine *N*-acetylation and regulatory engineering improved putrescine production by *Corynebacterium glutamicum*. J Biotechnol. 2015;201:75–85.
82. Nguyen AQ, Schneider J, Reddy GK, Wendisch VF. Fermentative production of the diamine putrescine: system metabolic engineering of *Corynebacterium glutamicum*. Metabolites. 2015;5:211–31.
83. Kawasaki N, Nakayama A, Yamano N, Takeda S, Kawata Y, Yamamoto N, Aiba S. Synthesis, thermal and mechanical properties and biodegradation of branched polyamide 4. Polymer. 2005;46:9987–93.
84. Lammens TM, Franssen MCR, Scott EL, Sanders JPM. Synthesis of biobased *N*-methylpyrrolidone by one-pot cyclization and methylation of γ-aminobutyric acid. Green Chem. 2010;12:1430–6.
85. Choi JW, Yim SS, Lee SH, Kang TJ, Park SJ, Jeong KJ. Enhanced production of gamma-aminobutyrate (GABA) in recombinant *Corynebacterium glutamicum* by expressing glutamate decarboxylase active in expanded pH range. Microb Cell Factories. 2015;14:21.
86. Okai N, Takahashi C, Hatada K, Ogino C, Kondo A. Disruption of *pknG* enhances production of gamma-aminobutyric acid by *Corynebacterium glutamicum* expressing glutamate decarboxylase. AMB Express. 2014;4:20.
87. Shi F, Jiang J, Li Y, Li Y, Xie Y. Enhancement of γ-aminobutyric acid production in recombinant *Corynebacterium glutamicum* by co-expressing two glutamate decarboxylase genes from Lactobacillus brevis. J Ind Microbiol Biotechnol. 2013;40:1285–96.
88. Takahashi C, Shirakawa J, Tsuchidate T, Okai N, Hatada K, Nakayama H, Tateno T, Ogino C, Kondo A. Robust production of gamma-amino butyric acid using recombinant *Corynebacterium glutamicum* expressing glutamate decarboxylase from *Escherichia coli*. Enzym Microb Technol. 2012;51:171–6.
89. Zhao Z, Ding JY, Ma WH, Zhou NY, Liu SJ. Identification and characterization of γ-aminobutyric acid uptake system GabPCg (NCgl0464) in *Corynebacterium glutamicum*. Appl Environ Microbiol. 2012;78:2596–601.

90. Jorge JM, Leggewie C, Wendisch VF. A new metabolic route for the production of gamma-aminobutyric acid by *Corynebacterium glutamicum* from glucose. Amino Acids. 2016;48(11):2519–31.
91. Lammens TM, Potting J, Sanders JPM, De Boer IJM. Environmental comparison of biobased chemicals from glutamic acid with their petrochemical equivalents. Environ Sci Technol. 2011;45:8521–8.
92. Le Nôtre J, Scott EL, Franssen MCR, Sandersa JPM. Biobased synthesis of acrylonitrile from glutamic acid. Green Chem. 2011;13:807–9.
93. Lammens TM, Le Nôtre J, Franssen MCR, Scott EL, Sanders JPM. Synthesis of biobased succinonitrile from glutamic acid and glutamine. ChemSusChem. 2011;4:785–91.

# Index





Z. Fang et al. (eds.), *Production of Platform Chemicals from Sustainable Resources*, Biofuels and Biorefineries 7, DOI 10.1007/978-981-10-4172-3

**B**

**C**

**Table 2.1** Selected metal cations pKa and hardness values

| Metal cation | pKa | Hardness value (eV) [54] |
|---|---|---|
| $Fe^{3+}$ | 2.46 | 13.1 |
| $Al^{3+}$ | 4.85 | 45.8 |
| $Cu^{2+}$ | 6.50 | 8.3 |
| $Fe^{2+}$ | 9.49 | 7.3 |
| $Ca^{2+}$ | 12.7 | 19.7 |

Adapted with permission from Kamireddy SR, Li J, Tucker M, Degenstein J, Ji Y Effects and mechanism of metal chloride salts on pretreatment and enzymatic digestibility of corn stover. Ind Eng Chem Res 52 (5):1775-1782. Copyright (2013) American Chemical Society [41].

metal chlorides, such as $FeCl_3$, $AlCl_3$, $CrCl_2$, have been widely used due to their ability to catalyse the dehydration of $C_5$-sugars into furfural [17, 53].

Liu and Wyman reported that aqueous solutions containing 0.8 wt. % $CaCl_2$ (pH = 5.59) or 0.8 wt. % $MgCl_2$ (pH = 5.87) are acidic and that the pH value of 0.8 wt. % $FeCl_3$ solutions is particularly low (pH = 1.86) [55]. Cai et al. reported the influence of several metal halides, $FeCl_3$, $AlCl_3$, $CrCl_3$, $CuCl_2$, and $ZrOCl_2$ in THF-water system, on the conversion of different lignocellulosic biomasses, such as maple wood and corn stover, into furfural [56]. All pH values of metal solutions were determined in 1:1 THF-water system. The $FeCl_3$ solution (pH = 1.90) allowed the highest theoretical furfural yield of 95 mol% from maple wood at 170 °C for 60 min of reaction time and at 4:1 THF-water system. Its high Brønsted character permitted slowing down the xylose conversion rates and enhancing the furfural selectivity at longer reaction times. On the other hand, $CuCl_2$ (pH = 2.78) and $AlCl_3$ (pH = 2.88) had a moderate effect resulting in furfural yields in the range of 81–89 mol% and 58–76 mol%, respectively, probably due to their moderate Brønsted acidity. The effect of $ZrOCl_2$ (pH = 1.65) on maple wood was characterised by a low furfural yield (44 mol%) and a poor selectivity (37 mol%), despite its strong Brønsted acidity. The $CrCl_3$ solution (pH = 3.13) allowed high xylose conversion (100% in 5 min), however it had a negative impact on furfural yield as furfural losses quickly overtook the furfural production. The authors proposed that the strong Lewis acid character of $CrCl_3$ played an important role in furfural losses [56].

## Furfural from $C_5$-Sugars and Lignocellulosic Feedstocks

### *Monophasic Aqueous and Non-aqueous Systems*

Converting xylan polymer to furfural is a critical step in the utilisation of lignocellulosic biomass, since β-D-xylose is the main constituent of hemicellulose. The ability to produce furfural with a high yield and selectivity implies a low production of humins as conversion of xylose to furfural is crucial to limit the production of side products and humins.

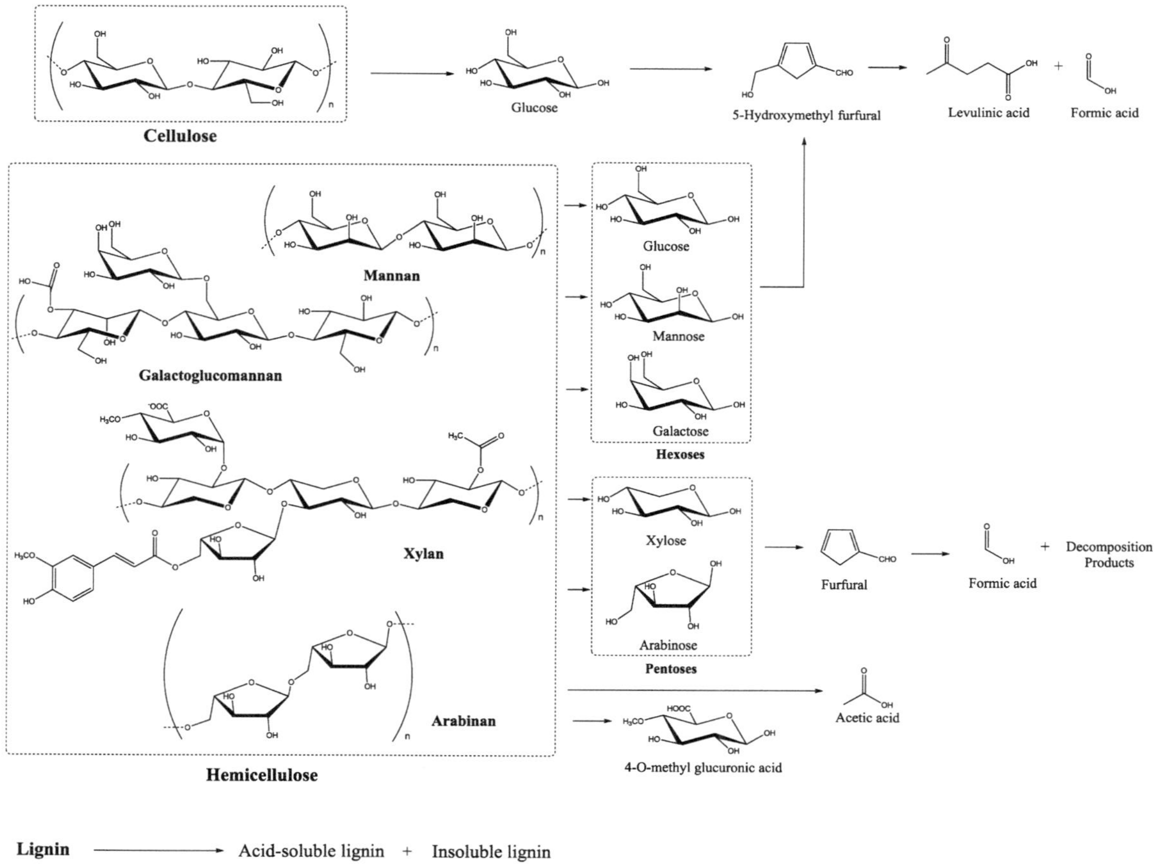


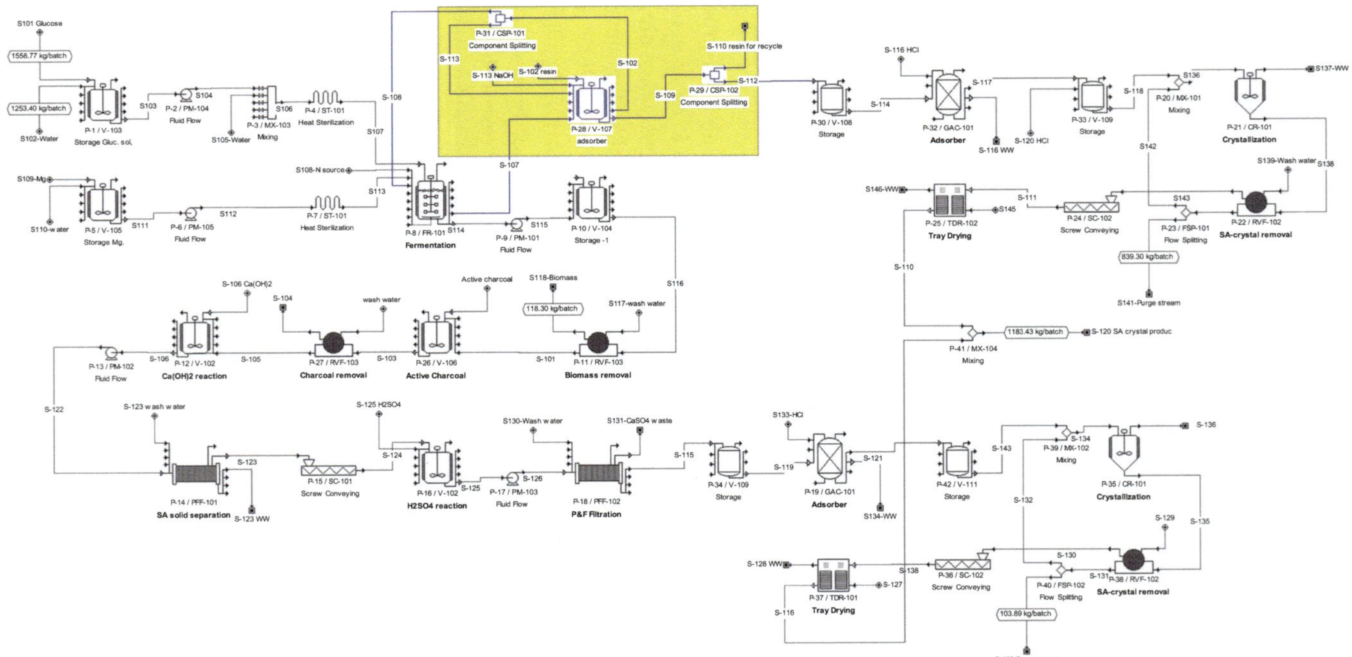


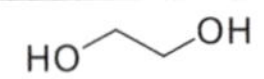 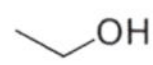 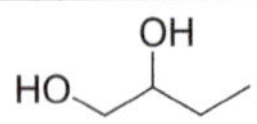 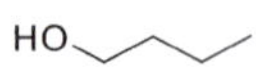 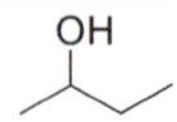 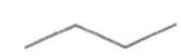 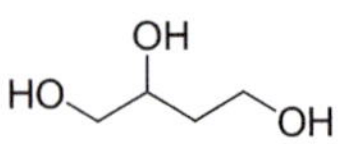 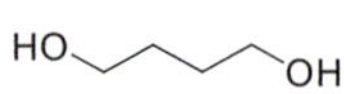 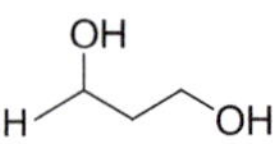 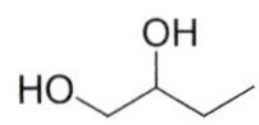 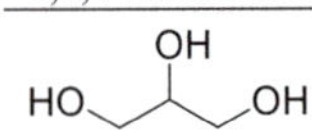 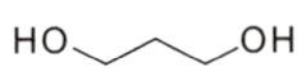 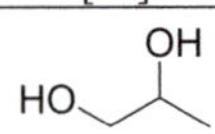 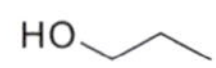 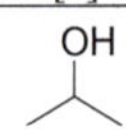 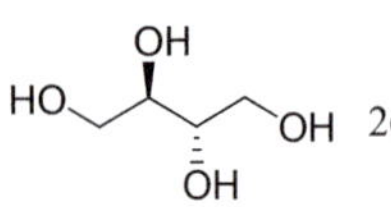 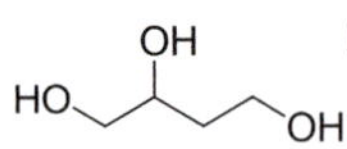 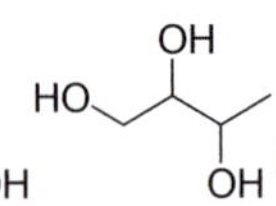 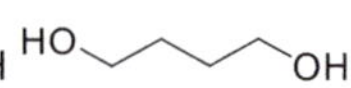 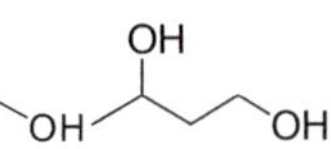 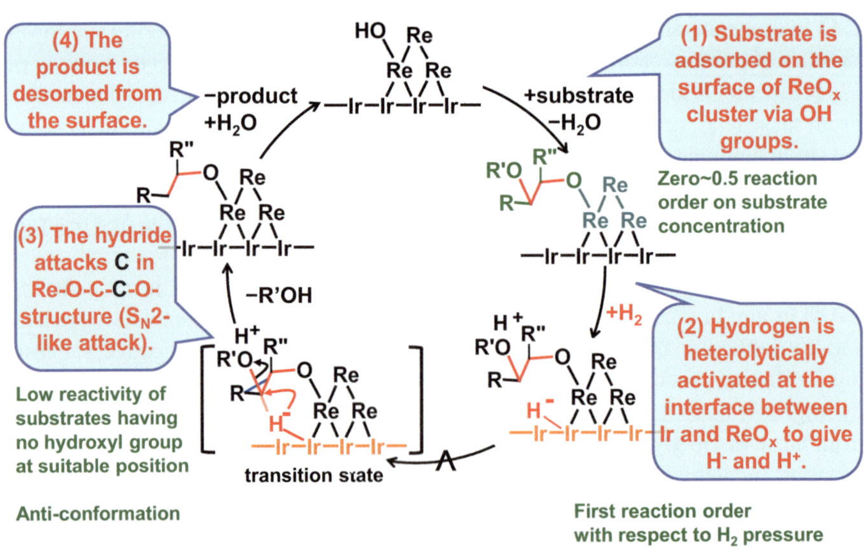


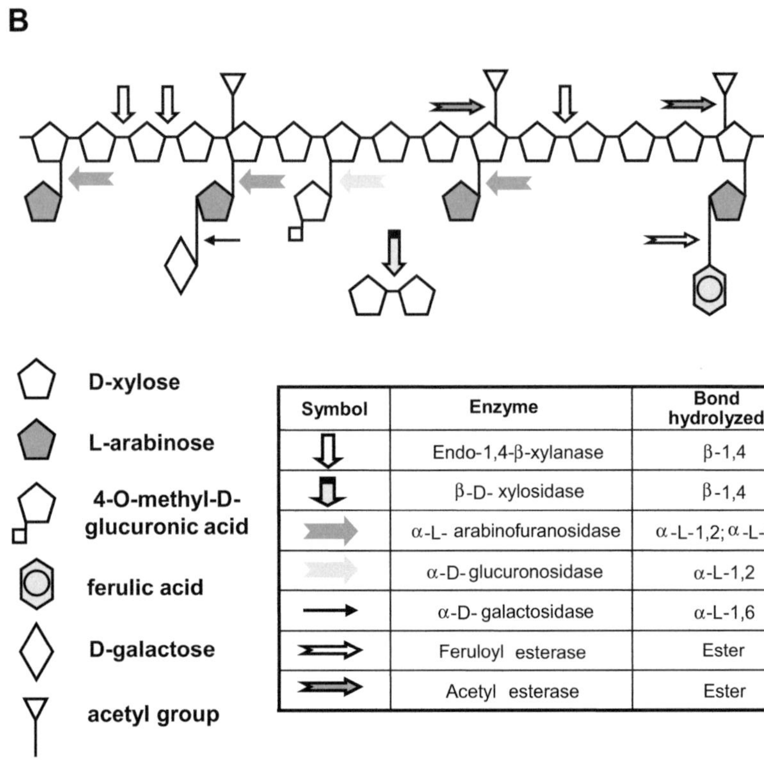


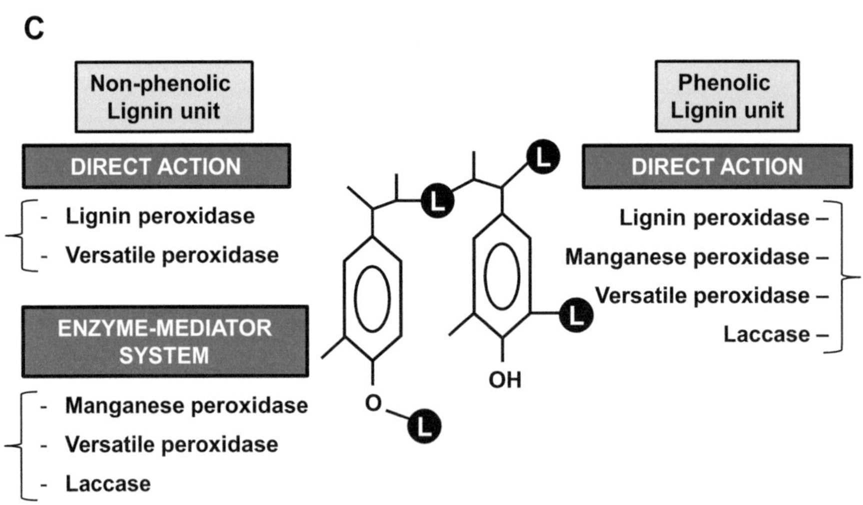


**Table 13.2** Production of γ-valerolactone from levulinic acid and carbohydrates using homogeneous catalysts[a]

| No. | Substrate | Catalyst | Solvent | $T$ (°C) | $H_2$ source | $t$ (h) | $Y_{GVL}$(%) | Ref. |
|---|---|---|---|---|---|---|---|---|
| 1 | LA | $RuCl_2(PPh_3)_3$ | $ScCO_2$ | 150 | 40 bar | 6 | 72.5 | [79] |
| 2 | LA | $(PPh_2)_3NC_3H_6$ + RuHCl $(PPh_3)_3$ | THF | 160 | 70 bar | 18 | 78 | [80] |
| 3 | LA | $Ru(acac)_3$ | THF | 160 | 70 bar | 18 | 99 | [80] |
| 4 | LA | $RuH_2(PPh_3)_4$ | THF | 160 | 70 bar | 18 | 95 | [80] |
| 5 | LA | $RuCl_2(PPh_3)_3 + (PPh_2)_3NC_3H_6$ | THF | 160 | 70 bar | 18 | 95 | [80] |
| 6 | LA | $Ru(acac)_3 + (PPh_2)_3 NC_3H_6$ | THF | 160 | 70 bar | 18 | 95 | [80] |
| 7 | LA | $Ru(acac)_3$ + TPPTS | $H_2O$ | 140 | 69 bar $H_2$ | 12 | 95 | [78] |
| 8 | LA | $[(\eta^6\text{-}C_6Me_6)\ Ru(bpy)(H_2O)][SO_4]$ | $H_2O$ | 70 | HCOONa | 18 | 25 | [78] |
| 9 | LA | $Ru(acac)_3 + PBu_3$ | – | 200 | 82.8 bar $H_2$ | 6 | 37 | [78] |
| 10 | LA | $Ru(acac)_3 + PnOct_3 + NH_4PF_6$ | – | 160 | 100 bar $H_2$ | 18 | >99 | [69] |
| 11 | LA | $RuCl_2(PPh_3)$ | – | 180 | 12 kg/cm² $H_2$ | 24 | 99 | [81] |
| 12 | LA | $RuH_2(PPh_3)_4$ | – | 180 | 12 kg/cm² $H_2$ | 24 | 58 | [81] |
| 13 | LA | $Ru(acac)_3$/ $RP(C_6H_4\text{-m-}SO_3Na)_2$[b] | – | 140 | 10 bar $H_2$ | 4.5 | 39–100 | [82, 83] |
| 14 | LA | $RuCl_3{\bullet}H_2O/PPh_3$ | – | 150 | HCOOH | 12 | 93 | [54] |
| 15 | Glucose | $Ru(CO)_4I_2$ + HI | $H_2O$ | 200 | 30 bar $H_2$ | 8 | 39.1 | [74] |
| 16 | Fructose | $HCO_2\text{-}CF_3$, Ru/C | $H_2O + HCO_2\text{-}CF_3$ | 180 | 94 bar $H_2$ | 8 | 62 | [84] |
| 17 | Glucose | $HCO_2\text{-}CF_3$, Ru/C | $H_2O + HCO_2\text{-}CF_3$ | 180 | 94 bar $H_2$ | 8 | 38 | [84] |
| 18 | Sucrose | $HCO_2\text{-}CF_3$, Ru/TPPTS | $H_2O + HCO_2\text{-}CF_3$ | 180 | 94 bar $H_2$ | 8 | 52 | [84] |

Partial Table 13.2 reprinted with permission from Ref [19]. Copyright © 2015, Elsevier

[a]Note: *T: temperature, t: time, Y: yield of GVL*

**Table 13.4.** (continued)

| No. | Substrate | Catalyst | Other reaction conditions | $Y_{GVL}$ (%) | Ref. |
|---|---|---|---|---|---|
| 31 | Furfural | Zr-beta +Amberlyst-70 | 2-butanol solvent, 120 °C, 24 h | 66 | [112] |
| 32 | EL | $Zr(OH)_4$ | 2-butanol solvent, 200 °C, 1 h | 81.3 | [13] |
| 33 | LA | β-$Mo_2C$ | Water, dioxane, 180 °C, 30 bar $H_2$ | >81 | [113] |

Adapted and reprinted with permission from Ref. [19]. Copyright © 2015, Elsevier

[a]Note: *Y*: yield of γ-valerolactone (GVL); [b]*ML: methyl levulinate*; [c]*EL: ethyl levulinate*; [d]*BL: butyl levulinate*

synthesis [115]. Besides, formic acid was often used to promote transfer hydrogenation reactions [116] due to the generation of $H_2$ and $CO_2$ from the decomposition based on a simple equation 13.1 [117]:

$$HCOOH \rightarrow CO_2 + H_2 \tag{13.1}$$

The selectivity of the product depends on the types of catalyst used, CO and water as side-products were often produced simultaneously following the formula (Eq. 13.2):

$$HCOOH \rightarrow CO + H_2O \tag{13.2}$$

recent investigations reported that several transition-metal complexes or Noble metal catalysts could act as catalysts to improve the selectivity of $H_2$ in the decomposition of formic acid. Laurenczy et al. [118] studied the decomposition of formic acid into $H_2$ and $CO_2$ in aqueous solutions catalyzed by $[Ru(H_2O)_6]^{2+}$, $[Ru(H_2O)_6]^{3+}$, and $RuCl_3$ $xH_2O$ in the presence of TPPTS (TPPTS = meta-trisulfonatedtriphenylphosphine). The decomposition of formic acid is slow, unless sodium formate is present as an initiator. Beller and co-workers studied the promotion effect through the addition of different amines (i.e. triethylamine, N,N-dimethylhexylamine, N,N-dimethylaminoethanol) in the catalytic decomposition of formic acid in the aqueous mixture of HCOOH and HCOONa [119]. Under such conditions, the reaction was observed even at room temperature over Ru-based catalysts including $[RuCl_2(p\text{-cymene})]_2$ and $[RuCl_2(Benzene)]_2$ $/(PPh_3)_3$. Ogo et al. studied a homogeneous Ru catalyst $[(\eta_6\text{-}C_6Me_6)Ru(bpy)(H_2O)][SO_4]$ able to promote the transfer hydrogenation of LA with HCOONa in liquid water solutions and the possible pathway was proposed in Fig. 13.8, where the GVL and 1,4-pentanediol were obtained in 25% yield [120]. Heeres et al. [84] studied a one-pot synthesis of GVL from the conversion of C6-sugars using trifluoroacetic acid, Ru/C catalyst, and formic acid as the hydrogen donor.

## I

## N

**Q**



**R**

MIX
Papier aus verantwortungsvollen Quellen
Paper from responsible sources
FSC® C105338

If you have any concerns about our products,
you can contact us on
**ProductSafety@springernature.com**

In case Publisher is established outside the EU,
the EU authorized representative is:
**Springer Nature Customer Service Center GmbH**
**Europaplatz 3, 69115 Heidelberg, Germany**

Printed by Libri Plureos GmbH
in Hamburg, Germany